Study Guide

Charles M. Seiger

Atlantic Community College

FUNDAMENTALS OF
Anatomy &
Physiology

FOURTH EDITION

MARTINI

PRENTICE HALL Upper Saddle River, NJ 07458

Executive Editor: *David Kendric Brake*
Acquisition Editor: *Linda Schreiber*
Senior Development Editor: *Laura J. Edwards*
Asst.Vice President, ESM Production & Manufacturing: *David W. Riccardi*
Special Projects Manager: *Barbara A. Murray*
Production Editor: *Veronica Schwartz, Dawn Blayer*
Supplement Cover Manager: *Paul Gourhan*
Manufacturing Manager: *Trudy Pisciotti*
Formatting: *Eric Trinadad*
Illustrations: *William C. Ober, M.D.; Claire W. Garrison, R.N.; Matthew S. Garrison*
Cover photograph: *Andrea Weber and Andrew Pacho*
Photo© *Lois Greenfield, 1993*

© 1998, 1995, 1992 by **PRENTICE-HALL, INC.**
Simon & Schuster/A Viacom Company
Upper Saddle River, NJ 07458

Printed in the United States of America

10 9 8 7 6 5 4 3

ISBN 0-13-751819-6

Prentice-Hall International (UK) Limited, *London*
Prentice-Hall of Australia Pty. Limited, *Sydney*
Prentice-Hall Canada, Inc., *Toronto*
Prentice-Hall Hispanoamericana, S.A., *Mexico*
Prentice-Hall of India Private Limited, *New Delhi*
Prentice-Hall of Japan, Inc., *Tokyo*
Simon & Schuster Asia Pte. Ltd., *Singapore*
Editora Prentice-Hall do Brasil, Ltda., *Rio de Janeiro*

Contents

Preface

This revision of the *Study Guide* has been shaped by many years of teaching and writing experience. In preparing it, I have tried to objectively examine my own classroom involvement, evaluate student input about the ways they learn, draw on my discussions with colleagues about the dynamics of teaching and learning, and pay close attention to the comments of the users and reviewers of my previous edition. The result is a resource that I hope, when used in conjunction with FUNDAMENTALS OF ANATOMY AND PHYSIOLOGY, by Frederic Martini, will not only excite students about this course but also provide valuable reinforcement of difficult concepts.

The sequence of topics within the *Study Guide* parallels that of the text and also incorporates the Three-Level Learning System that students encounter in the text's end of chapter material. This system will be familiar to those of you conversant with Bloom's Taxonomy. It provides a means for students to accurately gauge their level of comprehension of topics. The Level 1 questions are keyed directly to chapter objectives and offer a review of facts and terms. The Level 2 exercises encourage students to synthesize concepts, and Level 3 questions promote critical thinking. Each level contains appropriate question types including matching, multiple choice, completion, essay, labeling and concept mapping. Significant changes have been made in many exercises as a result of reviewer feedback.

I would like to acknowledge the students and colleagues who have had an impact on this work through their careful and thoughtful reviews. I am also grateful for the support provided by my editors, David Brake and Linda Schreiber, as well as the thoroughness of my development editor, Laura Edwards, and my production editor, Veronica Schwartz. Finally and most importantly, I thank my family for its continuing support of a very time intensive project.

Of course, any errors or omissions found by the reader are attributable to the author, rather than to the reviewers. Readers with comments, suggestions, relevant reprints, or corrections should contact me at the address below.

Charles Seiger
c/o Prentice Hall
1208 E. Broadway, Suite 200
Tempe, AZ 85282

Format and Features

The three level review system utilized in the Martini text and all of its supplements affords each student a logical framework for the progressive development of skill as he or she advances through material of increasing levels of complexity. The three levels of organization are designed to help students (1) learn basic facts, ideas, and principles; (2) increase their capacity for abstraction and concept integration; and (3) think critically in applying what they have learned to specific situations.

Level 1, *A review of Chapter Objectives,* consists of exercises that test the student's mastery of vocabulary and recall of information through objective-based multiple choice, completion, matching, and illustration labeling questions.

Level 2 focuses on *Concept Synthesis,* a process that actively involves the student in combining, integrating, and relating the basic facts and concepts mastered in Level 1. In addition to multiple choice, completion, and short answer/essay questions, this level incorporates a feature that is unique to this study guide: the development and completion of concept maps. The examples provided give students a framework of information that encourages them to complete the map. Once completed, these maps help the user to understand relationships between ideas and to differentiate related concepts. In addition, they have been designed to serve as models that will encourage students to develop new maps of their own.

Another feature of Level 2 is the use of Body Treks with Robo, the micro-robot. The organ systems of the human body are examined by combining fantasy with actual techniques that are currently being used experimentally to study anatomical features and monitor physiological activity. The treks are written to demonstrate the interrelationship of concepts in a unique and dynamic way.

Level 3 activities promote *Critical Thinking and Application* skills through the use of life experiences, plausible clinical situations, and common diagnostic problems. These techniques draw on the students' analytical and organizational powers, enabling them to develop associations and relationships that ultimately may benefit their decision-making ability in clinical situations and everyday life experiences.

An answer key for each section has been provided at the end of the study guide.

1

An Introduction to Anatomy and Physiology

■ Overview

Are you interested in knowing something about your body? Have you ever wondered what makes your heart beat or why and how muscles contract to produce movement? If you are curious to understand the what, why, and how of the human body, then the study of anatomy and physiology is essential for you. The term *anatomy* is derived from a Greek word which means to cut up, or anatomize (dissect) representative animals or human cadavers which serve as the basis for understanding the structure of the human body. *Physiology* is the science that attempts to explain the physical and chemical processes occurring in the body. Anatomy and Physiology provide the foundation for personal health and clinical applications.

Chapter 1 is an introduction to anatomy and physiology *citing* some of the basic functions of living organisms, *defining* various specialties of anatomy and physiology, *identifying* levels of organization in living things, *explaining* homeostasis and regulation, and *introducing* some basic anatomical terminology. The information in this chapter will provide the framework for a better understanding of anatomy and physiology, and includes basic concepts and principles necessary to get you started on a successful and worthwhile trek through the human body.

☐ LEVEL 1   REVIEW OF CHAPTER OBJECTIVES

1. Describe the basic functions of living organisms.
2. Define anatomy and physiology and describe the various specialties of each discipline.
3. Identify the major levels of organization in living organisms from the simplest to the most complex.
4. Identify the organ systems of the human body, their functions, and the major components of each system.
5. Explain the concept of homeostasis and its significance for living organisms.

6. Describe how positive and negative feedback are involved in homeostatic regulation.
7. Use anatomical terms to describe body sections, body regions and relative positions.
8. Identify the major body cavities and their subdivisions.

[L1] Multiple choice:

Place the letter corresponding to the correct answer in the space provided.

OBJ. 1   _____ 1. Creating subsequent generations of similar organisms describes the basic function of:

      a. growth and development
      b. growth and reproduction
      c. growth and assimilation
      d. a, b, and c

OBJ. 1   _____ 2. All of the basic functions in living things necessitate a common need for:

      a. getting rid of waste
      b. survival
      c. energy
      d. growth

OBJ. 1   _____ 3. All of the following selections include functions in living organisms except:

      a. oxygen, carbon dioxide
      b. growth, reproduction
      c. response, adaptability
      d. absorption, movement

OBJ. 2   _____ 4. Anatomy is the study of _____ and physiology is the study of _____.

      a. function, structure
      b. animals, plants
      c. cells, microorganisms
      d. structure, function

OBJ. 2   _____ 5. Systemic anatomy considers the structure of major _____, while surface anatomy refers to the study of _____.

      a. anatomical landmarks, organ systems
      b. organ systems, superficial markings
      c. superficial markings, macroscopic anatomy
      d. superficial external features, anatomical landmarks

OBJ. 2   _____ 6. The anatomical specialty which provides a bridge between the realms of macroscopic anatomy and microscopic anatomy is:

      a. gross anatomy
      b. regional anatomy
      c. developmental anatomy
      d. surgical anatomy

OBJ. 2 ____ 7. The specialized study which analyzes the structure of individual cells is:

      a. histology

      b. microbiology

      c. cytology

      d. pathology

OBJ. 2 ____ 8. The scientist who studies the effects of *diseases* on organ or system functions would be classified as a:

      a. histophysiologist

      b. cell physiologist

      c. system physiologist

      d. pathological physiologist

OBJ. 3 ____ 9. The smallest *living* units in the body are:

      a. elements

      b. sub-atomic particles

      c. cells

      d. molecules

OBJ. 3 ____ 10. The level of organization that reflects the interactions between organ systems is the:

      a. cellular level

      b. tissue level

      c. molecular level

      d. organism

OBJ. 4 ____ 11. The two regulatory systems in the human body include the:

      a. nervous and endocrine

      b. digestive and reproductive

      c. muscular and skeletal

      d. cardiovascular and lymphatic

OBJ. 5 ____ 12. *Homeostasis* refers to:

      a. the chemical operations under way in the body

      b. individual cells becoming specialized to perform particular functions

      c. changes in an organisms immediate environment

      d. the existence of a stable internal environment

OBJ. 6 ____ 13. When a variation outside of normal limits triggers an automatic response that corrects the situation, the mechanism is called:

      a. positive feedback

      b. crisis management

      c. negative feedback

      d. homeostasis

OBJ. 6 ____ 14. When the initial stimulus produces a response that exaggerates the stimulus, the mechanism is called:

      a. autoregulation

      b. negative feedback

      c. extrinsic regulation

      d. positive feedback

OBJ. 7   _____ 15. When a person is lying down face up in the anatomical position, the individual is said to be:

      a. prone

      b. rostral

      c. supine

      d. proximal

OBJ. 7   _____ 16. Moving from the wrist toward the elbow is an example of moving in a _____ direction.

      a. proximal

      b. distal

      c. medial

      d. lateral

OBJ. 7   _____ 17. RLQ is an abbreviation used as a reference to designate a specific:

      a. section of the vertebral column

      b. area of the cranial vault

      c. region of the pelvic girdle

      d. abdominopelvic quadrant

OBJ. 7   _____ 18. Making a sagittal section results in the separation of:

      a. anterior and posterior portions of the body

      b. superior and inferior portions of the body

      c. dorsal and ventral portions of the body

      d. right and left portions of the body

OBJ. 7   _____ 19. The process of choosing one sectional plane and making a series of sections at small intervals is called:

      a. parasagittal sectioning

      b. resonance imaging

      c. serial reconstruction

      d. sectional planing

OBJ. 8   _____ 20. The subdivisions of the dorsal body cavity include the:

      a. thoracic and abdominal cavity

      b. abdominal and pelvic cavity

      c. pericardial and pleural cavity

      d. cranial and spinal cavity

OBJ. 8   _____ 21. The subdivisions of the ventral body cavity include the:

      a. pleural and pericardial cavity

      b. thoracic and abdominopelvic cavity

      c. pelvic and abdominal cavity

      d. cranial and spinal cavity

OBJ. 8   _____ 22. The heart and the lungs are located in the _____ cavity.

      a. pericardial

      b. thoracic

      c. pleural

      d. abdominal

Level
-1-

OBJ. 8   _____ 23. The ventral body cavity is divided by a flat muscular sheet
called the:

        a. mediastinum

        b. pericardium

        c. diaphragm

        d. peritoneum

OBJ. 8   _____ 24. The procedure used to monitor circulatory pathways using radiodense dyes produces an X-ray image known as:

        a. an MRI

        b. a CT scan

        c. an echogram

        d. an angiogram

OBJ. 8   _____ 25. Checking for tumors or other tissue abnormalities is best accomplished by the use of:

        a. computerized tomography

        b. X-ray

        c. ultrasound

        d. magnetic resonance imaging

[L1]   Completion:

Using the terms below, complete the following statements.

digestion	transverse	responsiveness	embryology
autoregulation	endocrine	liver	positive feedback
organs	histologist	pericardial	medial
regulation	excretion	physiology	peritoneal
tissues	mediastinum	distal	molecules
integumentary	digestive	extrinsic	urinary

OBJ. 1   26. Moving your hand away from a hot stove is an example of a basic
function called _____.

OBJ. 1   27. In order for food to be utilized by cells in the human body, it must first
be broken down by the process of _____.

OBJ. 1   28. Harmful waste products are discharged into the environment by the
process of _____.

OBJ. 2   29. A person who specializes in the study of tissue is called a _____.

OBJ. 2   30. The study of early developmental processes is called _____.

OBJ. 2   31. The study of the *functions* of the living cell is called cell _____.

OBJ. 3   32. In complex organisms such as the human being, cells unite to form
_____ .

OBJ. 3   33. At the chemical level of organization, chemicals interact to form
complex _____.

OBJ. 3   34. The cardiovascular system is made up of structural units called
_____.

OBJ. 4   35. The kidneys, bladder, and ureters are organs which belong to the
_____ system.

OBJ. 4   36. The esophagus, large intestine, and stomach are organs which belong
to the _____ system.

Level
-1-

OBJ. 4   37. The organ system to which the skin belongs is the _____ system.

OBJ. 5   38. The term that refers to the adjustments in physiological systems that are responsible for the preservation of homeostasis is homeostatic _____.

OBJ. 5   39. The homeostatic control that results from the activities of the nervous or endocrine systems is called _____ regulation.

OBJ. 6   40. When the activities of a cell, tissue, organ, or system change automatically due to environmental variation the homeostatic mechanism which operates is called _____.

OBJ. 6   41. A response that is important in accelerating processes that must proceed to completion rapidly is called _____.

OBJ. 6   42. The two systems usually controlled by negative feedback mechanisms are the nervous and _____ system.

OBJ. 7   43. Tenderness in the right upper quadrant (RUQ) might indicate problems with the _____.

OBJ. 7   44. A term that means "close to the long axis of the body" is _____.

OBJ. 7   45. A term that means "away from an attached base" is _____ .

OBJ. 7   46. A plane that is perpendicular to the long axis of the body is a _____ section.

OBJ. 8   47. The subdivision of the thoracic cavity which houses the heart is the _____ cavity.

OBJ. 8   48. The large central mass of connective tissue that surrounds the pericardial cavity and separates the two pleural cavities is the _____.

OBJ. 8   49. The abdominopelvic cavity is also known as the _____ cavity.

[L1]   Matching:

Match the terms in column B with the terms in column A. Use letters for answers in the spaces provided.

PART I

		COLUMN A		COLUMN B
OBJ. 1	_____	50. excretion	A.	disease
OBJ. 1	_____	51. respiration	B.	organelles
OBJ. 2	_____	52. gross anatomy	C.	endocrine
OBJ. 2	_____	53. pathology	D.	oxygen, carbon dioxide
OBJ. 3	_____	54. internal cell structures	E.	cardiovascular
OBJ. 4	_____	55. heart	F.	waste elimination
OBJ. 4	_____	56. pituitary	G.	macroscopic

PART II

		COLUMN A		COLUMN B
OBJ. 5	_____	57. homeostasis	H.	skull
OBJ. 5	_____	58. automatic system change	I.	crisis management
OBJ. 6	_____	59. receptor	J.	steady state
OBJ. 6	_____	60. nervous system	K.	ventral body cavity
OBJ. 7	_____	61. cranial	L.	stimulus
OBJ. 7	_____	62. prone	M.	abdominopelvic
OBJ. 8	_____	63. peritoneal	N.	autoregulation
OBJ. 8	_____	64. coelom	O.	face down

Level
-1-

[L1]   Drawing/Illustration Labeling:

Identify each numbered structure by labeling the following figures:

OBJ. 7   *FIGURE 1.1*   Planes of the Body

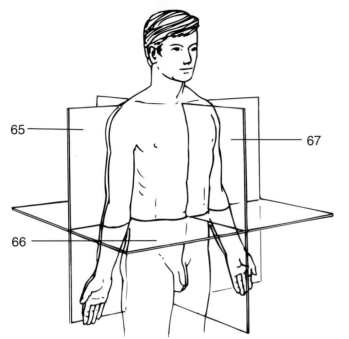

65 _frontal / coronal_
66 _transverse / horizontal_
67 _sagital_

OBJ. 7   *FIGURE 1.2*   Human Body Orientation & Direction

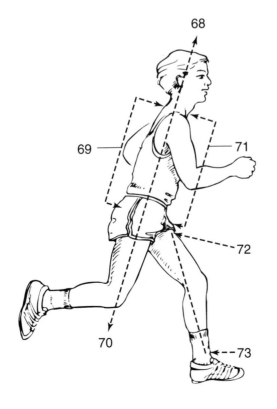

68 _____

69 _____

70 _____

71 _____

72 _____

73 _____

Level
-1-

OBJ. 7   *FIGURE 1.3*   Regional Body References

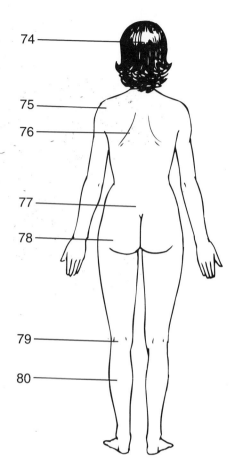

Posterior view
(dorsal)

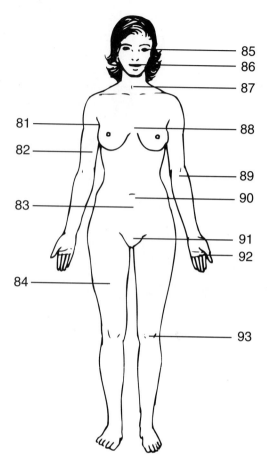

Anterior view
(ventral)

74 _____

75 _____

76 _____

77 _____

78 _____

79 _____

80 _____

81 _____

82 _____

83 _____

84 _____

85 _____

86 _____

87 _____

88 _____

89 _____

90 _____

91 _____

92 _____

93 _____

Level
-1-

OBJ. 8   *FIGURE 1.4*   Body Cavities – Sagittal View

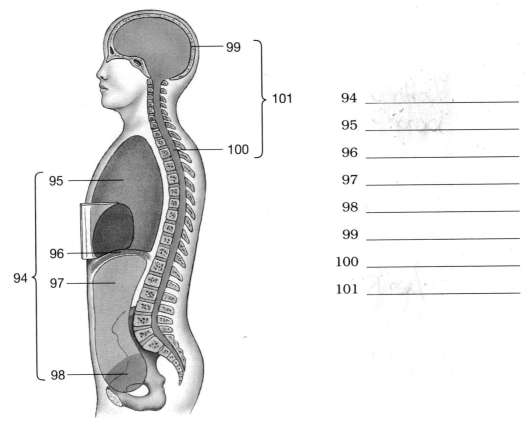

94 _____
95 _____
96 _____
97 _____
98 _____
99 _____
100 _____
101 _____

OBJ. 8   *FIGURE 1.5*   Body Cavities – Anterior View

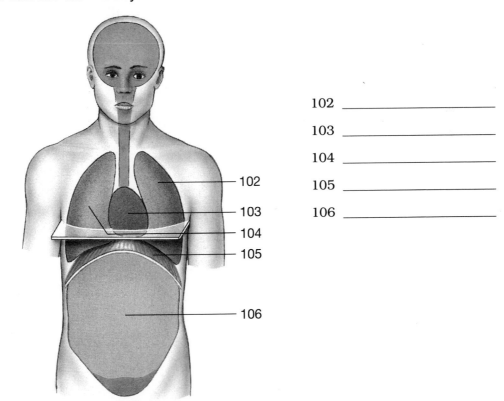

102 _____
103 _____
104 _____
105 _____
106 _____

When you have successfully completed the exercises in L1 proceed to L2.

Level
–1–

☐ LEVEL 2   CONCEPT SYNTHESIS

Concept Map I:

Using the following terms, fill in the circled, numbered, blank spaces to complete the concept map. Follow the numbers which comply with the organization of the map.

Surgical Anatomy   Regional Anatomy
Embryology   Cytology
Tissues   Macroscopic Anatomy
Structure of Organ Systems

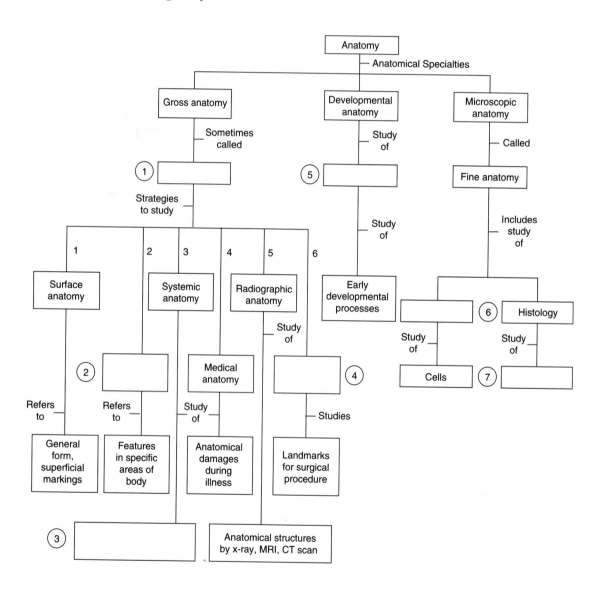

Concept Map II:

Using the following terms, fill in the circled, numbered, blank spaces to complete the concept map. Follow the numbers which comply with the organization of the map.

Pathological Physiology
Functions of Living Cells
Exercise Physiology
Functions of Anatomical
  Structures

Histophysiology
Specific Organ Systems
Body Function Response to Athletics
Body Function Response to Changes
  in Atmospheric Pressure

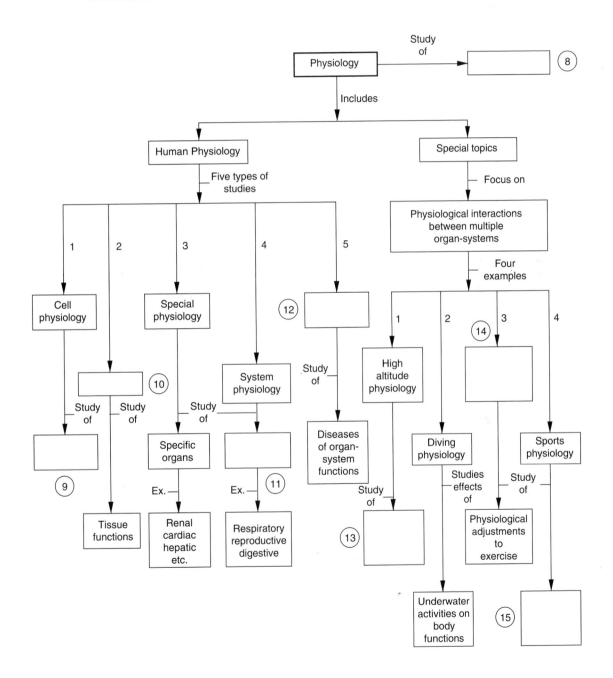

Concept Map III:

Using the following terms, fill in the circled, numbered, blank spaces to complete the concept map. Follow the numbers which comply with the organization of the concept map.

Pelvic Cavity   Spinal Cord   Cranial Cavity
Heart   Abdominopelvic Cavity   Two Pleural Cavities

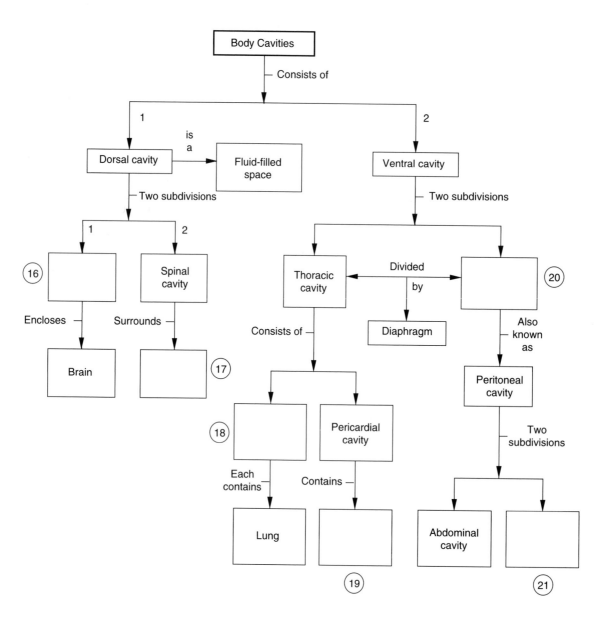

Concept Map IV:

Using the following terms, fill in the circled, numbered, blank spaces to complete the concept map. Follow the numbers which comply with the organization of the concept map.

Angiogram	High-energy radiation	CT Scans
Echogram	Radio Waves	Radiologist

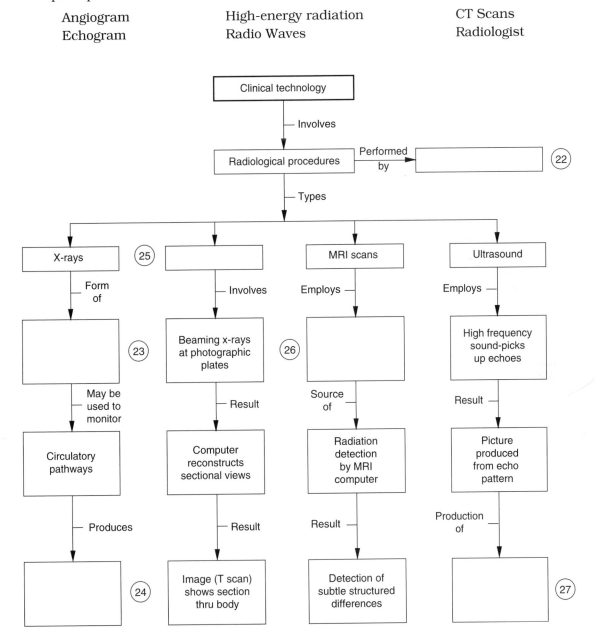

Body Trek:

Using the terms below, fill in the blanks to complete the trek through the *levels of organization* in the human body.

Tissues   Organelles   Cells   Organism   Subatomic particles
Protoplasm   Atoms   Systems   Molecules   Organs

    Robo, the micro-robot, is introduced into the body by way of the mouth where immediate contact is made with the lining of the mouth, which consists of a mucous epithelium. Immediate feedback to Mission Control gives information about the chemical interactions taking place, resulting in the formation of cells. Robo discloses that protons, neutrons, and electrons, which are (28) _____ , are combining in specific numbers and arrangements to form (29) _____ . There, forms, invisible to the naked eye but revealed by Robo's advanced detection system, seem to be sharing and/or giving and taking electrons and forming bonds which hold them together to make (30) _____. The chemical conglomeration of the bonded forms results in a complex living, somewhat colloidal, substance called (31) _____. The living matter contains some organized structures called (32) _____. The living matter is surrounded by a double phospholipid-protein layered enclosure known as a membrane. The enclosed living substance along with its organized microscopic forms comprises the makeup of the individual units of structure and function in all living things called (33) _____. As Robo's trek continues it is quite evident that there are many of the individual units which are combining with one another to form (34) _____. Four kinds are detected as Robo treks into other areas of the body. Epithelial was rather plentiful in the mouth, while other areas of the body include the presence of muscular, nervous, and connective types. The complex, multi-unit types form more organized and complex structural and functional units called (35) _____ , which, when performing in a similar capacity, make up the eleven body (36) _____. The complex, complete, living being is referred to as an (37) _____. With the completion of Robo's investigation, Mission Control programs a convenient exit by way of the mouth, and preparations will be made for the next body trek.

[L2]   Multiple Choice:

Place the letter corresponding to the correct answer in the space provided.

_____ 38. Because snakes are cold-blooded they seek warmth in the winter and coolness in the summer in order to survive temperature extremes. This is an example of:

       a. diversity

       b. conservation

       c. creativity

       d. adaptability

_____ 39. Beginning with cells and proceeding through increasing levels of complexity, the correct sequence is:

       a. cells, tissues, organs, system

       b. cells, organs, tissues, system

       c. cells, system, tissues, organs

       d. system, organs, tissues, cells

_____ 40. Damage at the cellular, tissue, or organ level often affects the entire system. This supports the view that:

       a. each level is totally independent of the others

       b. each level has its own specific function

       c. each level is totally dependent on the other

       d. the lower levels depend on the higher levels

Level
=2=

_____ 41. Anatomical position refers to a person standing erect, feet facing forward and:

      a. arms hanging to sides and palms of hands facing anteriorally and the thumbs located medially

      b. arms in a raised position and palms of hands facing forward with the thumbs to the outside

      c. arms hanging to sides and palms of hands facing forward with the thumbs to the outside

      d. arms in a raised position and palms of hands facing dorsally with the thumbs to the inside

_____ 42. From the following selections, identify the directional terms in _correct sequence_ which apply to the areas of the human body. (ventral, posterior, superior, inferior)

      a. anterior, dorsal, cephalic, caudal

      b. dorsal, anterior, caudal, cephalic

      c. caudal, cephalic, anterior, posterior

      d. cephalic, caudal, posterior, anterior

_____ 43. If an observer is facing someone in front of him, the left side of the subject is to the observer's:

      a. left

      b. right

      c. dorsal side

      d. proximal end

_____ 44. Resistance to X-ray penetration is called radiodensity. From the following selections, choose the one that correctly shows the order of _increasing_ radiodensity of materials in the human body.

      a. air, liver, fat, blood, bone, muscle

      b. air, fat, liver, blood, muscle, bone

      c. air, fat, blood, liver, muscle, bone

      d. air, liver, blood, fat, muscle, bone

_____ 45. From the organ systems listed below, select the organs in correct sequence which are found in each of the systems. (cardiovascular, digestive, endocrine, urinary, integumentary)

      a. blood vessels, pancreas, kidneys, lungs, nails

      b. heart, stomach, lungs, kidneys, hair

      c. heart, liver, pituitary gland, kidneys, skin

      d. lungs, gall bladder, ovaries, bladder, sebaceous glands

_____ 46. In a homeostatic system, the mechanism which triggers an automatic response that corrects the situation is:

      a. the presence of a receptor area and an effector area

      b. an exaggeration of the stimulus

      c. temporary repair to the damaged area

      d. a variation outside of normal limits

_____ 47. Suppose an individual's body temperature is 37.3° C, which is outside the "normal" range. This variation from the "normal" range may represent:

        a. an illness that has not been identified

        b. individual variation rather than a homeostatic malfunction

        c. the need to see a physician immediately

        d. a variability that is abnormal

_____ 48. If the temperature of the body climbs above 99° F, negative feedback is triggered by:

        a. increased heat conservation by restricted blood flow to the skin

        b. the individual experiences shivering

        c. activation of the positive feedback mechanism

        d. an increased heat loss through enhanced blood flow to the skin and sweating

_____ 49. The term medial surface refers to the area:

        a. close to the long axis of the body

        b. away from the long axis of the body

        c. toward an attached base

        d. away from an attached base

_____ 50. Diagnostic procedures have not changed significantly in a thousand years. The four basic components used by physicians when conducting a physical examination are:

        a. X-ray, CT scans, MRI scans, ultrasound

        b. inspection, palpation, percussion, auscultation

        c. blood pressure, urinalysis, X-ray, check reflexes

        d. use of otoscope, ophthalmoscope, blood pressure, urinalysis

_____ 51. In order for a hypothesis to be valid, the three necessary characteristics are that it will be:

        a. intuitive, theoretical, and conclusive

        b. modifiable, reliable, scientific

        c. testable, unbiased, and repeatable

        d. a, b, and c are correct

_____ 52. Using the scientific method to investigate a problem begins by:

        a. designing an experiment to conduct the investigation

        b. collecting and analyzing data

        c. evaluating data to determine its relevance and validity

        d. proposing a hypothesis

Level
2

[L2] Completion:

Using the terms below, complete the following statements.

appendicitis	adaptability	stethoscope
sternum	knee	elbow
nervous	lymphatic	extrinsic regulation
cardiovascular		

53. As a result of exposure to increased sunlight, the skin produces pigments that absorb damaging solar radiation and provide a measure of protection. This process is called _____.

54. The system responsible for internal transport of cells and dissolved materials, including nutrients, wastes, and gases, is the _____ system.

55. The system responsible for defense against infection and disease is the _____.

56. Activities of the nervous and endocrine systems to control or adjust the activities of many different systems simultaneously is _____.

57. The system which performs crisis management by directing rapid, short-term, and very specific responses is the _____ system.

58. The popliteal artery refers to the area of the body in the region of the _____.

59. Tenderness in the right lower quadrant of the abdomen may indicate _____.

60. Moving proximally from the wrist brings you to the _____.

61. Auscultation is a technique that employs the use of a _____.

62. If a surgeon makes a midsagittal incision in the inferior region of the thorax, the incision would be made through the _____.

[L2] Short Essay Answer:

Briefly answer the following questions in the spaces provided below.

63. Despite obvious differences, all living things perform the same basic functions. List at least six (6) functions that are active processes in living organisms.

64. Show your understanding of the *levels of organization* in a complex living thing by using arrows and listing in correct sequence from the simplest level to the most complex level.

65. Describe the position of the body when it is in the *anatomical position.*

66. What is the major difference between negative feedback and positive feedback?

67. References are made by the anatomist to the *front, back, head,* and *tail* of the human body. What directional reference terms are used to describe each one of these directions? Your answer should be in the order listed above.

68. What is the difference between a *sagittal* section and a *transverse* section?

69. What are the two (2) essential functions of body cavities in the human body?

70. List the basic sequence of the steps involved in the development of a scientific theory. (Use arrows to show stepwise sequence.)

When you have successfully completed the exercises in L2 proceed to L3.

Level
=2=

☐ LEVEL 3   CRITICAL THINKING/APPLICATION

Using principles and concepts learned in *Introduction to Anatomy and Physiology,* answer the following questions. Write your answers on a separate sheet of paper.

1. Unlike the abdominal viscera, the thoracic viscera is separated into two compartments by an area called the mediastinum. What is the clinical importance of this compartmental arrangement?

2. The events of childbirth are associated with the process of positive feedback. Describe the events which confirm this statement.

3. A radioactive tracer is induced into the heart to trace the possibility of a blockage in or around the uterus. Give the sequence of *body cavities* that would be included as the tracer travels in the blood from the heart through the aorta and the uterine artery.

4. Suppose autoregulation fails to maintain homeostasis in the body. What is the process called which takes over by initiating activity of both the nervous and endocrine systems? What are the results of this comparable homeostatic mechanism?

5. Monitoring fetal development may be dangerous for the fetus if improper diagnostic techniques are used. Why is ultrasound an effective means of monitoring fetal development?

6. Gastroenterologists use X-rays to check for ulcers or other stomach and upper digestive tract disorders. Before the X-rays are taken why is it necessary for the patient to drink large quantities of a solution that contains barium ions?

7. Body temperature is regulated by a control center in the brain that functions like a thermostat. Assuming a normal range of 98°–99° F, identify from the graph below what would happen if there was an increase or decrease in body temperature beyond the normal limits. Use the following selections to explain what would happen at no. 1 and no. 2 on the graph.

- body cools
- shivering
- increased sweating
- temperature declines

- body heat is conserved
- ↑ blood flow to skin
- ↓ blood flow to skin
- temperature rises

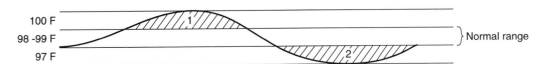

2

The Chemical Level of Organization

■ Overview

Technology within the last 40 to 50 years has allowed humans to "see and understand" the unseen within the human body. Today, instead of an "organ system view" of the body, we are able to look through the eyes of scientists using sophisticated technological tools and see the "ultramicro world" within us. Our technology has permitted us to progress from the "macro-view" to the "micro-view" and to understand that the human body is made up of atoms, and that the interactions of these atoms control the physiological processes within the body.

The study of anatomy and physiology begins at the most fundamental level of organization, namely individual atoms and molecules. The concepts in Chapter 2 provide a framework for understanding how simple components combine to make up the more complex forms that comprise the human body. Information is provided that shows how the chemical processes need to continue throughout life in an orderly and timely sequence if homeostasis is to be maintained.

The student activities in this chapter stress many of the important principles that make up the science of chemistry, both inorganic and organic. The tests are set up to measure your knowledge of chemical principles and to evaluate your ability to apply these principles to the structure and functions of organ systems within the human body.

☐ LEVEL 1 REVIEW OF CHAPTER OBJECTIVES

1. Describe an atom and how atomic structure affects interactions between atoms.
2. Compare the different ways in which atoms combine to form molecules and compounds.
3. Use chemical notation to symbolize chemical reactions.
4. Distinguish among the major types of chemical reactions that are important for studying physiology.
5. Describe the crucial role of enzymes in metabolism.
6. Distinguish between organic and inorganic compounds.
7. Explain how the chemical properties of water make life possible.
8. Discuss the importance of pH and the role of buffers in body fluids.
9. Describe the physiological roles of inorganic compounds.
10. Discuss the structure and function of carbohydrates, lipids, proteins, nucleic acids, and high-energy compounds.

[L1] Multiple choice:

Place the letter corresponding to the correct answer in the space provided.

OBJ. 1 _____ 1. The smallest chemical units of matter of which no chemical change can alter their identity are:

      a. electrons

      b. mesons

      c. protons

      d. atoms

OBJ. 1 _____ 2. The three subatomic particles that are stable constituents of atomic structure are:

      a. carbon, hydrogen, oxygen

      b. protons, neutrons, electrons

      c. atoms, molecules, compounds

      d. cells, tissues, organs

OBJ. 1 _____ 3. The protons in an atom are found only:

      a. outside the nucleus

      b. in the nucleus or outside the nucleus

      c. in the nucleus

      d. in orbitals

OBJ. 2 _____ 4. The unequal sharing of electrons in a molecule of water is an example of:

      a. an ionic bond

      b. a polar covalent bond

      c. a double covalent bond

      d. a strong covalent bond

OBJ. 2 _____ 5. The formation of cations and anions illustrates the attraction between:

      a. ionic bonds

      b. polar covalent bonds

      c. nonpolar covalent bonds

      d. double covalent bonds

OBJ. 3 _____ 6. The symbol 2H means:

      a. one molecule of hydrogen

      b. two molecules of hydrogen

      c. two atoms of hydrogen

      d. a, b, and c are correct

OBJ. 4 _____ 7. From the following choices, select the one which diagrams a typical *decomposition* reaction:

      a. $A + B \rightleftarrows AB$

      b. $AB + CD \rightarrow AD + CB$

      c. $AB \rightarrow A + B$

      d. $C + D \rightarrow CD$

OBJ. 4   _____ 8. A + B ⇌ AB is an example of a(n) _____ reaction.

       a. reversible

       b. synthesis

       c. decomposition

       d. a, b, and c are correct

OBJ. 4   _____ 9. AB + CD ⇌ AD + CB is an example of a(n) _____ reaction.

       a. reversible

       b. synthesis

       c. exchange

       d. a, b, and c are correct

OBJ. 5   _____ 10. The presence of an appropriate enzyme affects only the:

       a. rate of a reaction

       b. direction of the reaction

       c. products that will be formed from the reaction

       d. a, b, and c are correct

OBJ. 5   _____ 11. Organic catalysts made by a living cell to promote a specific reaction are called:

       a. nucleic acids

       b. buffers

       c. enzymes

       d. metabolites

OBJ. 6   _____ 12. The major difference between inorganic and organic compounds is that _inorganic_ compounds are usually:

       a. small molecules held together partially or completely by ionic bonds

       b. made up of carbon, hydrogen, and oxygen

       c. large molecules which are soluble in water

       d. easily destroyed by heat

OBJ. 6   _____ 13. The four major classes of organic compounds are:

       a. water, acids, bases, and salts

       b. carbohydrates, fats, proteins, and water

       c. nucleic acids, salts, bases, and water

       d. carbohydrates, lipids, proteins, and nucleic acids

OBJ. 7   _____ 14. The ability of water to maintain a relatively constant temperature and then prevent rapid changes in body temperature is due to its:

       a. solvent capacities

       b. molecular structure

       c. boiling and freezing point

       d. capacity to absorb and distribute heat

OBJ. 8 _____ 15. To maintain homeostasis in the human body, the normal pH range of the blood must remain at:

      a. 6.80 to 7.20

      b. 7.35 to 7.45

      c. 7.0

      d. 6.80 to 7.80

OBJ. 8 _____ 16. The human body generates significant quantities of acids that may promote a disruptive:

      a. increase in pH

      b. pH of about 7.40

      c. decrease in pH

      d. sustained muscular contraction

OBJ. 9 _____ 17. The ideal medium for the absorption and/or transport of inorganic or organic compounds is:

      a. oil

      b. water

      c. blood

      d. lymph fluid

OBJ. 9 _____ 18. A solute that dissociates to release hydrogen ions and causes a decrease in pH is:

      a. a base

      b. a salt

      c. an acid

      d. water

OBJ. 9 _____ 19. A solute that removes hydrogen ions from a solution is:

      a. an acid

      b. a base

      c. a salt

      d. a buffer

OBJ. 10 _____ 20. A carbohydrate molecule is made up of:

      a. carbon, hydrogen, oxygen

      b. monosaccharides, disaccharides, polysaccharides

      c. glucose, fructose, galactose

      d. carbon, hydrogen, nitrogen

OBJ. 10 _____ 21. Carbohydrates are most important to the body in that they serve as primary sources of:

      a. tissue growth and repair

      b. metabolites

      c. energy

      d. digestible forms of food

Level
—1—

OBJ. 10 ____ 22. The building blocks of proteins consist of chains of small molecules which are called:

       a. peptide bonds

       b. amino acids

       c. R groups

       d. amino groups

OBJ. 10 ____ 23. Special proteins that are involved in metabolic regulation are called:

       a. transport proteins

       b. contractile proteins

       c. structural proteins

       d. enzymes

OBJ. 10 ____ 24. The three basic components of a *single nucleotide* of a nucleic acid are:

       a. purines, pyrimidines, sugar

       b. sugar, phosphate group, nitrogen base

       c. guanine, cytosine, thymine

       d. pentose, ribose, deoxyribose

OBJ. 10 ____ 25. The most important high-energy compound found in the human body is:

       a. DNA

       b. UTP

       c. ATP

       d. GTP

[L1]   Completion:

Using the terms below, complete the following statements.

glucose	inorganic	solute	electrolytes
H_2	protons	endergonic	buffers
decomposition	acidic	mass number	solvent
salt	2H	carbonic acid	exergonic
isomers	ionic bond	organic	dehydration synthesis
catalysts	covalent bonds		

OBJ. 1   26. The atomic number of an atom is determined by the number of _____.

OBJ. 1   27. The total number of protons and neutrons in the nucleus is the _____.

OBJ. 2   28. Atoms that complete their outer electron shells by sharing electrons with other atoms result in molecules held together by _____.

OBJ. 2   29. When one atom loses an electron and another accepts that electron, the result is the formation of a(n) _____.

OBJ. 3   30. The chemical notation that would indicate "one molecule of hydrogen composed of two hydrogen atoms" would be _____.

OBJ. 3   31. The chemical notation that would indicate "two individual atoms of hydrogen" is _____.

OBJ. 4   32. A reaction that breaks a molecule into smaller fragments is called _____.

OBJ. 4   33. Reactions that release energy are said to be _____ , and reactions that absorb heat are called _____ reactions.

OBJ. 5   34. Compounds that accelerate chemical reactions without themselves being permanently changed are called _____.

OBJ. 6   35. Compounds that contain the elements carbon and hydrogen, and usually oxygen, are _____ compounds.

OBJ. 6   36. Acid, bases, and salts are examples of _____ compounds.

OBJ. 7   37. The fluid medium of a solution is called the _____ , and the dissolved substance is called the _____.

OBJ. 7   38. Soluble inorganic molecules whose ions will conduct an electric current in solution are _____.

OBJ. 8   39. A solution with a pH of 6.0 is _____.

OBJ. 8   40. Compounds in body fluids that maintain pH within normal limits are _____.

OBJ. 9   41. The interaction of an acid and a base in which the hydrogen ions of the acid are replaced by the positive ions of the base results in the formation of a(n) _____.

OBJ. 9   42. An example of a weak acid that serves as an effective buffer in the human body is _____.

OBJ. 10   43. The most important metabolic "fuel" in the body is _____.

OBJ. 10   44. The linking together of chemical units by the removal of water to create a more complex molecule is called _____.

OBJ. 10   45. Molecules that have the same molecular formula but different structural formulas are _____.

[L1] Matching:

Match the terms in column B with the terms in column A. Use letters for answers in the spaces provided.

PART I

		COLUMN A		COLUMN B
OBJ. 1	_____	46. electron	A.	two products; two reactants
OBJ. 2	_____	47. inert gases	B.	sodium chloride
OBJ. 2	_____	48. polar covalent bond	C.	negative electric charge
OBJ. 3	_____	49. NaCl	D.	inorganic acid
OBJ. 4	_____	50. exchange reaction	E.	stable
OBJ. 5	_____	51. enzyme	F.	unequal sharing of electrons
OBJ. 6	_____	52. HCl	G.	catalyst
OBJ. 6	_____	53. NaOH	H.	sodium hydroxide

PART II

		COLUMN A		COLUMN B
OBJ. 7	_____	54. ionization	I.	hydroxyl group
OBJ. 7	_____	55. hydrophilic	J.	uracil
OBJ. 8	_____	56. OH^-	K.	thymine
OBJ. 8	_____	57. H^+	L.	bicarbonate ion
OBJ. 9	_____	58. HCO_3^-	M.	hydrogen ion
OBJ. 10	_____	59. DNA-N base	N.	dissociation
OBJ. 10	_____	60. RNA-N base	O.	dissolve in water

Level —1—

[L1]   Drawing/Illustration Labeling:

Identify each numbered structure by labeling the following figures:

OBJ. 1   ***FIGURE 2.1***   Diagram of an Atom

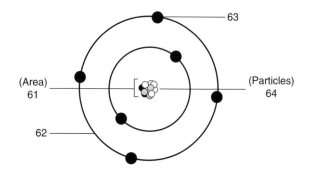

61 _____

62 _____

63 _____

64 _____

OBJ. 2   ***FIGURE 2.2***   Identification of Types of Bonds

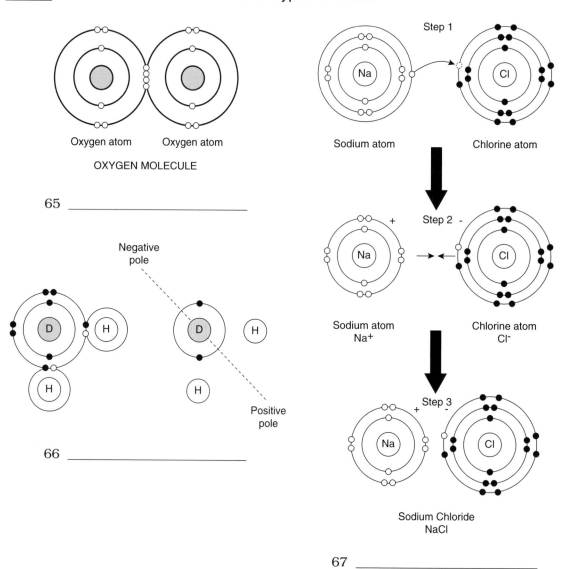

65 _____

66 _____

67 _____

OBJ. 10   *FIGURE 2.3*   Identification of Organic Molecules
(Select from the following terms to identify each molecule.)

polysaccharide   cholesterol   monosaccharide
amino acid   DNA   saturated fatty acid
disaccharide   polyunsaturated fatty acid

68 _____

69 _____

70 _____

71 _____

72 _____

73 _____

74 _____

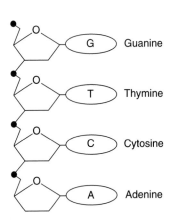

G   Guanine

T   Thymine

C   Cytosine

A   Adenine

75 _____

When you have successfully completed the exercises in L1 proceed to L2.

Level
–1–

☐ LEVEL 2   CONCEPT SYNTHESIS

Concept Map I:

Using the following terms, fill in the circled, numbered, blank spaces to complete the concept map. Follow the numbers to comply with the organization of the map.

Monosaccharides   Sucrose   Complex carbohydrates
Glucose   Glycogen   Bulk, Fiber
Disaccharides

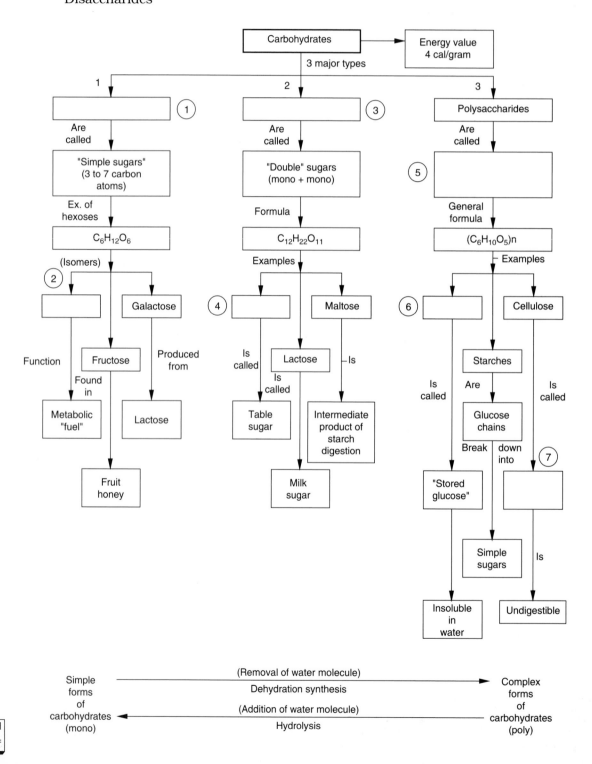

Concept Map II:

Using the following terms, fill in the circled, numbered, blank spaces to complete the concept map. Follow the numbers to comply with the organization of the map.

Local hormones Di-   Saturated

Glyceride   Phospholipid Carbohydrate + diglyceride

Steroids   Glycerol + fatty acids

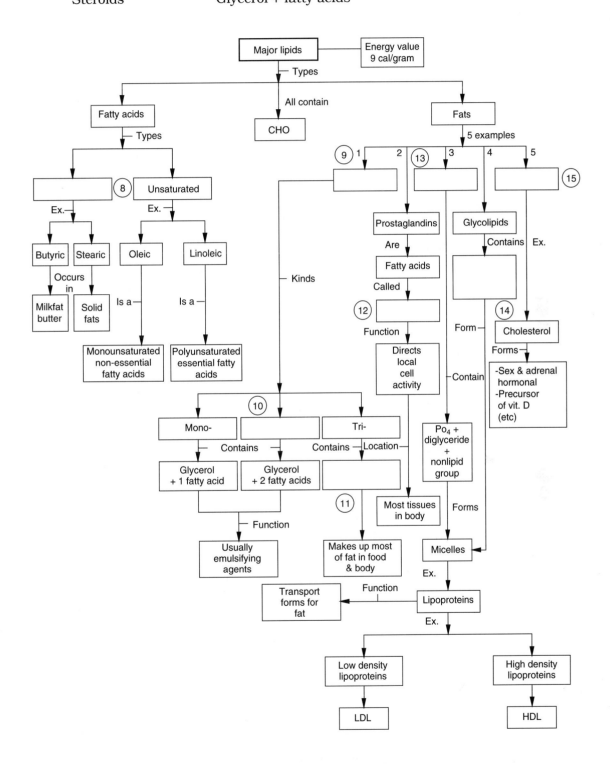

Concept Map III:

Using the following terms, fill in the circled, numbered, blank spaces to complete the concept map. Follow the numbers to comply with the organization of the concept map.

Variable group   –COOH   Globular proteins
Enzymes   Elastin   Quaternary
Structural proteins   Amino acids   Alpha helix
Amino group   Primary

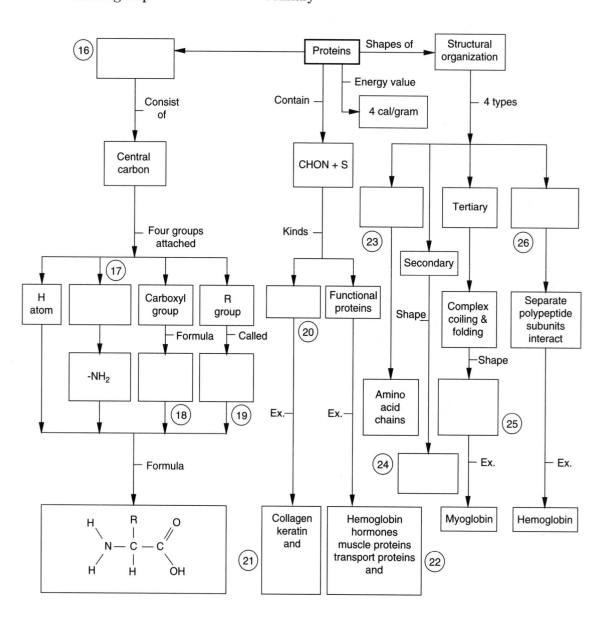

Concept Map IV:

Using the following terms, fill in the circled, numbered, blank spaces to complete the concept map. Follow the numbers to comply with the organization of the concept map.

Ribonucleic acid   Deoxyribose nucleic acid   N bases
Pyrimidines   Adenine   Thymine
Deoxyribose   Purines   Ribose

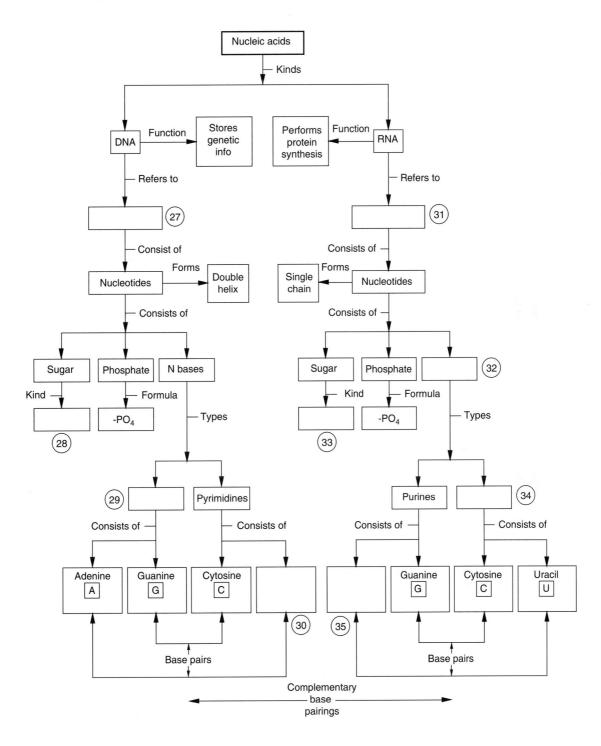

Body Trek:

Using the terms below, fill in the blanks to complete the trek through the chemical organization in the human body.

Heat capacity	Neutrons	Six	Polar covalent
Glycogen	Protein	Zero	Solvent
Polypeptide	Isotope	66	Water
Phosphorylation	Oxygen	ATP	Negatively
DNA	Deuterium	Eight	Electrons
Protons	Lowering	100	Glucose
Molecule	Hydrophobic	RNA	Monosaccharide
Hydrophilic	Oxygen gas	Orbitals	Hydrolysis
Nucleus	Dehydration synthesis	Double covalent bond	

Robo's task to monitor chemical activity begins with a trek outside the body. The robotic "engineers" have decided to place Robo in a cyclotron, an apparatus used in atomic research. The robotic sensors immediately detect an environment that has negative particles called (36) _____ , which are whirling around in cloud-like formations called (37) _____. The particles are travelling at a high rate of speed and appear to be circling stationary particles in a central region called the (38) _____. The central region contains positive charges called (39) _____ and, with the exception of hydrogen in its natural state, all the areas "observed" contain particles which are neutral called (40) _____. Robo's electronic systems appear to be overloaded and there is difficulty monitoring all the activity in the surrounding environment. Feedback to Mission Control reads, "Oops! There goes one that looks exactly like hydrogen, but it has two (2) neutrons looking like a heavier hydrogen called (41) _____, which is chemically identified as a(n) (42) _____ commonly used as a radioactive tracer in research laboratories." Robo signals the presence of an "area" that contains eight (8) positive charges, eight (8) neutral particles, and eight (8) negative charges, which is identified as a(n) (43) _____ atom. Atoms of this kind appear to be chemically active because the outermost cloud only has (44) _____ negative particles and a full complement of (45) _____ is necessary if the atom is to maintain chemical stability. In the atmosphere these "active atoms" are constantly in "search" of atoms of their own kind. When two of these atoms contact one another there is a chemical interaction evidenced by each one sharing two electrons and forming a(n) (46) _____ , resulting in one (47) _____ of (48) _____. When a single oxygen atom comes into contact with two hydrogen atoms a(n) (49) _____ bond is formed due to an unequal sharing of electrons. The result is a region around the oxygen atom that is (50) _____ charged and the region around the hydrogen atoms is positively charged.

After removal of the micro-robot from the cyclotron, preparation is made for Robo to be catheterized through the heart to begin the body trek. The robot is immediately swept from the heart into the large aorta where it is transported in a "sea" of liquid with suspended solid particles. As it treks along, it is obvious that the liquid portion of the surrounding solution is (51) _____ , which comprises about (52) _____% of the total body weight. Mission Control requests more information about this liquid medium. An immediate response confirms that the medium can withstand extremes of cold temperatures reaching a freezing point of (53) _____°C and a boiling point of (54) _____°C. The molecules of this medium have a high (55) _____ , which prevents rapid changes in body temperature, and the heat absorbed during evaporation makes perspiration an effective means of losing heat and (56) _____ body temperature. The robot senses the dissolving of numerous organic and inorganic compounds establishing the (57) _____ property of the aqueous medium. The readily dissolved molecules are said to be (58) _____ , and some of the body's fat deposits, which are insoluble, are said to be (59) _____. Robo's adventure through the bloodstream is briefly interrupted by a trek through the liver. The robot's sensors are fully activated to cope with all the information available in the liver. Robo is surrounded by a sweet substance, (60) _____ , a (61) _____ , which is the body's most important metabolic fuel. The molecules of this sweet stuff are "sticking" together because water is "leaking out" and

molecules of (62) _____ are forming due to the process of (63) _____ .
What a chemical show! Robo detects a signal that indicates additional sugar is needed for
energy in other parts of the body and the molecules formed by the "sticking" together of
the sugar molecules need to be separated so other cells can use this fuel for energy. The
liver responds by engaging in the process of (64) _____ and sending the "ready
fuel" by way of the bloodstream to where it is needed. Other cells are involved in protein
synthesis. As the (65) _____ in the nucleus transcribes onto (66) _____ ,
a message will translate into the production of (67) _____ chains resulting in the
formation of large (68) _____ molecules that have both structural and functional
roles in the body. The process of (69) _____ in the mitochondria of the liver cells
is producing a high-energy compound called (70) _____ . The end of Robo's trek
is imminent and an exit route is facilitated via the hepatic vein with eventual arrival in
the brachial vein where the process of bloodletting will release the robot. What a trek!

[L2]   Multiple Choice:

Place the letter corresponding to the correct answer in the space provided.

_____ 71. The chemical properties of every element are determined by:

    a. the number and arrangement of electrons in the outer
       energy level

    b. the number of protons in the nucleus

    c. the number of protons and neutrons in the nucleus

    d. the atomic weight of the atom

_____ 72. Whether or not an atom will react with another atom will be determined
primarily by:

    a. the number of protons present in the atom

    b. the number of electrons in the outermost energy level

    c. the atomic weight of the atom

    d. the number of subatomic particles present in the atom

_____ 73. In the formation of *nonpolar* covalent bonds there is:

    a. equal sharing of protons and electrons

    b. donation of electrons

    c. unequal sharing of electrons

    d. equal sharing of electrons

_____ 74. The symbol $2H_2O$ means that two identical molecules of water are each
composed of:

    a. 4 hydrogen atoms and 2 oxygen atoms

    b. 2 hydrogen atoms and 2 oxygen atoms

    c. 4 hydrogen atoms and 1 oxygen atom

    d. 2 hydrogen atoms and 1 oxygen atom

_____ 75. The reason water is particularly effective as a solvent is:

    a. polar molecules are formed due to the closeness of
       hydrogen atoms

    b. cations and anions are produced by hydration

    c. hydrophobic molecules have many polar covalent bonds

    d. it has a high heat capacity which dissolves molecules

Level
=2=

_____ 76. The action of a buffer to maintain pH with normal limits consists primarily of:

      a. removing or replacing hydrogen ions

      b. taking Rolaids or Tums

      c. the replacement of bicarbonate ions

      d. the addition of weak acids to weak bases

_____ 77. A _salt_ may best be described as:

      a. an organic molecule created by chemically altering an acid or base

      b. an inorganic molecule that buffers solutions

      c. an inorganic molecule created by the reaction of an acid and a base

      d. an organic molecule used to flavor food

_____ 78. The chemical makeup of a lipid molecule is different from a carbohydrate in that the lipid molecule:

      a. contains much less oxygen than a carbohydrate having the same number of carbon atoms

      b. contains twice as much oxygen as the carbohydrate

      c. contains equal amounts of carbon and oxygen in its molecular structure

      d. the chemical makeup is the same

_____ 79. Lipid deposits are important as _energy reserves_ because:

      a. they appear as fat deposits on the body

      b. they are readily broken down to release energy

      c. the energy released from lipids is metabolized quickly

      d. lipids provide twice as much energy as carbohydrates

_____ 80. Proteins differ from carbohydrates in that they:

      a. are not energy nutrients

      b. do not contain carbon, hydrogen, and oxygen

      c. always contain nitrogen

      d. are inorganic compounds

_____ 81. Compared to the other major organic compounds, nucleic acids are unique in that they:

      a. contain nitrogen

      b. store and process information at the molecular level

      c. are found only in the nuclei of cells

      d. control the metabolic activities of the cell

_____ 82. Isotopes of an element are:

      a. atoms whose nuclei contain different numbers of protons

      b. atoms whose nuclei have equal numbers of protons and neutrons

      c. atoms with an equal number of protons, neutrons, and electrons

      d. atoms whose nuclei contain different numbers of neutrons

Level
=2=

_____ 83. From the selections that follow, choose the one that represents the symbols for each of the following elements in the correct order. (carbon, sodium, phosphorus, iron, oxygen, nitrogen, sulfur)

      a. C, So, Ph, I, O, Ni, S

      b. C, Na, P, I, O, N, S

      c. C, Na, P, Fe, O, N, S

      d. C, Na, P, Fe, O_2, N_2, S

_____ 84. If an atom has an atomic number of 92 and its atomic weight is 238, how many protons does the atom have?

      a. 238

      b. 92

      c. 146

      d. 54

_____ 85. If the second energy level of an atom has one (1) electron, how many more does it need to fill it to its maximum capacity?

      a. 1

      b. 2

      c. 5

      d. 7

_____ 86. The atomic structure of hydrogen looks like which one of the following?

  a.

    1 p+
    1 n
    1e⁻

  c.

    1 p+
    1 n
    2e⁻

  b.

    1 n
    2e⁻

  d,

    1 p+
    1e⁻

_____ 87. The number of neutrons in $_8O^{17}$ is:

      a. 8

      b. 9

      c. 7

      d. 17

_____ 88. Which one of the following selections represents the pH of the _weakest_ acid?

      a. 7.0

      b. 1.3

      c. 3.2

      d. 6.7

_____ 89. The type of bond that has the most important effects on the properties of water and the shapes of complex molecules is the:

      a. hydrogen bond

      b. ionic bond

      c. covalent bond

      d. polar covalent bond

_____ 90. If oxygen has an atomic weight of 16, what is the molecular weight of an oxygen molecule?

        a. 16

        b. 8

        c. 32

        d. 2

_____ 91. The chemical reaction A + B $\rightleftarrows$ AB is an example of a(n):

        a. synthesis reaction

        b. decomposition reaction

        c. reversible reaction

        d. a, b, and c are correct

_____ 92. If the concentration of hydrogen ions is (0.000001) what is the pH?

        a. 1

        b. 5

        c. 6

        d. 7

_____ 93. Two simple sugars joined together form a disaccharide. The reaction involved for this to occur necessitates:

        a. the removal of water to create a more complex molecule

        b. the addition of water to create a more complex molecule

        c. the presence of cations and anions to initiate electrical attraction

        d. the disassembling of molecules through hydrolysis

_____ 94. The presence of a _carboxylic acid group_ at the end of a carbon chain demonstrates a characteristic common to all:

        a. amino acids

        b. inorganic acids

        c. nucleic acids

        d. a, b, and c are correct

_____ 95. The reason the hemoglobin molecule qualifies as a quaternary protein structure is:

        a. it is fibrous and generally insoluble in body fluids

        b. it has four interacting globular subunits

        c. it has four major functions in the human body

        d. its four subunits react independently of each other

[L2]   Completion:

Using the terms below, complete the following statements.

nucleic acids	isomers	molecular weight
peptide bond	ions	hydrolysis
ionic bonds	hydrophobic	inorganic compounds
mole	saturated	dehydration synthesis
molecule	alkaline	alpha particles

96. The helium nucleus, consisting of two protons and two neutrons, describes the characteristics of _____.

97. For every element, a quantity that has a mass in grams equal to the atomic weight will contain the same number of atoms. The name given to this quantity is a(n) _____.

98. A chemical structure containing more than one atom is a(n) _____.

99. Atoms or molecules that have positive or negative charges are called _____.

100. Electrical attraction between opposite charges produces a strong _____.

101. The sum of the atomic weight of the components in a compound is the _____.

102. Small molecules held together partially or completely by ionic bonds are _____.

103. Molecules that have few if any polar covalent bonds and do not dissolve in water are _____.

104. If the pH is above 7 with hydroxyl ions in the majority, the solution is _____.

105. Of the four major classes of organic compounds, the one responsible for storing genetic information is the _____.

106. Molecules that have the same molecular formula but different structural formulas are _____.

107. The process that breaks a complex molecule into smaller fragments by the addition of a water molecule is _____.

108. Glycogen, a branched polysaccharide composed of interconnected glucose molecules, is formed by the process of _____.

109. Butter, fatty meat, and ice cream are examples of sources of fatty acids that are said to be _____.

110. The attachment of a carboxylic acid group of one amino acid to the amino group of another forms a connection called a(n) _____.

[L2]   Short Essay:

Briefly answer the following questions in the spaces provided below.

111. Suppose an atom has 8 protons, 8 neutrons and 8 electrons. Construct a diagram of the atom and identify the subatomic particles by placing them in their proper locations.

112. In chemical samples and processes, why are relationships expressed in moles rather than grams?

Level =2=

113. Why are the elements helium, argon, and neon called *inert gases?*

114. In a water (H_2O) molecule the unequal sharing of electrons creates a *polar covalent bond.* Why?

115. Compute the molecular weight (MW) of one molecule of glucose ($C_6H_{12}O_6$). [Note: atomic weights C = 12; H = 1; O = 16]

116. List six important characteristics of water that make life possible.

117. List the four major classes of organic compounds found in the human body and give an example for each one.

118. Differentiate between a saturated and an unsaturated fatty acid.

119. Using the four kinds of nucleotides that make up a DNA molecule, construct a model that will show the correct arrangement of the components which make up each nucleotide. *Name each nucleotide.*

120. What are the three components that make up one nucleotide of ATP?

When you have successfully completed the exercises in L2 proceed to L3.

☐ LEVEL 3   CRITICAL THINKING/APPLICATION

Using principles and concepts learned in Chapter 2, answer the following questions. Write your answers on the answer sheet provided.

1. Using the letters AB and CD, show how each would react in an exchange reaction.

2. Why might "baking soda" be used to relieve excessive stomach acid?

3. Using the glucose molecule ($C_6H_{12}O_6$), demonstrate your understanding of dehydration synthesis by writing an equation to show the formation of a molecule of sucrose ($C_{12}H_{22}O_{11}$). [Make sure the equation is balanced.]

4. Even though the recommended dietary intake for carbohydrates is 55–60 percent of the daily caloric intake, why do the carbohydrates account for less than 3 percent of our total body weight?

5. Why can a drug test detect the use of marijuana for days after the drug has been used?

6. Why is it potentially dangerous to take excessive amounts of vitamins A, D, E, and K?

7. You are interested in losing weight so you decide to eliminate your intake of fats completely. You opt for a fat substitute such as *Olestra*, which contains compounds that cannot be used by the body. Why might this decision be detrimental to you?

8. A friend of yours is a bodybuilder who takes protein supplements with the idea that this will increase the body's muscle mass. What explanation might you give your friend to convince him/her that the purchase of protein supplements is a waste of money?

3

The Cellular Level of Organization: Cell Structure

■ Overview

Cells are highly organized basic units of structure and function in all living things. In the human body all cells originate from a single fertilized egg and additional cells are produced by the division of the preexisting cells. These cells become specialized in a process called differentiation, forming tissues and organs that perform specific functions. The "roots" of the study of anatomy and physiology are established by understanding the basic concepts of cell biology. This chapter provides basic principles related to the structure of cell organelles and the vital physiological functions that each organelle performs.

The following exercises relating to cell biology will reinforce the student's ability to learn the subject matter and to synthesize and apply the information in meaningful and beneficial ways.

❑ LEVEL 1   REVIEW OF CHAPTER OBJECTIVES

1. List the functions of the cell membrane and the structural features that enable it to perform those functions.
2. Specify the routes by which different ions and molecules can enter or leave a cell and the factors that may restrict such movement.
3. Describe the various transport mechanisms that cells use to facilitate the absorption or removal of specific substances.
4. Explain the origin and significance of the transmembrane potential.
5. Describe the organelles of a typical cell and indicate the specific functions of each.
6. Explain the functions of the cell nucleus.
7. Discuss the nature and importance of the genetic code.
8. Summarize the process of protein synthesis.
9. Describe the stages of the cell life cycle.
10. Describe the process of mitosis and explain its significance.
11. Define differentiation and explain its importance.

[L1]   Multiple choice:

Place the letter corresponding to the correct answer in the space provided.

OBJ. 1 _____ 1. The outer boundary of the intracellular material is called the:

   a. cytosol
   b. extracellular fluid
   c. cell membrane
   d. cytoplasm

OBJ. 1 _____ 2. The major components of the cell membrane are:

   a. carbohydrates, fats, proteins, water
   b. carbohydrates, lipids, ions, vitamins
   c. amino acids, fatty acids, carbohydrates, cholesterol
   d. phospholipids, proteins, glycolipids, cholesterol

OBJ. 1 _____ 3. Most of the communication between the interior and exterior of the cell occurs by way of:

   a. the phospholipid bilayer
   b. the peripheral proteins
   c. intergral protein channels
   d. receptor sites

OBJ. 2 _____ 4. Ions and other small water-soluble materials cross the cell membrane only by passing through:

   a. ligands
   b. membrane anchors
   c. a channel
   d. a receptor protein

OBJ. 2 _____ 5. The mechanism by which glucose can enter the cytoplasm without expending ATP is via:

   a. a carrier protein
   b. a glycocalyx
   c. a catalyzed reaction
   d. a recognition protein

OBJ. 2 _____ 6. Passive or leakage channels within the cell membrane:

   a. open or close to regulate ion passage
   b. permit water and ion movement at all times
   c. require ATP as an energy source
   d. a, b, and c are correct

OBJ. 3 _____ 7. All transport through the cell membrane can be classified as either:

   a. active or passive
   b. diffusion or osmosis
   c. pinocytosis or phagocytosis
   d. permeable or impermeable

Level
-1-

OBJ. 3 ———— 8. The major difference between diffusion and bulk flow is that when molecules move by *bulk flow* they move:

      a. at random

      b. as a unit in one direction

      c. as individual molecules

      d. slowly over long distances

OBJ. 3 ———— 9. The rate that solute molecules are filtered depends on:

      a. their size

      b. the force of the hydrostatic pressure

      c. the rate at which water passes through the membrane

      d. a, b, and c are correct

OBJ. 4 ———— 10. The transmembrane potential is important in maintaining the integrity of a cell because:

      a. it greatly increases the cell's sensitivity to its environment

      b. it adds strength and helps to stabilize the shape of the cell

      c. it provides mechanical strength and blocks the passage of water

      d. it serves to balance the negative charges of the cell's proteins

OBJ. 4 ———— 11. The factor(s) that interact(s) to create and maintain the transmembrane potential is (are) the:

      a. membrane permeability for sodium

      b. membrane permeability for potassium

      c. presence of the Na-K exchange pump

      d. a, b, and c are correct

OBJ. 5 ———— 12. Cytosol contains a high concentration of _____ whereas extracellular fluid contains a high concentration of _____.

      a. sodium, potassium

      b. carbohydrates, proteins

      c. fatty acids, amino acids

      d. potassium, sodium

OBJ. 5 ———— 13. The primary components of the cytoskeleton, which gives the cell strength and rigidity and anchors the position of major organelles, are:

      a. microvilli

      b. microtubules

      c. microfilaments

      d. thick filaments

OBJ. 5   _____ 14. Of the following selection, the one that contains only *membranous* organelles is:

     a. cytoskeleton, microvilli, centrioles, cilia, ribosomes

     b. centrioles, lysosomes, nucleus, endoplasmic reticulum, cilia, ribosomes

     c. mitochondria, nucleus, endoplasmic reticulum, Golgi apparatus, lysosomes, peroxisomes

     d. nucleus, mitochondria, lysosomes, centrioles, micro villi

OBJ. 5   _____ 15. Approximately 95 percent of the energy needed to keep a cell alive is generated by the activity of the:

     a. mitochondria

     b. ribosomes

     c. nucleus

     d. microtubules

OBJ. 5   _____ 16. *Nucleoli* are nuclear organelles that:

     a. contain the chromosomes

     b. are responsible for producing DNA

     c. control nuclear operations

     d. synthesize the components of ribosomes

OBJ. 5   _____ 17. The three major functions of the *endoplasmic reticulum* are:

     a. hydrolysis, diffusion, osmosis

     b. detoxification, packaging, modification

     c. synthesis, storage, transport

     d. pinocytosis, phagocytosis, storage

OBJ. 5   _____ 18. The functions of the *Golgi apparatus* include:

     a. synthesis, storage, alteration, packaging

     b. isolation, protection, sensitivity, organization

     c. strength, movement, control, secretion

     d. neutralization, absorption, assimilation, secretion

OBJ. 5   _____ 19. Peroxisomes, which are smaller than lysosomes, are primarily responsible for:

     a. the control of lysosomal activities

     b. absorption and neutralization of toxins

     c. synthesis and packaging of secretions

     d. renewal and modification of the cell membrane

OBJ. 6   _____ 20. The *major* factor that allows the nucleus to control cellular operations is through its:

     a. ability to communicate chemically through nuclear pores

     b. location within the cell

     c. regulation of protein synthesis

     d. "brain-like" sensory devices that monitor cell activity

Level
-1-

OBJ. 6

_____ 21. Ribosomal proteins and RNA are produced primarily in the:

      a. nucleus

      b. nucleolus

      c. cytoplasm

      d. mitochondria

OBJ. 7

_____ 22. Along the length of the DNA strand, information is stored in the sequence of:

      a. the sugar-phosphate linkages

      b. nitrogen bases

      c. the hydrogen bonds

      d. the amino acid chain

OBJ. 7

_____ 23. A sequence of three nitrogen bases can specify the identity of:

      a. a specific gene

      b. a single DNA molecule

      c. a single amino acid

      d. a specific peptide chain

OBJ. 8

_____ 24. The process where RNA polymerase uses the genetic information to assemble a strand of mRNA is:

      a. translation

      b. transcription

      c. initiation

      d. elongation

OBJ. 8

_____ 25. If the DNA triplet is TAG, the corresponding codon on the mRNA strand will be:

      a. AUC

      b. AGC

      c. ATC

      d. ACT

OBJ. 8

_____ 26. If the mRNA has the codons (GGG) – (GCC) – (AAU), it will bind to the tRNAs with anticodons:

      a. (CCC) – (CGG) – (TTA)

      b. (CCC) – (CGG) – (UUA)

      c. (CCC) – (AUC) – (UUA)

      d. (CCC) – (UAC) – (TUA)

OBJ. 9

_____ 27. The correct sequence of the cell cycle beginning with interphase is:

      a. G_1, G_0, G_2, S_1, G_m

      b. G_0, G_1, S, G_2, G_m

      c. G_1, G_0, S_1, G_m, G_2

      d. G_0, S_1, G_1, G_m, G_2

OBJ. 9

_____ 28. The process of mitosis begins when the cell enters the:

      a. S phase

      b. G_m phase

      c. G_1 phase

      d. G_0 phase

Level
—1—

OBJ. 10   _____ 29. The four stages of mitosis in correct sequence are:

        a. prophase, anaphase, metaphase, telophase

        b. prophase, metaphase, telophase, anaphase

        c. prophase, metaphase, anaphase, telophase

        d. prophase, anaphase, telophase, metaphase

OBJ. 11   _____ 30. The process of differentiation resulting in the appearance of characteristic cell specializations involves:

        a. irreversible alteration in protein structure

        b. the presence of an oncogene

        c. gene activation or repression

        d. the process of fertilization

[L1]   Completion:

Using the terms below, complete the following statements.

transmembrane	gene	fixed ribosomes
translation	nuclear pores	interphase
endocytosis	phospholipid bilayer	anaphase
ribosomes	cytoskeleton	resting
nucleus	differentiation	

OBJ. 1   31. Structurally, the cell membrane is called a(n) _____.

OBJ. 2   32. Intracellular membrane proteins are bound to a network of supporting filaments called the _____.

OBJ. 3   33. The packaging of extracellular materials in a vesicle at the cell surface for importation into the cell is called _____.

OBJ. 4   34. The separation of positive and negative ions by the cell membrane produces a potential difference called the _____ potential.

OBJ. 4   35. In an undisturbed cell the transmembrane potential is a(n) _____ potential.

OBJ. 5   36. The cell organelles responsible for the synthesis of proteins using information provided by nuclear DNA are _____.

OBJ. 5   37. The areas of the endoplasmic reticulum that are called the rough ER contain _____.

OBJ. 6   38. Chemical communication between the nucleus and cytosol occurs through the _____.

OBJ. 6   39. The organelle that determines the structural and functional characteristics of the cell is the _____.

OBJ. 7   40. Triplet codes needed to produce a specific peptide chain comprise the makeup of a(n) _____.

OBJ. 8   41. The construction of a functional polypeptide using the information provided by an mRNA strand is called _____.

OBJ. 9   42. Somatic cells spend the majority of their functional lives in a state known as _____.

OBJ. 10   43. The phase of mitosis in which the chromatid pairs separate and the daughter chromosomes move toward opposite ends of the cells is _____.

OBJ. 11   44. The specialization process that causes a cell's functional abilities to become more restricted is called _____.

Level
-1-

[L1]   Matching:

Match the terms in column B with the terms in column A. Use letters for answers in the spaces provided.

PART I   **COLUMN A**   **COLUMN B**

OBJ. 1	_____	45.	cell membrane	A.	glycoprotein
OBJ. 1	_____	46.	integral proteins	B.	slightly negatively charged
OBJ. 1	_____	47.	receptor proteins	C.	protein "factories"
OBJ. 2	_____	48.	recognition protein	D.	carrier-mediated
OBJ. 3	_____	49.	osmosis	E.	slightly positively charged
OBJ. 3	_____	50.	active transport	F.	membrane turnover
OBJ. 4	_____	51.	inside cell membrane	G.	transmembrane potentials
OBJ. 4	_____	52.	cell membrane exterior	H.	phospholipid bilayer
OBJ. 5	_____	53.	ribosomes	I.	ligand sensitivity
OBJ. 5	_____	54.	golgi apparatus	J.	net diffusion of water

PART II   **COLUMN A**   **COLUMN B**

OBJ. 6	_____	55.	nucleus	K.	RNA-nitrogen base
OBJ. 6	_____	56.	DNA strands	L.	DNA replication
OBJ. 7	_____	57.	thymine	M.	normal cell functions
OBJ. 7	_____	58.	uracil	N	cell control center
OBJ. 8	_____	59.	mRNA formation	O.	nuclear division
OBJ. 8	_____	60.	produces peptide chain	P.	cell specialization
OBJ. 9	_____	61.	S phase	Q.	chromosomes
OBJ. 9	_____	62.	G_0 phase	R.	elongation
OBJ. 10	_____	63.	mitosis	S.	transcription
OBJ. 11	_____	64.	differentiation	T.	DNA-nitrogen base

☐ LEVEL 2   CONCEPT SYNTHESIS

Concept Map I:

Using the following terms, fill in the circled, numbered, blank spaces to complete the concept map. Follow the numbers which comply with the organization of the map.

Ribosomes   Nucleolus   Membranous   Centrioles
Lysosomes   Lipid bilayer   Proteins   Organelles
Fluid component

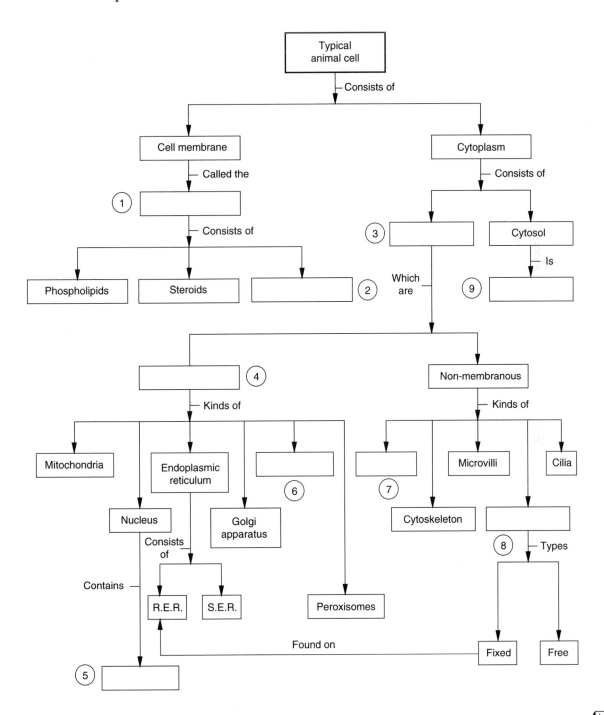

Concept Map II:

Using the following terms, fill in the circled, numbered, blank spaces to complete the concept map. Follow the numbers which comply with the organization of the map.

Require energy   Lipid solubility   Osmosis   Ion pumps
Cell eating   Pinocytosis   Carrier proteins Ligands
Filtration

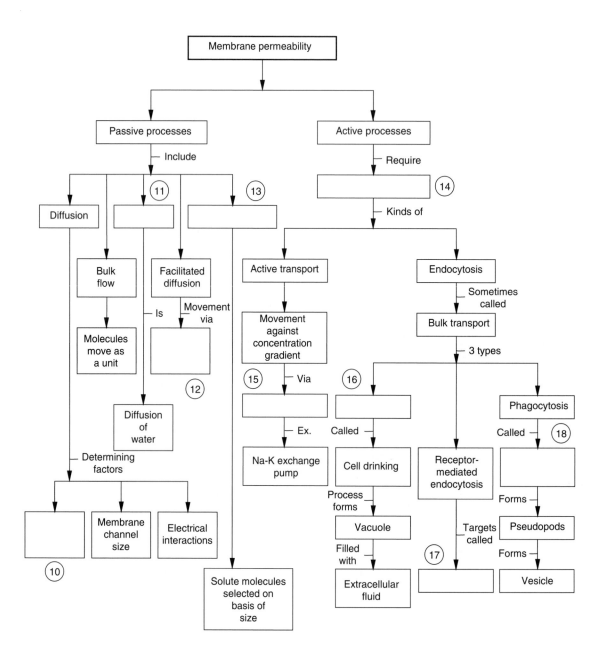

Concept Map III:

Using the following terms, fill in the circled numbered, blank spaces to complete the concept map. Follow the numbers which comply with the organization of the concept map.

Metaphase   Somatic cells   Telophase   Cytokinesis

DNA replication G_2 phase   Mitosis G_1 phase

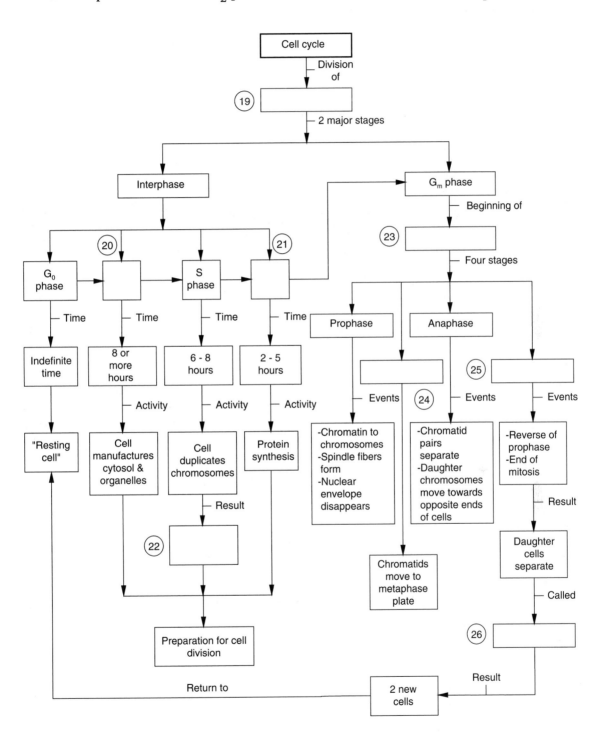

Body Trek:

Using the terms below, fill in the blanks to complete the trek through the chemical organization in the human body.

Integral proteins	Ions	Nonmembranous	Matrix
Cell membrane	Cytosol	Extracellular fluid	Cilia
Intracellular fluid	ATP	Peripheral proteins	Proteins
Phospholipid	Saccules	Protein synthesis	Organelles
Cell division	Lysosomes	Respiratory enzymes	Nucleus
Mitochondrion	Cristae	Golgi apparatus	Ribosomes
Chromosomes	Channels	Cytoskeleton	Nucleoli
Rough endoplasmic reticulum	Nuclear	Nuclear envelope	Nucleoplasm
		Endoplasmic reticulum	

Robo is inhaled into the body via a deep inspiration while its human host is sleeping. The micro-robot immediately lodges in the trachea among a group of pseudostratified, ciliated, columnar epithelial cells. Robo's initial maneuver is to get into a position for entry into one of the cells. The robot radios control command, "Might need a life jacket." There appears to be a watery "moat" around each cell, most likely the (27) _____ . The robots location seems to be in jeopardy because of the presence of small "finger-like" projections, the (28) _____ , which are swaying back and forth and threatening to "wash" the robot farther down into the respiratory tract. The robot's mechanical arm extends and grabs hold of the outer boundary of a cell, the (29) _____ . Robo is instructed to "look for" (30) _____ which form (31) _____ large enough for the robot to gain entry into the cell. As it passes through the opening, its chemical sensors pick up the presence of a (32) _____ bilayer with integral proteins embedded in the membrane and (33) _____ attached to the inner membrane surface. Once inside, the trek through the cytosol or (34) _____ begins. Resistance to movement is greater inside the cell than outside because of the presence of dissolved nutrients, (35) _____ , soluble and insoluble (36) _____ , and waste products. The first "observation" inside the cell is what looks like a protein framework, the (37) _____ , which gives the cytoplasm strength and flexibility. Formed structures, the (38) _____ , are in abundance, some attached and others freely floating in the (39) _____. Robo's contact with some of the organelles is easy because they are (40) _____ while others may be difficult to get into because of the presence of a membrane. Structures such as the ribosomes, which are involved in (41) _____ , and the centrioles, which direct strands of DNA during (42) _____ , are noted quite easily as Robo treks through the cytosol. A small "cucumber-shaped" structure, a(n) (43) _____ , is "sighted"; however, the robot's entry capabilities are fully taxed owing to the unusual double membrane. After penetrating the outer membrane, an inner membrane containing numerous folds called (44) _____ blocks further entry. The inner membrane serves to increase the surface area exposed to the fluid contents or (45) _____. The presence of (46) _____ attached to the folds would indicate that this is where (47) _____ is generated by the mitochondria. Robo senses the need for a quick "diffusion" through the outer membrane and back into the cytosol. The moving cytosol carries the micro-robot close to the center of the cell where it contacts the cell's control center, the (48) _____. Robo is small enough to pass through the double membrane, the (49) _____ , because it contains (50) _____ . Once inside, the environment is awesome! Nuclear organelles called (51) _____ engage in activities which would indicate that they synthesize the components of the (52) _____ since RNA and ribosomal proteins are in abundance. The fluid content of the nucleus, the (53) _____ , contains ions, enzymes, RNA, DNA, and their nucleotides. The DNA strands form complex structures called (54) _____ , which contain information to synthesize thousands of different proteins and control the synthesis of RNA. Robo leaves the nucleus by way of the ER or (55) _____ where it senses the manufacture of proteins at specific sites along the membrane, the (56) _____ , or RER. Some of the synthesized molecules are stored, others will be transported along with the robot by transport vesicles that will deliver them to the (57) _____ , a system of flattened membrane discs called (58) _____ . Robo relays the message to control command, "Looks like a stack of dinner plates in which synthesis and packaging of secretions along with cell membrane renewal and modification is taking place." Robo's exit from the cell occurs through the Golgi saccules since they communicate with the ER and the cell surface. Robo "boards" a secretory vesicle and is on its way out of the cell when it senses the presence of "suicide packets" called (59) _____ along the way. The trek ends with the robot's entrance into the extracellular fluid and passage into the lumen of the trachea where it waits for a "cough" from its host to exit the respiratory tract and return to Mission Control.

[L2] Multiple Choice:

Place the letter corresponding to the correct answer in the space provided.

_____ 60. Homeostasis at the tissue, organ, system, and individual levels reflects:

    a. the combined and coordinated actions of many cells

    b. a combined effort of a specific cell population

    c. maintenance of all organelles within the cell

    d. cell differentiation during embryonic development

_____ 61. Structurally, the cell membrane is best described as a:

    a. phospholipid layer integrated with peripheral proteins

    b. phospholipid bilayer interspersed with proteins

    c. protein bilayers interspersed with phospholipids

    d. protein layer interspersed with peripheral phospholipids

_____ 62. Isolating the cytoplasm from the surrounding fluid environment by the cell membrane is important because:

    a. the cell organelles lose their shape if the membrane is destroyed

    b. the nucleus needs protection to perform its vital functions

    c. cytoplasm has a composition different from the extracellular fluid and the differences must be maintained

    d. the cytoplasm contains organelles that need to be located in specific regions in order to function properly

_____ 63. Solutes cannot cross the lipid portion of a cell membrane because:

    a. the lipid tails of phospholipid molecules are highly hydrophobic and will not associate with water molecules

    b. most solutes are too large to get through the channels formed by integral proteins

    c. the hydrophilic heads are at the membrane surface and the hydrophobic tails are on the inside

    d. communication between the interior of the cell and the extracellular fluid is cut off

_____ 64. Regulation of exchange with the environment is an important function of the cell membrane because it controls:

    a. most of the activity occurring in the extracellular fluid

    b. the entry of ions and nutrients, and the elimination of wastes

    c. the activity that is occurring in the intracellular fluid

    d. alterations that might occur in the extracellular fluid

_____ 65. Membranous organelles differ from nonmembranous organelles in that membranous organelles are:

      a. always in contact with the cytosol

      b. unable to perform functions essential to normal cell maintenance

      c. usually found close to the nucleus of the cell

      d. surrounded by lipid membranes that isolate them from the cytosol

_____ 66. The major functional difference between flagella and cilia is that flagella:

      a. move fluid past a stationary cell

      b. move fluids or secretions across the cell surface

      c. move a cell through the surrounding fluid

      d. move DNA molecules during cell division

_____ 67. The smooth ER (SER) has a variety of functions that center around the synthesis of:

      a. lipids and carbohydrates

      b. proteins and lipids

      c. glycogen and proteins

      d. carbohydrates and proteins

_____ 68. The reason lysosomes are sometimes called "cellular suicide packets" is:

      a. the lysosome fuses with the membrane of another organelle

      b. lysosomes fuse with endocytic vesicles with solid materials

      c. the breakdown of lysosomal membranes can destroy a cell

      d. lysosomes have structures which penetrate other cells

_____ 69. The energy-producing process in the mitochondria involves a series of reactions in which _____ is consumed and _____ is generated.

      a. carbon dioxide; oxygen

      b. water; oxygen

      c. carbon dioxide; water

      d. oxygen; carbon dioxide

_____ 70. The most notable characteristic of the G_0 phase of an interphase cell is that:

      a. the cell is manufacturing cell organelles

      b. it is not in preparation for mitosis

      c. the cell duplicates its chromosomes

      d. DNA polymerase binds to nitrogen bases

_____ 71. The replication of DNA occurs primarily during the:

      a. S phase

      b. G_1 phase

      c. G_2 phase

      d. G_m phase

_____ 72. The process of *cytokinesis* refers to:

    a. the constriction of chromosomes along the metaphase plate

    b. the formation of spindle fibers between the centriole pairs

    c. the reorganization of the nuclear contents

    d. the physical separation of the daughter cells

_____ 73. The *passive* factor which helps to maintain the transmembrane potential is:

    a. K^+ diffuse out of the cell faster than Na^+ can enter and the interior of the cell develops an excess of negative charges

    b. Na^+ diffuse out of the cell faster than K^+ can enter and the exterior of the cell develops a negative charge

    c. K^+ and Na^+ diffuse equally across the cell membrane and the interior of the cell develops a negative charge

    d. the presence of proteins inside the cell and the presence of Cl^- outside the cell

_____ 74. The reason that dead skin cells are usually shed in thick sheets rather than individually is because of:

    a. the presence of the intercellular cement holding them together

    b. the strength of the links of the desmosomes

    c. the fusion of the cell membranes forming tight junctions

    d. the interlocking of the membrane proteins forming gap junctions

_____ 75. The reason that *water-soluble* ions and molecules cannot enter certain regions of the cell membrane is because of:

    a. the presence of hydrophilic ends exposed to the solution

    b. the presence of channels too small for ions and molecules to enter

    c. the presence of hydrophobic tails on the interior of the membrane

    d. the presence of gated channels which close when ions are present

_____ 76. The effect of diffusion in body fluids is that it:

    a. tends to increase the concentration gradient of the fluid

    b. tends to scatter the molecules and inactivate them

    c. tends to repel like charges and attract unlike charges

    d. tends to eliminate local concentration gradients

_____ 77. During *osmosis* water will always flow across a membrane toward the solution that has the:

    a. highest concentration of solvents

    b. highest concentration of solutes

    c. equal concentrations of solute

    d. equal concentrations of solvents

Level
=2=

_____ 78. A solution that is hypotonic to cytoplasm has:

      a. a solute concentration lower than that of the cytoplasm

      b. a solute concentration higher than that of the cytoplasm

      c. a solute concentration that is equal to that of the cytoplasm

      d. an osmotic concentration higher than that of the intra-cellular fluid

_____ 79. Red blood cells are _hemolyzed_ when the cells are placed in contact with:

      a. a hypotonic solution

      b. a hypertonic solution

      c. an isotonic solution

      d. a salt solution

_____ 80. An injection of a concentrated salt solution into the circulatory system would result in:

      a. little or no effect on the red blood cells

      b. hemolysis of the red blood cells

      c. a slight increase in cellular volume

      d. crenation of the red blood cells

_____ 81. _Facilitated diffusion_ differs from ordinary diffusion in that:

      a. ATP is expended during facilitated diffusion

      b. molecules move against a concentration gradient

      c. carrier proteins are involved

      d. it is an active process utilizing carriers

_____ 82. One of the great advantages of moving materials by _active transport_ is:

      a. carrier proteins are not necessary

      b. the process is not dependent on a concentration gradient

      c. the process has no energy cost

      d. receptor sites are not necessary for the process to occur

_____ 83. In the human body, the process of _phagocytosis_ is illustrated by:

      a. air expelled from the lungs

      b. a specific volume of blood expelled from the left ventricle

      c. vacuolar digestion of a solvent

      d. a white blood cell engulfing a bacterium

_____ 84. Epsom salts exert a laxative effect due to the process of:

      a. osmosis

      b. diffusion

      c. diarrhea

      d. phagocytosis

_____ 85. The formation of a malignant tumor indicates that

  a. the cells are remaining within a connective tissue capsule
  b. the tumor cells resemble normal cells, but they are dividing faster
  c. mitotic rates of cells are no longer responding to normal control mechanisms
  d. metastasis is necessary and easy to control

[L2]   Completion:

Using the terms below, complete the following statements.

cilia	hydrophobic	peroxisomes	microvilli
exocytosis	rough ER	phagocytosis	channels
diffusion	endocytosis	isotonic	cytokinesis
hormones	permeability	mitosis	

86. Ions and water-soluble compounds cannot cross the lipid portion of a cell membrane because the lipid tails of the phospholipid molecules are highly _____.

87. Water molecules, small water-soluble compounds, and ions pass in and out of cells through _____.

88. Cells that are actively engaged in _absorbing_ materials from the extracellular fluid, such as the cells of the digestive tract and the kidneys, contain _____.

89. In the respiratory tract, sticky mucus and trapped dust particles are moved toward the throat and away from delicate respiratory surface because of the presence of _____.

90. Pancreatic cells that manufacture digestive _enzymes_ contain an extensive _____.

91. Glycoproteins synthesized by the Golgi apparatus that cover most cell surfaces are usually released by the process of _____.

92. Cells may remove bacteria, fluids, and organic debris from their surroundings in vesicles at the cell surface by the process of _____.

93. Toxins such as alcohol or hydrogen peroxide that are absorbed from the extracellular fluid or generated by chemical reactions in the cytoplasm are absorbed and neutralized by _____.

94. The property that determines the cell membrane's effectiveness as a barrier is its _____.

95. A drop of ink spreading to color an entire glass of water demonstrates the process of _____.

96. If a solution has the same solute concentration as the cytoplasm and will not cause a net movement in or out of the cells, the solution is said to be _____.

97. An important means of coordinating carrier protein activity throughout the body is provided by _____.

98. Pseudopodia are cytoplasmic extensions that function in the process of _____.

99. The division of somatic cells followed by the formation of two daughter cells is a result of the process of _____.

100. The _end_ of a cell division is marked by the completion of _____.

[L2]   Short Essay:

Briefly answer the following questions in the spaces provided below.

101. Confirm your understanding of cell *specialization* by citing five (5) systems in the human body and naming a specialized cell found in each system.

102. List four (4) general functions of the cell membrane.

103. List three (3) ways in which the cytosol differs chemically from the extracellular fluid.

104. What *organelles* would be necessary to construct a functional "typical" cell? (Assume the presence of cytosol and a cell membrane.)

105. What are the *functional* differences among centrioles, cilia, and flagella?

106. What three (3) major factors determine whether a substance can diffuse across a cell membrane?

☐ LEVEL 3   CRITICAL THINKING/APPLICATION

Using principles and concepts learned in Chapter 3, answer the following questions. Write your answers on a separate sheet of paper.

1. A friend tells you that he has been diagnosed as having Kartagener's syndrome. What is this disorder and how does it affect your friend?

2. An instructor at the fitness center tells you that bodybuilders have the potential for increased supplies of energy and improved muscular performance because of increased numbers of mitochondria in their muscle cells. You ask, "Why?"

3. How does a transmembrane potential increase a cell's sensitivity to its environment?

4. Using the principles of *tonicity*, explain why a lifeguard at an ocean beach will have more of a chance to save a drowning victim than a lifeguard at an inland freshwater lake.

5. One remedy for constipation is a saline laxative such as Epsom salts ($MgSO_4$). Why do such salts have a laxative effect?

6. An advertisement currently appearing on television suggests that after strenuous exercise "Gatorade gives your body what it thirsts for." What is the benefit derived from ingesting such a drink after strenuous exercise?

7. In a hospital, a nurse gave a patient recovering from surgery a transfusion of 5% salt solution by mistake instead of a transfusion of physiological saline (0.9% salt). The patient almost immediately went into shock and soon after died. What caused the patient to enter into a state of shock?

8. If a prudent homemaker is preparing a tossed salad in the afternoon for the evening meal, the vegetables to be used will be placed in a bowl of cold water in order to keep these vegetables crisp. Osmotically speaking, explain why the vegetables remain crisp.

Level
≡**3**≡

4

The Tissue Level of Organization

■ Overview

Have you ever thought what it would be like to live on an island all alone? Your survival would depend on your ability to perform all the activities necessary to remain healthy and alive. In today's society surviving alone would be extremely difficult because of the important interrelationships and interdependence we have with others to support us in all aspects of life inherent in everyday living. So it is with individual cells in the multi-cellular body.

Individual cells of similar structure and function join together to form groups called *tissues*, which are identified on the basis of their origin, location, shape, and function. Many of the tissues are named according to the organ system in which they are found or the function which they perform such as neural tissue in the nervous system, muscle tissue in the muscular system or connective tissue that is involved with the structural frame-work of the body and supporting, surrounding, and interconnecting other tissue types.

The activities in Chapter 4 introduce the discipline of *histology*, the study of tissues, with emphasis on the four primary types: epithelial, connective, muscle,and neural tis-sue. The questions and exercises are designed to help you organize and conceptualize the interrelationships and interdependence of individual cells that extend to the tissue level of cellular organization.

❑ LEVEL 1   REVIEW OF CHAPTER OBJECTIVES

1. Identify the four major tissues of the body and their roles.
2. Describe the types and functions of epithelial cells.
3. Discuss the relationship between form and function for each epithelial type.
4. Compare the structures and functions of the various types of connective tissues.
5. Explain how epithelial and connective tissues combine to form four different types of membranes and specify the functions and locations of each.
6. Describe how connective tissue establishes the framework of the body.
7. Describe the three types of skeletal muscle tissue and their special structural features.
8. Discuss the basic structure and role of neural tissue.
9. Describe how nutrition and aging affect tissues of the body.

[L1]   Multiple choice:

Place the letter corresponding to the correct answer in the space provided.

OBJ. 1   _____   1. The four primary tissue types found in the human body are:

        a. squamous, cuboidal, columnar, glandular

        b. adipose, elastic, reticular, cartilage

        c. skeletal, cardiac, smooth, muscle

        d. epithelial, connective, muscle, neural

OBJ. 2   _____   2. The type of tissue that covers exposed surfaces and lines internal passageways and body cavities is:

        a. muscle

        b. neural

        c. epithelial

        d. connective

OBJ. 2   _____   3. The two types of *layering* recognized in epithelial tissues are:

        a. cuboidal and columnar

        b. squamous and cuboidal

        c. columnar and stratified

        d. simple and stratified

OBJ. 2   _____   4. The types of cells that form glandular epithelium that secrete enzymes and buffers in the pancreas and salivary glands are:

        a. simple squamous epithelium

        b. simple cuboidal epithelium

        c. stratified cuboidal epithelium

        d. transitional epithelium

OBJ. 2   _____   5. The type of epithelial tissue found only along the ducts that drain sweat glands is:

        a. transitional epithelium

        b. simple squamous epithelium

        c. stratified cuboidal epithelium

        d. pseudostratified columnar epithelium

OBJ. 3   _____   6. A single layer of epithelium covering a basement membrane is termed:

        a. simple epithelium

        b. stratified epithelium

        c. squamous epithelium

        d. cuboidal epithelium

OBJ. 3   _____   7. Simple epithelial cells are characteristic of regions where:

        a. mechanical or chemical stresses occur

        b. support and flexibility are necessary

        c. padding and elasticity are necessary

        d. secretion and absorption occur

Level
-1-

OBJ. 3 _____ 8. From a surface view, cells that look like fried eggs laid side by side are:

      a. squamous epithelium

      b. simple epithelium

      c. cuboidal epithelium

      d. columnar epithelium

OBJ. 3 _____ 9. Stratified epithelium has several cell layers above the basement membrane and is usually found in areas where:

      a. secretion and absorption occur

      b. mechanical or chemical stresses occur

      c. padding and elasticity are necessary

      d. storage and secretion occur

OBJ. 3 _____ 10. Cells that form a neat row with nuclei near the center of each cell and that appear square in typical sectional views are:

      a. stratified epithelium

      b. squamous epithelium

      c. cuboidal epithelium

      d. columnar epithelium

OBJ. 3 _____ 11. The major structural difference between columnar epithelia and cuboidal epithelia is that the *columnar epithelia*:

      a. are hexagonal and the nuclei are near the center of each cell

      b. consist of several layers of cells above the basement membrane

      c. are thin and flat and occupy the thickest portion of the membrane

      d. are taller and slender and the nuclei are crowded into a narrow band close to the basement membrane

OBJ. 3 _____ 12. Simple squamous epithelium would be found in the following areas of the body:

      a. urinary tract and inner surface of circulatory system

      b. respiratory surface of lungs

      c. lining of body cavities

      d. a, b, and c are correct

OBJ. 3 _____ 13. Stratified columnar epithelia provide protection along portions of the following systems:

      a. skeletal, muscular, endocrine, integumentary

      b. lymphatic, cardiovascular, urinary, reproductive

      c. nervous, skeletal, muscular, endocrine

      d. reproductive, digestive, respiratory, urinary

OBJ. 3 _____ 14. Glandular epithelia contain cells that produce:

      a. exocrine secretions only

      b. exocrine or endocrine secretions

      c. endocrine secretions only

      d. secretions released from goblet cells only

OBJ. 4   _____ 15. The three basic components of all connective tissues are:

     a. free exposed surface, exocrine secretions, endocrine secretions

     b. fluid matrix, cartilage, osteocytes

     c. specialized cells, extracellular protein fibers, ground substance

     d. satellite cells, cardiocytes, osteocytes

OBJ. 4   _____ 16. The three classes of connective tissue based on structure and function are:

     a. fluid, supporting, and connective tissue proper

     b. cartilage, bone, and blood

     c. collagenic, reticular, and elastic

     d. adipose, reticular, and ground

OBJ. 4   _____ 17. The two *major* cell populations found in connective tissue proper are:

     a. fibroblasts and adipocytes

     b. mast cells and lymphocytes

     c. melanocytes and mesenchymal cells

     d. fixed cells and wandering cells

OBJ. 4   _____ 18. Most of the volume in loose connective tissue is made up of:

     a. elastic fibers

     b. ground substance

     c. reticular fibers

     d. collagen fibers

OBJ. 4   _____ 19. The major purposes of adipose tissue in the body are:

     a. strength, flexibility, elasticity

     b. support, connection, conduction

     c. padding, cushioning, insulating

     d. absorption, compression, lubrication

OBJ. 4   _____ 20. Reticular tissue forms the basic framework and organization for several organs that have:

     a. a complex three-dimensional structure

     b. tightly packed collagen and elastic fibers

     c. adipocytes that are metabolically active

     d. relative proportions of cells, fibers, and ground substance

OBJ. 4   _____ 21. Tendons are cords of dense regular connective tissue that:

     a. cover the surface of a muscle

     b. connect one bone to another

     c. attach skeletal muscles to bones

     d. surround organs such as skeletal muscle tissue

OBJ. 4   _____ 22. Ligaments are bundles of elastic and collagen fibers that:

     a. connect one bone to another bone

     b. attach skeletal muscle to bones

     c. connect one muscle to another muscle

     d. cover the surface of a muscle

Level
-1-

OBJ. 4 _____ 23. The three major subdivisions of the extracellular fluid in the body are:

      a. blood, water, and saliva

      b. plasma, interstitial fluid, and lymph

      c. blood, urine, and saliva

      d. spinal fluid, cytosol, and blood

OBJ. 5 _____ 24. The type of tissue that fills internal spaces and provides structural support and a framework for communication within the body is:

      a. connective

      b. epithelial

      c. muscle

      d. neural

OBJ. 5 _____ 25. The mucous membranes that are lined by simple epithelia perform the functions of:

      a. digestion and circulation

      b. respiration and excretion

      c. absorption and secretion

      d. a, b, and c are correct

OBJ. 5 _____ 26. The mesothelium of serous membranes is very thin, a structural characteristic that makes them:

      a. subject to friction

      b. relatively waterproof and usually dry

      c. resistant to abrasion and bacterial attack

      d. extremely permeable

OBJ. 6 _____ 27. The two types of supporting connective tissues found in the body are:

      a. regular and irregular connective tissue

      b. collagen and reticular fibers

      c. proteoglycans and chondrocytes

      d. cartilage and bone

OBJ. 6 _____ 28. The three major types of cartilage found in the body are:

      a. collagen, reticular, and elastic cartilage

      b. regular, irregular, and dense cartilage

      c. hyaline, elastic, and fibrocartilage

      d. interstitial, appositional, and calcified

OBJ. 6 _____ 29. The flap (pinna) of the outer ear is extremely resilient and flexible because it contains:

      a. dense, irregular connective tissue

      b. collagen fibers

      c. fibrocartilage

      d. elastic cartilage

OBJ. 6 _____ 30. The pads that lie between the vertebrae of the spinal column contain:

      a. elastic fibers

      b. fibrocartilage

      c. hyaline cartilage

      d. dense, regular connective tissue

Level
-1-

OBJ. 6 _____ 31. Bone cells found in the lacunae within the matrix are called:

     a. osteocytes

     b. chondrocytes

     c. adipocytes

     d. stroma

OBJ. 7 _____ 32. Muscle tissue has the ability to:

     a. provide a framework for communication within the body

     b. carry impulses from one part of the body to another

     c. cover exposed surfaces of the body

     d. contract and produce active movement

OBJ. 7 _____ 33. The three types of muscle tissue found in the body are:

     a. elastic, hyaline, fibrous

     b. striated, nonstriated, fibrous

     c. voluntary, involuntary, nonstriated

     d. skeletal, cardiac, smooth

OBJ. 7 _____ 34. Skeletal muscle fibers are very unusual because they may be:

     a. a foot or more in length, and each cell contains hundreds of nuclei

     b. subject to the activity of pacemaker cells, which establish contraction rate

     c. devoid of striations, spindle-shaped with a single nucleus

     d. unlike smooth muscle cells capable of division

OBJ. 8 _____ 35. Neural tissue is specialized to:

     a. contract and produce movement

     b. carry electrical impulses from one part of the body to another

     c. provide structural support and fill internal spaces

     d. line internal passageways and body cavities

OBJ. 8 _____ 36. The major function of *neurons* in neural tissue is:

     a. to provide a supporting framework for neural tissue

     b. to regulate the composition of the interstitial fluid

     c. to act as phagocytes that defend neural tissue

     d. to transmit signals that take the form of changes in the transmembrane potential

OBJ. 8 _____ 37. Structurally, neurons are unique because they are the only cells in the body that have:

     a. lacunae and canaliculi

     b. axons and dendrites

     c. satellite cells and neuroglia

     d. soma and stroma

Level
-1-

OBJ. 9   _____ 38. The restoration of homeostasis after an injury involves two related processes, which are:

        a. necrosis and fibrous

        b. infection and immunization

        c. inflammation and regeneration

        d. isolation and reconstruction

OBJ. 9   _____ 39. The release of histamine by mast cells at an injury site produces the following responses:

        a. redness, warmth, and swelling

        b. bleeding, clotting, healing

        c. necrosis, fibrosis, scarring

        d. hematoma, shivering, retraction

OBJ. 9   _____ 40. One of the major effects of aging on connective tissues is:

        a. it becomes thinner and the individual bruises easily

        b. cartilage and bone soften and lose the ability to provide support

        c. cartilage becomes stiffer and less resilient and bones become brittle

        d. it becomes thicker and the tissue becomes harder

OBJ. 9   _____ 41. The three major factors that play a role in age-related reduction in bone strength in women are:

        a. disease, injury, genetics

        b. posture, injury, inactivity

        c. lordosis, scyphosis, osteoporosis

        d. inactivity, inadequate diet, decreased estrogen

OBJ. 9   _____ 42. To maintain normal bone structure throughout life, it is necessary for women to subscribe to a program that includes:

        a. hormonal replacement therapy

        b. exercise

        c. calcium supplements

        d. a, b, and c are necessary

[L1]   Completion:

Using the terms below, complete the following statements.

areolar	exocytosis	epithelial
necrosis	skeletal	recticular
endothelium	abscess	aponeuroses
genetic	collagen	neuroglia
neural	connective	mesothelium
lamina propria		

OBJ. 1   43. The four primary tissue types found in the body are connective, muscle, epithelial, and _____.

OBJ. 2   44. The type of tissue that makes up the surface of the skin is _____.

OBJ. 3   45. The epithelium that lines the body cavity is the _____.

OBJ. 3   46. The lining of the heart and blood vessels is called a(n) _____.

Level
-1-

OBJ. 3   47. In merocrine secretion, the product is released through _____.

OBJ. 4   48. Of the four primary types, the tissue that stores energy in bulk quantities is _____.

OBJ. 4   49. The most common fibers in connective tissue proper are _____ fibers.

OBJ. 4   50. Connective tissue fibers forming a branching, interwoven framework that is tough but flexible describes _____ fibers.

OBJ. 4   51. The least specialized connective tissue in the adult body is _____.

OBJ. 5   52. The loose connective tissue of a mucous membrane is called the _____.

OBJ. 6   53. Collagenous bands that cover the surface of a muscle and assist in attaching the muscle to another structure are called _____.

OBJ. 7   54. The only type of muscle tissue that is under voluntary control is _____.

OBJ. 8   55. Neural tissue contains several different kinds of supporting cells called _____.

OBJ. 9   56. The death of cells or tissues from disease or injury is referred to as _____.

OBJ. 9   57. An accumulation of pus in an enclosed tissue space is called an _____.

OBJ. 9   58. The most difficult age-related tissue changes to treat are those that are _____.

[L1]   Matching:

Match the terms in column B with the terms in column A. Use letters for answers in the spaces provided.

PART I

		COLUMN A	COLUMN B
OBJ. 1	____ 59.	histology	A. trachea
OBJ. 2	____ 60.	covering epithelia	B. absorption
OBJ. 2	____ 61.	glandular epithelia	C. fixed cells
OBJ. 3	____ 62.	microvilli	D. incomplete cellular layer
OBJ. 3	____ 63.	cilia	E. study of tissues
OBJ. 4	____ 64.	fibroblasts	F. wandering cells
OBJ. 4	____ 65.	mast cells	G. exocrine
OBJ. 5	____ 66.	synovial membrane	H. epidermis

PART II

		COLUMN A	COLUMN B
OBJ. 6	____ 67.	elastic ligaments	I. neuroeffector junction
OBJ. 7	____ 68.	muscle tissue	J. dendrites
OBJ. 7	____ 69.	skeletal muscle tissue	K. movement
OBJ. 8	____ 70.	synapse	L. tissue homeostasis
OBJ. 8	____ 71.	neuron	M. interconnects vertebrae
OBJ. 9	____ 72.	histamine	N. dilation
OBJ. 9	____ 73.	fibrin	O. clot
OBJ. 9	____ 74.	good nutrition	P. satellite cells, voluntary

[L1]   Drawing/Illustration Labeling:

Identify each numbered structure by labeling the following figures:

OBJ. 2   **FIGURE 4.1**   Types of Epithelial Tissue

75 _____

79 _____

76 _____

80 _____

77 _____

81 _____

78 _____

OBJ. 4   *FIGURE 4.2*   Types of Connective Tissue (proper)

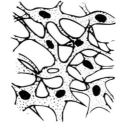

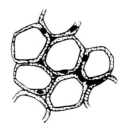

82 _____   83 _____   84 _____

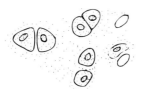

85 _____   86 _____

87 _____   88 _____   89 _____

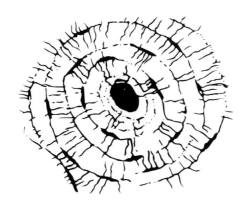

90 _____   91 _____

Level
—1—

OBJ. 7   *FIGURE 4.3*   Types of Muscle Tissue

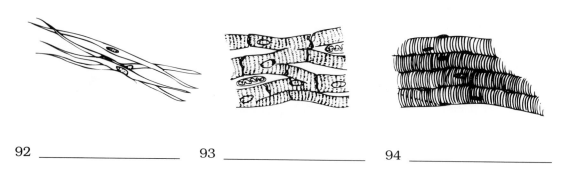

92 _____   93 _____   94 _____

OBJ. 8   *FIGURE 4.4*   Identify the Type of Tissue.

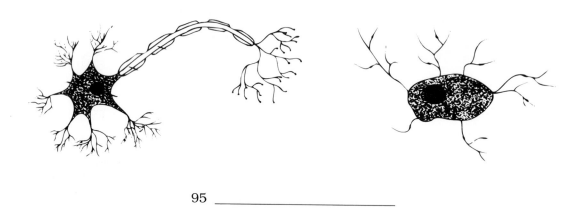

95 _____

Level
—1—   **When you have successfully completed the exercises in L1 proceed to L2.**

☐ LEVEL 2   CONCEPT SYNTHESIS

Concept Map I:

Using the following terms, fill in the circled, numbered blank spaces to complete the concept map. Follow the numbers that comply with the organization of the map.

Connective   Muscle   Neural   Epithelial

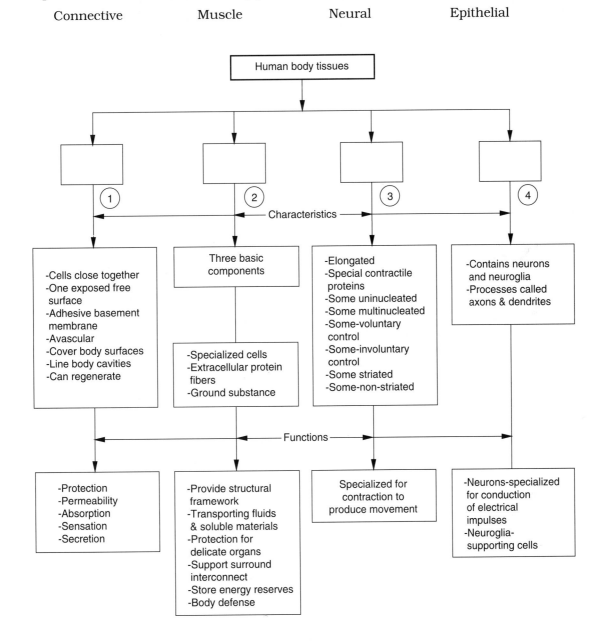

Concept Map II:

Using the following terms, fill in the circled, numbered, blank spaces to complete the concept map. Follow the numbers that comply with the organization of the map.

Transitional
Male urethra mucosa
Stratified squamous

Stratified cuboidal
Exocrine glands
Pseudostratified columnar

Lining
Endocrine glands

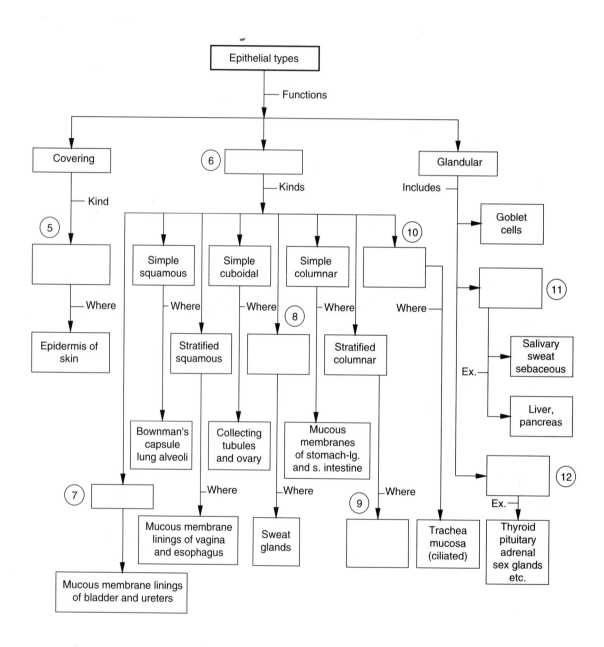

Concept Map III:

Using the following terms, fill in the circled, numbered, blank spaces to complete the concept map. Follow the numbers that comply with the organization of the concept map.

Ligaments   Loose connective tissue   Regular
Bone   Chondrocytes in lacunae   Tendons
Adipose   Fluid connective tissue   Blood
Hyaline

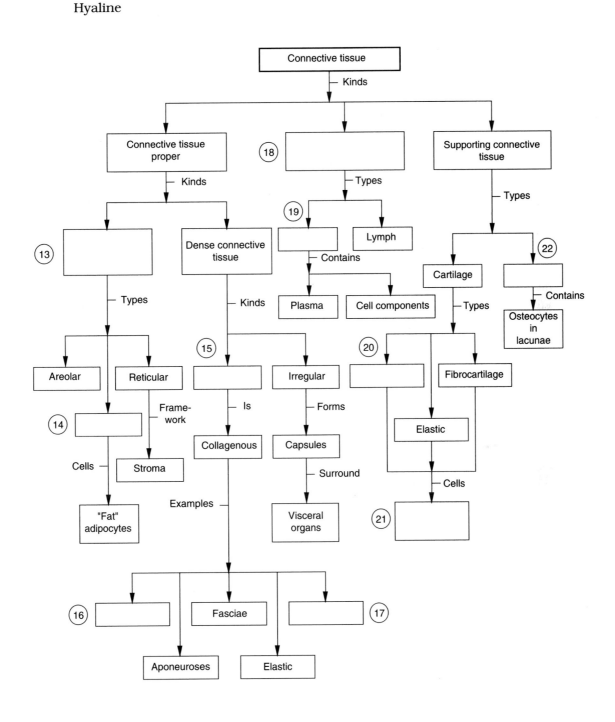

Concept Map IV:

Using the following terms, fill in the circled, numbered, blank spaces to complete the concept map. Follow the numbers that comply with the organization of the concept map.

Ground substance   Phagocytosis   Collagen
Mast cells   Fixed
Adipose cells   Branching interwoven framework

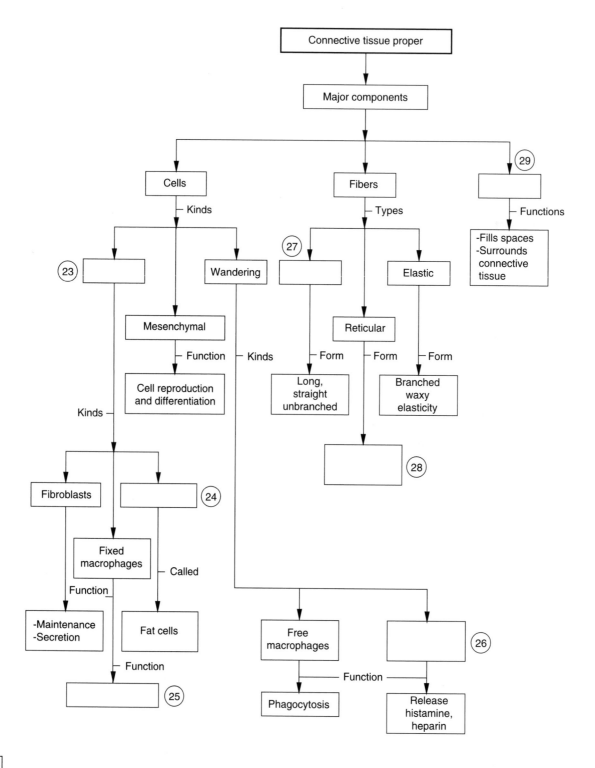

Concept Map V:

Using the following terms, fill in the circled, numbered, blank spaces to complete the concept map. Follow the numbers that comply with the organization of the concept map.

Multinucleated   Cardiac   Viscera
Involuntary   Nonstriated   Voluntary

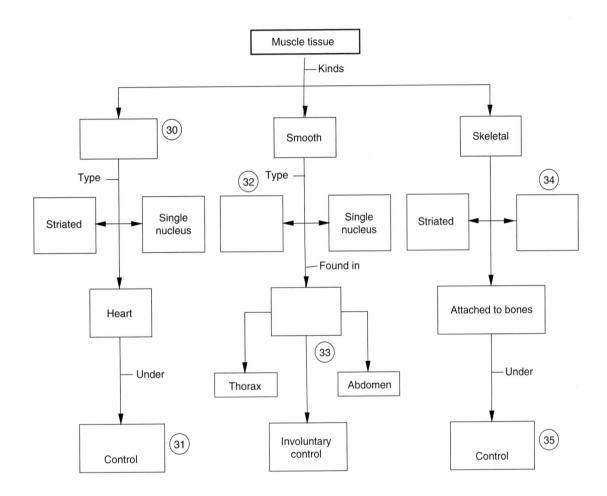

Concept Map VI:

Using the following terms, fill in the circled, numbered, blank spaces to complete the concept map. Follow the numbers that comply with the organization of the concept map.

Soma   Neuroglia   Dendrites

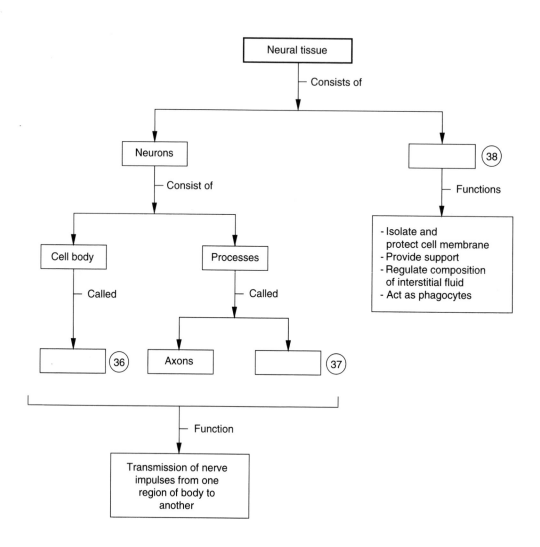

Concept Map VII:

Using the following terms fill in the circled, numbered, blank spaces to complete the concept map. Follow the numbers that comply with the organization of the map.

Thick, waterproof, dry
Pericardium
Synovial

Mucous
No basement membrane
Fluid formed on membrane
    surface

Goblet cells
Skin
Phagocytosis

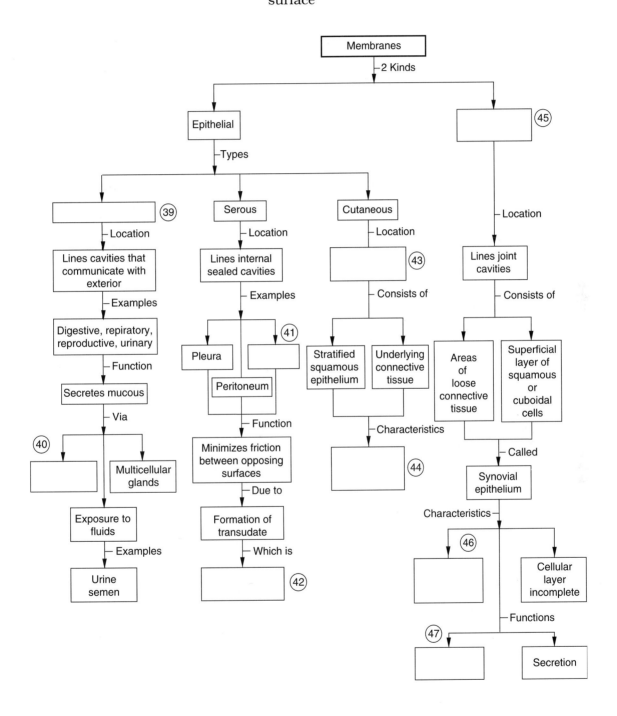

Body Trek:

To complete the body trek to study tissues, the micro-robot will be used in an experimental procedure by a pathologist to view and collect tissue samples from a postmortem examination. Robo is equipped with a mini camera to scan body cavities and organs and will use its tiny arm with a blade to retrieve tissue samples for study. The procedure avoids the necessity of severe invasive activity but allows a complete "tissue autopsy" to determine the ultimate cause of death. The tissue samples will be collected by the robot, taken to the laboratory for preparation, microscopically analyzed, and a report will be written and filed by the pathologist and *you*. All descriptions of the tissues will be designated as normal or abnormal. The report will be categorized as follows:

Body location; tissue type; description/appearance; N; A.

Using the terms listed below, complete the report on pages 78 – 79 relating to the body tissues by entering your responses in the blank spaces. The letter *N* refers to *normal*; the letter *A* to *abnormal*.

Epithelia	*Connective*	*Muscle*	*Neural*
Simple cuboidal	Chondrocytes in lacunae	Heart	Neurons–Axons
Trachea Mucosa (ciliated)	Tendons; Ligaments	Skeletal	Dendrites
Stratified squamous	Cardiovascular system	Nonstriated	Neuroglia
Layers of column-like cells	Irregular dense fibrous	Uninucleated	Support cells
Transitional	External ear; Epiglottis		
Simple squamous	Hyaline cartilage		
	Bone or Osseous		
	Adipose		

Body Location	Tissue Type	Description/Appearance	N	A
EPITHELIAL				
Mucous Membrane Lining of Mouth & Esophagus	(48)	Multiple Layers of Flattened Cells	X	
Mucosa of Stomach & Large Intestine	Simple Columnar	Single Rows of Column-Shaped Cells	X	
(49)	Pseudostratified Columnar	One Cell Layer – Cells Rest on Basement Membrane – (Evidence of Decreased Number of Cilia)		X
Respiratory Surface of Lungs	(50)	Single Layer of Flattened Cells (Excessive Number of Cells & Abnormal Chromosomes Observed)		X
Sweat Glands	Stratified Cuboidal	Layers of Hexagonal or Cube-like Cells	X	
Collecting Tubules of Kidney	(51)	Hexagonal Shape Neat Row of Single Cells	X	
Mucous Membrane Lining of Urinary Bladder	(52)	Cells with Ability to Slide Over One Another, Layered Appearance	X	
Male Urethra	Stratified Columnar	(53)	X	

Level =2=

Body Location	Tissue Type	Description/Appearance	N	A
CONNECTIVE				
Subcutaneous Tissue; Around Kidneys; Buttocks, Breasts	(54)	Closely Packed Fat Cells	X	
Widely Distributed Packages Organs; Forms Basement Membrane of Epithelial	Areolar (loose)	Three Types of Fibers; Many Cell Types	X	
(55)	Regular Dense Fibrous	Fibroblasts in Matrix, Parallel Collagenic and Elastic Fibers	X	
Dermis; Capsules of Joints; Fascia of Muscles	(56)	Fibroblast in Matrix, Irregularly Arranged Fibers	X	
Ends of Long Bones; Costal Cartilages of Ribs; Support Nose, Trachea, Larynx	(57)	Chondrocytes in Lacunae, Groups 2-4 Cells	X	
Intervertebral Disks Disks of Knee Joints	Fibro Cartilage	(58)	X	
(59)	Elastic Cartilage	Chondrocytes in Lacunae	X	
Skeleton	(60)	Osteocytes in Lacunae, Vascularized	X	
(61)	Blood	Liquid – Plasma RBC, WBC, Platelets	X	
MUSCLE				
Attached to Bones	(62)	Long; Cylindrical; Multinucleate	X	
(63)	Cardiac	Cardiocytes; Intercalated Disks	X	
Walls of Hollow Organs; Blood Vessels	Smooth	(64)	X	
NEURAL				
Brain; Spinal Cord; Peripheral Nervous System	Neural	(65)	X	

(End of Report)

This report confirms that death was due to metaplasia and anaplasia caused by excessive smoking. The tumor cancer cells in the lungs had extensive abnormal chromosomes.

Level
=2=

[L2]   Multiple Choice:

Place the letter corresponding to the correct answer in the space provided.

_____ 66. If epithelial cells are classified according to their cell *shape*, the classes would include:

  a. simple, stratified, pseudostratified
  b. squamous, cuboidal, columnar
  c. simple, squamous, stratified
  d. pseudostratified, stratified, columnar

_____ 67. If epithelial cells are classified according to their *function*, the classes would include those involved with:

  a. support, transport, storage
  b. defense, support, storage
  c. lining, covering, secreting
  d. protection, defense, transport

_____ 68. Epithelial tissues are differentiated from the other tissue types because they:

  a. always have a free surface exposed to the environment or to some internal chamber or passageway
  b. have few extracellular materials between adjacent epithelial cells
  c. do not contain blood vessels
  d. a, b, and c are correct

_____ 69. Certain epithelial cells are called *pseudostratified* columnar epithelium because:

  a. they have a layered appearance but all the cells contact the basement membrane
  b. they are stratified and all the cells do not contact the basement membrane
  c. their nuclei are all located the same distance from the cell surface
  d. they are a mixture of cell types

_____ 70. Three methods used by glandular epithelial cells to release secretions are:

  a. serous, mucous, and mixed secretions
  b. alveolar, acinar, tubuloacinar secretions
  c. merocrine, apocrine, holocrine secretions
  d. simple, compound, tubular secretions

_____ 71. Milk production in the breasts and underarm perspiration occur through:

  a. holocrine secretion
  b. apocrine secretion
  c. merocrine secretion
  d. tubular secretion

_____ 72. Holocrine secretions differ from other methods of secretion because:

  a. cytoplasm is lost as well as the secretory product
  b. the secretory product is released through exocytosis
  c. the product is released but the cell is destroyed
  d. the secretions leave the cell intact

_____ 73. Examples of exocrine glands that secrete onto some internal or external surface are:

        a. pituitary and thyroid

        b. thymus and salivary

        c. pancreas and pituitary

        d. serous and mucous

_____ 74. The two _fluid connective tissues_ found in the human body are:

        a. mucous and matrix

        b. blood and lymph

        c. ground substance and hyaluronic acid

        d. collagen and plasma

_____ 75. _Supporting connective tissues_ found in the body are:

        a. muscle and bone

        b. mast cells and adipocytes

        c. cartilage and bone

        d. collagen and reticular fibers

_____ 76. The common factor shared by the three connective tissue fiber types is that all three types are:

        a. formed through the aggregation of protein subunits

        b. abundant in all major organs in the body

        c. resistant to stretching due to the presence of ground substance

        d. springy, resilient structures capable of extensive stretching

_____ 77. During a weight loss program when nutrients are scarce, adipocytes:

        a. differentiate into mesenchymal cells

        b. are normally destroyed and disappear

        c. tend to enlarge and eventually divide

        d. deflate like collapsing balloons

_____ 78. Hyaline cartilage would serve to:

        a. support the pinna of the outer ear

        b. connect the ribs to the sternum

        c. support the epiglottis

        d. support the vocal cords

_____ 79. Summarizing the structural and functional properties of skeletal muscle tissues, it can be considered:

        a. nonstriated involuntary muscle

        b. nonstriated voluntary muscle

        c. striated voluntary muscle

        d. striated involuntary muscle

_____ 80. The major identifying feature characteristic of _mucous membranes_ is:

        a. they line cavities that communicate with the exterior

        b. they line the sealed, internal cavities of the body

        c. they minimize friction between opposing surfaces

        d. enclosed organs of the body are in close contact at all times

Level
2

_____ 81. Mucous membranes would be found primarily in the following systems:

      a. skeletal, muscular, endocrine, circulatory
      b. integumentary, lymphatic, nervous, endocrine
      c. digestive, respiratory, reproductive, urinary
      d. skeletal, lymphatic, circulatory, muscular

_____ 82. The pleura, peritoneum, and pericardium are examples of:

      a. mucous membranes
      b. body cavities
      c. visceral organs
      d. serous membranes

_____ 83. The primary function of a _serous_ membrane is to:

      a. provide nourishment and support to the body lining
      b. reduce friction betwee the parietal and visceral surfaces
      c. establish boundaries between internal organs
      d. line cavities that communicate with the exterior

_____ 84. In contrast to serous or mucous membranes, the _cutaneous_ membrane is:

      a. thin, permeable to water, and usually moist
      b. lubricated by goblet cells found in the epithelium
      c. thick, relatively waterproof, and usually dry
      d. covered with a specialized connective tissue, the lamina propria

_____ 85. The two factors that distinguish synovial epithelium from other types of epithelia are:

      a. there is a basement membrane and the cells are in rows next to each other
      b. there is no basement membrane and the cells are in rows next to each other
      c. there is a basement membrane and small spaces exist between adjacent cells
      d. there is no basement membrane and small spaces exist between adjacent cells

_____ 86. A component that synovial fluid and ground substance have in common is the presence of:

      a. hyaluronic acid
      b. phagocytes
      c. ascites
      d. satellite cells

_____ 87. The capsules that surround most organs such as the kidneys and organs in the thoracic and peritoneal cavities are components of the:

      a. superficial fascia
      b. deep fascia
      c. subserous fascia
      d. subcutaneous layer

Level
=2=

_____ 88. The two primary requirements for maintaining tissue homeostasis over time are:

   a. exercise and supplements

   b. hormonal therapy and adequate nutrition

   c. metabolic turnover and adequate nutrition

   d. supplements and hormonal therapy

_____ 89. The *loss* of cilia over time by epithelial cells is referred to as:

   a. dysplasia

   b. fibrosis

   c. anaplasia

   d. metaplasia

_____ 90. Of the following, the one that best defines *inflammation* is:

   a. the secretion of histamine to increase blood flow to the injured area

   b. a defense which involves the coordinated activities of several tissues

   c. a restoration process to heal the injured area

   d. the stimulation of macrophages to defend injured tissue

[L2]   Completion:

Using the terms below, complete the following statements.

stroma	dense regular connective tissue	adhesions
cancer	chemotherapy	remission
anaplasia	lamina propria	alveolar
caution	goblet cells	fibrosis
transudate	subserous fascia	dysplasia

91. The loose connective tissue component of a mucous membrane is called the _____.

92. The fluid formed on the surfaces of a serous membrane is called a(n) _____.

93. Restrictive fibrous connections caused by damaged serous membranes attracting fibroblasts which bind opposing membranes together with collagen fibers are referred to as _____.

94. The layer of loose connective tissue that lies between the deep fascia and the serous membranes that line body cavities is the _____.

95. Oncologists are physicians who specialize in the identification and treatment of _____.

96. The major goal of cancer treatment is to achieve _____.

97. The seven warning signs of cancer can be remembered by using the first letter of each warning phrase, which is _____.

98. Methods that involve the administration of drugs that kill cancerous tissues or prevent mitotic divisions are called _____,

99. In smokers, an *irreversible* form of lung cancer in which tissue organization breaks down is _____.

100. In smokers, a *reversible* condition in which the cilia of the trachea are first paralyzed is _____.

101. The only example of unicellular exocrine glands in the body is that of _____.

102. Glands made up of cells in a blind pocket are _____.

103. The basic framework of reticular tissue found in the liver, spleen, lymph nodes, and bone marrow is the _____.

104. Scar tissue is a dense collagenous framework produced by the process of _____.

105. Tendons, aponeuroses, fascia, elastic tissue, and ligaments are all examples of _____.

[L2]   Short Essay:

Briefly answer the following questions in the spaces provided below.

106. What are the four primary tissue types in the body?

107. Summarize the four essential functions of epithelial tissue.

108. What is the functional difference between microvilli and cilia on the exposed surfaces of epithelial cells?

109. How do the processes of merocrine, apocrine, and holocrine secretions differ?

110. List the types of exocrine glands in the body and identify their secretions.

111. What three basic components are found in all connective tissues?

112. What three classifications are recognized to classify connective tissues?

113. What three basic types of fibers are found in connective tissue?

114. What four kinds of membranes consisting of epithelial and connective tissues that cover and protect other structures and tissues are found in the body?

115. What are the three types of muscle tissue?

116. What two types of cell populations make up neural tissue, and what is the primary function of each type?

☐ LEVEL 3   CRITICAL THINKING/APPLICATION

Using principles and concepts learned about the tissue level of organization, answer the following questions. Write your answers on a separate sheet of paper.

1. Suppose you work in a hospital laboratory as a specialist in the *Histology* department. How do you perceive yourself as being an important part of a team of Allied Health Professionals involved in health care?

2. The processes of cornification and keratinization are associated with stratified squamous epithelial tissue. How do the processes differ and what is their value to the body?

3. Exocrine glands secrete products that reach the surface by means of excretory *ducts*. What exocrine glands are found in the integumentary and digestive systems?

4. How are connective tissues associated with body immunity?

5. Why is the bacterium *Staphylococcus aureus* particularly dangerous when infection occurs in the connective tissues of the body?

6. The knee joint is quite susceptible to injury involving the tearing of cartilage pads within the knee joint. In most cases, why is surgery needed?

7. The recommended dietary intake of vitamin C for an adult is 60 mg daily. What relationship does an optimum amount of vitamin C have to tissue development in the body?

8. Mast cells are connective tissue cells often found near blood vessels. What chemicals do they release and how are the chemicals used following an injury or infection?

9. Continuous exposure to cigarette smoke causes progressive deterioration of tissues in the respiratory tract. Using the terms *dysplasia, metaplasia,* and *anaplasia,* describe the progressive deterioration that affects the histological organization of tissues and organs in the respiratory tract, which eventually leads to death.

10. You are asked to give a talk about cancer in speech class. How might you use the word "caution" to identify the warning signs that might make early detection possible?

5

The Integumentary System

■ Overview

The integumentary system consists of the skin and associated structures including hair, nails, and a variety of glands. The four primary tissue types making up the skin comprise what is considered to be the largest structurally integrated organ system in the human body.

Because the skin and its associated structures are readily seen by others, a lot of time is spent caring for the skin, to enhance its appearance and prevent skin disorders that may alter desirable structural features on and below the skin surface. The integument manifests many of the functions of living matter, including protection, excretion, secretion, absorption, synthesis, storage, sensitivity, and temperature regulation. Studying the important structural and functional relationships in the integument provides numerous examples which demonstrate patterns that apply to tissue interactions in other organ systems.

☐ LEVEL 1   REVIEW OF CHAPTER OBJECTIVES

1. Describe the main structural features of the epidermis and explain their functional significance.
2. Explain what accounts for individual and racial differences in skin, such as skin color.
3. Discuss the effects of ultraviolet radiation on the skin and the role played by melanocytes.
4. Describe the structure and functions of the dermis.
5. Explain the mechanisms that produce hair and determine hair texture and color.
6. Discuss the various kinds of glands found in the skin and their secretions.
7. Explain how the sweat glands of the integumentary system play a major role in regulating body temperature.
8. Describe the anatomical structure of nails and how they are formed.
9. Explain how the skin responds to injury and repairs itself.
10. Summarize the effects of the aging process on the skin.

[L1]   Multiple choice:

Place the letter corresponding to the correct answer in the space provided.

OBJ. 1   _____   1. The two functional components of the integument include:

      a. dermis and epidermis

      b. hair and skin

      c. cutaneous membrane and accessory structures

      d. eleidin and keratin

OBJ. 1   _____   2. The layers of the epidermis, beginning with the deepest layer and proceeding outwardly, include the stratum:

      a. corneum, granulosum, spinosum, germinativum

      b. granulosum, spinosum, germinativum, corneum

      c. spinosum, germinativum, corneum, granulosum

      d. germinativum, spinosum, granulosom, corneum

OBJ. 1   _____   3. The layers of the epidermis where mitotic divisions occur are:

      a. germinativum and spinosum

      b. corneum and germinativum

      c. pinosum and corneum

      d. mitosis occurs in all the layers

OBJ. 1   _____   4. For a cell to move from the stratum germinivatum to the stratum corneum, it takes approximately:

      a. 6 weeks

      b. 7 days

      c. 1 month

      d. 14 days

OBJ. 1   _____   5. Epidermal cells in the strata spinosum and germinativum function as a chemical factory in that they can convert:

      a. steroid precursors to vitamin D when exposed to sunlight

      b. eleidin to keratin

      c. keratohyalin to eleidin

      d. a and c

OBJ. 2   _____   6. Differences in skin color between individuals and races reflect distinct:

      a. numbers of melanocytes

      b. melanocyte distribution patterns

      c. levels of melanin synthesis

      d. U.V. responses and nuclear activity

OBJ. 2   _____   7. The two basic factors interacting to produce skin color are:

      a. sunlight and ultraviolet radiation

      b. the presence of carotene and melanin

      c. melanocyte production and oxygen supply

      d. circulatory supply and pigment concentration and composition

OBJ. 3   _____ 8. Skin exposure to *small amounts* of ultraviolet radiation serves to:

        a. produce a tan that is beneficial to the skin

        b. convert a steroid related to cholesterol into vitamin D

        c. induce growth of cancerous tissue in the skin

        d. induce melanocyte production

OBJ. 3   _____ 9. Excessive exposure of the skin to U.V. radiation may cause redness, edema, blisters, and pain. The presence of blisters classifies the burn as:

        a. first degree

        b. second degree

        c. third degree

        d. none of these

OBJ. 4   _____ 10. The two major components of the dermis are:

        a. capillaries and nerves

        b. dermal papillae and a subcutaneous layer

        c. sensory receptors and accessory structures

        d. papillary and deep reticular layers

OBJ. 4   _____ 11. From the following selections, choose the one that identifies what the dermis contains to communicate with other organ systems.

        a. blood vessels

        b. lymphatics

        c. nerve fibers

        d. a, b, and c

OBJ. 4   _____ 12. Special smooth muscles in the dermis that, when contracted, produce "goose bumps" are called:

        a. tissue papillae

        b. arrector pili

        c. root sheaths

        d. cuticular papillae

OBJ. 5   _____ 13. Hair production occurs in the:

        a. reticular layers of the dermis

        b. papillary layer of the dermis

        c. hypodermis

        d. stratum germinativum of the epidermis

OBJ. 5   _____ 14. Except for red hair, the natural factor responsible for varying shades of hair color is:

        a. number of melanocytes

        b. amount of carotene production

        c. the type of pigment present

        d. a, b, and c

OBJ. 6   _____ 15. Accessory structures of the skin include:

        a. dermis, epidermis, hypodermis

        b. cutaneous and subcutaneous layers

        c. hair follicles, sebaceous and sweat glands

        d. blood vessels, macrophages, neurons

OBJ. 6   _____ 16. Sensible perspiration released by the eccrine sweat glands serves to:

        a. cool the surface of the skin

        b. reduce body temperature

        c. dilute harmful chemicals

        d. a, b, and c

OBJ. 7   _____ 17. When the body temperature becomes abnormally high, thermo-regulatory homeostasis is maintained by:

        a. an increase in sweat gland activity and blood flow to the skin

        b. a decrease in blood flow to the skin and sweat gland activity

        c. an increase in blood flow to the skin and a decrease in sweat gland activity

        d. an increase in sweat gland activity and a decrease in blood flow to the skin

OBJ. 8   _____ 18. Nail production occurs at an epithelial fold not visible from the surface called the:

        a. eponychium

        b. cuticle

        c. nail root

        d. lunula

OBJ. 9   _____ 19. The immediate response by the skin to an injury is:

        a. bleeding occurs and mast cells trigger an inflammatory response

        b. the epidermal cells are immediately replaced

        c. fibroblasts in the dermis create scar tissue

        d. the formation of a scab

OBJ. 9   _____ 20. The practical limit to the healing process of the skin is the formation of inflexible, fibrous, noncellular:

        a. scabs

        b. skin grafts

        c. ground substance

        d. scar tissue

OBJ. 10   _____ 21. Dangerously high body temperatures occur sometimes in the elderly due to:

        a. reduction in the number of Langerhans cells

        b. decreased blood supply to the dermis

        c. decreased sweat gland activity

        d. b and c

Level
-1-

OBJ. 10 _____ 22. A factor which causes increased skin damage and infection in the elderly is:

     a. decreased sensitivity of the immune system

     b. decreased vitamin D production

     c. a decline in melanocyte activity

     d. a decline in glandular activity

OBJ. 10 _____ 23. Hair turns gray or white due to:

     a. a decline in glandular activity

     b. a decrease in the number of Langerhans cells

     c. decreased melanocyte activity

     d. decreased blood supply to the dermis

OBJ. 10 _____ 24. Sagging and wrinkling of the integument occurs from:

     a. the decline of germinative cell activity in the epidermis

     b. a decrease in the elastic fiber network of the dermis

     c. a decrease in vitamin D production

     d. deactivation of sweat glands

[L1]   Completion:

Using the terms below, complete the following statements.

apocrine	Langerhans cells	follicle	eponychium
cyanosis	decrease	sebaceous glands	vellus
connective	blisters	glandular	iron
melanocyte	stratum corneum	sebum	MSH
eccrine glands	stratum lucidum	contraction	melanin

OBJ. 1   25. In areas where the skin is thick, such as the palms of the hands and the soles of the feet, the cells are flattened, densely packed, and filled with eleidin. This layer is called the _____.

OBJ. 1   26. Keratin, a fibrous protein, would be found primarily in the _____.

OBJ. 1   27. Mobile macrophages that are a part of the immune system and found scattered among the deeper cells of the epidermis are called _____.

OBJ. 2   28. The peptide secreted by the pituitary gland which increases the rate of melanin synthesis is _____.

OBJ. 3   29. The pigment which absorbs ultraviolet radiation before it can damage mitochondrial DNA is _____.

OBJ. 4   30. The type of tissue which comprises most of the dermis is _____.

OBJ. 5   31. The fine "peach fuzz" hairs found over much of the body surface are called _____.

OBJ. 5   32. Hair follicles are often associated with _____.

OBJ. 5   33. Hair develops from a group of epidermal cells at the base of a tube-like depression called a(n) _____.

OBJ. 5   34. Red hair contains a(n) _____ pigment that does not occur in hair of any other color.

OBJ. 6   35. The secretion which lubricates and inhibits the growth of bacteria on the skin is called _____.

OBJ. 6

36. The glands in the skin which become active when the body temperature rises above normal are the _____.

OBJ. 7

37. The sweat glands that communicate with hair follicles are called _____.

OBJ. 7

38. If the body temperature drops below normal, heat is conserved by a(n) _____ in the diameter of dermal blood vessels.

OBJ. 8

39. The stratum corneum that covers the exposed nail closest to the root is the _____.

OBJ. 9

40. During a sustained reduction in circulatory supply, the skin takes on a bluish coloration called _____.

OBJ. 9

41. A second-degree burn is readily identified by the appearance of _____.

OBJ. 9

42. An essential part of the healing process during which the edges of a wound are pulled closer together is called _____.

OBJ. 10

43. In older Caucasians, the skin becomes very pale because of a decline in _____ activity.

OBJ. 10

44. In older adults, dry and scaly skin is usually a result of a decrease in _____ activity.

[L1]   Matching:

Match the terms in column B with the terms in column A. Use letters for answers in the spaces provided.

	COLUMN A		COLUMN B
OBJ. 1	_____ 45. epidermis	A.	vitamin D
OBJ. 2	_____ 46. melanin	B.	1st degree burn
OBJ. 3	_____ 47. ultraviolet radiation	C.	club hair
OBJ. 4	_____ 48. dermis	D.	eccrine gland
OBJ. 5	_____ 49. inactive follicle	E.	nails
OBJ. 6	_____ 50. body odor	F.	loss of elastin
OBJ. 7	_____ 51. thermoregulation	G.	blood vessel and nerve supply
OBJ. 8	_____ 52. accessory structures	H.	stratified squamous epithelium
OBJ. 9	_____ 53. erythema	I.	apocrine sweat glands
OBJ. 10	_____ 54. wrinkles	J.	skin pigment

Level
-1-

[L1]   Drawing/Illustration Labeling:

Identify each numbered structure by labeling the following figures:

OBJ. 1   ***FIGURE 5.1***   Organization of the Integument
OBJ. 4

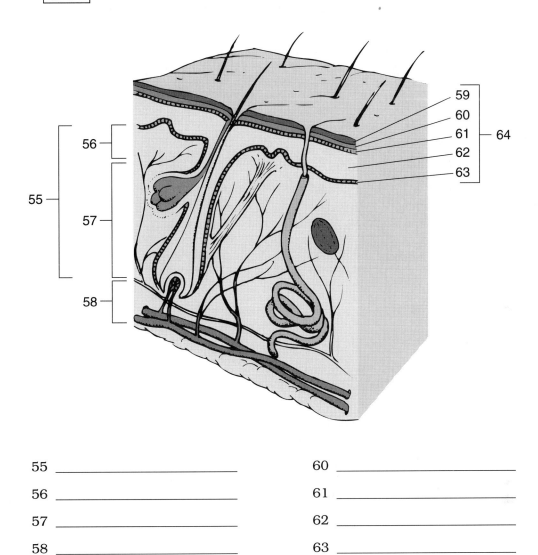

55 _____   60 _____

56 _____   61 _____

57 _____   62 _____

58 _____   63 _____

59 _____   64 _____

Level
–1–

OBJ. 2 ***FIGURE 5.2*** Hair Follicle, Sebaceous Gland Arrector Pili Muscle

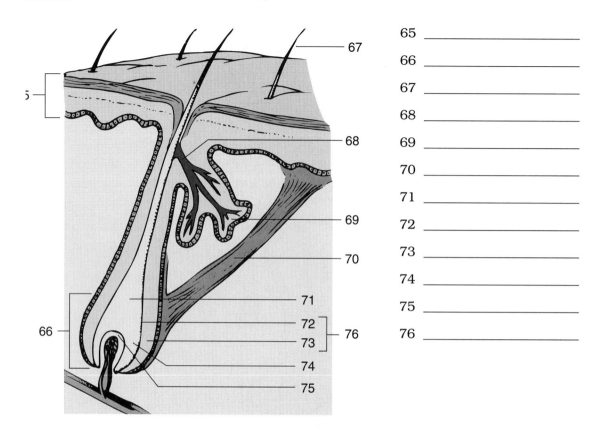

65 _____
66 _____
67 _____
68 _____
69 _____
70 _____
71 _____
72 _____
73 _____
74 _____
75 _____
76 _____

OBJ. 8 ***FIGURE 5.3*** Nail Structure (a) Nail Surface (b) Sectional View

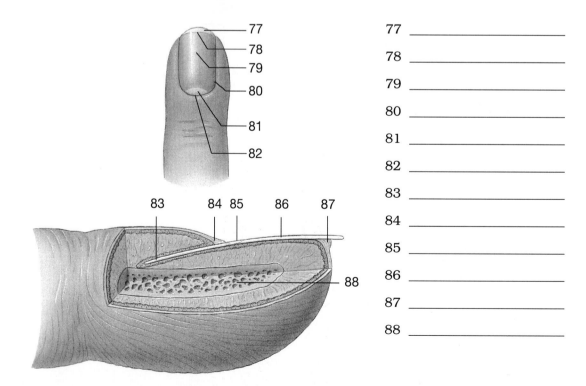

77 _____
78 _____
79 _____
80 _____
81 _____
82 _____
83 _____
84 _____
85 _____
86 _____
87 _____
88 _____

Level
–1–

When you have successfully completed the exercises in L1 proceed to L2.

☐ LEVEL 2   CONCEPT SYNTHESIS

Concept Map I:

Using the terms below, fill in the circled, numbered blank spaces to complete the concept map. Follow the numbers which comply with the organization of the map.

Sensory Reception   Vitamin D Synthesis   Produce Secretions
Lubrication   Dermis   Exocrine Glands

Concept Map II:

Using the terms below, fill in the circled, numbered blank spaces to complete the concept map. Follow the numbers which comply with the organization of the map.

Nerves   Epidermis   Collagen
Skin   Hypodermis   Connective
Fat   Granulosum   Papillary Layer

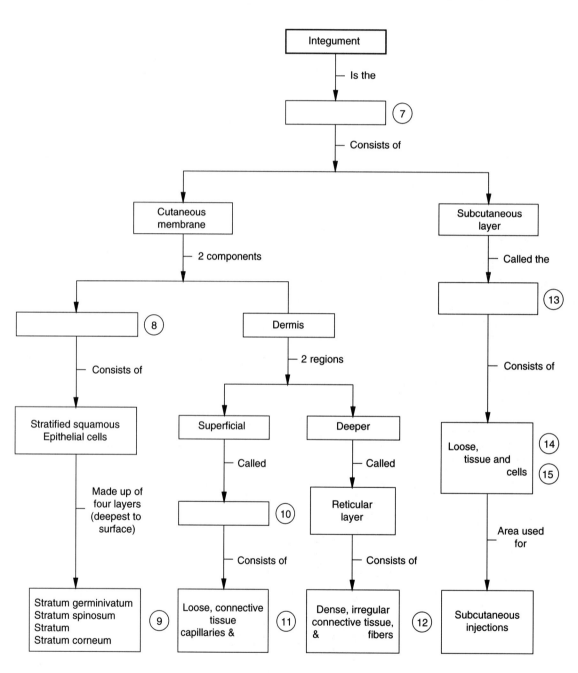

Concept Map III:

Using the terms below, fill in the circled, numbered blank spaces to complete the concept map. Follow the numbers complying with the organization of the concept map.

Merocrine or "Eccrine"
Thickened Stratum Corneum
Cerumen ("Ear Wax")
Odors

"Peach Fuzz"
Arms & Legs
Terminal
Sebaceous

Glands
Lunula
Cuticle

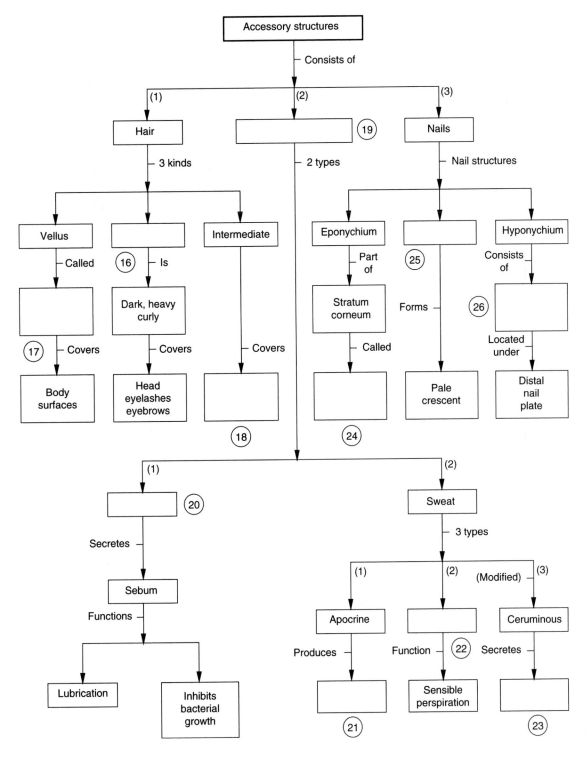

Body Trek:

Using the terms below, fill in the blanks to complete the trek through the integument.

Stratum lucidum	Stratum granulosum	Sebaceous
Collagen	Hyponichium	Desmosomes
Stem cells	Epidermal ridges	Eleidin
Mitosis	Papillary	Cuticle
Elastic	Stratum spinosum	Keratin
Accessory	Dermal papillae	Keratohyalin
Lunula	Hypodermis	Reticular layer

The body trek begins as the micro-robot is placed on the nail plate of the thumbnail, a(n) (27) _____ part of the integument. Looking over the free edge of the nail, Robo observes a thickened stratum corneum, the (28) _____. Turning around and advancing toward the proximal end of the nail plate, a pale crescent known as the (29) _____ comes into view. This area serves as the "entrance" to a part of the stratum corneum which is folded and extends over the exposed nail nearest the root, forming the (30) _____, or eponychium. Robo expresses confusion as the message relayed to Mission Control says, "I thought I would be treking on a living surface, but I sense nothing but layers of dead, flattened cells." A chemical analysis confirms the presence of a fibrous protein called (31) _____ , which makes the cells durable and water resistant and allows them to exist for about two weeks until they are shed or washed away. Robo treks along the "dead" surface until it arrives at the palm of the hand, a "thick" area of the skin. The robot's descent is carefully controlled because of the presence of a "glassy" (32) _____, a clear layer of flattened, densely packed cells filled with (33) _____ , a protein derived from (34) _____ which is produced in great quantities in the "grainy" layers below, the (35) _____. Below these strata the layers are several cells thick and bound together by (36) _____. Some of the cells act as if they are in the process of dividing. Due to the arrangement and activity of these cells, Robo confirms that its location is the (37) _____. The deeper Robo probes, the more lively the activity becomes. In the deepest stratum of the epidermis, the dominant cell population consists of (38) _____ , which are actively involved in the process of (39) _____. The deeper layers of the epidermis that Robo has trekked through form (40) _____, which extend into the dermis, increasing the area of contact between the two regions and providing a route for the trek into the dermis. The dermis consists of a (41) _____ layer of loose connective tissue, and the first appearance of capillaries and nerves is evident. The region derives its name from nipple-shaped mounds called (42) _____, which project between the epidermal ridges. Probing deeper, Robo senses an area of dense, irregular connective tissue, bundles of collagen fibers, lymphatics, fat cells, muscle cells, and accessory sweat and (43) _____ glands. These structures give rise to a layer known as the (44) _____. The (45) _____ provides strength, and scattered (46) _____ fibers give the dermis the ability to stretch and contract during normal movements. The fibers of the deepest dermal layer are continuous with a subcutaneous layer, the (47) _____ , an area of loose connective tissue and an abundance of fat cells. Sensory receptors in the dermal and deep epidermal layers interfere with Robo's movements and the search for an exit is imminent. Mission Control instructs Robo to find a hair follicle on the dorsal side of the hand, mount the hair shaft, and proceed to the distal end of the hair where the robot is removed and re-energized for the next trek.

[L2] Multiple Choice:

Place the letter corresponding to the correct answer in the space provided.

_____ 48. According to the "rule of nines" if the left arm and left leg of an adult are burned, the percentage of surface area affected would be:

      a. 9%

      b. 18%

      c. 27%

      d. 54%

Level
=2=

_____ 49. A hospital report listed an injury to the skin as a jagged, irregular surface tear, produced by solid impact or an irregular object. The term used to describe this injury is:

      a. a puncture

      b. an avulsion

      c. an abrasion

      d. a laceration

_____ 50. Psoriasis is a skin disorder in which there is abnormal increased mitotic activity in the:

      a. stratum spinosum

      b. stratum lucidum

      c. stratum germinativum

      d. stratum corneum

_____ 51. Third degree burns differ from first and second degree burns in that:

      a. the epidermis, dermis, and hypodermis are destroyed

      b. they are more painful

      c. fluid accumulates between the dermis and epidermis

      d. the burn is restricted to the superficial layers of the skin

_____ 52. Wounds are classified as open or closed. From the following choices, select the one which does not include only open wounds.

      a. abrasions, avulsions, incisions, lacerations

      b. lacerations, incisions, avulsions, contusions

      c. punctures, abrasions, incisions, avulsions

      d. avulsions, lacerations, incisions, punctures

_____ 53. Because freshwater is hypotonic to body fluids, sitting in a freshwater bath causes:

      a. water to leave the epidermis and dehydrate the tissue

      b. water from the interstitial fluid to penetrate the surface and evaporate

      c. water to enter the epidermis and cause the epithelial cells to swell

      d. complete cleansing because the bacteria on the surface drown

_____ 54. Malignant melanomas are extremely dangerous and life threatening because:

      a. they develop in the germinative layer of the epidermis

      b. they form tumors which interfere with circulation

      c. metastasis is restricted to the dermis and epidermis

      d. the melanocytes grow rapidly and metastasize through the lymphatic system

_____ 55. A "boil" or furuncle develops when:

      a. apocrine glands cease to function

      b. the duct of a sebaceous gland becomes blocked

      c. eccrine gland malfunctions cause an inflammation

      d. bacteria invade the sweat glands

Level
=2=

_____ 56. Ceruminous glands are modified sweat glands located in the:

        a. reticular layer of the dermis

        b. stratum spinosum of the epidermis

        c. nasal passageways

        d. external auditory canal

_____ 57. Yellow fingernails are indicated in patients who have:

        a. chronic respiratory disorders

        b. thyroid gland disorders

        c. AIDS

        d. a, b, or c

_____ 58. An abscess is best described as:

        a. an accumulation of pus in an enclosed tissue space

        b. a widespread inflammation of the dermis caused by bacterial infection

        c. a necrosis occurring because of inadequate circulation

        d. a sore which affects the skin near a joint or projecting bone

[L2]   Completion:

Using the terms below, complete the following statements.

keloid	Langerhans cells
Merkel cells	alopecia areata
hemangiomas	hirsutism
contusions	seborrheic dermatitis
hives	papillary region of dermis

59. If a doctor diagnoses a skin inflammation as urticaria, he suspects that the patient has _____.

60. Growth of hair on women in patterns usually characteristic of men is called _____.

61. Black and blue marks, or "black eyes," are familiar examples of _____.

62. The dendrites of sensory neurons and specialized touch receptor cells provide information about objects touching the skin. These specialized cells are called _____.

63. Dermatitis is an inflammation of the skin that primarily involves the _____.

64. Benign tumors which usually occur in the dermis are called _____.

65. A localized hair loss that can affect either sex is called _____.

66. "Cradle cap" in infants and "dandruff" in adults result from an inflammation around abnormally active sebaceous glands called _____.

67. A thick, flattened mass of scar tissue that begins at the injury site and grows into the surrounding tissue is called the _____.

68. Due to the onset of aging, increased damage and infection is apparent because of a decrease in the number of _____.

Level
=2=

[L2]   Short Essay:

Briefly answer the following questions in the spaces provided below.

69. A friend says to you, "Don't worry about what you say to her; she is thick-skinned." Anatomically speaking, what areas of the body would your friend be referring to? Why are these areas thicker?

70. Two females are discussing their dates. One of the girls says, "I liked everything about him except he had body odor." What is the cause of body odor?

71. A hypodermic needle is used to introduce drugs into the loose connective tissue of the hypodermis. Beginning on the surface of the skin in the region of the thigh, list, in order, the layers of tissue the needle would penetrate to reach the hypodermis.

72. The general public associates a tan with good health. What is wrong with this assessment?

73. Many shampoo advertisements list the ingredients, such as honey, kelp extracts, beer, vitamins and other nutrients as being beneficial to the hair. Why could this be considered false advertisement?

74. Two teenagers are discussing their problems with acne. One says to the other, "Sure wish I could get rid of these whiteheads." The other replies, "At least you don't have blackheads like I do." What is the difference between a "whitehead" and a "blackhead"?

Level
=2=

75. You are a nurse in charge of a patient who has decubitus ulcers, or "bedsores." What causes bedsores, and what should have been done to prevent skin tissue degeneration or necrosis which has occurred?

76. After a first date, a young man confides to a friend, "Every time I see her, I get 'goose bumps.'" What is happening in this young man's skin?

When you have successfully completed the exercises in L2 proceed to L3.

☐ LEVEL 3   CRITICAL THINKING/APPLICATION

Using principles and concepts learned about the integumentary system, answer the following questions. Write your answers on a separate sheet of paper.

1. Even though the stratum corneum is water resistant, it is not waterproof. When the skin is immersed in water, osmotic forces may move water in or out of the epithelium. Long-term exposure to seawater endangers survivors of a shipwreck by accelerating dehydration. How and why does this occur?

2. A young Caucasian girl is frightened during a violent thunderstorm, during which lightning strikes nearby. Her parents notice that she is pale, in fact, she has "turned white." Why has her skin color changed to this "whitish" appearance?

3. Tretinoin (Retin-A) has been called the anti-aging cream. Since it is applied topically, how does it affect the skin?

4. The cornification process is quite specialized in hair production. Using the terms below, briefly describe the process. Underline each term as you describe the process.

basal cells	medulla	papilla
matrix	cortex	hard keratin
keratinization	soft keratin	cuticle

5. Someone asks you, "Is hair really important to the human body?" What responses would you give to show the functional necessities of hair?

6. Suppose that a hair in the scalp grows at a rate of around 0.33 mm/day. What would the length of the hair be in inches after a period of three years? (Note: 2.54 cm = 1 in.)

7. Individuals who participate in endurance sports must continually provide the body with fluids. Explain why this is necessary.

8. Why do calluses form on the palms of the hands when doing manual labor?

9. Bacterial invasion of the superficial layers of the skin is quite common. Why is it difficult to reach the underlying connective tissues? (Cite at least six (6) features of the skin that help protect the body from invasion by bacteria.)

6

Osseous Tissue and Skeletal Structure

■ Overview

Osteology is a specialized science that is the study of bone (osseous) tissue and skeletal structure and function. The skeletal system consists of bones and related connective tissues which include cartilage, tendons, and ligaments. Chapter 6 addresses the topics of bone development and growth, histological organization, bone classification, the effects of nutrition, hormones, and exercise on the skeletal system, and the homeostatic mechanisms that operate to maintain skeletal structure and function throughout the lifetime of an individual.

The exercises in this chapter are written to show that although bones have common microscopic characteristics and the same basic dynamic nature, each bone has a characteristic pattern of ossification and growth, a characteristic shape, and identifiable surface features that reflect its integrated functional relationship to other bones and other systems throughout the body. It is important for the student to understand that even though bone tissue is structurally stable, it is living tissue and is functionally dynamic.

☐ LEVEL 1   REVIEW OF CHAPTER OBJECTIVES

1. Describe the functions of the skeletal system.
2. Identify the cell types found in bone and list their major functions.
3. Compare the structures and functions of compact and spongy bone.
4. Compare the mechanisms of intramembranous and endochondral ossification.
5. Discuss the timing of bone development and growth and account for the differences in the internal structure of adult bones.
6. Describe the remodeling and homeostatic mechanisms of the skeletal system.
7. Discuss the effects of nutrition, hormones, exercise, and aging on bone development and the skeletal system.
8. Describe the different types of fractures and explain how they heal.
9. Classify bones according to their shapes and give examples of each type.

[L1] Multiple choice:

Place the letter corresponding to the correct answer in the space provided.

OBJ. 1 _____ 1. The function(s) of the skeletal system is (are):

      a. it is a storage area for calcium and lipids

      b. it is involved in blood cell formation

      c. it provides structural support for the entire body

      d. a, b, and c are correct

OBJ. 1 _____ 2. Storage of lipids that represent an important energy reserve in bone occur in areas of:

      a. red marrow

      b. yellow marrow

      c. bone matrix

      d. ground substance

OBJ. 2 _____ 3. Mature bone cells found in lacunae are called:

      a. osteoblasts

      b. osteocytes

      c. osteoclasts

      d. osteoprogenitors

OBJ. 2 _____ 4. Giant multinucleated cells involved in the process of osteolysis are:

      a. osteocytes

      b. osteoblasts

      c. osteoclasts

      d. osteoprogenitor cells

OBJ. 3 _____ 5. One of the basic histological differences between compact and spongy bone is that in *compact bone*:

      a. the basic functional unit is the Haversian system

      b. there is a lamellar arrangement

      c. there are plates or struts called trabeculae

      d. osteons are not present

OBJ. 3 _____ 6. Spongy or cancellous bone, unlike compact bone, resembles a network of bony struts separated by spaces that are normally filled with:

      a. osteocytes

      b. lacunae

      c. bone marrow

      d. lamella

OBJ. 3 _____ 7. Spongy bone is found primarily at the _____ of long bones:

      a. bone surfaces, except inside joint capsules

      b. expanded ends of long bones, where they articulate with other skeletal elements

      c. axis of the diaphysis

      d. exterior region of the bone shaft to withstand forces applied at either end

Level
-1-

OBJ. 3 _____ 8. Compact bone is usually found where:

    a. bones are not heavily stressed

    b. stresses arrive from many directions

    c. trabeculae are aligned with extensive cross-bracing

    d. stresses arrive from a limited range of directions

OBJ. 4 _____ 9. During intramembranous ossification the developing bone grows outward from the ossification center in small struts called:

    a. spicules

    b. lacunae

    c. the osteogenic layer

    d. dermal bones

OBJ. 4 _____ 10. When osteoblasts begin to differentiate with a connective tissue, the process is called:

    a. endochondral ossification

    b. osteoprogenation

    c. osteolysis

    d. intramembranous ossification

OBJ. 4 _____ 11. The process during which bones begin development as cartilage models and the cartilage is later replaced by bone is called:

    a. intramembranous ossification

    b. endochondral ossification

    c. articular ossification

    d. secondary ossification

OBJ. 4 _____ 12. The region known as the *metaphysis* is the area where:

    a. secondary ossification centers are located

    b. the epiphysis begins to calcify at birth

    c. cartilage is being replaced by bone

    d. collagen fibers become cemented into lamella by osteoblasts

OBJ. 5 _____ 13. The bony skeleton begins to form about _____ after fertilization, and usually does not stop growing until about age _____.

    a. 6 weeks; 25

    b. 3 weeks; 18

    c. 3 days; 12

    d. 5 months; 35

OBJ. 5 _____ 14. The process of replacing other tissues with bone is called:

    a. calcification

    b. ossification

    c. remodeling

    d. osteoprogenesis

Level -1-

OBJ. 6   _____ 15. Of the following selections, the one that describes a homeostatic mechanism of the skeleton is:

         a. as one osteon forms through the activity of osteoblasts, another is destroyed by osteoclasts

         b. mineral absorption from the mother's bloodstream during prenatal development

         c. Vitamin D stimulating the absorption and transport of calcium and phosphate ions

         d. a, b, and c are correct

OBJ. 6   _____ 16. The condition that produces a reduction in _bone mass_ sufficient to compromise normal function is:

         a. osteopenia

         b. osteitis deformans

         c. osteomyelitis

         d. osteoporosis

OBJ. 7   _____ 17. The major effect that exercise has on bones is:

         a. it provides oxygen for bone development

         b. it enhances the process of calcification

         c. it serves to maintain and increase bone mass

         d. it accelerates the healing process when a fracture occurs

OBJ. 7   _____ 18. Growth hormone from the pituitary gland and thyroxine from the thyroid gland maintain normal bone growth activity at the:

         a. epiphyseal plates

         b. diaphysis

         c. periosteum

         d. endosteum

OBJ. 8   _____ 19. In a _greenstick_ fracture:

         a. only one side of the shaft is broken and the other is bent

         b. the shaft bone is broken across its long axis

         c. the bone protrudes through the skin

         d. the bone is shattered into small fragments

OBJ. 8   _____ 20. A Pott's fracture, which occurs at the ankle and affects both bones of the lower leg, is identified primarily by:

         a. a transverse break in the bone

         b. dislocation

         c. a spiral break in the bone

         d. nondisplacement

OBJ. 9   _____ 21. Of the following selections, the one which correctly identifies a long bone is:

         a. rib

         b. sternum

         c. humerus

         d. patella

Level
–1–

OBJ. 9   _____ 22. Bones forming the roof of the skull and the scapula are referred to as:

    a. irregular bones

    b. flat bones

    c. short bones

    d. sesamoid bones

[L1]   Completion:

Using the terms below, complete the following statements.

osteoclasts	compound	osteocytes
osteon	minerals	calcitriol
ossification	support	epiphysis
calcium	irregular	communited
intramembranous	Wormian	endochondral
osteoblasts	yellow marrow	remodeling

OBJ. 1   23. The storage of lipids in bones occurs in the _____.

OBJ. 1   24. Of the five major functions of the skeleton, the two that depend on the dynamic nature of bone are storage and _____.

OBJ. 2   25. Cuboidal cells that synthesize the organic components of the bone matrix are _____.

OBJ. 2   26. In adults, the cells responsible for maintaining the matrix in osseous tissue are the _____.

OBJ. 3   27. The basic functional unit of compact bone is the _____.

OBJ. 3   28. The expanded region of a long bone consisting of spongy bone is called the _____.

OBJ. 4   29. When osteoblasts differentiate within a mesenchymal or fibrous connective tissue, the process is called _____ ossification.

OBJ. 4   30. The type of ossification that begins with the formation of a hyaline cartilage model is _____.

OBJ. 5   31. The process which refers specifically to the formation of bone is _____.

OBJ. 5   32. The major mineral associated with the development and mineralization of bone is _____.

OBJ. 6   33. The organic and mineral components of the bone matrix are continually being recycled and renewed through the process of _____.

OBJ. 6   34. During bone renewal, as one osteon forms through the activity of osteoblasts, another is destroyed by _____.

OBJ. 7   35. The ability of bone to adapt to new stresses results from the turnover and recycling of _____.

OBJ. 7   36. The hormone synthesized in the kidneys which is essential for normal calcium and phosphate ion absorption in the digestive tract is _____.

OBJ. 8   37. Fractures which shatter the affected area into a multitude of bony fragments are called _____ fractures.

OBJ. 8   38. Fractures which project through the skin are called _____ fractures.

Level
—1—

OBJ. 9   39. Bones which have complex shapes with short, flat, notched, or ridged surfaces are termed _____.

OBJ. 9   40. Sutural bones which are small, flat, odd-shaped bones found between the flat bones of the skull are referred to as _____ bones.

[L1]   Matching:

Match the terms in column B with the terms in column A. Use letters for answers in the spaces provided.

PART I

		COLUMN A	COLUMN B
OBJ. 1	_____ 41.	blood cell formation	A. synthesize osteoid
OBJ. 2	_____ 42.	osteoprogenitor cells	B. spicules
OBJ. 2	_____ 43.	osteoblasts	C. Haversian System
OBJ. 3	_____ 44.	spongy bones	D. interstitial and appositional growth
OBJ. 3	_____ 45.	osteon	
OBJ. 4	_____ 46.	intramembraneous ossification	E. bone formation
			F. cancellous bone
OBJ. 4	_____ 47.	endochondral ossification	G. red bone marrow
OBJ. 5	_____ 48.	osteogenesis	H. fracture repair

PART II

		COLUMN A	COLUMN B
OBJ. 6	_____ 49.	bone Maintenance	I. Vitamin D deficiency
OBJ. 7	_____ 50.	thyroxine	J. bony fragments
OBJ. 7	_____ 51.	rickets	K. broken "wrist"
OBJ. 8	_____ 52.	colles fracture	L. long bone
OBJ. 8	_____ 53.	comminuted fracture	M. patella
OBJ. 9	_____ 54.	humerus	N. stimulates bone growth
OBJ. 9	_____ 55.	patella	O. remodeling

Level
–1–

[L1]   Drawing/Illustration Labeling:

Identify each numbered structure by labeling the following figures:

OBJ. 3   **FIGURE 6.1**   Structural Organization of Bone

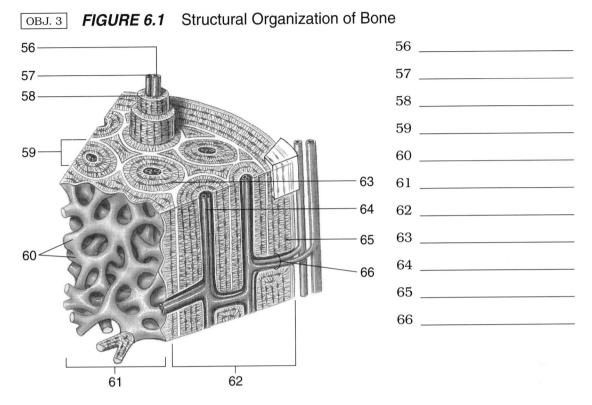

56 _____

57 _____

58 _____

59 _____

60 _____

61 _____

62 _____

63 _____

64 _____

65 _____

66 _____

OBJ. 3   **FIGURE 6.2**   Structure of a Long Bone — L.S., X.S. and 3-D Views

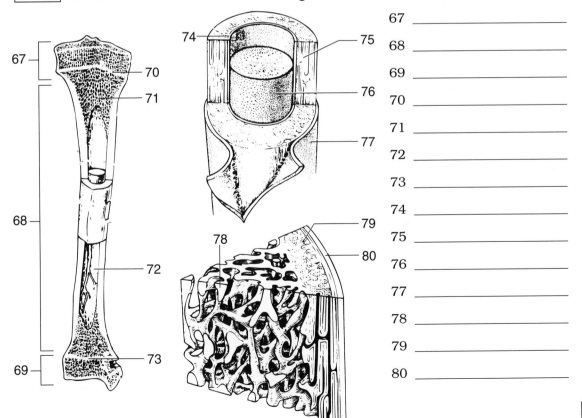

67 _____

68 _____

69 _____

70 _____

71 _____

72 _____

73 _____

74 _____

75 _____

76 _____

77 _____

78 _____

79 _____

80 _____

Level
-1-

OBJ. 8   *FIGURE 6.3*   Types of Fractures

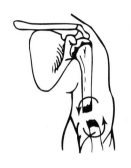

81 _____

84 _____

82 _____

85 _____

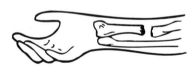

83 _____

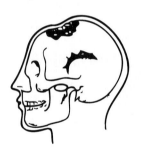

86 _____

When you have successfully completed the exercises in L1 proceed to L2.

Level
-1

☐ LEVEL 2   CONCEPT SYNTHESIS

Concept Map I:

Using the following terms, fill in the circled, numbered, blank spaces to complete the concept map. Follow the numbers that comply with the organization of the map.

Osteocytes   Collagen   Intramembranous ossification
Periosteum   Hyaline cartilage

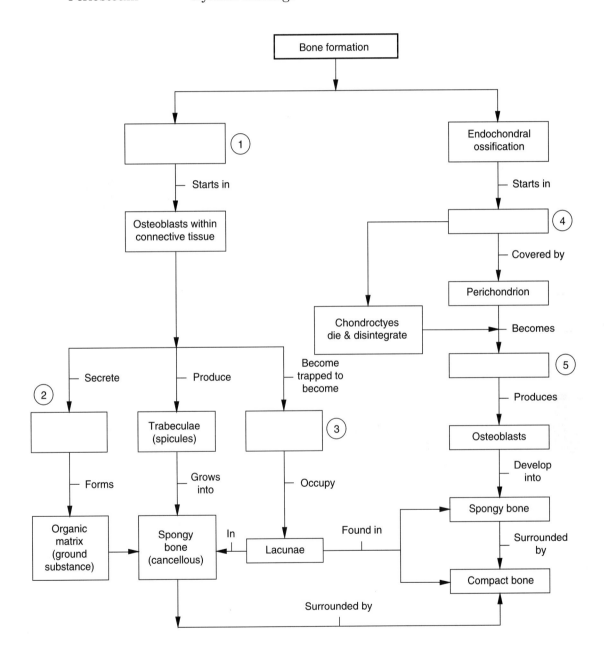

Concept Map II:

Using the following terms, fill in the circled, numbered, blank spaces to complete the concept map. Follow the numbers that comply with the organization of the map.

↓ Calcium level   Releases stored Ca from bone   Parathyroid
Homeostasis   ↓ Intestinal absorption of Ca   Calcitonin

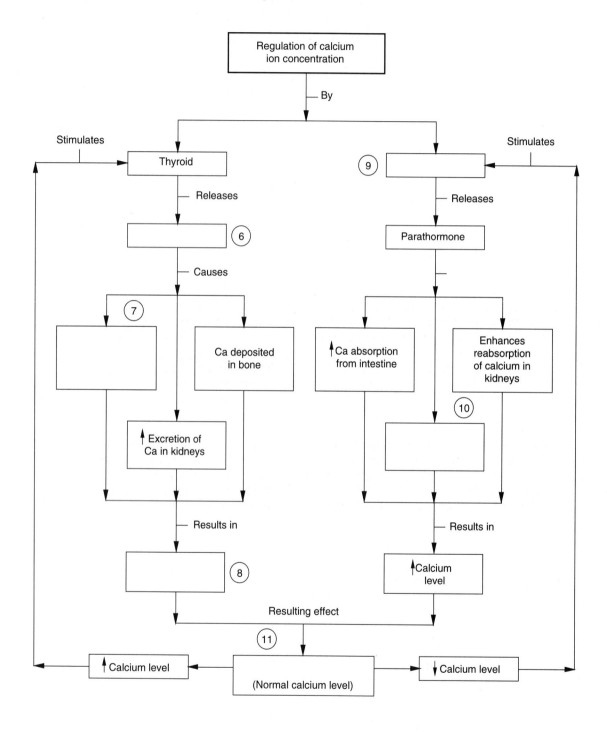

Body Trek:

Using the terms below, fill in the blanks to complete the trek through a long bone in the upper arm, the humerus.

Red marrow	Endosteum	Lacuna
Trabeculae	Canaliculi	Volkmann's canal
Osteoclasts	Blood vessels	Compact bone
Compound	Osteon	Osteocytes
Lamella	Yellow marrow	Haversian canal
Periosteum	Red blood cells	Cancellous or spongy

For this trek Robo will enter the interior of the humerus, the long bone in the upper part of the arm. The entry point is accessible due to a (12)_____ fracture in which the bone has projected through the skin at the distal end of the shaft. The robot proceeds to an area of the bone that is undisturbed by the trauma occurring in the damaged region. The micro-robot enters the medullary cavity which contains (13) _____, and moves proximally through a "sea" of fat to a region where it contacts the lining of the cavity, the (14) _____. After passing through the lining, Robo senses an area that projects images of an interlocking network of long plates or beams riddled with holes or spaces, which are characteristic of (15) _____ bone. The structural forms of this network are called (16) _____, which consist of a bony matrix, the (17) _____, with bone cells, the (18) _____, located between the layers. The bone cells communicate with other bone cells through small channels called (19) _____. The "holes" or spaces have a reddish glow and appear to be actively involved in producing disk-shaped cells or (20) _____, which establish the robot's position in a cavity containing (21) _____. Robo's extended arm grabs onto one of the "bony beams" and, after moving along the beam for a short distance, contact is made with a large canal located at a right angle to the bone's shaft. This canal, called (22) _____, is the major communicating pathway between the bone's interior and exterior surface, the (23) _____. Advancing through the canal, the robot's sensors are signaling dense tissue surrounding the canal indicating that this is the region of (24) _____. Suddenly, Robo arrives at an intersection where the canal dead-ends; however, another large tube-like canal runs parallel to the long axis of the bone. This tube, the (25) _____, contains nerves, (26) _____, and lymphatic vessels. This canal, with its contents and associated concentric lamellae and osteocytes, is referred to as a(n) (27) _____. The robot's visit to an osteocyte located in a(n) (28) _____ is accomplished by trekking from the large canal into smaller canaliculi which form a dense transportation network connecting all the living cells of the bony tissue to the nutrient supply. The giant osteocytes with dark nuclei completely fill the lumen at the bone cell sites located throughout the lamella. Around the bone sites specialized bone digesting cells, the (29) _____, are liquefying the matrix, making the area insensitive to the robot's electronic devices, terminating the effectiveness of the signals transmitted to Mission Control. The exit program is relayed to the robot and the "reverse" trek begins through the bone's canal "system" and a return to the fracture site for removal and preparation for the next trek.

[L2]   Multiple Choice:

Place the letter corresponding to the correct answer in the space provided.

_____ 30. Changing the magnitude and direction of forces generated by skeletal muscles is an illustration of the skeletal function of:

a. protection

b. leverage

c. energy reserve in bones

d. storage capability of bones

Level
=2=

_____ 31. A sesamoid bone would most often be found:

        a. in-between the flat bones of the skull

        b. in the spinal vertebrae

        c. near joints at the knee, the hands, and the feet

        d. in the forearm and the lower leg

_____ 32. The outer surface of the bone, the periosteum:

        a. isolates the bone from surrounding tissues

        b. provides a route for circulatory and nervous supply

        c. actively participates in bone growth and repair

        d. a, b, and c are correct

_____ 33. Osteolysis is an important process in the regulation of:

        a. calcium and phosphate concentrations in body fluids

        b. organic components in the bone matrix

        c. the production of new bone

        d. the differentiation of osteoblasts into osteocytes

_____ 34. The calcification of cartilage results in the production of:

        a. spongy bone

        b. ossified cartilage

        c. compact bone

        d. calcified cartilage

_____ 35. In human beings, the _major_ factor determining the _size_ and proportion of the body is:

        a. the growth of the skeleton

        b. the amount of food eaten

        c. the size of the musculature

        d. thyroid metabolism

_____ 36. Healing of a fracture, even after severe damage, depends on the survival of:

        a. the bone's mineral strength and its resistance to stress

        b. the circulatory supply and the cellular components of the endosteum and periosteum

        c. osteoblasts within the spongy and compact bone

        d. the external callus which protects the fractured area

_____ 37. The three (3) organs regulating the calcium ion concentration in body fluids are:

        a. heart, liver, lungs

        b. liver, kidneys, stomach

        c. pancreas, heart, lungs

        d. bones, intestinal tract, kidneys

_____ 38. When the calcium ion concentration of the blood rises above normal, secretion of the hormone _calcitonin_:

        a. promotes osteoclast activity

        b. increases the rate of intestinal absorption

        c. increases the rate of calcium ion excretion

        d. activates the parathyroid gland to release parathormone

Level
=2=

_____ 39. When cartilage is produced at the epiphyseal side of the metaphysis at the same rate as bone is deposited on the opposite side, bones:

      a. grow wider

      b. become shorter

      c. grow longer

      d. become thicker

_____ 40. The major advantage(s) for bones to undergo continual remodeling is (are):

      a. it may change the shape of a bone

      b. it may change the internal structure of a bone

      c. it may change the total amount of minerals deposited in the bones

      d. a, b, and c are correct

_____ 41. The fibers of _tendons_ intermingle with those of the periosteum, attaching:

      a. skeletal muscles to bones

      b. the end of one bone to another bone

      c. the trabecular framework to the periosteum

      d. articulations with the trabeculae

_____ 42. Giant cells, called _osteoclasts_, with 50 or more nuclei serve to:

      a. synthesize the organic components of the bone matrix

      b. form the trabecular framework which protects cells of the bone marrow

      c. line the inner surfaces of the central canals

      d. secrete acids which dissolve the bony matrix and release the stored minerals

_____ 43. After fertilization occurs, the skeleton begins to form in about:

      a. 2 weeks

      b. 4 weeks

      c. 6 weeks

      d. 3 months

_____ 44. The circulating hormones that stimulate bone growth are:

      a. calcitonin and parathormone

      b. growth hormone and thyroxine

      c. oxytocin and secretin

      d. epinepherine and relaxin

_____ 45. Appositional bone growth at the outer surface results in:

      a. an increase in the diameter of a growing bone

      b. an increase in the overall length of a bone

      c. a thickening of the cartilages that support the bones

      d. an increased hardening of the periosteum

_____ 46. The vitamins that are specifically required for normal bone growth are:

      a. Vitamins B1, B2, B3

      b. Vitamins C and B complex

      c. Vitamins A, D, E, K

      d. Vitamins A, C, D

Level
=2=

_____ 47. A gradual deformation of the skeleton is a condition known as:

         a. osteomyelitis

         b. Paget's disease

         c. osteomalacia

         d. osteopenia

_____ 48. Hypersecretion of parathyroid hormone would produce changes in the bone similar to those associated with:

         a. rickets

         b. osteomyelitis

         c. osteomalacia

         d. osteitis

_____ 49. A fracture that projects through the skin is a:

         a. compound fracture

         b. comminuted fracture

         c. Colles' fracture

         d. greenstick fracture

_____ 50. A fracture in which one side of the shaft is broken and the other side is bent is a:

         a. compound fracture

         b. greenstick fracture

         c. transverse fracture

         d. comminuted fracture

[L2]   Completion:

Using the terms below, complete the following statements.

osteomalacia	osteopenia	epiphyseal plates	depressed
diaphysis	endochondral	rickets	canaliculi
osteoblasts	intramembranous		

51. The communication pathways from the lacunae, which connect the osteocytes with one another and with the blood vessels of the Haversian canal, are _____.

52. The condition in which an individual develops a bowlegged appearance as the leg bones bend under the weight of the body is _____.

53. The type of cells responsible for the production of new bone are _____.

54. Dermal bones, such as several bones of the skull, the lower jaw, and the collarbone, are a result of _____ ossification.

55. Limb bone development is a good example of the process of _____ ossification.

56. Long bone growth during childhood and adolescence is provided by persistence of the _____.

57. The condition that can occur in children or adults whose diet contains inadequate levels of calcium or vitamin D is _____.

58. Fragile limbs, a reduction in height, and the loss of teeth are a part of the aging process referred to as _____.

59. A common type of skull fracture is a _____ fracture.

60. The location of compact bone in an adult's bone is the _____.

Level
=2=

[L2]   Short Essay:

Briefly answer the following questions in the spaces provided below.

61. What five (5) major functions is the skeletal system responsible for in the human body?

62. What are the primary histological *differences* between compact and spongy bone?

63. What is the difference between the *periosteum* and the *endosteum*?

64. How does the process of *calcification* differ from *ossification*?

65. Differentiate between the *beginning stage* of intramembranous and endochondral ossification.

66. The conditions of *gigantism* and *pituitary dwarfism* are extreme opposites. What effect does hormonal regulation of bone growth have on each condition?

67. What six (6) broad classifications are used to divide the 206 bones of the body into categories? Give at least one example of each classification.

68. What are the fundamental relationships between the skeletal system and other body systems?

When you have successfully completed the exercises in L2 proceed to L3.

☐ LEVEL 3 CRITICAL THINKING/APPLICATION

Using principles and concepts learned in Chapter 6, answer the following questions. Write your answers on a separate sheet of paper .

1. Why is an individual who experiences premature puberty not as tall as expected at age 18?

2. How might a cancerous condition in the body be related to the development of a severe reduction in bone mass and excessive bone fragility?

3. During a conversation between two grandchildren, one says to the other, "Grandpa seems to be getting shorter as he gets older." Why is this a plausible observation?

4. Good nutrition and exercise are extremely important in bone development, growth, and maintenance. If you were an astronaut, what vitamin supplements and what type of exercise would you need to be sure that the skeletal system retained its integrity while in a weightless environment in space?

5. A recent sports article stated, "High-tech equipment is designed to pack power, but it also might leave you packing ice." How might this statement affect you if, as a "week-end" tennis player who has used a regular-sized wooden racket for years, you decide to purchase a new graphite wide-body racket whose frame has doubled in thickness?

6. Stairmasters (stair-climbing machines) have been popular in health clubs and gyms in the last few years. Since climbing stairs involves continuous excessive force on the knees, what tissues in the knee joint are subject to damage?

7. The skeletal remains of a middle-aged male have been found buried in a shallow grave. As the pathologist assigned to the case, what type of anatomical information could you provide by examining the bones to determine the identity of the individual?

7

The Axial Skeleton

■ Overview

The skeletal system in the human body is composed of 206 bones, 80 of which are found in the axial division, and 126 of which make up the appendicular division. Chapter 7 includes a study of the bones and associated parts of the axial skeleton, located along the body's longitudinal axis and center of gravity. They include 22 skull bones, 6 auditory ossicles, 1 hyoid bone, 26 vertebrae, 24 ribs, and 1 sternum. This bony framework protects and supports the vital organs in the dorsal and ventral body cavities. In addition, the bones serve as areas for muscle attachment, articulate at joints for stability and movement, assist in respiratory movements, and stabilize and position elements of the appendicular skeleton.

   The student study and review for this chapter includes the identification and location of bones, the identification and location of bone markings, the functional anatomy of bones, and the articulations that comprise the axial skeleton.

◻ LEVEL 1   REVIEW OF CHAPTER OBJECTIVES

1. Identify the bones of the axial skeleton and specify their functions.
2. Identify the bones of the cranium and face and explain the significance of the markings on the individual bones.
3. Describe the structure of the nasal complex and the functions of the individual bones.
4. Explain the function of the paranasal sinuses.
5. Describe key structural differences in the skulls of infants, children, and adults.
6. Identify and describe the curvatures of the spinal column and their functions.
7. Identify the vertebral regions and describe the distinctive structural and functional characteristics of each vertebral group.
8. Explain the significance of the articulations between the thoracic vertebrae and ribs, and between the ribs and sternum.

[L1] Multiple choice:

Place the letter corresponding to the correct answer in the space provided.

OBJ. 1 _____ 1. The axial skeleton can be recognized because it:

    a. includes the bones of the arms and legs

    b. forms the longitudinal axis of the body

    c. includes the bones of the pectoral and pelvic girdles

    d. a, b, and c are correct

OBJ. 1 _____ 2. What percentage of the bones in the body comprise the axial skeleton?

    a. 60%

    b. 80%

    c. 20%

    d. 40%

OBJ. 1 _____ 3. Of the following selections, the one that includes bones found exclusively in the axial skeleton is:

    a. ear ossicles, scapula, clavicle, sternum, hyoid

    b. vertebrae, ischium, ilium, skull, ribs

    c. skull, vertebrae, ribs, sternum, hyoid

    d. sacrum, ear ossicles, skull, scapula, ilium

OBJ. 1 _____ 4. The axial skeleton creates a framework that supports and protects organ systems in:

    a. the dorsal and ventral body cavities

    b. the pleural cavity

    c. the abdominal cavity

    d. the pericardial cavity

OBJ. 2 _____ 5. The bones of the *cranium* that exclusively represent *single, unpaired* bones are:

    a. occipital, parietal, frontal, temporal

    b. occipital, frontal, sphenoid, ethmoid

    c. frontal, temporal, parietal, sphenoid

    d. ethmoid, frontal, parietal, temporal

OBJ. 2 _____ 6. The *paired* bones of the cranium are:

    a. ethmoid and sphenoid

    b. frontal and occipital

    c. occipital and parietal

    d. parietal and temporal

OBJ. 2 _____ 7. The *associated* bones of the skull include the:

    a. mandible and maxilla

    b. nasal and lacrimal

    c. hyoid and auditory ossicles

    d. vomer and palatine

OBJ. 2   _____   8. The *single, unpaired* bones that make up the skeletal part of the face are the:

      a. mandible and vomer

      b. nasal and lacrimal

      c. mandible and maxilla

      d. nasal and palatine

OBJ. 2   _____   9. The *sutures* that articulate the bones of the skull are:

      a. parietal, occipital, frontal, temporal

      b. calvaria, foramen, condyloid, lacerum

      c. posterior, anterior, lateral, dorsal

      d. lambdoidal, sagittal, coronal, squamosal

OBJ. 2   _____   10. The bones that make up the *eye socket* or *orbit* include:

      a. lacrimal, zygomatic, maxilla

      b. ethmoid, temporal, zygomatic

      c. lacrimal, ethmoid, sphenoid

      d. temporal, frontal, sphenoid

OBJ. 2   _____   11. *Foramina*, located on the bones of the skull, serve primarily as passageways for:

      a. airways and ducts for secretions

      b. sound and sight

      c. nerves and blood vessels

      d. muscle fibers and nerve tissue

OBJ. 2   _____   12. The lines, tubercles, crests, ridges, and other processes on the bones represent areas which are used primarily for:

      a. attachment of muscles to bones

      b. attachment of bone to bone

      c. joint articulation

      d. increasing the surface area of the bone

OBJ. 2   _____   13. Areas of the head that are involved in the formation of the skull are called:

      a. fontanels

      b. craniocephalic centers

      c. craniulums

      d. ossification centers

OBJ. 3   _____   14. The *sinuses* or internal chambers in the skull are found in:

      a. sphenoid, ethmoid, vomer, lacrimal bones

      b. sphenoid, frontal, ethmoid, maxillary bones

      c. ethmoid, frontal, lacrimal, maxillary bones

      d. lacrimal, vomer, ethmoid, frontal bones

OBJ. 3   _____   15. The nasal complex consists of the:

      a. frontal, sphenoid, and ethmoid bones

      b. maxilla, lacrimal and ethmoidal concha

      c. inferior concha

      d. a, b, and c are correct.

Level
-1-

OBJ. 4 ____ 16. The air-filled chambers that communicate with the nasal cavities are the:

      a. condylar processes
      b. paranasal sinuses
      c. maxillary foramina
      d. mandibular foramina

OBJ. 4 ____ 17. The mucus membrane of the paranasal sinuses responds to environmental stress by:

      a. breaking up air flow in the nasal cavity
      b. creating swirls and eddies in the sinuses
      c. accelerating the production of mucus
      d. a, b, and c are correct

OBJ. 3 ____ 18. The reason the skull can be distorted without damage during birth is:

      a. fusion of the ossification centers is completed
      b. the brain is large enough to support the skull
      c. fibrous connective tissue connects the cranial bones
      d. shape and structure of the cranial elements are elastic

OBJ. 5 ____ 19. At birth, the bones of the skull can be distorted without damage because of the:

      a. cranial foramina
      b. fontanels
      c. alveolar process
      d. cranial ligaments

OBJ. 5 ____ 20. The most significant growth in the skull occurs before age five (5) because:

      a. the brain stops growing and cranial sutures develop
      b. brain development is incomplete until maturity
      c. the cranium of a child is larger than that of an adult
      d. the ossification and articulation process is completed.

OBJ. 6 ____ 21. The primary spinal curves that appear late in fetal development:

      a. help shift the trunk weight over the legs
      b. accommodate the lumbar and cervical regions
      c. become accentuated as the toddler learns to walk
      d. accommodate the thoracic and abdominopelvic viscera.

OBJ. 6 ____ 22. An abnormal lateral curvature of the spine is called:

      a. kyphosis
      b. lordosis
      c. scoliosis
      d. amphiarthrosis

Level
-1-

OBJ. 5   _____ 23. The vertebrae that indirectly effect changes in the volume of the rib cage are the:

    a. cervical vertebrae

    b. thoracic vertebrae

    c. lumbar vertebrae

    d. sacral vertebrae

OBJ. 7   _____ 24. The most massive and least mobile of the vertebrae are the:

    a. thoracic

    b. cervical

    c. lumbar

    d. sacral

OBJ. 7   _____ 25. Of the following selections, the one which correctly identifies the sequence of the vertebra from superior to inferior is:

    a. thoracic, cervical, lumbar, coccyx, sacrum

    b. cervical, lumbar, thoracic, sacrum, coccyx

    c. cervical, thoracic, lumbar, sacrum, coccyx

    d. cervical, thoracic, sacrum, lumbar, coccyx

OBJ. 7   _____ 26. When identifying the vertebra, a numerical shorthand is used such as C_3. The C refers to:

    a. the region of the vertebrae

    b. the position of the vertebrae in a specific region

    c. the numerical order of the vertebrae

    d. the articulating surface of the vertebrae

OBJ. 7   _____ 27. C_1 and C_2 have specific names, which are the:

    a. sacrum and coccyx

    b. atlas and axis

    c. cervical and costal

    d. sacrum and coccyx

OBJ. 7   _____ 28. The sacrum consists of five fused elements which afford protection for:

    a. reproductive, digestive, and excretory organs

    b. respiratory, reproductive, and endocrine organs

    c. urinary, respiratory, and digestive organs

    d. endocrine, respiratory, and urinary organs

OBJ. 7   _____ 29. The primary purpose of the coccyx is to provide:

    a. protection for the urinary organs

    b. protection for the anal opening

    c. an attachment site for leg muscles

    d. an attachment site for a muscle that closes the anal opening

OBJ. 8

_____ 30. The first seven pairs of ribs are called true ribs, while the lower five pairs are called *false* ribs because:

    a. the fused cartilages merge with the costal cartilage
    b. they do not attach directly to the sternum
    c. the last two pair have no connection with the sternum
    d. they differ in shape from the true ribs

OBJ. 8

_____ 31. The skeleton of the chest or thorax consists of:

    a. cervical vertebrae, ribs, and sternum
    b. cervical vertebrae, ribs, thoracic vertebrae
    c. cervical vertebrae, ribs, pectoral girdle
    d. thoracic vertebrae, ribs, sternum

OBJ. 8

_____ 32. The three components of the adult sternum are the:

    a. pneumothorax, hemothorax, tuberculum
    b. manubrium, body, xiphoid process
    c. head, capitulum, tuberculum
    d. angle, body, shaft

[L1] Completion:

Using the terms below, complete the following statements.

centrum	muscles	floating
axial	costal	capitulum
fontanels	cranium	microcephaly
mucus	cervical	foramen magnum
compensation	paranasal	inferior concha

OBJ. 1

33. The part of the skeletal system that forms the longitudinal axis of the body is the _____ division.

OBJ. 1

34. The bones of the skeleton provide an extensive surface area for the attachment of _____.

OBJ. 2

35. The part of the skull that provides protection for the brain is the _____.

OBJ. 2

36. The opening that connects the cranial cavity with the canal enclosed by the spinal column is the _____.

OBJ. 3

37. The paired scroll-like bones located on each side of the nasal septum are the _____.

OBJ. 3

38. The airspaces connected to the nasal cavities are the _____ sinuses.

OBJ. 4

39. Irritants are flushed off the walls of the nasal cavities because of the presence of _____.

OBJ. 5

40. At birth, the cranial bones are connected by areas of fibrous connective tissues called _____.

OBJ. 5

41. An undersized head caused by a cessation of brain enlargement and skull growth is called _____.

OBJ. 6

42. The spinal curves that assist in allowing a child to walk and run are called _____ curves.

Level
–1–

OBJ. 7   43. The medium, heart-shaped, flat face which serves as a facet for rib articulation on the *thoracic* vertebrae is called the _____.

OBJ. 7   44. The vertebrae that stabilize relative positions of the brain and spinal cord are the _____ vertebrae.

OBJ. 8   45. The cartilaginous extensions that connect the ribs to the sternum are the _____ cartilages.

OBJ. 8   46. A typical rib articulates with the vertebral column at the area of the rib called the _____.

OBJ. 8   47. The last two pairs of ribs that do not articulate with the sternum are called _____ ribs.

[L1] Matching:

Match the terms in column B with the terms in column A. Use letters for answers in the spaces provided.

PART I

		COLUMN A	COLUMN B
OBJ 1	_____	48. hyoid bone	A. calvaria
OBJ 1	_____	49. respiratory movement	B. premature closure of fontanel
OBJ. 2	_____	50. skullcap	C. infant skull
OBJ. 2	_____	51. sphenoid bone	D. vomer
OBJ. 3	_____	52. nasal septum	E. paranasal sinuses
OBJ. 4	_____	53. air-filled chambers	F. sella turcica
OBJ. 5	_____	54. fontanel	G. elevation of rib cage
OBJ. 5	_____	55. craniostenosis	H. stylohyoid ligaments

PART II

		COLUMN A	COLUMN B
OBJ. 6	_____	56. primary curves	I. lower back
OBJ. 7	_____	57. cervical vertebrae	J. ribs 8–10
OBJ. 7	_____	58. lumbar vertebrae	K. C_1
OBJ. 7	_____	59. atlas	L. C_2
OBJ. 7	_____	60. axis	M. ribs 1–7
OBJ. 8	_____	61. vertebrosternal ribs	N. accommodation
OBJ. 8	_____	62. vertebrochondral ribs	O. neck

[L1]   Drawing/Illustration Labeling:

Identify each numbered structure by labeling the following figures:

OBJ. 1   ***FIGURE 7.1***   Bones of the Axial Skeleton

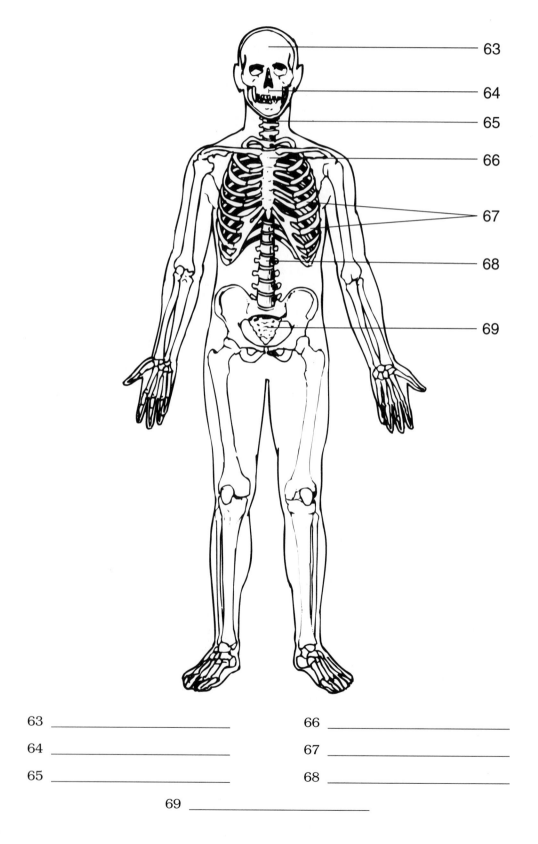

63 _____   66 _____

64 _____   67 _____

65 _____   68 _____

69 _____

OBJ. 2
OBJ. 3

FIGURE 7.2   Anterior View of the Skull

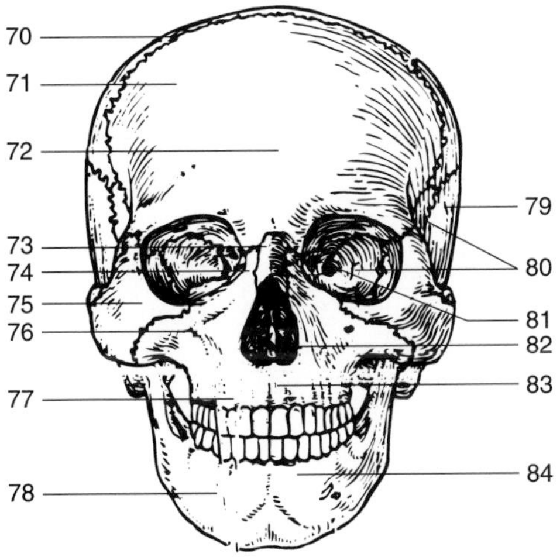

70 _____
71 _____
72 _____
73 _____
74 _____
75 _____
76 _____

77 _____
78 _____
79 _____
80 _____
81 _____
82 _____
83 _____

84 _____

OBJ. 2
OBJ. 3

FIGURE 7.3   Lateral View of the Skull

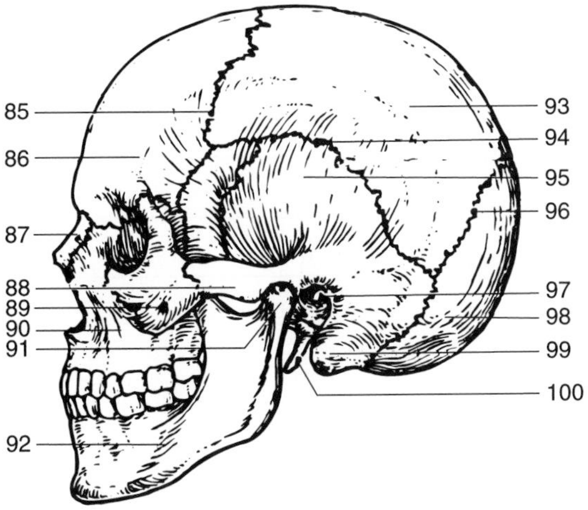

85 _____
86 _____
87 _____
88 _____
89 _____
90 _____
91 _____
92 _____

93 _____
94 _____
95 _____
96 _____
97 _____
98 _____
99 _____
100 _____

OBJ. 2
OBJ. 3

FIGURE 7.4   Inferior View of the Skull

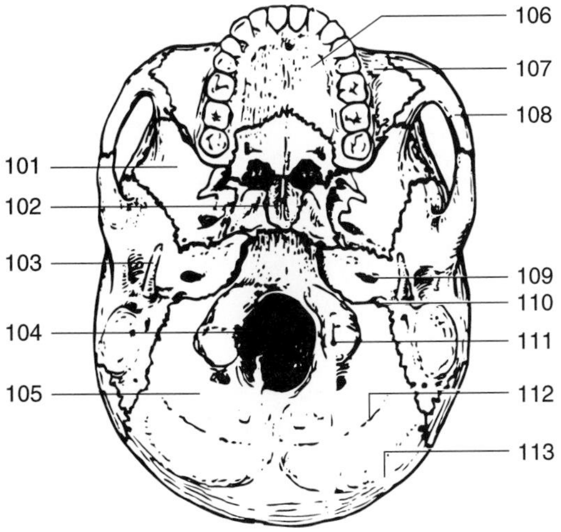

101 _____
102 _____
103 _____
104 _____
105 _____
106 _____

107 _____
108 _____
109 _____
110 _____
111 _____
112 _____

113 _____

Level
-1-

OBJ. 4   *FIGURE 7.5*   Paranasal Sinuses

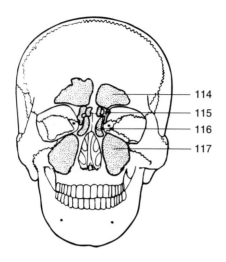

114 _____

115 _____

116 _____

117 _____

OBJ. 5   *FIGURE 7.6*   Fetal Skull – Lateral View

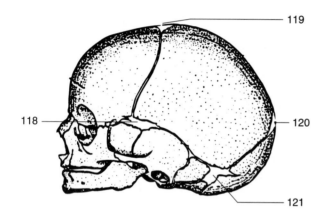

118 _____

119 _____

120 _____

121 _____

OBJ. 5   *FIGURE 7.7*   Fetal Skull – Superior View

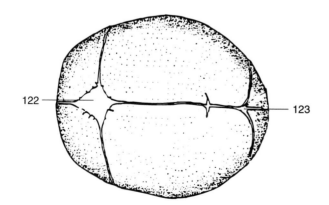

122 _____

123 _____

Level
—1—

OBJ. 7   **FIGURE 7.8**   The Vertebral Column

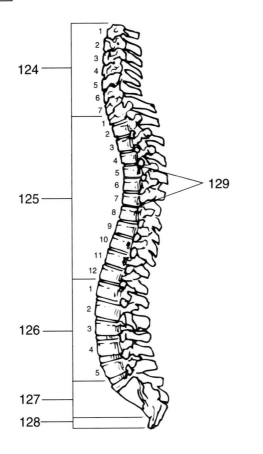

124 _____

125 _____

126 _____

127 _____

128 _____

129 _____

OBJ. 8   **FIGURE 7.9**   The Ribs

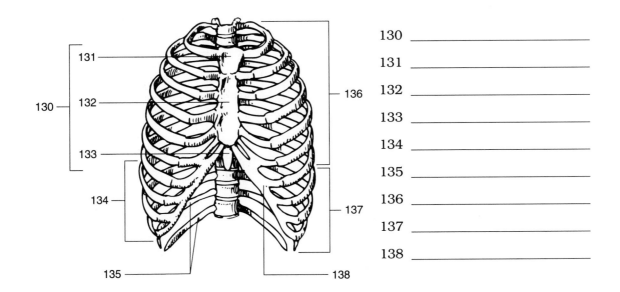

130 _____

131 _____

132 _____

133 _____

134 _____

135 _____

136 _____

137 _____

138 _____

When you have successfully completed the exercises in L1 proceed to L2.

Level
_1_

☐ LEVEL 2   CONCEPT SYNTHESIS

Concept Map I:

Using the following terms, fill in the circled, numbered, blank spaces to complete the concept map. Follow the numbers that comply with the organization of the map.

Floating ribs, 2 pair	Temporal	Sutures
Hyoid	Sacral	Vertebral Column
Lacrimal	Xiphoid process	Occipital
Sternum	Skull	Mandible
Thoracic	Coronal	

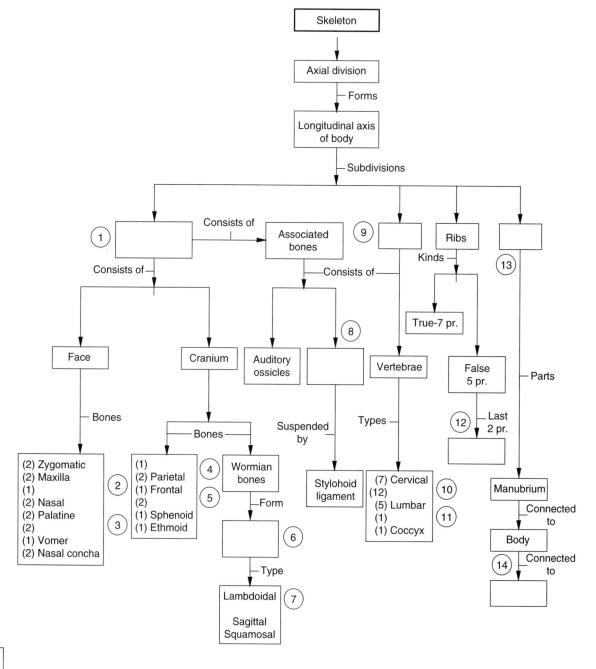

Body Trek:

Using the terms below, fill in the blanks to complete the study of the axial skeleton.

Mandible	Thoracic	Sternum	Xiphoid process
Hyoid	Sacrum	Skull	Cranium
True	Occipital	Parietal	Zygomatic
Ribs	Vertebrae	Lumbar	Lacrimal
Axial	Sphenoid	False	Nasal
Sutures	Cervical	Manubrium	Sagittal
Floating			

Because of the difficulties involved with treking from bone to bone inside the body, Robo is placed in an anatomy laboratory to monitor the activities of students who are using a skeleton to locate and name the bones of the axial skeleton. The students are initially unaware of the robot's presence. It is programmed to offer sounds of encouragement as the study proceeds.

The students are quick to recognize that the bones of the (15) _____ division of the skeleton form the longitudinal axis of the body. A suggestion is made to identify all the components of the division first and then study each part separately. [Robo: "Wow, what an idea!"] The most superior part, the (16) _____ , consists of the (17) _____ , the face, and associated bones. Just below the neck, and looking anteriorly at the skeleton, the students immediately identify the twelve pairs of (18) _____ and the (19) _____ , a broad-surfaced bone with three individual articulated parts. Turning the skeleton around for a posterior view, the twenty-six artistically, articulated (20) _____ are exposed. After identifying the component parts, the class is divided into small groups to study each part. Members of each group share with the entire class what their study has revealed.

The first group has chosen the skull. Having identified the face as a part of the skull, group members notice the presence of six paired bones, which include the (21) _____ , maxilla, (22) _____ , palatine, (23) _____ , and the nasal concha. The (24) _____ and the vomer bone comprise the single bones of the face. The cranial part of the skull includes four unpaired bones, the (25) _____ , frontal, (26) _____ , and the ethmoid. The paired bones of the cranium are the (27) _____ and the temporal. [Robo: "Good job!"] The students laugh and are curious to know who or what is making the muffled sounds. [Robo: "Continue."] The first group continues by locating the Wormian bones, which appear in (28) _____ of the skull. These immovable joints have names such as lambdoidal, coronal, squamosal, and (29) _____ . The associated skull bones include the ear ossicles and the (30) _____ , the only bone in the body not articulated with another bone. [Robo: "Excellent observation!"]

The students chuckle as the second group notices the uniqueness of the vertebral column. Twenty-six vertebrae are counted and categorized into regions according to their location. The first seven make up the (31) _____ area; the next twelve the (32) _____ region; the five that follow, the (33) _____ ; and the last two fused areas, the (34) _____ and the coccyx. [Robo: "Very interesting!"] The students ignore the comment.

The third group, which is studying the ribs, is debating which ones are the true ribs and which ones are the false ribs. [Robo: "Count them!"] Students look around, seem confused, but decide to count. They conclude that one through seven are the (35) _____ ribs, and eight through twelve are the (36) _____ ribs. [Robo: "Easy, wasn't it?"] Students laugh. Someone in the group notices that the last two pairs do not attach to the cartilage above. Another responds, "That is the reason they are called the (37) _____ ribs."

The fourth group appears intrigued with the structure of the sternum owing to its three-part articulated form, which includes the superiorly located (38) _____ , the body, and the (39) _____ located at the inferior end. [Robo: "Easy job, but good work!"]

The lab instructor finally removes the mirco-robot from its location inside the skull of the skeleton and informs the students that Robo will return to the lab when they study the appendicular division of the skeleton.

[L2]   Multiple Choice:

Place the letter corresponding to the correct answer in the space provided.

_____ 40. Brain growth, skull growth, and completed cranial suture development occur:

     a. before birth

     b. right after birth

     c. before age 5

     d. between age 5 and 10

_____ 41. The area of the greatest degree of flexibility along the vertebral column is found from:

     a. $C_3 - C_7$

     b. $T_1 - T_6$

     c. $T_7 - T_{12}$

     d. $L_1 - L_5$

_____ 42. After a hard fall, compression fractures or compression/dislocation fractures most often involve the:

     a. last thoracic and first two lumbar vertebrae

     b. first two cervical vertebrae

     c. sacrum and coccyx

     d. fifth lumbar and the sacrum

_____ 43. Intervertebral discs are found in between all the vertebrae except:

     a. between C_1 and T_1, and T_{12} and L_1; sacrum and coccyx

     b. the sacrum and the coccyx

     c. between L_5 and the sacrum

     d. between C_1 and C_2 and the sacrum and coccyx

_____ 44. Part of the loss in _height_ that accompanies aging results from:

     a. degeneration of osseous tissue in the diaphysis of long bones

     b. degeneration of skeletal muscles attached to bones

     c. the decreasing size and resiliency of the intervertebral discs

     d. the reduction in the number of vertebrae due to aging

_____ 45. The skull articulates with the vertebral column at the:

     a. foramen magnum

     b. occipital condyles

     c. lambdoidal sutures

     d. C_1 and C_2

_____ 46. The long framework of the sphenoid bone that houses the pituitary gland is the:

     a. crista galli

     b. styloid process

     c. sella turcica

     d. frontal squama

Level
=2=

_____ 47. The growth of the cranium is usually associated with:

       a. the expansion of the brain

       b. the development of the fontanels

       c. the closure of the sutures

       d. the time of birth

_____ 48. Beginning at the superior end of the vertebral canal and proceeding inferiorally:

       a. the diameter of the cord and the size of the neural arch increase

       b. the diameter of the cord increases and the size of the neural arch decreases

       c. the diameter of the cord decreases and the size of the neural arch increases

       d. the diameter of the cord and the size of the neural arch decrease

_____ 49. The vertebrae that are directly articulated with the ribs are:

       a. cervical and thoracic

       b. thoracic only

       c. cervical only

       d. thoracic and lumbar

_____ 50. During CPR, proper positioning of the hands is important so that an excessive pressure will not break the:

       a. manubrium

       b. xiphoid process

       c. body of the sternum

       d. costal cartilages

[L2]   Completion:

Using the terms below, complete the following statements.

kyphosis	mental foramina	pharyngotympanic
auditory ossicles	lordosis	alvelolar processes
metopic	tears	scoliosis
compensation		

51. The structure that ends inside the mass of the temporal bone which connects the airspace of the middle ear with the pharynx is the _____ tube.

52. At birth the two frontal bones which have not completely fused are connected at the _____ suture.

53. The lacrimal bones house the structures that are associated with the production and release of _____.

54. The associated skull bones of the middle ear that conduct sound vibrations from the tympanum to the inner ear are the _____.

55. The oral margins of the maxillae that provide the sockets for the teeth are the

    _____.

56. Small openings which serve as nerve passageways on each side of the body of the mandible are the _____.

57. The lumbar and cervical curves which appear several months after birth and help to position the body weight over the legs are known as _____ curves.

58. A normal thoracic curvature which becomes exaggerated, producing a "roundback" appearance, is a _____.

59. An exaggerated lumbar curvature or "swayback" appearance is a _____.

60. An abnormal lateral curvature which usually appears in adolescence during periods of rapid growth is _____.

[L2]   Short Essay:

Briefly answer the following questions in the spaces provided below.

61. What are the four (4) primary functions of the axial skeleton?

62. A. List the *paired bones* of the *cranium.*
    B. List the *unpaired* (single) bones of the *cranium.*

63. A. List the *paired* bones of the *face.*
    B. List the *unpaired* (single) bones of the *face.*

64. Why are the auditory ossicles and hyoid bone referred to as associated bones of the skull?

65. What is *craniostenosis* and what are the results of this condition?

66. What is the difference between a *primary curve* and a *secondary curve* of the spinal column?

67. Distinguish among the abnormal spinal curvature distortions of kyphosis, lordosis, and scoliosis.

68. What is the difference between the *true* ribs and the *false* ribs?

When you have successfully completed the exercises in L2 proceed to L3.

☐ LEVEL 3   CRITICAL THINKING/APPLICATION

Using principles and concepts learned about the axial skeleton, answer the following questions. Write your answers on a separate sheet of paper.

1. What senses are affected by the cranial and facial bones that protect and support the sense organs?

2. A friend of yours tells you that she has been diagnosed as having TMJ. What bones are involved with this condition, how are they articulated, and what symptoms are apparent?

3. Your nose has been crooked for years and you have had a severe sinus condition to accompany the disturbing appearance of the nose. You suspect there is a relationship between the crooked nose and the chronic sinus condition. How do you explain your suspicion?

4. During a car accident you become a victim of *whiplash*. You experience pains in the neck and across the upper part of the back. Why?

5. During a child's routine physical examination it is discovered that one leg is shorter than the other. Relative to the spinal column, what might be a plausible explanation for this condition?

6. A clinical diagnosis has been made that substantiates the presence of a herniated disc and a severe case of sciatica. What is the relationship between the two conditions?

8

The Appendicular Division

■ Overview

How many things can you think of that require the use of your hands? Your arms? Your legs? Your feet? If you are an active person, the list would probably be endless. There are few daily activities, if any, that do not require the use of the arms and/or legs performing in a dynamic, coordinated way to allow you to be an active, mobile individual. This intricately articulated framework of 126 bones comprises the appendicular division of the skeleton. The bones consist of the pectoral girdle and the upper limbs, and the pelvic girdle and the lower limbs. The bones in this division provide support, are important as sites for muscle attachment, articulate at joints, are involved with movement and mobility, and are utilized in numerous ways to control the environment that surrounds you every second of your life.

The student study and review for Chapter 8 includes the identification and location of bones and bone markings, and the functional anatomy of the bones and articulations that comprise the appendicular skeleton.

❑ LEVEL 1   REVIEW OF CHAPTER OBJECTIVES

1. Identify each bone of the appendicular skeleton.
2. Identify the bones that form the pectoral girdle, their functions, and their superficial features.
3. Identify the bones of the upper limb, their functions, and their superficial features.
4. Identify the bones that form the pelvic girdle, their functions, and their superficial features.
5. Identify the bones of the lower limb, their functions, and their superficial features.
6. Discuss structural and functional differences between the pelvis of a female and that of a male.
7. Explain how study of the skeleton can reveal significant information about an individual.
8. Summarize the skeletal differences between males and females.
9. Briefly describe how the aging process affects the skeletal system.

[L1]   Multiple choice:

Place the letter corresponding to the correct answer in the space provided.

OBJ. 1   _____   1. The bones that make up the appendicular division of the skeleton consist of:

    a. the bones which form the longitudinal axis of the body

    b. the rib cage and the vertebral column

    c. the skull and the arms and legs

    d. the pectoral and pelvic girdles, and the upper and lower limbs

OBJ. 1   _____   2. One of the major functional differences between the appendicular and axial divisions is that the appendicular division:

    a. serves to adjust the position of the head, neck, and trunk

    b. protects organ systems in the dorsal and ventral body cavities

    c. makes you an active, mobile individual

    d. assists directly in respiratory movements

OBJ. 1   _____   3. A composite structure that includes portions of both the appendicular and axial skeleton is the:

    a. pelvis

    b. pectoral girdle

    c. pelvic girdle

    d. a, b, and c are correct

OBJ. 2   _____   4. The bones of the *pectoral* girdle include:

    a. clavicle and scapula

    b. ilium and ischium

    c. humerus and femur

    d. ulna and radius

OBJ. 2   _____   5. The large posterior process on the scapula is the:

    a. coracoid process

    b. acromion process

    c. olecranon fossa

    d. styloid process

OBJ. 3   _____   6. The parallel bones that support the forearm are the:

    a. humerus and femur

    b. ulna and radius

    c. tibia and fibula

    d. scapula and clavicle

OBJ. 3   _____   7. The process of the humerus that articulates with the scapula is the:

    a. greater tubercle

    b. lesser tubercle

    c. head

    d. ball

OBJ. 4   _____ 8. The bones of the *pelvic* girdle include:

   a. tibia and fibula
   b. ilium, pubis, ischium
   c. ilium, ischium, acetabulum
   d. coxa, patella, acetabulum

OBJ. 4   _____ 9. The primary function of the *pectoral* girdle is to:

   a. protect the organs of the thorax
   b. provide areas for articulation with the vertebral column
   c. position the shoulder joint and provide a base for arm movement
   d. support and maintain the position of the skull

OBJ. 4   _____ 10. The amphiarthrotic articulation which limits movements between the two pubic bones is the:

   a. pubic symphysis
   b. obturator foramen
   c. greater sciatic notch
   d. pubic tubercle

OBJ. 5   _____ 11. The large medial bone of the lower leg is the:

   a. femur
   b. fibula
   c. tibia
   d. humerus

OBJ. 5   _____ 12. A prominent deviation that runs along the center of the posterior surface of the femur, which serves as an attachment site for muscles that abduct the femur, is the:

   a. greater trochanter
   b. trochanteric crest
   c. trochanteric lines
   d. linea aspera

OBJ. 6   _____ 13. The general appearance of the pelvis of the female compared to the male is that the female pelvis is:

   a. heart-shaped
   b. robust, heavy, rough
   c. relatively deep
   d. broad, light, smooth

OBJ. 6   _____ 14. The shape of the pelvic inlet in the female is:

   a. heart-shaped
   b. triangular
   c. oval to round
   d. somewhat rectangular

OBJ. 7 ____ 15. Of the following selections, the one that would be used to esti-
mate muscular development and body weight is:

    a. sex and age

    b. degenerative changes in the normal skeletal system

    c. normal timing of skeletal development

    d. development of various ridges and general bone mass

OBJ. 8 ____ 16. The two specific *areas* of the skeleton that are generally used to
identify significant differences between a male and female are:

    a. arms and legs

    b. ribs and vertebral column

    c. skull and pelvis

    d. a, b, and c are correct

OBJ. 8 ____ 17. The two important *skeletal elements* that are generally used to
determine sex and age are:

    a. teeth and healed fractures

    b. presence of muscular and fatty tissue

    c. condition of the teeth and muscular mass

    d. bone weight and bone markings

OBJ. 9 ____ 18. In determining the age of a skeleton, which of the following
selections would be used?

    a. the presence/absence of epiphyseal plates

    b. the size and roughness of bone markings

    c. the mineral content of the bones

    d. a, b, and c are correct

[L1]   Completion:

Using the terms below, complete the following statements.

knee	styloid	glenoid fossa	pelvis
pubic symphysis	clavicle	acetabulum	teeth
malleolus	age	childbearing	upper & lower extremities
coxae	wrist	pectoral girdle	

OBJ. 1   19. The appendicular skeleton includes the bones of the pectoral and
pelvic girdles and the _____.

OBJ. 2   20. The only direct connection between the pectoral girdle and the axial
skeleton is the _____.

OBJ. 2   21. The shoulder area and its component bones comprise a region referred
to as the _____.

OBJ. 2   22. The scapula articulates with the proximal end of the humerus at the
_____.

OBJ. 3   23. The ulna and the radius both have long shafts that contain like processes
called _____ processes.

OBJ. 3   24. The radiocarpal articulations and the intercarpal articulations are
responsible for the movements in the region of the _____.

OBJ. 4   25. The pelvic girdle consists of six bones collectively referred to as the
_____.

OBJ. 4   26. Ventrally the coxae are connected by a pad of fibrocartilage at the _____.

OBJ. 5   27. The process that the tibia and fibula have in common that acts as a shield for the ankle is the _____.

OBJ. 5   28. At the hip joint to either side, the head of the femur articulates with the _____.

OBJ. 5   29. The popliteal ligaments are responsible for reinforcing the back of the _____.

OBJ. 6   30. An enlarged pelvic outlet in the female is an adaptation for _____.

OBJ. 7   31. A major means of determining the medical history of a person is to examine the condition of the individual's _____.

OBJ. 8   32. Skeletal differences between males and females are usually identified axially on the skull and/or the appendicular aspects of the _____.

OBJ. 9   33. Reduction in mineral content of the bony matrix is an example of a skeletal change related to _____.

[L1]   Matching:

Match the terms in column B with the terms in column A. Use letters for answers in the spaces provided.

		COLUMN A		**COLUMN B**
OBJ. 1	_____	34. carpals, tarsals	A.	lower arm bones
OBJ. 2	_____	35. pectoral girdle	B.	involved with running
OBJ. 2	_____	36. shoulder joint	C.	hip bones
OBJ. 3	_____	37. radius, ulna	D.	lower leg bones
OBJ. 4	_____	38. pelvic girdle	E.	oval to round
OBJ. 5	_____	39. patella	F.	wrist and ankle bones
OBJ. 5	_____	40. tibia, fibula	G.	ball and socket
OBJ. 6	_____	41. female pelvic inlet	H.	male characteristic
OBJ. 6	_____	42. male pelvic inlet	I.	smooth bone markings
OBJ. 7	_____	43. talar arch	J.	shoulder bones
OBJ. 8	_____	44. prominent bone markings	K.	heart-shaped
OBJ. 9	_____	45. process of aging	L.	kneecap

Level
-1-

[L1]   Drawing/Illustration Labeling:

Identify each numbered structure by labeling the following figures:

OBJ. 1   **_FIGURE 8.1_**   Appendicular Skeleton

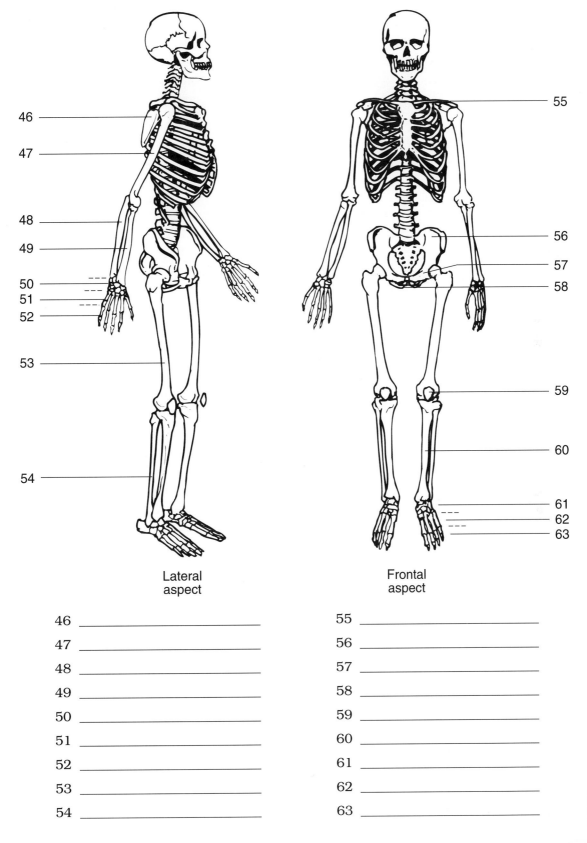

Lateral aspect

Frontal aspect

46 _____

47 _____

48 _____

49 _____

50 _____

51 _____

52 _____

53 _____

54 _____

55 _____

56 _____

57 _____

58 _____

59 _____

60 _____

61 _____

62 _____

63 _____

OBJ. 2   *FIGURE 8.2*   The Scapula

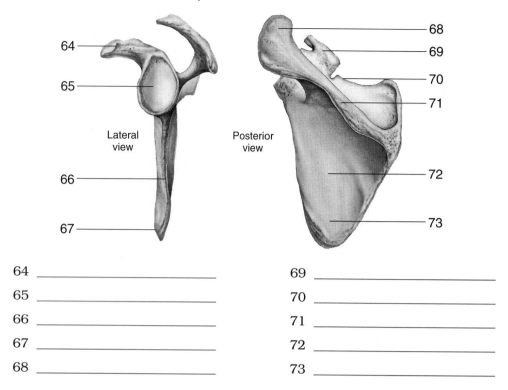

Lateral view

Posterior view

64	_____	69	_____
65	_____	70	_____
66	_____	71	_____
67	_____	72	_____
68	_____	73	_____

OBJ. 3   *FIGURE 8.3*   The Humerus

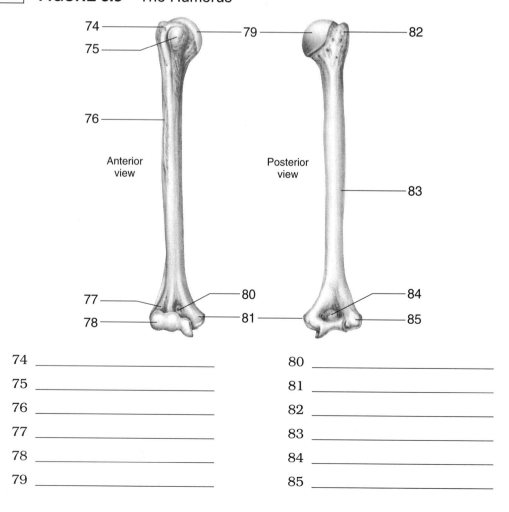

Anterior view

Posterior view

74	_____	80	_____
75	_____	81	_____
76	_____	82	_____
77	_____	83	_____
78	_____	84	_____
79	_____	85	_____

Level
-1-

OBJ. 3   *FIGURE 8.4*   The Radius and Ulna

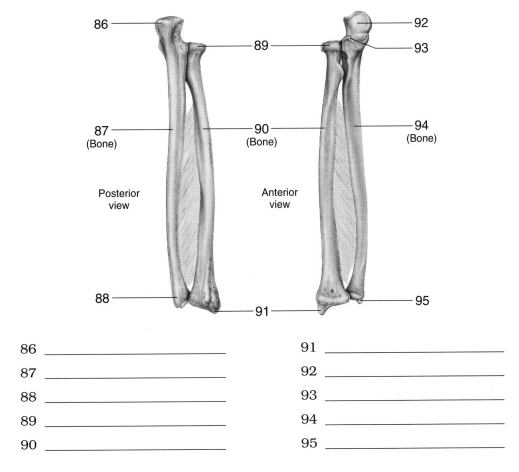

Posterior
view

Anterior
view

86 _____

87 _____

88 _____

89 _____

90 _____

91 _____

92 _____

93 _____

94 _____

95 _____

OBJ. 3   *FIGURE 8.5*   Bones of the Wrist and Hand

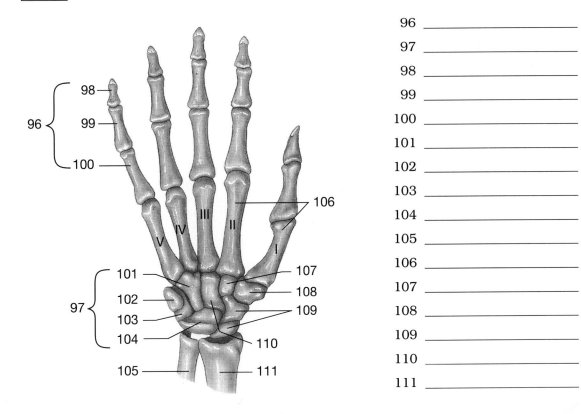

96 _____

97 _____

98 _____

99 _____

100 _____

101 _____

102 _____

103 _____

104 _____

105 _____

106 _____

107 _____

108 _____

109 _____

110 _____

111 _____

OBJ. 4   **FIGURE 8.6**   The Pelvis (anterior view)

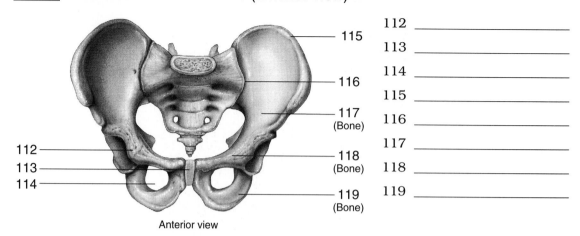

Anterior view

112	_____
113	_____
114	_____
115	_____
116	_____
117	_____
118	_____
119	_____

OBJ. 4   **FIGURE 8.7**   The Pelvis (lateral view)

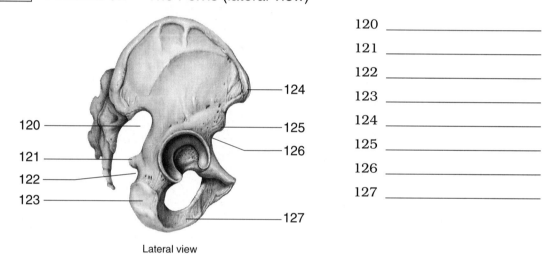

Lateral view

120	_____
121	_____
122	_____
123	_____
124	_____
125	_____
126	_____
127	_____

OBJ. 5   **FIGURE 8.8**   The Femur (anterior and posterior views)

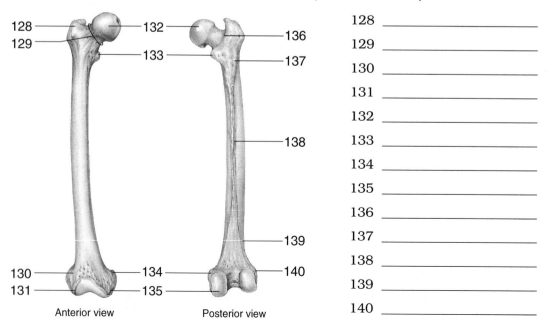

Anterior view   Posterior view

Level
-1-

128	_____
129	_____
130	_____
131	_____
132	_____
133	_____
134	_____
135	_____
136	_____
137	_____
138	_____
139	_____
140	_____

OBJ. 5   *FIGURE 8.9*   The Tibia and Fibula

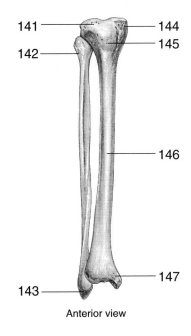

Anterior view

141 _____

142 _____

143 _____

144 _____

145 _____

146 _____

147 _____

OBJ. 5   *FIGURE 8.10*   Bones of the Ankle & Foot

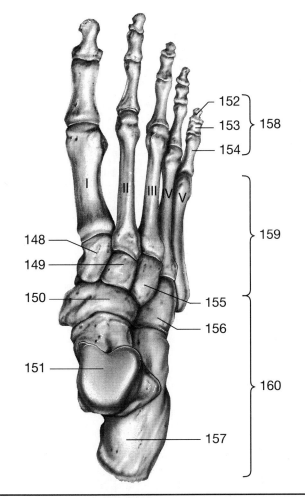

148 _____

149 _____

150 _____

151 _____

152 _____

153 _____

154 _____

155 _____

156 _____

157 _____

158 _____

159 _____

160 _____

When you have successfully completed the exercises in L1 proceed to L2.

Level
–1–

☐ LEVEL 2   CONCEPT SYNTHESIS

Concept Map I:

Using the following terms, fill in the circled, numbered, blank spaces to complete the concept map. Follow the numbers that comply with the organization of the map.

Fibula   Ischium   Femur
Humerus   Pectoral girdle   Metacarpals
Phalanges   Radius   Tarsals

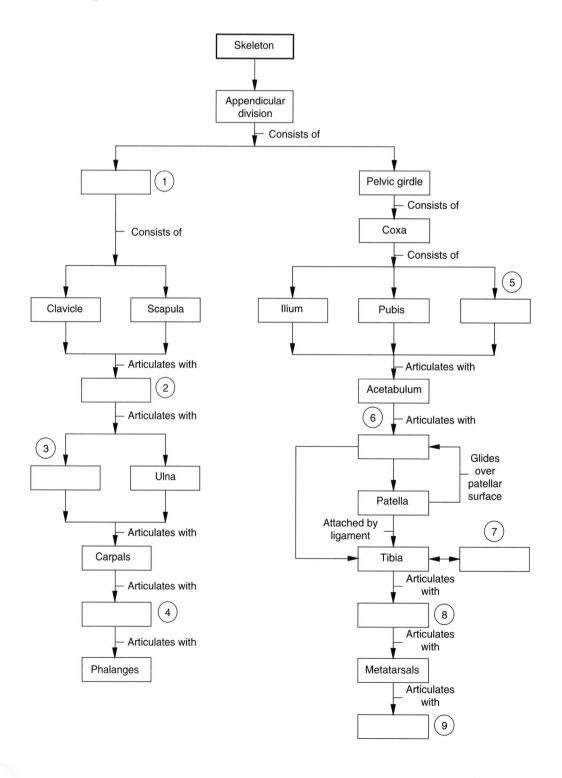

Body Trek:

Using the terms below, fill in the blanks to complete the study of the appendicular skeleton.

Patella	Tibia	Metatarsals	Pubis
Metacarpals	Ilium	Ulna	Femur
Tarsals	Hinge	Fibula	Radius
Clavicle	Pectoral	Brachium	Scapula
Carpals	Pelvic	Ball and socket	Ischium
Phalanges	Humerus	Knee	Shoulder
Finger bones	Acetabulum	Elbow	

Robo is returned to the lab to monitor the activities of the students as they use a skeleton to locate and name the bones and joints of the appendicular skeleton. The students are aware of the robot's presence, but unsure of its location. They look forward to the "verbal" encouragement Robo offers. The four groups that studied the axial skeleton remain intact, and each group is assigned an area of the appendicular skeleton to identify and study. Group 1 is assigned the pectoral girdle and the shoulder joint; group 2 the bones of the upper and lower arm and the elbow joint; group 3 the pelvic girdle and hip joint; and group 4 the bones of the upper and lower leg and the knee joint.

Group 1 locates and identifies the scapula and the (10) _____ , the bones that make up the (11) _____. Group members notice that the shoulder joint is shaped like a (12) _____ because the rounded convex surface of the upper arm bone fits into a cup-like socket of the (13) _____ . [Robo: "Just like I would have done it!"] Students can't believe that a robot can talk.

Group 2 notices that there is only one long bone in the upper arm, the (14) _____ , and two long bones in the lower arm, the (15) _____ , which is always located on the thumb side, and the (16) _____. The long bone of the (17) _____ , or upper arm, is joined with the two long bones of the lower arm by the elbow joint, which functions like a (18) _____. The distal ends of the lower arm bones are articulated with the wrist bones, the (19) _____ , which join with the hand bones, the (20) _____ , which are connected to the phalanges or (21) _____. [Robo: "Wish I had arms and hands!"] A student shouts, "As small as you are we wouldn't be able to see them." Others laugh.

Settling down, group 3 has difficulty differentiating among the largest bone of the pelvic girdle, the (22) _____ , from the (23) _____ and the (24) _____ , which forms a joint at the inferior part of the pelvis. Group members observe that the pelvic socket, the (25) _____ , is comprised of all three bones of the (26) _____ girdle. [Robo: "What a job! Can't wait to hear the next group get 'it' all together."]

Group 4 responds enthusiastically to Robo's comment by locating the longest and heaviest bone in the body, in the upper leg, the (27) _____ , which has a ball-like head that fits into the pelvic socket and forms a joint that functions like the (28) _____ joint. Students point out that the large bone articulates at the distal end with a long thin bone, the (29) _____ , and the larger and thicker (30) _____ , both of which make up the bones of the lower leg. The (31) _____ , a bone that glides across the articular surface of the femur, is embedded within a tendon in the region of the (32) _____ joint, which functions like the (33) _____ joint. [Robo: "I'm impressed!"] Students laugh. Distally, the long bones of the lower leg articulate with the ankle bones called (34) _____ , which join with the bones of the foot, the (35) _____ , which are connected to the toe bones called (36) _____. [Robo: "Wish I had legs and feet so I could escape from this acetabulum."] Students are amused and raise questions about the technology involved in producing micro-robots and their potential uses for monitoring activity within the human body. Robo is removed from its location and returned to Mission Control to begin preparation for the next trek.

[L2]   Multiple Choice:

Place the letter corresponding to the correct answer in the space provided.

_____ 37. The unique compromise of the articulations in the appendicular skeleton is:

    a. the stronger the joint, the less restricted the range of motion

    b. the weaker the joint, the more restricted the range of motion

    c. the stronger the joint, the more restricted the range of motion

    d. the strength of the joint and range of motion are unrelated

_____ 38. If you run your fingers along the superior surface of the shoulder joint, you will feel a process called the:

    a. coracoid

    b. acromion

    c. coronoid

    d. styloid

_____ 39. In a shoulder separation, the:

    a. acromioclavicular joint undergoes partial or complete dislocation

    b. clavicle and the scapula separate

    c. the head of the femur separates from the clavicle

    d. the muscles attached to the clavicle and scapula are torn

_____ 40. The process on the humerus located near the head that establishes the contour of the shoulder is the:

    a. intertubercular groove

    b. deltoid tuberosity

    c. lateral epicondyle

    d. greater tubercle

_____ 41. The four _proximal_ carpals are:

    a. trapezium, trapezoid, capitate, hamate

    b. trapezium, triquetal, capitate, lunate

    c. scaphoid, trapezium, lunate, capitate

    d. scaphoid, lunate, triquetal, pisiform

_____ 42. The _distal_ carpals are:

    a. scaphoid, lunate, triquetal, pisiform

    b. trapezium, trapezoid, capitate, hamate

    c. trapezium, scaphoid, trapezoid, lunate

    d. scaphoid, capitate, triquetal, hamate

_____ 43. Structures found in synovial joints that reduce friction where large muscles and tendons pass across the joint capsule are referred to as:

    a. joint capsule

    b. capsular ligaments

    c. bursae

    d. articular condyles

Level =2=

_____ 44. The amphiarthrotic articulation which limits movement between the two pubic bones is the:

      a. pubic symphysis

      b. iliopectineal line

      c. iliac tuberosity

      d. obturator foramen

_____ 45. The only ankle bone that articulates with the tibia and the fibula is the:

      a. calcaneus

      b. talus

      c. navicular

      d. cuboid

_____ 46. Severe fractures of the femoral neck have the highest complication rate of any fracture because:

      a. primary limits are imposed by the surrounding muscles

      b. of the restrictions imposed by ligaments and capsular fibers

      c. of the thickness and length of the bone

      d. the blood supply to the region is relatively limited

_____ 47. A complete dislocation of the knee is extremely unlikely because of:

      a. the pair of cartilaginous pads that surround the knee

      b. the presence of the medial and lateral menisci

      c. the presence of bursae which serve to reduce friction

      d. the seven major ligaments that stabilize the knee joint

_____ 48. The mechanism that allows standing for prolonged periods without continually contracting the extensor muscles is:

      a. the popliteal ligaments extend between the femur and the heads of the tibia and fibula

      b. the limitations of the ligaments and the movement of the femur

      c. the patellar ligaments providing support for the front of the knee, helping to maintain posture

      d. a slight lateral rotation of the tibia tightens the anterior cruciate ligament and joins the meniscus between the tibia and femur

_____ 49. The reason opposition can occur between the hallux (first toe) and the first metatarsal is:

      a. the articulation is a hinge rather than a saddle joint

      b. the articulation is a saddle rather than a hinge joint

      c. the articulation is a gliding diarthrosis

      d. the articulation is like a ball-and-socket joint

_____ 50. When a ligament is stretched to the point where some of the collagen fibers are torn, the injury is called a:

      a. dislocation

      b. strain

      c. sprain

      d. dancer's fracture

Level
=2=

_____ 51. Abnormalities that affect the bones can have a direct effect on the muscles because:

   a. when a muscle gets larger, the bones become stronger
   b. most of the body's calcium is tied up in the skeleton
   c. when a bone gets more massive, the muscle enlarges proportionately
   d. a, b, and c are correct.

[L2]   Completion:

Using the terms below, complete the following statements.

calcaneus	hallux	bursitis	pollex
clavicle	fibula	ilium	metacarpals
femur	arthroscopy	pelvis	bursae
ulna	scapula	acetabulum	

52. The bone which cannot resist strong forces and provides the only fixed support for the pectoral girdle is the _____.

53. At its proximal end, the round head of the humerus articulates with the _____.

54. The bone that forms the medial support of the forearm is the _____.

55. The bones which form the palm of the hand are the _____.

56. Chambers lined by synovial membrane and filled with synovial fluid are the _____.

57. Propping your head above a desk while struggling through your A & P textbook may result in "student's elbow," which is a form of _____.

58. At the proximal end of the femur, the head articulates with the curved surface of the _____.

59. The largest coxal bone is the _____.

60. The coxae, the sacrum, and the coccyx form a composite structure called the _____.

61. The longest and heaviest bone in the body is the _____.

62. The bone in the lower leg which is completely excluded from the knee joint is the _____.

63. The large heel bone which receives weight transmitted to the ground by the inferior surface of the talus is the _____.

64. The thumb is anatomically referred to as the _____.

65. The first toe is anatomically referred to as the _____.

66. The technique which uses fiber optics to permit exploration of a joint without major surgery is _____.

[L2]   Short Essay:

Briefly answer the following questions in the spaces provided below.

67. What are the primary functions of the appendicular skeleton?

68. What are the basic components of the appendicular skeleton?

69. What bones comprise the pectoral girdle? The pelvic girdle?

70. What are the structural and functional similarities and differences between the ulna and the radius?

71. Functionally, what is the relationship between the elbow joint and the knee joint?

72. What are the structural and functional similarities and differences between the tibia and the fibula?

73. Functionally, what is the commonality between the shoulder joint and the hip joint?

74. How are the articulations of the carpals of the wrist comparable to those of the tarsals of the ankle?

75. How are muscles and bones physiologically linked?

When you have successfully completed the exercises in L2 proceed to L3.

☐ LEVEL 3   CRITICAL THINKING/APPLICATION

Using principles and concepts learned about the appendicular skeleton, answer the following questions. Write your answers on a separate sheet of paper.

1. The carpals are arranged in two rows of four each. In order for tendons, nerves, and blood vessels to pass into the hand from the wrist a ligament stretches across the wrist which forms a tunnel on the anterior surface of the wrist called the *carpal tunnel.* Tingling, burning, and numbness in the hand are symptoms that result from *carpal tunnel syndrome.* What causes these symptoms to occur?

2. A recent headline on the sports page of the newspaper read, "Football player dies awaiting leg surgery." The player was expected to be sidelined for a matter of weeks with a broken leg but died suddenly due to lung hemorrhaging and inability to oxygenate his blood. What is the relationship between a fractured leg and pulmonary hemorrhaging that caused the death of this young athlete?

3. What structural characteristics of the female pelvis, compared to those of the male, make it adaptable for delivery of the newborn?

4. What is the association between the metabolic disorder known as *gout*, which affects the joints, and damage to the kidney?

5. From time to time it is fashionable to wear pointed shoes. How might a decision to wear pointed shoes contribute to the formation of a bunion?

6. What is the commonality and what are the differences of "housemaid's knee," "weaver's bottom" and "student's elbow"?

9

Articulations

■ Overview

Ask the average individual what things are necessary to maintain life and usually the response will be oxygen, food, and water. Few people relate the importance of movement as one of the factors necessary to maintain life, yet the human body doesn't survive very long without the ability to produce body movements. One of the functions of the skeletal system is to permit body movements; however, it is not the rigid bones that allow movement, but the articulations, or joints, between the bones.

The elegant movements of the ballet dancer or gymnast, as well as the everyday activities of walking, talking, eating, and writing, require the coordinated activity of all the joints, some of which are freely movable and flexible while others remain rigid to stabilize the body and serve to maintain balance.

Chapter 9 classifies the joints of the body according to structure and function. Classification by structure is based on the type of connective tissue between the articulating bones, while the functional classification depends on the degree of movement permitted within the joint.

The exercises in this chapter are designed to allow you to organize and conceptualize the principles of kinetics and biomechanics in the science of arthrology—the study of joints. Demonstrating the various movements permitted at each of the movable joints will facilitate and help you to visualize the adaptive advantages as well as the limitations of each type of movement.

☐ LEVEL 1   REVIEW OF CHAPTER OBJECTIVES

1. Contrast the major categories of joints and explain the relationship between structure and function for each category.

2. Describe the basic structure of a synovial joint, identifying possible accessory structures and their functions.

3. Describe the dynamic movements of the skeleton.

4. List the different types of synovial joints and discuss how the characteristic motions of each type is related to its anatomical structure.

5. Describe the articulations between the vertebrae of the vertebral column.

6. Describe the structure and function of the shoulder, elbow, hip, and knee joints.

7. Explain the relationship between joint strength and mobility, using specific examples.

[L1]   Multiple choice:

Place the letter corresponding to the correct answer in the space provided.

OBJ. 1   _____   1. Joints, or articulations, are classified on the basis of their degree of movement. From the following selections choose the one which identifies, in correct order, the following joints on the basis of: no movement; slightly movable; freely movable.

      a. amphiarthrosis, diarthrosis, synarthrosis

      b. diarthrosis, synarthrosis, amphiarthrosis

      c. amphiarthrosis, synarthrosis, diarthrosis

      d. synarthrosis, amphiarthrosis, diarthrosis

OBJ. 1   _____   2. The amphiarthrotic articulation which limits movements between the two pubic bones is the:

      a. pubic symphysis

      b. obturator foramen

      c. greater sciatic notch

      d. pubic tubercle

OBJ. 1   _____   3. The type of synarthrosis that binds each tooth to the surrounding bony socket is a:

      a. synchondrosis

      b. syndesmosis

      c. gomphosis

      d. symphysis

OBJ. 1   _____   4. Which joint is correctly matched with the types of joint indicated?

      a. symphisis pubis – fibrous

      b. knee – synovial

      c. sagittal suture – cartilaginous

      d. intervetebral disc – synovial

OBJ. 2   _____   5. The function(s) of synovial fluid that fills the joint cavity is (are):

      a. It nourishes the chondrocytes.

      b. It provides lubrication.

      c. It acts as a shock absorber.

      d. a, b, and c are correct.

OBJ. 2   _____   6. The primary function(s) of menisci in synovial joints is (are):

      a. to subdivide a synovial cavity

      b. to channel the flow of synovial fluid

      c. to allow for variations in the shapes of the articular surfaces

      d. a, b, and c are correct

OBJ. 3   _____   7. Flexion is defined as movement that:

      a. increases the angle between articulating elements

      b. decreases the angle between articulating elements

      c. moves a limb from the midline of the body

      d. moves a limb toward the midline of the body

Level
-1-

OBJ. 3   _____ 8. The movement that allows you to gaze at the ceiling is:

      a. rotation

      b. circumduction

      c. hyperextension

      d. elevation

OBJ. 3   _____ 9. If an articulation permits only angular movement in the forward/backward plane, or prevents any movement other than rotation around its longitudinal axis, it is:

      a. monaxial

      b. biaxial

      c. triaxial

      d. polyaxial

OBJ. 4   _____ 10. The reason that the elbow and knee are called hinge joints is:

      a. the articulator surfaces are able to slide across one another

      b. all combinations of movement are possible

      c. they permit angular movement in a single plane

      d. sliding and rotation are prevented, and angular motion is restricted to two directions

OBJ. 4   _____ 11. The type of joint that connects the fingers and toes with the metacarpals and metatarsals is:

      a. an ellipsoidal joint

      b. a biaxial joint

      c. a synovial joint

      d. a, b, and c are correct

OBJ. 4   _____ 12. The shoulder and hip joints are examples of:

      a. hinge joints

      b. ball and socket joints

      c. pivot joints

      d. gliding joints

OBJ. 5   _____ 13. The part(s) of the vertebral column that do not contain intervetebral discs is (are):

      a. the sacrum

      b. the coccyx

      c. first and second cervical vertebrae

      d. a, b, and c are correct

OBJ. 5   _____ 14. Movements of the vertebra column are limited to:

      a. flexion and extension

      b. lateral flexion

      c. rotation

      d. a, b, and c are correct

OBJ. 6   _____ 15. The joint that permits the greatest range of motion of any joint in the body is:

         a. hip joint
         b. shoulder joint
         c. elbow joint
         d. knee joint

OBJ. 6   _____ 16. The elbow joint is quite stable because:

         a. the bony surfaces of the humeris and ulna interlock
         b. the articular capsule is very thick
         c. the capsule is reinforced by stout ligaments
         d. a, b, and c are correct

OBJ. 6   _____ 17. In the hip joint, the arrangement that keeps the head of the femur from moving away from the acetabulum is:

         a. the formation of a complete bony socket
         b. the presence of fat pads covered by synovial membranes
         c. the articular capsule encloses the femoral head and neck
         d. the acetabular bones and the femoral head fit tightly

OBJ. 6   _____ 18. The knee joint functions as a:

         a. hinge joint
         b. ball and socket joint
         c. saddle joint
         d. gliding joint

OBJ. 6   _____ 19. The reason the points of contact in the knee joint are constantly changing is:

         a. there is no single unified capsule or a common synovial cavity
         b. the menisci conform to the shape of the surface of the femur
         c. the rounded femoral condyles roll across the top of the tibia
         d. a, b, and c are correct

OBJ. 7   _____ 20. The popliteal ligaments extend between the femur and the heads of the tibia and fibula reinforcing the:

         a. anterior surface of the knee joint
         b. lateral surface of the knee joint
         c. inside of the joint capsule
         d. back of the knee joint

OBJ. 7   _____ 21. The iliofemoral, pubofemoral, and ischiofemoral ligaments are involved in reinforcement and stabilization of the:

         a. shoulder joint
         b. hip joint
         c. knee joint
         d. elbow joint

Level
—1—

OBJ. 7   _____ 22. The subacromial bursa and the subcoracoid bursa reduce friction at the:

      a. elbow joint

      b. hip joint

      c. shoulder joint

      d. knee joint

OBJ. 7   _____ 23. The ligamentum flavum, interspinous and supraspinous ligament bind together and stabilize the:

      a. scapulohumeral joint

      b. vertebral column

      c. humero-ulnar joint

      d. olecranal joint

[L1]   Completion:

Using the terms below, complete the following statements.

synovial	symphysis	hip
bursae	elbow	synostosis
scapulohumeral	anulus fibrosus	knee
supination	suture	accessory ligaments
ellipsoidal	flexion	gliding

OBJ. 1   24. A synarthrotic joint found only between the bones of the skull is a _____.

OBJ. 1   25. A totally rigid immovable joint resulting from fusion of bones is a _____.

OBJ. 1   26. The amphiarthrotic joint where bones are separated by a wedge or pad of fibrocartilage is a _____.

OBJ. 2   27. Localized thickenings of joint capsule are called _____.

OBJ. 2   28. Small, synovial-filled pockets that form where a tendon or ligament rubs against other tissues are called _____.

OBJ. 3   29. A movement that reduces the angle between the articulating elements is _____.

OBJ. 3   30. Movement in the wrist and hand in which the palm is turned forward is _____.

OBJ. 4   31. Diarthrotic joints that permit a wide range of motion are called _____ joints.

OBJ. 4   32. The type of joint that connects the fingers and toes with the metacarpals and metatarsals is an _____.

OBJ. 5   33. The joints between the superior and inferior articulations of adjacent vertebrae are _____.

OBJ. 5   34. The tough outer layer of fibrocartilage on intervertebral discs is the _____.

OBJ. 6   35. The joint that permits the greatest range of motion of any joint in the body is the _____ joint.

OBJ. 6   36. The extremely stable joint that is almost completely enclosed in a bony socket is the _____ joint.

Level
—1—

OBJ. 7   37. The joint that resembles three separate joints with no single unified capsule or common synovial cavity is the _____.

OBJ. 7   38. The radial collateral, annular, and ulnar collateral ligaments provide stability for the _____ joint.

[L1]   Matching:

Match the terms in column B with the terms in column A. Use letters for answers in the spaces provided.

PART I

			COLUMN A	COLUMN B
OBJ. 1	____	39.	synarthrosis	A. ball and socket
OBJ. 1	____	40.	amphiarthrosis	B. forearm bone movements
OBJ. 2	____	41.	articular discs	C. thumb movement for grasping
OBJ. 2	____	42.	acetabulum; head of femur	D. biaxial movement
OBJ. 3	____	43.	pronation-supination	E. monoaxial movement
OBJ. 3	____	44.	opposition	F. slightly movable joint
OBJ. 4	____	45.	elbow	G. menisci
OBJ. 4	____	46.	wrist	H. immovable joint

PART II

			COLUMN A	COLUMN B
OBJ. 5	____	47.	pivot joint	I. knee joint capsule
OBJ. 6	____	48.	shoulder joint	J. knee
OBJ. 6	____	49.	hinge joint	K. intervertebral discs
OBJ. 6	____	50.	tendons	L. triaxial movement
OBJ. 6	____	51.	reinforce knee joint	M. between atlas and axis
OBJ. 7	____	52.	cruciate ligaments	N. muscle to bone attachment
OBJ. 7	____	53.	nucleus pulposus	O. popliteal ligaments

[L1] Drawing/Illustration Labeling:

Identify each numbered structure by labeling the following figures:

OBJ. 2 **FIGURE 9.1** Structure of a Synovial Joint

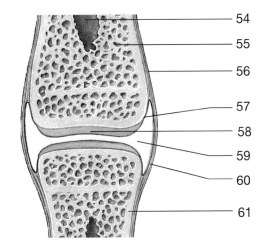

54 _____   58 _____

55 _____   59 _____

56 _____   60 _____

57 _____   61 _____

OBJ. 6 **FIGURE 9.2** Diarthrosis: Structure of the ShoulderJoint

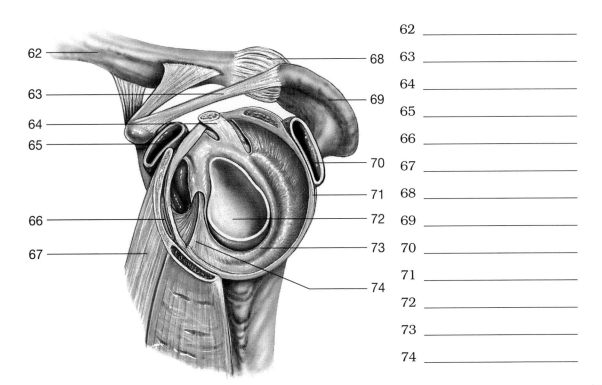

62 _____

63 _____

64 _____

65 _____

66 _____

67 _____

68 _____

69 _____

70 _____

71 _____

72 _____

73 _____

74 _____

Level
-1-

OBJ. 3 | *FIGURE 9.3* Movements of the Skeleton

Identify each skeletal movement.

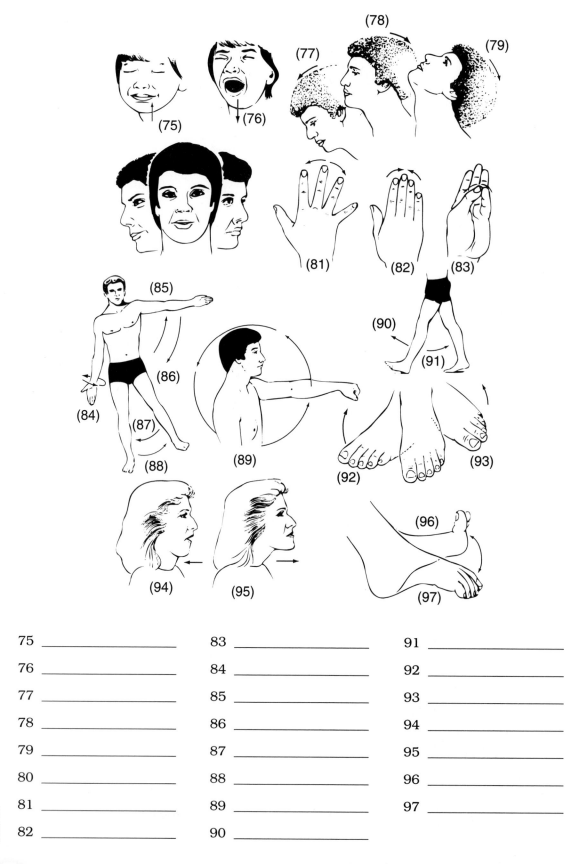

75 _____	83 _____	91 _____
76 _____	84 _____	92 _____
77 _____	85 _____	93 _____
78 _____	86 _____	94 _____
79 _____	87 _____	95 _____
80 _____	88 _____	96 _____
81 _____	89 _____	97 _____
82 _____	90 _____	

Level
-1-

Instructions:

Match the joint movements at the top of the page with the appropriate type of joint in Figure 9.4. Use letters for answers in spaces labeled *movement.*

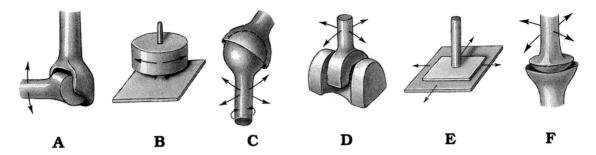

| A | B | C | D | E | F |

Instruction:

Identify the *types* of joints in Figure 9.4. Place the answer in the space labeled *type.* (Synarthrosis, amphiarthrosis, diarthrosis)

OBJ. 4   ***FIGURE 9.4***   Types of Diarthrotic Joints

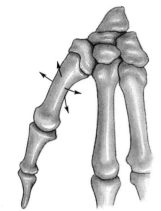

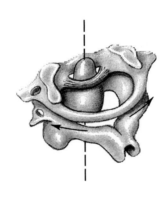

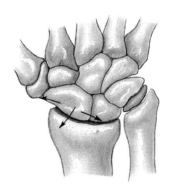

(Type) 98 _____

(Movement) 99 _____

(Type) 100 _____

(Movement) 101 _____

(Type) 102_____

(Movement) 103 _____

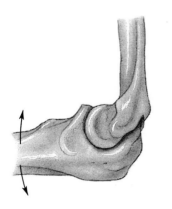

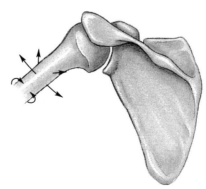

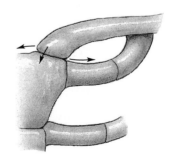

(Type) 104 _____

(Movement) 105 _____

(Type) 106 _____

(Movement) 107 _____

(Type) 108_____

(Movement) 109 _____

OBJ. 6   **FIGURE 9.5**   Sectional View of the Knee Joint

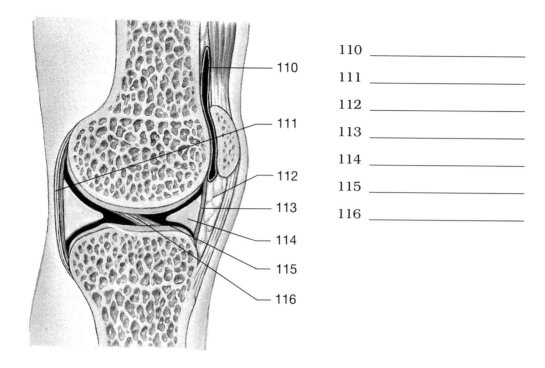

110 _____

111 _____

112 _____

113 _____

114 _____

115 _____

116 _____

OBJ 7   **FIGURE 9.6**   Anterior View of a Flexed Knee

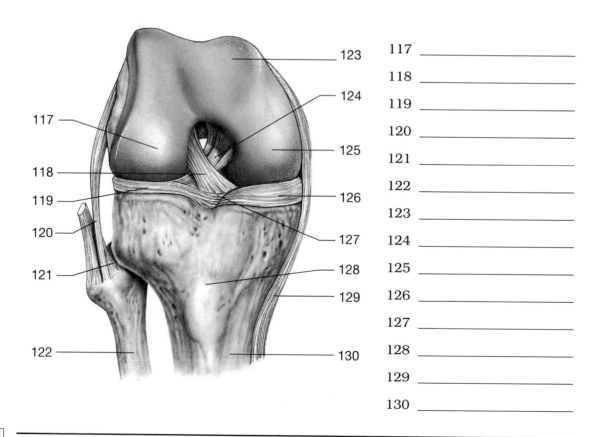

117 _____

118 _____

119 _____

120 _____

121 _____

122 _____

123 _____

124 _____

125 _____

126 _____

127 _____

128 _____

129 _____

130 _____

Level
–1–   **When you have successfully completed the exercises in L1 proceed to L2.**

☐ LEVEL 2   CONCEPT SYNTHESIS

Concept Map I:

Using the following terms, fill in the circled, numbered, blank spaces to complete the concept map. Follow the numbers that comply with the organization of the concept map.

Amphiarthrosis	Sutures	Wrist	Monoaxial
Symphysis	Cartilaginous	Synovial	Fibrous
No movement			

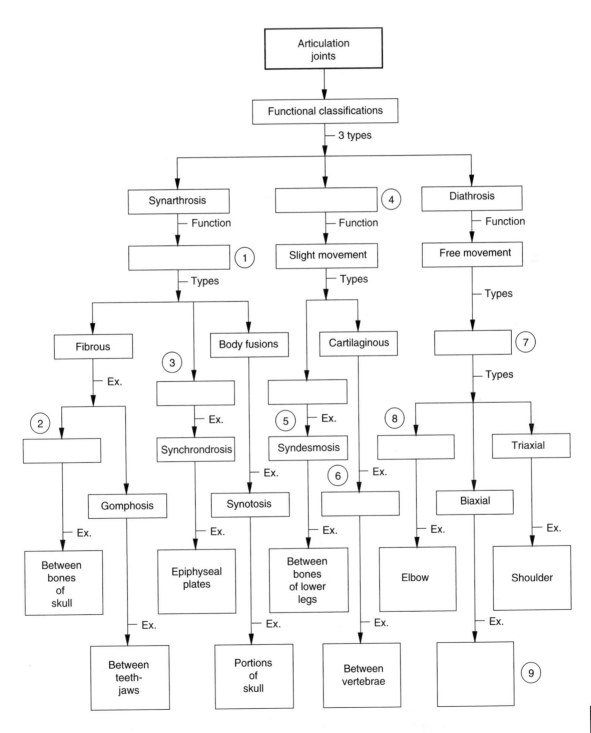

Body Trek:

Using the terms below, fill in the blanks to complete the roboscopic examination of a typical, normal knee joint.

capsule	synovial fluid	condyles	tibia
femur	menisci	cushions	femoral
stability	patellar	synovial	tibial collateral
fibula	cruciate	popliteal	fibular collateral
tibial	seven		

The ultra-tiny "roboscope" attached to Robo is inserted into the knee joint and immediately a signal is relayed to mission control, "lights on please, it's dark in here!" The immediate site resembles an area that looks like three separate compartments, two between the femur and tibia, and one between the patella and the patellar surface of the femur. Robo responds, "disoriented and complicated; need direction. No single unified (10) _____, and no common (11) _____ cavity." The cavity is filled with a clear, viscous fluid, the (12) _____ , which appears to make the surfaces of the medial and lateral (13) _____ of the (14) _____ and the articulating surface of the (15) _____ shiny and slippery. Movement of Robo is somewhat restricted because of a pair of fibrocartilage pads, the medial and lateral (16) _____ which lie between the femoral and tibial surfaces. Even though movement is restricted, the pads act like (17) _____ , making the "visit" to the area quite comfortable and soft. It appears that these pads provide some lateral (18) _____ to the joint. Meanwhile back in mission control, the pictures shown by the roboscope reveal the presence of (19) _____ major ligaments which stabilize the knee joint. The (20) _____ ligament provides support to the anterior surface of the knee joint, while two (21) _____ ligaments extend between the femur and the heads of the tibia and (22) _____ , providing reinforcement to the back of the knee joint. Probing inside the joint capsule, the roboscope reveals the crossed anterior and posterior (23) _____ ligaments, limiting movement of the femur and maintaining alignment of the (24) _____ and (25) _____ condyles. Due to the tightness of the (26) _____ ligament reinforcing the medial surface of the knee joint and the (27) _____ ligament reinforcing the lateral surface, the leg must be fully extended and stabilized allowing Robo to remain in an upright position while "scoping" the region free from being crushed or damaged due to knee flexion. The mission completed, the signal is given for "Robo retrieval" confirming the success of the "joint" probe with Robo "scoping" a knee using the roboscope—Yea!

[L2] Multiple Choice:

Place the letter corresponding to the correct answer in the space provided.

_____ 28. Small, fluid-filled pockets in connective tissue lined by synovial membranes are:

      a. bursae

      b. menisci

      c. fat pads

      d. articular capsules

_____ 29. A movement away from the longitudinal axis of the body in the frontal plane is:

      a. adduction

      b. hyperextension

      c. abduction

      d. circumduction

Level
=2=

_____ 30. The opposing movement of supination is:

    a. rotation

    b. pronation

    c. protraction

    d. opposition

_____ 31. The special movement of the thumb that enables it to grasp and hold an object is:

    a. supination

    b. inversion

    c. circumduction

    d. opposition

_____ 32. Twiddling your thumbs during a lecture demonstrates the action that occurs at a:

    a. hinge joint

    b. ball and socket joint

    c. saddles joint

    d. gliding joint

_____ 33. A characteristic decrease in height with advanced age may result from:

    a. decreased water content of the nucleus pulposus in an intervertebral disc

    b. a decrease in the anulus fibrosus of an intervertebral disc

    c. a decrease in the collagen fibers in the bodies of the vertebrae

    d. a, b, and c are correct

_____ 34. When the nucleus pulposus breaks through the anulus and enters the vertebral canal, the result is a (an):

    a. slipped disc

    b. anulus fibrosus

    c. herniated disc

    d. ligamentous nuchae

_____ 35. Contraction of the biceps brachii muscle produces:

    a. pronation of the forearm and extension of the elbow

    b. supination of the forearm and extension of the elbow

    c. supination of the forearm and flexion of the elbow

    d. pronation of the forearm and flexion of the elbow

_____ 36. The unique compromise of the articulations in the appendicular skeleton is:

    a. the stronger the joint, the less restricted the range of motion

    b. the weaker the joint, the more restricted the range of motion

    c. the stronger the joint, the more restricted the range of motion

    d. the strength of the joint and range of motion are unrelated

Level =2=

_____ 37. Even though the specific cause may vary, arthritis always involves damage to the:

>   a. articular cartilages
>   b. bursae
>   c. accessory ligaments
>   d. epiphyseal disks

_____ 38. In a shoulder separation the:

>   a. clavicle and scapula separate
>   b. head of the femur separates from the clavicle
>   c. muscles attached to the clavicle and scapula are torn
>   d. acromioclavicular joint undergoes partial or complete separation

_____ 39. If you run your fingers along the superior surface of the shoulder joint, you will feel a process called the:

>   a. coracoid
>   b. acromion
>   c. coronoid
>   d. styloid

_____ 40. Structures found in synovial joints that reduce friction where large muscles and tendons pass across the joint capsule are referred to as:

>   a. capsular ligaments
>   b. menisci
>   c. bursae
>   d. articular condyles

_____ 41. A complete dislocation of the knee is extremely unlikely because of:

>   a. the pair of cartilaginous pads that surround the knee
>   b. the presence of the media and lateral muscle
>   c. the presence of bursae which serve to reduce friction
>   d. the seven major ligaments that stabilize the knee joint

_____ 42. The mechanism that allows standing for prolonged periods without continually contracting the extensor muscles is:

>   a. the popliteal ligaments extend between the femur and the heads of the tibia and fibula
>   b. the limitations of the ligaments and the movement of the femur
>   c. the patellar ligaments providing support for the front of the knee, helping to maintain posture
>   d. a slight lateral rotation of the tibia tightens the anterior cruciate ligament and joins the meniscus between the tibia and femur

_____ 43. The reason opposition can occur between the hallux (first toe) and the first metatarsal is:

>   a. the articulation is a hinge rather than a saddle joint
>   b. the articulation is a saddle rather than a hinge joint
>   c. the articulation is a gliding diarthrosis
>   d. the articulation is like a ball-and-socket joint

_____ 44. When a ligament is stretched to the point where some of the collagen fibers are torn, the injury is called a:

        a. dislocation

        b. strain

        c. sprain

        d. dancer's fracture

[L2]   Completion:

Using the terms below, complete the following statements.

gomphosis	synchondrosis	arthritis
hyperextension	fat pads	articular cartilage
menisci	amykylosis	syndesmosis
tendons	luxation	rheumatism

45. An abnormal fusion between articulating bones in response to trauma and friction is referred to as an _____.

46. Rheumatic diseases that affect synovial joints result in the development of _____.

47. The synarthrosis that binds each tooth to the surrounding bony socket is a _____.

48. A rigid cartilaginous connection such as an epiphyseal plate is called a _____.

49. The amphiarthrotic distal articulation between the tibia and fibula is a _____.

50. The joint accessory structures which may subdivide a synovial cavity are _____.

51. The accessory structures which provide protection for the articular cartilages are the _____.

52. Arthritis always involves damage to the _____.

53. A movement that allows you to gaze at the ceiling is _____.

54. A general term that indicates pain and stiffness affecting the skeletal and/or muscular systems is _____.

55. The structures that pass across or around a joint that may limit the range of motion and provide mechanical support are _____.

56. When articulating surfaces are forced out of position the displacement is called a _____.

[L2]   Short Essay

57. What three major categories of joints are identified based on the range of motion permitted?

58. What is the difference between a _synostosis_ and a _symphysis_?

59. What is the functional difference between a ligament and a tendon?

60. What are the structural and functional differences between: (a) a bursa and (b) a meniscus?

61. What is the functional role of synovial fluid in a diarthrotic joint?

62. Give examples of the types of joints and their location in the body that exhibit the following types of movements: (a) monoaxial   (b) biaxial   (c) triaxial

63. What six (6) types of synovial joints are found in the human body? Give an example of each type.

64. Functionally, what is the relationship between the elbow and the knee joint?

65. Functionally, what is the commonality between the shoulder joint and the hip joint?

66. How are the articulations of the carpals of the wrist comparable to those of the tarsals of the ankle?

67. Based on an individual in anatomical position, what types of movements can be used to illustrate angular motion?

68. Why is the scapulohumeral joint the most frequently dislocated joint in the body?

69. What regions of the vertebral column do not contain intervertebral discs? Why are they unnecessary in these regions?

70. Identify the unusual types of movements which apply to the following examples:
    (1) twisting motion of the foot that turns the sole inward
    (2) grasping and holding an object with the thumb
    (3) standing on tiptoes
    (4) crossing the arms
    (5) opening the mouth
    (6) shrugging the shoulders

71. How does the stability of the elbow joint compare with that of the hip joint? Why?

72. What structural damage would most likely occur if a "locked knee" was struck from the side while in a standing position?

When you have successfully completed the exercises in L2 proceed to L3.

☐ LEVEL 3 CRITICAL THINKING/APPLICATION

Using principles and concepts learned about articulations, answer the following questions. Write your answers on a separate sheet of paper.

1. As a teacher of anatomy what structural characteristics would you identify for the students to substantiate that the hip joint is stronger and more stable than the shoulder joint?

2. A high school football player notices swelling in the knee joint. He decides he'd better tell the coach who responds by telling him, "You have water on the knee." As a student of anatomy, explain what the coach is talking about.

3. Stairmasters (stair-climbing machines) are popular in health clubs and gyms for cardiovascular exercise. Since climbing involves continuous excessive force on the knees, what tissues in the knee joint are subject to damage?

4. Greg is a pitcher on the high school baseball team. He spends many hours practicing to improve his pitching skills. Recently, he has been complaining about persistent pain beneath his right shoulder blade (scapula). What do you think is causing the pain?

5. Steve injured his right knee during a basketball game when he jumped to rebound the ball and landed off-balance on the right leg. He has been experiencing pain and limited mobility of the knee joint. What type of injury do you think Steve has? What techniques would be used to explore the extent of the damage?

6. Two patients sustain hip fractures. In one case, a pin is inserted into the joint and the injury heals well. In the other, the fracture fails to heal. Identify the types of fractures that are probably involved. Why did the second patient's fracture fail to heal and what steps can be taken to restore normal function?

7. Joan is an executive secretary for a large corporation. Her normal attire for work includes high-heeled shoes sometimes with pointed toes, resulting in the formation of bunions on her big toes. What are bunions and what causes them to form?

Level
3

Muscle Tissue

■ Overview

It would be impossible for human life to exist without muscle tissue. This is because many functional processes and all our dynamic interactions internally and externally with our environment depend on movement. In the human body, muscle tissue is necessary for movement to take place. The body consists of various types of muscle tissue, including cardiac, which is found in the heart; smooth, which forms a substantial part of the walls of hollow organs; and skeletal, which is attached to the skeleton. Smooth and cardiac muscles are involuntary and are responsible for transport of materials within the body. Skeletal muscle is voluntary and allows us to maneuver and manipulate in the environment; it also supports soft tissues, guards entrances and exits, and serves to maintain body temperature.

The review exercises in this chapter present the basic structural and functional characteristics of skeletal muscle tissue. Emphasis is placed on muscle cell structure, muscle contraction and muscle mechanics, the energetics of muscle activity, muscle performance, and integration with other systems.

☐ LEVEL 1   REVIEW OF CHAPTER OBJECTIVES

1. Describe the characteristics and functions of muscle tissue.
2. Describe the organization of muscle at the tissue level.
3. Explain the unique characteristics of skeletal muscle fibers.
4. Identify the structural components of a sarcomere.
5. Identify the components of the neuromuscular junction and summarize the events involved in the neural control of skeletal muscle function.
6. Explain the key steps involved in the contraction of a skeletal muscle fiber.
7. Compare the different types of muscle contractions.
8. Describe the mechanisms by which muscle fibers obtain the energy to power contractions.
9. Relate the types of muscle fibers to muscle performance.
10. Distinguish between aerobic and anaerobic endurance and explain their implications for muscular performance.

11. Specify the effects of exercise and aging on muscles.

12. Identify the structural and functional differences between skeletal, cardiac, and smooth muscle fibers.

13. Discuss the role that smooth muscle tissue plays in systems throughout the body.

[L1] Multiple Choice:

Place the letter corresponding to the correct answer in the space provided.

OBJ. 1 _____ 1. Muscle tissue consists of cells that are highly specialized for the function of:

    a. excitability

    b. contractibility

    c. extensibility

    d. a, b, and c are correct

OBJ. 1 _____ 2. The three types of muscle tissue are:

    a. epimysium, perimysium, endomysium

    b. skeletal, cardiac, smooth

    c. elastic, collagen, fibrous

    d. voluntary, involuntary, resting

OBJ. 1 _____ 3. Skeletal muscles move the body by:

    a. using the energy of ATP to form ADP

    b. activation of the excitation-coupling reaction

    c. means of neural stimulation

    d. pulling on the bones of the skeleton

OBJ. 2 _____ 4. Skeletal muscles are often called voluntary muscles because:

    a. ATP activates skeletal muscles for contraction

    b. the skeletal muscles contain myoneural junctions

    c. they contract when stimulated by motor neurons of the central nervous system

    d. connective tissue harnesses generated forces voluntarily

OBJ. 3 _____ 5. The smallest functional unit of the muscle fiber is:

    a. thick filaments

    b. thin filaments

    c. Z line

    d. sarcomere

OBJ. 3 _____ 6. Nerves and blood vessels are contained within the connective tissues of the:

    a. epimysium and endomysium

    b. the endomysium only

    c. epimysium and perimysium

    d. the perimysium only

OBJ. 4   _____   7. The *thin* filaments consist of:

         a. a pair of protein strands together to form chains of actin molecules

         b. a helical array of actin molecules

         c. a pair of protein strands wound together to form chains of myosin molecules

         d. a helical array of myosin molecules

OBJ. 4   _____   8. The *thick* filaments consist of:

         a. a pair of protein strands wound together to form chains of myosin molecules

         b. a helical array of myosin molecules

         c. a pair of protein strands wound together to form chains of actin molecules

         d. a helical array of actin molecules

OBJ. 5   _____   9. All of the muscle fibers controlled by a single motor neuron constitute a:

         a. motor unit

         b. sarcomere

         c. myoneural junction

         d. cross-bridge

OBJ. 5   _____   10. The tension in a muscle fiber will vary depending on:

         a. the structure of individual sarcomeres

         b. the initial length of muscle fibers

         c. the number of cross-bridge interactions within a muscle fiber

         d. a, b, and c are correct

OBJ. 5   _____   11. The reason there is *less* precise control over leg muscles compared to the muscles of the eye is:

         a. single muscle fibers are controlled by many motor neurons

         b. many muscle fibers are controlled by many motor neurons

         c. a single muscle fiber is controlled by a single motor neuron

         d. many muscle fibers are controlled by a single motor neuron

OBJ. 5   _____   12. The amount of tension produced when a skeletal muscle contracts depends on:

         a. the frequency of stimulation and the number of motor units involved

         b. the size of the motor units

         c. the degree to which wave summation occurs

         d. a, b, and c are correct

Level
—1—

OBJ. 6 ____ 13. The *sliding filament theory* explains that the *physical* change that takes place during contraction is:

    a. the thick filaments are sliding toward the center of the sarcomere alongside the thin filaments

    b. the thick and thin filaments are sliding toward the center of the sarcomere together

    c. the Z lines are sliding toward the H zone

    d. the thin filaments are sliding toward the center of the sarcomere alongside the thick filaments

OBJ. 6 ____ 14. Troponin and tropomyosin are two proteins that can prevent the contractile process by:

    a. combining with calcium to prevent active site binding

    b. causing the release of calcium from the sacs of the sarcoplasmic reticulum

    c. covering the active site and blocking the actin-myosin interaction

    d. inactivating the myosin to prevent cross-bridging

OBJ. 6 ____ 15. In order for an active site to be available for the cross-bridge binding, the presence of _____ is required.

    a. actin

    b. myosin

    c. actin and myosin

    d. calcium

OBJ. 7 ____ 16. In an *isotonic* contraction:

    a. the tension in the muscle varies as it shortens

    b. the muscle length doesn't change due to the resistance

    c. the cross-bridges must produce enough tension to overcome the resistance

    d. tension in the muscle decreases as the resistance increases

OBJ. 7 ____ 17. In an *isometric* contraction:

    a. tension rises but the length of the muscle remains constant

    b. the tension rises and the muscle shortens

    c. the tension produced by the muscle is greater than the resistance

    d. the tension of the muscle increases as the resistance decreases

OBJ. 8 ____ 18. A high blood concentration of the enzyme creatine phosphokinase (CPK) usually indicates:

    a. the release of stored energy

    b. serious muscle damage

    c. an excess of energy is being produced

    d. the mitochondria are malfunctioning

Level
–1–

OBJ. 8 _____ 19. Mitochondrial activities are relatively efficient, but their rate of ATP generation is limited by:

      a. the presence of enzymes

      b. the availability of carbon dioxide and water

      c. the energy demands of other organelles

      d. the availability of oxygen

OBJ. 9 _____ 20. The three major types of skeletal muscle fibers in the human body are:

      a. slow, fast resistant, fast fatigue

      b. slow, intermediate, fast

      c. SO, FR, FF

      d. a, b, and c are correct

OBJ. 9 _____ 21. Extensive blood vessels, mitochondria, and myoglobin are found in the greatest concentration in:

      a. fast fibers

      b. slow fibers

      c. intermediate fibers

      d. Type II fibers

OBJ. 10 _____ 22. The length of time a muscle can continue to contract while supported by mitochondrial activities is referred to as:

      a. anaerobic endurance

      b. aerobic endurance

      c. hypertrophy

      d. recruitment

OBJ. 10 _____ 23. Altering the characteristics of muscle fibers and improving the performance of the cardiovascular system results in improving:

      a. thermoregulatory adjustment

      b. hypertrophy

      c. anaerobic endurance

      d. aerobic endurance

OBJ. 11 _____ 24. Fibrosis in aging skeletal muscle tissue makes the muscle:

      a. harder and results in increased tolerance for exercise

      b. less flexible, and the collagen fibers can restrict movement and circulation

      c. more elastic but less flexible

      d. more flexible and more subject to injury

OBJ. 11 _____ 25. As a result of declining numbers of satellite cells due to aging:

      a. the amount of fibrous tissue increases

      b. repair capabilities are limited

      c. the muscles fatigue more rapidly

      d. there is a reduction in muscular strength

OBJ. 12   _____ 26. The property of cardiac muscle that allows it to contract without neural stimulation is:

   a. intercalation
   b. automaticity
   c. plasticity
   d. pacesetting

OBJ. 12   _____ 27. The type of muscle tissue that does not contain sarcomeres is:

   a. cardiac
   b. skeletal
   c. smooth
   d. a and c are correct

OBJ. 13   _____ 28. Smooth muscle contractions in the respiratory passageways would cause:

   a. increased resistance to air flow
   b. decreased resistance to air flow
   c. immediate death
   d. resistance to air flow will not be affected

OBJ. 13   _____ 29. Layers of smooth muscle in the reproductive tract of the female are important in:

   a. movement of ova
   b. movement of sperm if present
   c. expelling of the fetus at delivery
   d. a, b, and c are correct

[L1]   Completion:

Using the terms below, complete the following statements.

endurance	epimysium	flaccid	action potential
twitch	white muscles	ATP	sarcolemma
tendon	sarcomeres	fascicles	red muscles
contraction	atrophy	treppe	recruitment
hypertrophy	T tubules	oxygen debt	cross-bridges
troponin	pacemaker	plasticity	

OBJ. 1   30. The cells that make up muscle tissues are highly specialized for the process of _____.

OBJ. 2   31. The dense layer of collagen fibers surrounding a muscle is called the _____.

OBJ. 2   32. Bundles of muscle fibers are called _____.

OBJ. 2   33. The dense regular connective tissue that attaches skeletal muscle to bones is known as a _____.

OBJ. 3   34. The cell membrane that surrounds the cytoplasm of a muscle fiber is called the _____.

OBJ. 3   35. Structures that help distribute the command to contract throughout the muscle fiber are called _____.

OBJ. 3   36. Muscle cells contain contractible units called _____.

OBJ. 4   37. Because they connect thick and thin filaments, the myosin heads are also known as _____.

OBJ. 5   38. The conducted charge in the transmembrane potential is called a(n) _____.

OBJ. 5   39. The smooth but steady increase in muscular tension produced by increasing the number of active motor units is called _____.

OBJ. 6   40. Active site exposure during the contraction process occurs when calcium binds to _____.

OBJ. 7   41. The "staircase" phenomenon during which the peak muscle tension rises in stages is called _____.

OBJ. 7   42. A single stimulus-contraction-relaxation sequence in a muscle fiber is a _____.

OBJ. 8   43. When muscles are actively contracting, the process requires large amounts of energy in the form of _____.

OBJ. 9   44. Muscles dominated by fast fibers are sometimes referred to as _____.

OBJ. 9   45. Muscles dominated by slow fibers are sometimes referred to as _____.

OBJ. 10   46. The amount of oxygen used in the recovery period to restore normal pre-exertion conditions is referred to as _____.

OBJ. 10   47. The amount of time for which the individual can perform a particular activity is referred to as _____.

OBJ. 11   48. A reduction in muscle size, tone, and power is called _____.

OBJ. 11   49. The loss in muscle tone and mass that results from a lack of stimulation of skeletal muscles causes the muscles to become _____.

OBJ. 11   50. Muscle enlargement due to repeated stimulation to near-maximal tension is referred to as _____.

OBJ. 12   51. The timing of contractions in cardiac muscle tissues is determined by specialized muscle fibers called _____ cells.

OBJ. 13   52. The ability of smooth muscle to function over a wide range of lengths is called _____.

[L1]   Matching:

Match the terms in column B with the terms in column A. Use letters for answers in the spaces provided.

	COLUMN A	COLUMN B
OBJ. 1 _____	53. skeletal muscle	A. thick filaments
OBJ. 2 _____	54. fascicles	B. resting tension
OBJ. 3 _____	55. striations	C. synaptic cleft
OBJ. 4 _____	56. myosin	D. cardiac muscle fibers
OBJ. 5 _____	57. neuromuscular junction	E. muscle bundles
		F. lactic acid
OBJ. 6 _____	58. cross-bridging	G. red muscles
OBJ. 7 _____	59. muscle tone	H. fibrosis
OBJ. 8 _____	60. energy reserve	I. smooth muscle cell
OBJ. 9 _____	61. slow fibers	J. produce skeletal movements
OBJ. 10 _____	62. anaerobic glycolysis	
OBJ. 11 _____	63. aging	K. actin-myosin interaction
OBJ. 12 _____	64. intercalated discs	L. skeletal muscle cell
OBJ. 13 _____	65. no striations	M. creatine phosphate

[L1] Drawing/Illustration Labeling:

Identify each numbered structure by labeling the following figures:

OBJ. 2 **FIGURE 10.1** Organization of Skeletal Muscles

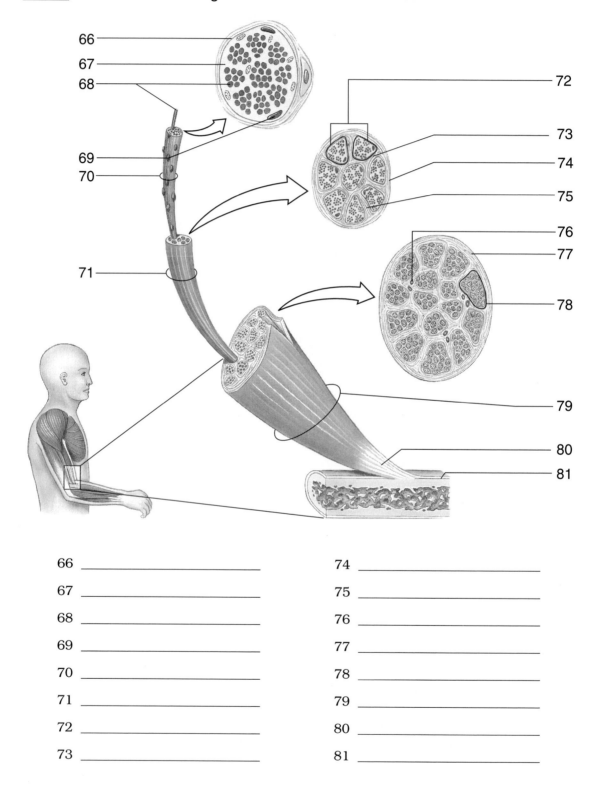

66 _____	74 _____	
67 _____	75 _____	
68 _____	76 _____	
69 _____	77 _____	
70 _____	78 _____	
71 _____	79 _____	
72 _____	80 _____	
73 _____	81 _____	

OBJ. 3   **FIGURE 10.2**   The Histological Organization of Skeletal Muscles

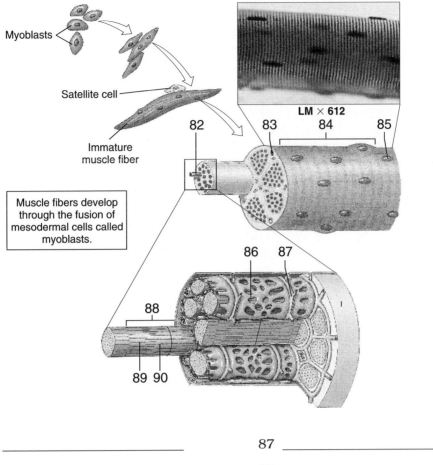

Myoblasts

Satellite cell

Immature
muscle fiber

LM × 612

82   83   84   85

Muscle fibers develop
through the fusion of
mesodermal cells called
myoblasts.

86   87

88

89   90

82 _____   87 _____

83 _____   88 _____

84 _____   89 _____

85 _____   90 _____

86 _____

OBJ. 3   **FIGURE 10.3**   Types of Muscle Tissue

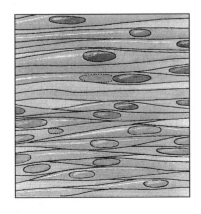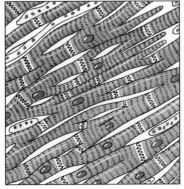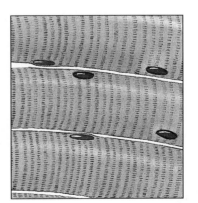

91 _____   92 _____   93 _____

Level
–1–

OBJ. 5   *FIGURE 10.4*   Structure of a Sarcomere

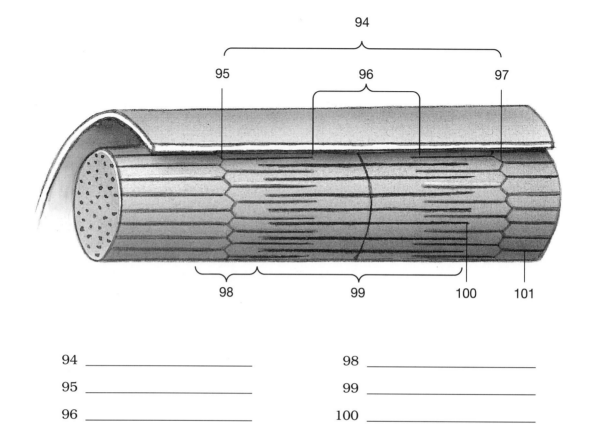

94	_____	98	_____
95	_____	99	_____
96	_____	100	_____
97	_____	101	_____

OBJ. 5   *FIGURE 10.5*   Neuromuscular Junction

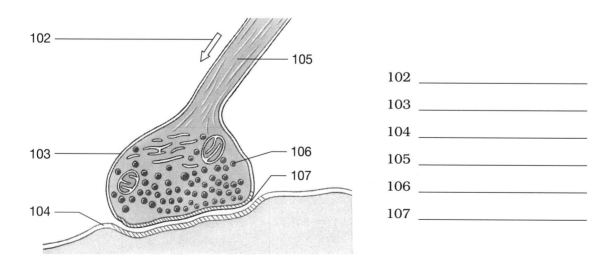

102	_____
103	_____
104	_____
105	_____
106	_____
107	_____

When you have successfully completed the exercises in L1 proceed to L2.

☐ LEVEL 2   CONCEPT SYNTHESIS

Concept Map I:

Using the following terms, fill in the circled, numbered, blank spaces to complete the concept map. Follow the numbers that comply with the organization of the map.

Smooth	Involuntary	Striated
Multinucleated	Bones	Non-striated
Heart		

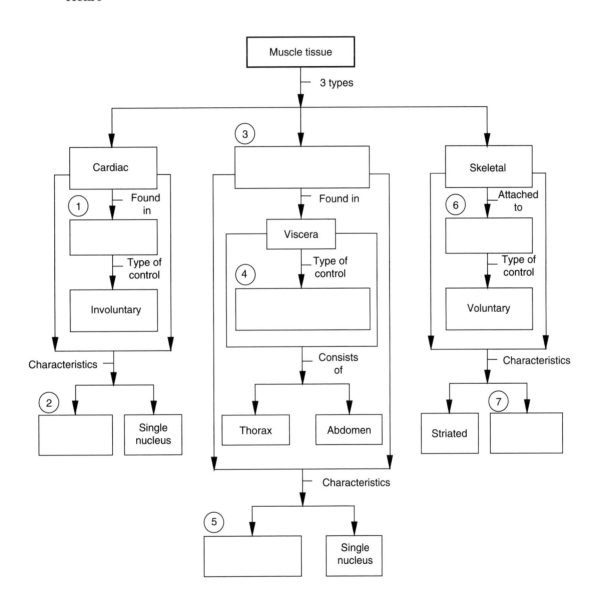

Concept Map II:

Using the following terms, fill in the circled, numbered, blank spaces to complete the concept map. Follow the numbers that comply with the organization of the map.

Z lines   Muscle bundles (fascicles)   Myofibrils
Actin   Thick filaments   Sarcomeres
H zone

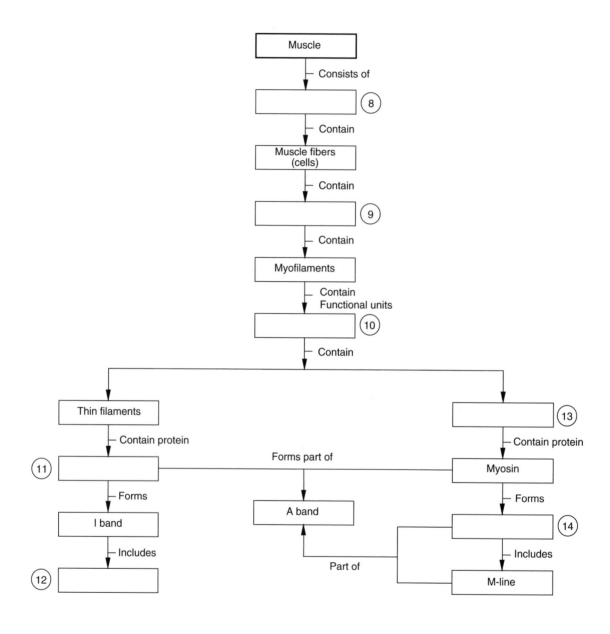

Concept Map III:

Using the following terms and/or phrases, fill in the numbered, blank spaces to complete the concept map on *muscle contraction*. Follow the numbers that comply with the organization of the map.

Cross-bridging (heads of myosin attach to turned-on thin filaments)

Energy + ADP + phosphate

Release of Ca++ from sacs of sarcoplasmic reticulum

Shortening, i.e., contraction of myofibrils and muscle fibers they compose

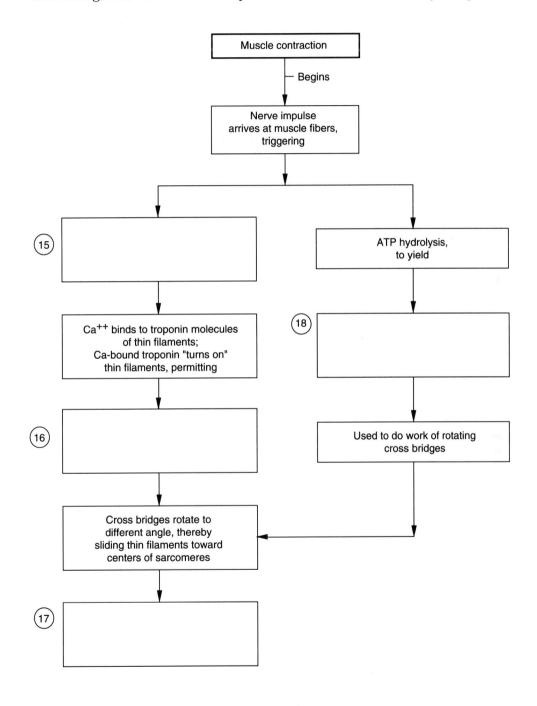

Body Trek:

Using the terms below, fill in the blanks to complete the trek through the muscle tissue.

thick	nuclei	fascicles	actin
contraction	Z line	endomysium	A band
myosin	sarcomeres	sarcolemma	T tubules
myofibrils	I band	myofilaments	perimysium
satellite	epimysium	sliding filament	

Robo's programming for the study of muscle tissue will take the micro-robot on a trek into a skeletal muscle. A large syringe is used to introduce the robot via a subcutaneous injection into the deep fascia until it comes into contact with a muscle that is surrounded by a dense layer of collagen fibers known as the (19) _____ . Once inside the muscle, the robot senses the appearance of a series of compartments consisting of collagen, elastic fibers, blood vessels, and nerves, all surrounded by the (20) _____ . The individual compartments are referred to as (21) _____ . Probing deeper, there appears to be a network of loose connective tissue with numerous reticular fibers, the (22) _____ , which surrounds each muscle fiber contained within the muscle bundle. Between this surrounding layer and the muscle cell membrane, (23) _____ cells, the "doctors" that function in the repair of damaged muscle tissue, are sparsely scattered. Robo's entrance into the muscle by way of the muscle cell membrane, called the (24) _____ , is facilitated by a series of openings to "pipe-like" structures that are regularly arranged and that project into the muscle fibers. These "pipes" are the (25) _____ that serve as the transport system between the outside of the cell and the cell interior. Just inside the membrane, the robot picks up the presence of many (26) _____ , a characteristic common to skeletal muscle fibers. Robo's trek from the pipe-like environment into the sarcoplasm exposes neatly arranged thread-like structures, the (27) _____ , which extend from one end of the muscle cell to the other. Mission Control receives information regarding the presence of smaller filaments called (28) _____ , consisting of large protein molecules some of which are arranged into thin filaments or (29) _____ , while others reveal a thicker appearance, the thick filaments or (30) _____ molecules. These thick and thin filaments are organized as repeating functional units called (31) _____ . Each functional unit extends from one (32) _____ to another, forming a filamentous network of disklike protein structures for the attachment of actin molecules. Robo senses a banded or striated appearance due to the areas of the unit where only thin filaments exist, called the (33) _____ , while a greater portion of the unit consists of overlapping thick and thin filaments called the (34) _____ . In the center of this area, only (35) _____ filaments are apparent. The interactions between the filaments result in the sliding of the thin filaments toward the center of the unit, a process referred to as the (36) _____ theory, and ultimately producing a function unique to muscle cells, the process of (37) _____ . As Robo's trek is completed, the robot's exit plan is to re-enter the "pipe-like" system within the muscle cell and leave the cell via the openings within the cell membrane. After retrieving Robo, Mission Control returns the robot to the control center for re-programming for the next task.

[L2]   Multiple Choice:

Place the letter corresponding to the correct answer in the space provided.

_____ 38. Muscle contraction occurs as a result of:

    a. interactions between the thick and thin filaments of the sarcomere

    b. the interconnecting filaments that make up the Z lines

    c. shortening of the A band, which contains thick and thin filaments

    d. shortening of the I band, which contains thin filaments only

Level =2=

_____ 39. The area of the A band in the sarcomere consists of:

     a. Z line, H band, M line

     b. M line, H band, zone of overlap

     c. thin filaments only

     d. overlapping thick and thin filaments

_____ 40. The process of cross-bridging, which occurs at an active site, involves a series of sequential-cyclic reactions that include:

     a. attach, return, pivot, detach

     b. attach, pivot, detach, return

     c. attach, detach, pivot, return

     d. attach, return, detach, pivot

_____ 41. Excitation-contraction coupling forms the link between:

     a. the release of Ca++ to bind with the troponin molecule

     b. depolarization and repolarization

     c. electrical activity in the sarcolemma and the initiation of a contraction

     d. the neuromuscular junction and the sarcoplasmic reticulum

_____ 42. When Ca^{++} binds to troponin, it produces a change by:

     a. initiating activity at the neuromuscular junction

     b. causing the actin-myosin interaction to occur

     c. decreasing the calcium concentration at the sarcomere

     d. exposing the active site on the thin filaments

_____ 43. The phases of a single twitch in sequential order include:

     a. contraction phase, latent phase, relaxation phase

     b. latent period, relaxation phase, contraction phase

     c. latent period, contraction phase, relaxation phase

     d. relaxation phase, latent phase, contraction phase

_____ 44. After contraction, a muscle fiber returns to its original length through:

     a. the active mechanism for fiber elongation

     b. elastic forces and the movement of opposing muscles

     c. the tension produced by the initial length of the muscle fiber

     d. involvement of all the sarcomeres along the myofibrils

_____ 45. A muscle producing peak tension during rapid cycles of contraction and relaxation is said to be in:

     a. complete tetanus

     b. incomplete tetanus

     c. treppe

     d. recruitment

_____ 46. The process of reaching *complete tetanus* is obtained by:

    a. applying a second stimulus before the relaxation phase has ended

    b. decreasing the concentration of calcium ions in the cytoplasm

    c. activation of additional motor units

    d. increasing the rate of stimulation until the relaxation phase is completely eliminated.

_____ 47. The total force exerted by a muscle as a whole depends on:

    a. the rate of stimulation

    b. how many motor units are activated

    c. the number of calcium ions released

    d. a, b, and c are correct

_____ 48. The primary energy reserves found in skeletal muscle cells are:

    a. carbohydrates, fats, proteins

    b. DNA, RNA, ATP

    c. ATP, creatine phosphate, glycogen

    d. ATP, ADP, AMP

_____ 49. The two mechanisms used to generate ATP from glucose are:

    a. aerobic respiration and anaerobic glycolysis

    b. ADP and creatine phosphate

    c. cytoplasm and mitochondria

    d. a, b, and c are correct

_____ 50. In *anaerobic glycolysis*, glucose is broken down to pyruvic acid, which is converted to:

    a. glycogen

    b. lactic acid

    c. acetyl-CoA

    d. citric acid

_____ 51. The maintenance of normal body temperature is dependent upon:

    a. the temperature of the environment

    b. the pH of the blood

    c. the production of energy by muscles

    d. the amount of energy produced by anaerobic glycolysis

_____ 52. Growth hormone from the pituitary gland and the male sex hormone, testosterone, stimulate:

    a. the rate of energy consumption by resting and active skeletal muscles

    b. muscle metabolism and increased force of contraction

    c. synthesis of contractile proteins and the enlargement of skeletal muscles

    d. the amount of tension produced by a muscle group

_____ 53. The hormone responsible for stimulating muscle metabolism and increasing the force of contraction during a sudden crisis is:

      a. epinephrine

      b. thyroid hormone

      c. growth hormone

      d. testosterone

_____ 54. The type of skeletal muscle fibers that have low fatigue resistance are:

      a. fast fibers

      b. slow fibers

      c. intermediate fibers

      d. Type I fibers

_____ 55. An example of an activity that requires _anaerobic endurance_ is:

      a. a 50-yard dash

      b. a 3-mile run

      c. a 10-mile bicycle ride

      d. running a marathon

_____ 56. Athletes training to develop anaerobic endurance perform:

      a. few, long, relaxing workouts

      b. a combination of weight training and marathon running

      c. frequent, brief, intensive workouts

      d. stretching, flexibility, and relaxation exercises

_____ 57. The major support that the muscular system gets from the cardiovascular system is:

      a. a direct response by controlling the heart rate and the respiratory rate

      b. constriction of blood vessels and decrease in heart rate for thermoregulatory control

      c. nutrient and oxygen delivery and carbon dioxide removal

      d. decreased volume of blood and rate of flow for maximal muscle contraction

[L2]   Completion:

Using the terms below, complete the following statements.

satellite	myoblasts	fatigue
isotonic	tetanus	motor unit
rigor	muscle tone	absolute refractory periods
A bands		

58. Specialized cells that function in the repair of damaged muscle tissue are called _____ cells.

59. During development, groups of embryonic cells that fuse together to create individual muscle fibers are called _____.

60. Resting tension in a skeletal muscle is called _____.

61. A rise of a few degrees in temperature within a muscle may result in a severe contraction called _____.

62. The time when a muscle cell cannot be stimulated because repolarization is occurring is the _____.

63. The condition that results when a muscle is stimulated but cannot respond is referred to as _____.

64. In a sarcomere, the dark bands (anisotropic bands) are referred to as _____.

65. A single cranial or spinal motor neuron and the muscle fibers it innervates comprise a _____.

66. At sufficiently high electrical frequencies, the overlapping twitches result in one strong, steady contraction referred to as _____.

67. When the muscle shortens but its tension remains the same, the contraction is _____.

[L2]   Short Essay:

Briefly answer the following questions in the spaces provided below.

68. What are the five functions performed by skeletal muscles?

69. What are the three layers of connective tissue that are part of each muscle?

70. Draw an illustration of a sarcomere and label the parts according to the unit organization.

71. Cite the five interlocking steps involved in the contraction process.

72. Describe the major events that occur at the neuromuscular junction to initiate the contractive process.

Level
=2=

73. Identify the types of muscle contractions illustrated at points A, B, and C on the diagram below.

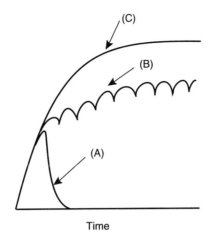

Time

74. What are the differences between an isometric and an isotonic contraction?

75. What is the relationship among fatigue, anaerobic glycolysis, and oxygen debt?

76. Why do fast fibers fatigue more rapidly than slow fibers?

When you have successfully completed the exercises in L2 proceed to L3.

☐ LEVEL 3   CRITICAL THINKING/APPLICATION

Using principles and concepts learned about skeletal muscle tissue, answer the following questions. Write your answers on a separate sheet of paper.

1. When a person drinks an excessive amount of alcohol, that individual becomes drunk or intoxicated, a condition resulting in a lack of muscular control and coordination. How does the alcohol cause this condition?

2. Suppose you are assigned the responsibility of developing training programs tailored to increase both aerobic and anaerobic endurance. What types of activities would be necessary to support these training programs?

3. During a lecture on death and dying, a student asks, "Is rigor mortis the same as physiological contracture?" How would you as the lecturer describe the similarities/ differences between these two conditions?

4. An individual who has been on a diet that eliminates dairy products complains of muscular spasms and nervousness. Why might these symptoms result from such a diet?

5. What does the bacterium *Clostridium botulinum* and the *poliovirus* have in common that ultimately affects the condition of muscles in the human body?

Level
≡3≡

11

The Muscular System

■ Overview

The study of the muscular system includes the skeletal muscles that make up about 40 percent of the body mass and that can be controlled voluntarily. It is virtually impossible to master all the facts involved with the approximately 700 muscles that have been identified in the human body. A representative number of muscles from all parts of the body are selected and surveyed. The survey includes the gross anatomy of muscles, anatomical arrangements, muscle attachments, and muscular performance relative to basic mechanical laws.

The Study Guide exercises for Chapter 11 focus on organizing the muscles into small numbers of anatomical and functional groups, the general appearance of the muscles, and the factors that interact to determine the effects of muscle contraction. This is a chapter that requires a great deal of memorization; therefore, arranging the information in an organized way and learning to recognize clues will make the task easier to manage and ultimately result in facilitating the memorization process.

❑ LEVEL 1 REVIEW OF CHAPTER OBJECTIVES

1. Describe the arrangement of fascicles in the various types of muscles, and explain the resulting functional differences.
2. Describe the different classes of levers and how they make muscles more efficient.
3. Predict the actions of a muscle on the basis of the relative positions of its origin and insertion.
4. Explain how muscles interact to produce or oppose movements.
5. Explain how the name of a muscle can help identify its location, appearance, and/or function.
6. Identify the principal axial muscles of the body, together with their origins, insertions, actions, and innervation.
7. Identify the principal appendicular muscles of the body, together with their origins, insertions, actions, and innervation.
8. Compare the major muscle groups of the upper and lower extremities, and relate their differences to their functional roles.

[L1]   Multiple Choice:

Place the letter corresponding to the correct answer in the space provided.

OBJ. 1   _____   1. The four types of muscles identified by different patterns of organization are:

    a. skeletal, smooth, cardiac, visceral

    b. movers, synergists, antagonists, agonists

    c. parallel, convergent, pennate, circular

    d. flexors, extensors, adductors, abductors

OBJ. 1   _____   2. In a convergent muscle the muscle fibers are:

    a. parallel to the long axis of the muscle

    b. based over a broad area, but all the fibers come together at a common attachment site

    c. arranged to form a common angle with the tendon

    d. arranged concentrically around an opening or recess

OBJ. 2   _____   3. A first class lever is one in which:

    a. the resistance is located between the applied force and the fulcrum

    b. a force is applied between the resistance and the fulcrum

    c. speed and distance traveled are increased at the expense of the force

    d. the fulcrum lies between the applied force and the resistance

OBJ. 2   _____   4. The effect of an arrangement where a force is applied between the resistance and the fulcrum illustrates the principles operating:

    a. first-class levers

    b. second-class levers

    c. third-class levers

    d. fourth-class levers

OBJ. 3   _____   5. The movable attachment of muscle to bone or other connective tissue is referred to as the:

    a. origin

    b. insertion

    c. rotator

    d. joint

OBJ. 4   _____   6. A muscle whose contraction is chiefly responsible for producing a particular movement is called:

    a. a synergist

    b. an antagonist

    c. an originator

    d. a prime mover

OBJ. 4 ____ 7. Muscles are classified functionally as synergists when:

    a. muscles perform opposite tasks and are located on opposite sides of the limb

    b. muscles contract together and are coordinated in affecting a particular movement.

    c. a muscle is responsible for a particular movement

    d. the movement involves flexion and extension

OBJ. 5 ____ 8. Extrinsic muscles are those that:

    a. operate within an organ

    b. position or stabilize an organ

    c. are prominent and can be easily seen

    d. are visible at the body surface

OBJ. 5 ____ 9. The reason we use the word "bicep" to describe a particular muscle is:

    a. there are two areas in the body where biceps are found

    b. there are two muscles in the body with the same characteristics

    c. there are two tendons of origin

    d. the man who named it was an Italian by the name of Biceppe Longo

OBJ. 6 ____ 10. From the following selections choose the one that includes only muscles of *facial expression*:

    a. lateral rectus, medial rectus, hypoglossus, stylohyoideus

    b. splenius, masseter, scalenes, platymsa

    c. procerus, capitis, cervicis, zygomaticus

    d. buccinator, orbicularis oris, risorius, frontalis

OBJ. 6 ____ 11. The muscles that maximize the efficient use of teeth during mastication are:

    a. temporalis, pterygoid, masseter

    b. procerus, capitis, zygomaticus

    c. mandibular, maxillary, zygomaticus

    d. glossus, platymsa, risorius

OBJ. 6 ____ 12. The names of the muscles of the tongue are readily identified because their descriptive names end in:

    a. genio

    b. pollicus

    c. glossus

    d. hallucis

OBJ. 6 ____ 13. The superficial muscles of the spine are identified by *subdivisions* that include:

    a. cervicis, thoracis, lumborum

    b. iliocostalis, longissimus, spinalis

    c. longissimus, transversus, longus

    d. capitis, splenius, spinalis

OBJ. 6   _____ 14. The *oblique* series of muscles located between the vertebral spine and the ventral midline include:

        a. cervicis, diaphragm, rectus abdominis, external obliques

        b. subclavius, trapezius, capitis, cervicis

        c. scalenes, intercostals, obliques, transversus

        d. internal obliques, thoracis, lumborum

OBJ. 6   _____ 15. The muscular floor of the pelvic cavity is formed by muscles that make up the:

        a. urogenital and anal triangle

        b. sacrum and the coccyx

        c. ischium and the pubis

        d. ilium and the ischium

OBJ. 7   _____ 16. From the following selections choose the one that includes only muscles that move the *shoulder girdle*:

        a. teres major, deltoid, pectoralis major, triceps

        b. procerus, capitis, pterygoid, brachialis

        c. trapezius, sternocleidomastoideus, pectoralis minor, subclavius

        d. internal oblique, thoracis, deltoid, pectoralis minor

OBJ. 7   _____ 17. From the following selections choose the one that includes only muscles that move the *upper arm*:

        a. deltoid, teres major, latissimus dorsi, pectoralis major

        b. trapezius, pectoralis minor, subclavius, triceps

        c. rhomboideus, serratus anterior, subclavius, trapezius

        d. brachialis, brachioradialis, pronator, supinator

OBJ. 8   _____ 18. The muscles that arise on the humerus and the forearm and rotate the radius without producing either flexion or extension of the elbow are the:

        a. pronator teres and supinator

        b. brachialis and brachioradialis

        c. triceps and biceps brachii

        d. carpi ulnaris and radialis

OBJ. 8   _____ 19. The terms that describe the *actions* of the muscles that move the palm and fingers are:

        a. depressor, levator, pronator, rotator

        b. tensor, supinator, levator, pronator

        c. rotator, tensor, extensor, adductor

        d. adductor, abductor, extensor, flexor

OBJ. 8   _____ 20. The muscle *groups* that are responsible for movement of the thigh include:

        a. abductor, flexor, extensor

        b. adductor, gluteal, lateral rotator

        c. depressor, levator, rotator

        d. procerus, capitis, pterygoid

Level
−1−

OBJ. 8

_____ 21. The *flexors* that move the lower leg commonly known as the hamstrings include:

    a. rectus femoris, vastus intermedius, vastus lateralis, vastus medialis

    b. sartorius, rectus femoris, gracilis, vastus medialis

    c. piriformis, lateral rotators, obturator, sartorius

    d. biceps femoris, semimembranosus, semitendinosus

OBJ. 8

_____ 22. The *extensors* that move the lower leg commonly known as the *quadriceps* include:

    a. piriformis, lateral rotator, obturator, sartorius

    b. semimembranosus, semitendinosus, gracilis, sartorius

    c. popliteus, gracilis, rectus femoris, biceps femoris

    d. rectus femoris, vastus intermedius, vastus lateralis, vastus medialis

OBJ. 8

_____ 23. Major muscles that produce plantar flexion involved with movement of the lower leg are the:

    a. tibialis anterior, calcaneal, popliteus

    b. flexor hallucis, obturator, gracilis

    c. gastrocnemius, soleus, tibialis posterior

    d. sartorius, soleus, flexor hallicus

OBJ. 8

_____ 24. The actions that the arm muscles produce that are not evident in the action of the leg muscles are:

    a. abduction and adduction

    b. flexion and extension

    c. pronation and supination

    d. rotation and adduction

OBJ. 8

_____ 25. The most common functional role of the muscles of both the forearm and the upper leg involves the action of:

    a. flexion and extension

    b. adduction and abduction

    c. rotation and supination

    d. pronation and supination

[L1]   Completion:

Using the terms below, complete the following statements.

cervicis   origin   popliteus   diaphragm
third-class   deltoid   biceps brachii   second-class
rectus femoris   sphincters   synergist

OBJ. 1   26. An example of a parallel muscle with a central body or belly is the
_____.

OBJ. 1   27. Circular muscles that guard entrances and exits of internal passageways are called _____.

OBJ. 2   28. The most common levers in the body are classified as _____
levers.

OBJ. 2   29. The type of lever in which a small force can balance a larger weight is
classified as a _____ lever.

OBJ. 3   30. The stationary, immovable, or less movable attachment of a muscle is
the _____.

OBJ. 4   31. A muscle that assists the prime mover in performing a particular
action is a _____.

OBJ. 5   32. The term that identifies the region of the body behind the knee is
_____.

OBJ. 5   33. The term that identifies the neck region of the body is _____.

OBJ. 6   34. The muscle that separates the thoracic and abdominopelvic cavities
is the _____.

OBJ. 7   35. The major abductor of the arm is the _____.

OBJ. 8   36. The large quadricep muscle that extends the leg and flexes the thigh
is the _____.

[L1]   Matching:

Match the terms in column B with the terms in column A. Use letters for answers in the
spaces provided.

PART I   COLUMN A   COLUMN B

OBJ. 1   _____ 37. slender band of collagen   A. wheelbarrow
fibers   B. agonist

OBJ. 1   _____ 38. tendon branches within   C. antagonist
muscles   D. raphe

OBJ. 2   _____ 39. first-class lever   E. insertion

OBJ. 2   _____ 40. second-class lever   F. see-saw

OBJ. 3   _____ 41. stationary muscle attachment   G. multipennate

OBJ. 3   _____ 42. movable muscle attachment   H. origin

OBJ. 4   _____ 43. prime mover

OBJ. 4   _____ 44. oppose action of prime mover

PART II   COLUMN A   COLUMN B

OBJ. 5   _____ 45. long and round muscles   I. hyoids

OBJ. 5   _____ 46. tailor's muscle   J. flexes the arm

OBJ. 6   _____ 47. oculomotor muscles   K. teres

OBJ. 6   _____ 48. extrinsic laryngeal muscles   L. rotator cuff

OBJ. 7   _____ 49. pectoralis major   M. gastrocnemius

OBJ. 7   _____ 50. support shoulder joint   N. eye position

OBJ. 8   _____ 51. "calf" muscle   O. sartorius

Level
-1-

[L1] Drawing/Illustration Labeling:

Identify each numbered structure by labeling the following figures:

OBJ. 6
OBJ. 7

FIGURE 11.1 Major Superficial Skeletal Muscles (anterior view).

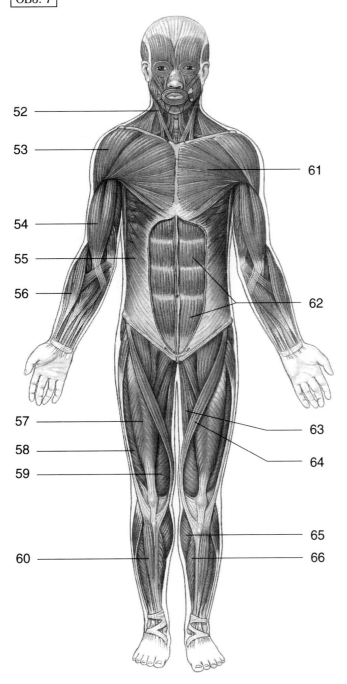

Anterior view

52 _____

53 _____

54 _____

55 _____

56 _____

57 _____

58 _____

59 _____

60 _____

61 _____

62 _____

63 _____

64 _____

65 _____

66 _____

Level
–1–

OBJ. 6
OBJ. 7

FIGURE 11.2   Major Superficial Skeletal Muscles (posterior view).

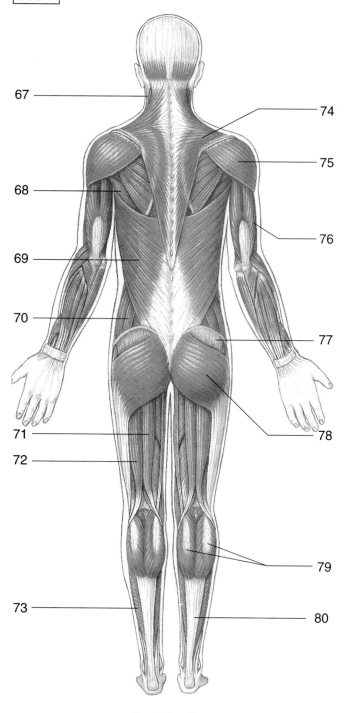

Posterior view

67 _____
68 _____
69 _____
70 _____
71 _____
72 _____
73 _____
74 _____
75 _____
76 _____
77 _____
78 _____
79 _____
80 _____

Level
-1-

OBJ. 6   ***FIGURE 11.3***   Superficial View of Facial Muscles (lateral view)

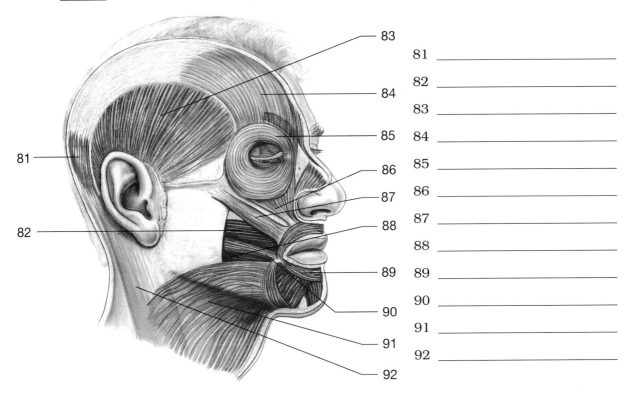

81 _____

82 _____

83 _____

84 _____

85 _____

86 _____

87 _____

88 _____

89 _____

90 _____

91 _____

92 _____

OBJ. 6   ***FIGURE 11.4***   Muscles of the Neck (anterior view)

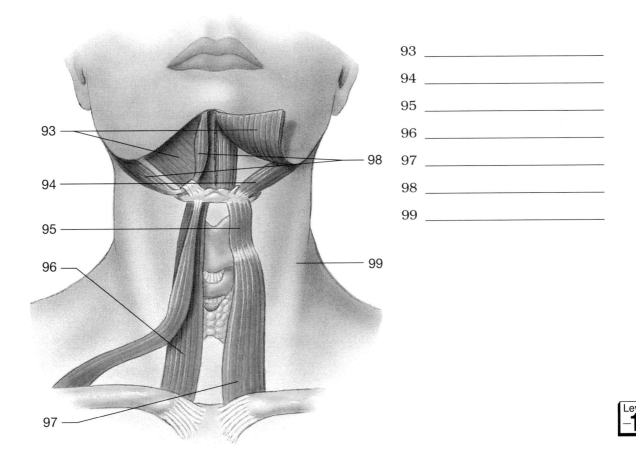

93 _____

94 _____

95 _____

96 _____

97 _____

98 _____

99 _____

Level
–1–

OBJ. 8   *FIGURE 11.5*   Superficial Muscles of the Forearm (anterior view)

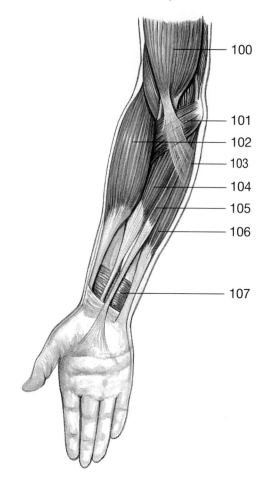

100 _____

101 _____

102 _____

103 _____

104 _____

105 _____

106 _____

107 _____

OBJ. 8   *FIGURE 11.6*   Muscles of the Hand (palmar group)

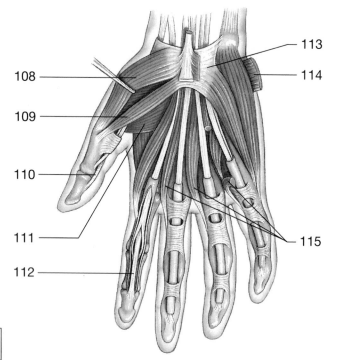

108 _____

109 _____

110 _____

111 _____

112 _____

113 _____

114 _____

115 _____

Level
–1–

OBJ. 8   **FIGURE 11.7**   Superficial Muscles of Thigh (anterior view)

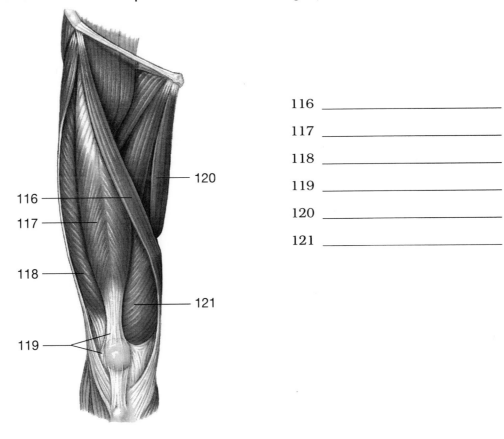

116 _____

117 _____

118 _____

119 _____

120 _____

121 _____

OBJ. 8   **FIGURE 11.8**   Superficial Muscles of Thigh (posterior view)

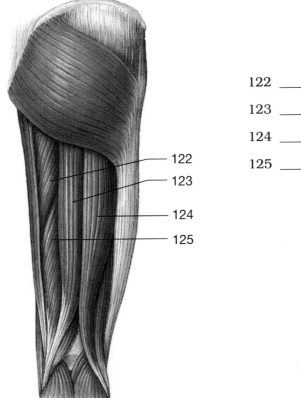

122 _____

123 _____

124 _____

125 _____

OBJ. 8   **FIGURE 11.9**   Superficial Muscles of Lower Leg (lateral view)

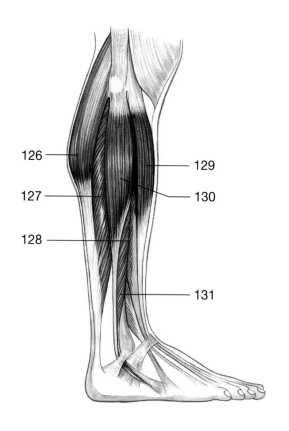

126 _____

127 _____

128 _____

129 _____

130 _____

131 _____

OBJ. 8   **FIGURE 11.10**   Superficial Muscles of Foot (plantar view)

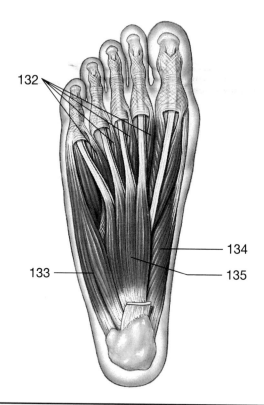

132 _____

133 _____

134 _____

135 _____

Level
-1-
   When you have successfully completed the exercises in L1 proceed to L2.

☐ **LEVEL 2   CONCEPT SYNTHESIS**

Concept Map I:

Using the following terms, fill in the circled, numbered, blank spaces to complete the concept map. Follow the numbers that comply with the organization of the map.

Bulbocavernosus   Capitis   Urethral sphincter
Splenius   Masseter   External obliques
Buccinator   Thoracis   Oculomotor muscles
Rectus abdominis   Diaphragm   Sternohyoid
Scalenes   External anal sphincter   External-internal
Hypoglossus   Stylohyoid   Intercostals

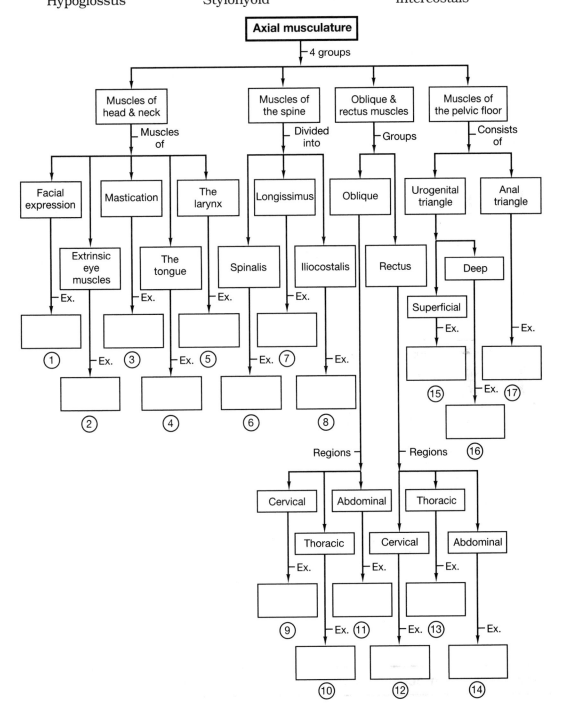

Concept Map II:

Using the following terms, fill in the circled, numbered, blank spaces to complete the concept map. Follow the numbers that comply with the organization of the map.

Psoas
Triceps
Obturators
Deltoid
Gastrocnemius

Flexor carpi radialis
Flexor digitorum longus
Gluteus maximus
Extensor carpi ulnaris
Extensor digitorum

Trapezius
Biceps brachii
Biceps femoris
Rectus femoris

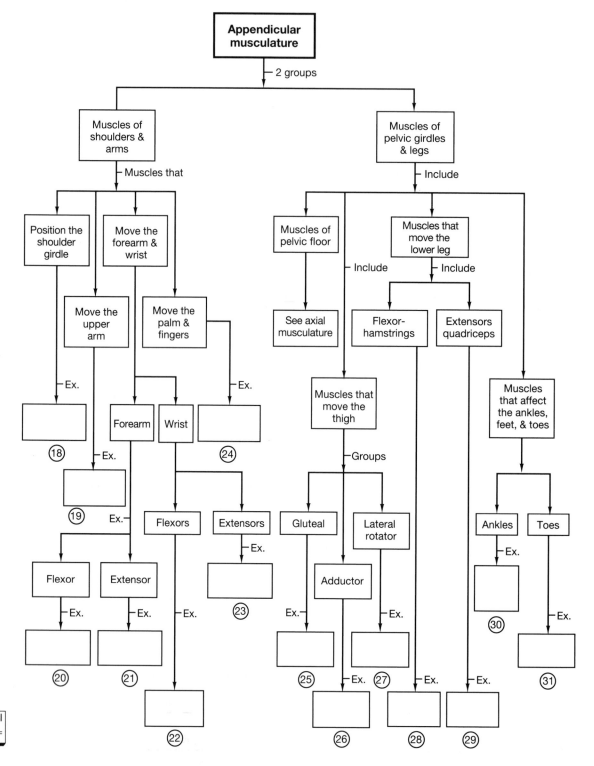

Body Trek:

For the study of origins and insertions, the use of a muscular model that can be dissected along with a skeleton to locate and identify the muscles and the bone processes where they originate and insert is recommended. Charts, diagrams, or photographs of a bodybuilder to support the study will facilitate observation and identification. From the selections below, complete the report as the origins and insertions are located for each superficial muscle listed. The origins are randomly listed in Column 1 and the insertions in Column 2.

COLUMN 1 – Origins

ischium, femur, pubis, ilium
medial margin of orbit
scapula
ribs
ribs (8–12), thoracic-lumbar vertebrae
humerus-lateral epicondyle
femoral condyles
manubrium and clavicle
lower eight ribs

COLUMN 2 – Insertions

tibia
lips
deltoid tuberosity of humerus
humerus-greater tubercle
symphysis pubis
ulna-olecranon process
first metatarsal
iliotibial tract and femur
clavicle, scapula
skin of eyebrow, bridge of nose

Origin – Insertion Report

MUSCLE – LOCATION	ORIGIN	INSERTION
Frontalis (Forehead)	Galea Aponeurotica	(32)
Orbicularis Oculi (Eye)	(33)	Skin Around Eyelids
Orbicularis Oris (Mouth)	Maxilla & Mandible	(34)
Sternocleidomastoid (Neck)	(35)	Mastoid – Skull
Trapezius (Back – Neck)	Occupital Bone Thoracic Vertebrae	(36)
Latissimus Dorsi (Back)	(37)	Lesser Tubercle of Humerus
Deltoid (Shoulder)	Clavicle, Scapula	(38)
Biceps Brachii (Upper Arm)	(39)	Tuberosity of Radius
Triceps (Upper Arm – Posterior)	Humerus	(40)
Brachioradialis (Lower Arm)	(41)	Radius – Styloid Process
Pectoralis Major (Chest)	Ribs 2 – 6, Sternum, Clavicle	(42)
Intercostals (Ribs)	(43)	Ribs
Rectus Abdominis – (Abdomen – Anterior)	Xiphoid Process– Costal Cartilages (5 – 7)	(44)
External Obliques Abdomen – (Lateral – Anterior)	(45)	Linea Alba, Iliac Crest
Gluteus Maximus – (Posterior)	Ilium, Scarum, Coccyx	(46)
Hamstrings (Thigh – Posterior)	(47)	Fibula, Tibia
Quadriceps – (Thigh – Anterior)	Ilium, Femur (Linea Aspera)	(48)
Gastrocnemius – (Posterior – Lower Leg)	(49)	Calcaneous via Achilles Tendon
Tibialis Anterior (Anterior – Lower Leg)	Tibia	(50)

[L2]   Multiple Choice:

Place the letter corresponding to the correct answer in the space provided.

_____ 51. The two factors that interact to determine the effects of *individual* skeletal muscle contraction are:

       a. the degree to which the muscle is stretched and the amount of tension produced

       b. the anatomical arrangement of the muscle fibers and the way the muscle attaches to the skeletal system

       c. the strength of the stimulus and the speed at which the stimulus is applied

       d. the length of the fibers and the metabolic condition of the muscle

_____ 52. When a muscle is stretched too far, the muscle loses power because:

       a. the overlap between myofilaments is reduced

       b. the amount of tension is increased due to over-stretching

       c. the sarcomere is stretched and the tension increases

       d. a, b, and c are correct

_____ 53. When a muscle contracts and its fibers shorten:

       a. the origin moves toward the insertion

       b. the origin and the insertion move in opposite directions

       c. the origin and insertion move in the same direction

       d. the insertion moves toward the origin

_____ 54. The *rectus* muscles, which lie between the vertebral spines and the ventral midline, are important:

       a. sphincters of the rectum

       b. rotators of the spinal column

       c. flexors of the spinal column

       d. extensors of the spine

_____ 55. Attaching a muscle to a lever can change:

       a. the direction of an applied force

       b. the distance and speed of movement applied by force

       c. the strength of a force

       d. a, b, and c are correct

_____ 56. The structural commonality of the *rectus femoris* and *rectus abdominis* is that they are:

       a. parallel muscles whose fibers run along the long axis of the body

       b. both found in the region of the rectum

       c. muscles whose fibers are arranged to control peristalsis in the rectum

       d. attached to bones involved in similar functional activities

Level
2

_____ 57. The origin and insertion of the *genioglossus* muscle is:

     a. inserts at chin, originates in tongue

     b. inserts and originates at chin

     c. originates at chin, inserts at tongue

     d. inserts and originates at tongue

_____ 58. The muscles that are synergistic with the diaphragm during inspiration are the:

     a. internal intercostals

     b. rectus abdominis

     c. pectoralis minor

     d. external intercostals

_____ 59. When playing an instrument such as a trumpet, the muscle used to purse the lips and blow forcefully is the:

     a. masseter

     b. buccinator

     c. pterygoid

     d. stylohyoid

_____ 60. The most powerful and important muscle used when chewing food is the:

     a. buccinator

     b. pterygoid

     c. masseter

     d. zygomaticus

_____ 61. If you are engaging in an activity in which the action involves the use of the *levator scapulae* you are:

     a. shrugging your shoulders

     b. raising your hand

     c. breathing deeply

     d. looking up toward the sky

_____ 62. The muscular elements that provide substantial support for the loosely built shoulder joint are collectively referred to as the:

     a. levator scapulae

     b. coracobrachialis

     c. pronator quadratus

     d. rotator cuff

_____ 63. The biceps muscle makes a prominent bulge when:

     a. extending the forearm pronated

     b. flexing the forearm supinated

     c. extending the forearm supinated

     d. flexing the forearm pronated

_____ 64. The carpal tunnels that are associated with the wrist bones are formed by the presence of:

     a. muscles which extend from the lower arm

     b. ligaments which attach the arm bones to the wrist

     c. tendon sheaths crossing the surface of the wrist

     d. muscular attachments to the phalanges

_____ 65. If an individual complains because of shin splints, the affected muscles are located over the:

> a. anterior surface of the leg
> b. posterior surface of the leg
> c. anterior and posterior surfaces of the leg
> d. anterior surface of the thigh

[L2]   Completion:

Using the terms below, complete the following statements.

quadriceps	innervation	sartorius
cervicis	biomechanics	strain
perineum	hamstrings	
charley horse	risorius	

66. The analysis of biological systems in mechanical terms is the study of _____.
67. The term that identifies the neck region of the body is _____.
68. The term used to refer to the identity of the nerve that controls a muscle is _____.
69. The muscle that is active when crossing the legs is the _____.
70. The facial muscle that is active when laughing is the _____.
71. The muscles of the pelvic floor that extend between the sacrum and pelvic girdle form the muscular _____.
72. The flexors of the legs are commonly referred to as the _____.
73. The extensors of the legs are commonly referred to as the _____.
74. A break or tear in a muscle is referred to as a _____.
75. A painful tear and contusion of a muscle often involving the quadricep is called a _____.

[L2]   Short Essay:

Briefly answer the following questions in the spaces provided below.

76. List the four types of skeletal muscles based on the pattern of organization within a fascicle. Give an example of each type.

77. What is the primary functional difference between an origin and an insertion?

78. What are the three (3) primary actions used to group muscles?

79. What are the four (4) groups of muscles that comprise the axial musculature?

80. What two (2) major groups of muscles comprise the appendicular musculature?

81. List the three (3) classes of levers, give a practical example of each, and identify a place where the action occurs in the body.

82. What three (3) muscles are included in the hamstrings?

83. What four (4) muscles are included in the quadriceps?

When you have successfully completed the exercises in L2 proceed to L3.

☐ LEVEL 3   CRITICAL THINKING/APPLICATION

Using principles and concepts learned about the muscles, answer the following questions. Write your answers on a separate sheet of paper.

1. What effects does a lever have on modifying the contraction of a muscle?
2. Baseball players, especially pitchers, are susceptible to injuries that affect the *rotator cuff* muscles. In what area of the body are these muscles located and what specific muscles may be affected?
3. One of the most common leg injuries in football involves pulling or tearing the "hamstrings." In what area of the body does this occur and what muscles may be damaged?
4. How does damage to a musculoskeletal compartment produce the condition of ischemia or "blood starvation"?
5. Your responsibility as a nurse includes giving intramuscular (IM) injections. (a) Why is an IM preferred over an injection directly into circulation? (b) What muscles are best suited as sites for IM injections?

Level
≡3≡

CHAPTER

12

Neural Tissue

■ Overview

The nervous system is the control center and communication network of the body, and its overall function is the maintenance of homeostasis. The nervous system and the endocrine system acting in a complementary way regulate and coordinate the activities of the body's organ systems. The nervous system generally affects short-term control, whereas endocrine regulation is slower to develop and the general effect is long-term control.

In this chapter the introductory material begins with an overview of the nervous system and the cellular organization in neural tissue. The emphasis for the remainder of the chapter concerns the structure and function of neurons, information processing, and the functional patterns of neural organization.

The integration and interrelation of the nervous system with all the other body systems is an integral part of understanding many of the body's activities, which must be controlled and adjusted to meet changing internal and external environmental conditions.

☐ LEVEL 1   REVIEW OF CHAPTER OBJECTIVES

1. List the two major anatomical subdivisions of the nervous system and describe the characteristics of each.
2. Describe the locations and functions of neuroglia in the central nervous system.
3. Sketch and label the structure of a typical neuron and describe the function of each component.
4. Classify neurons on the basis of their structure and function.
5. Explain how the resting potential is created and maintained.
6. Describe the events involved in the initiation and conduction of an action potential.
7. Discuss the factors that affect the speed with which action potentials are transmitted.
8. Describe the structure of a synapse and explain the mechanism involved in synaptic transmission.
9. Describe the major kinds of neurotransmitters and neuromodulators and discuss their effects on postsynaptic membranes.

10. Discuss the interactions that make possible the processing of information in neural tissue.

11. List the factors that affect neural function and explain the basis for their effects on neural activity.

12. Describe the patterns of interaction between neurons that are involved in the processing of information at higher levels.

[L1]   Multiple Choice:

Place the letter corresponding to the correct answer in the space provided.

OBJ. 1   _____   1. The two major anatomical subdivisions of the nervous system are:

    a. central nervous system (CNS) and peripheral nervous system (PNS)

    b. somatic nervous system and autonomic nervous system

    c. neurons and neuroglia

    d. afferent division and efferent division

OBJ. 1   _____   2. The central nervous system (CNS) consists of:

    a. afferent and efferent division

    b. somatic and visceral division

    c. brain and spinal cord

    d. autonomic and somatic division

OBJ. 1   _____   3. The primary function(s) of the nervous system include:

    a. providing sensation of the internal and external environments

    b. integrating sensory information

    c. regulating and controlling peripheral structures and systems

    d. a, b, and c are correct

OBJ. 2   _____   4. The two major cell populations of neural tissue are:

    a. astrocytes and oligodendrocytes

    b. microglia and ependymal cells

    c. satellite cells and Schwann cells

    d. neurons and neuroglia

OBJ. 2   _____   5. The types of glial cells in the central nervous system are:

    a. astrocytes, oligodendrocytes, microglia, and ependymal cells

    b. unipolar, bipolar, multipolar cells

    c. efferent, afferent, association cells

    d. motor, sensory, interneuron cells

OBJ. 2   _____   6. The white matter of the CNS represents a region dominated by the presence of:

    a. astrocytes

    b. oligodendrocytes

    c. neuroglia

    d. unmyelinated axons

Level
-1-

OBJ. 3 _____ 7. Neurons are responsible for:

  a. creating a three-dimensional framework for the CNS
  b. performing repairs in damaged neural tissue
  c. information transfer and processing in the nervous system
  d. controlling the interstitial environment

OBJ. 3 _____ 8. The region of a neuron with voltage-gated sodium channels is the:

  a. soma
  b. dendrite
  c. axon hillock
  d. perikaryon

OBJ. 4 _____ 9. Neurons are classified on the basis of their *structure* as:

  a. astrocytes, oligodendrocytes, microglia, ependymal
  b. efferent, afferent, association, interneurons
  c. motor, sensory, association, interneurons
  d. anaxonic, unipolar, bipolar, multipolar

OBJ. 4 _____ 10. Neurons are classified on the basis of their *function* as:

  a. unipolar, bipolar, multipolar
  b. motor, sensory, association
  c. somatic, visceral, autonomic
  d. central, peripheral, somatic

OBJ. 5 _____ 11. Depolarization of the membrane will shift the membrane potential toward:

  a. –90 mV
  b. – 85 mV
  c. –70 mV
  d. 0 mV

OBJ. 5 _____ 12. The resting membrane potential (RMP) of a typical neuron is:

  a. – 85 mV
  b. – 60 mV
  c. –70 mV
  d. 0 mV

OBJ. 6 _____ 13. If resting membrane potential is –70 mV and the threshold is – 60 mV, a membrane potential of – 62 mV will:

  a. produce an action potential
  b. depolarize the membrane to 0 mV
  c. repolarize the membrane to – 80 mV
  d. not produce an action potential

Level
–1–

OBJ. 6   _____ 14. At the site of an action potential the membrane contains:

      a. an excess of negative ions inside and an excess of negative ions outside

      b. an excess of positive ions inside and an excess of negative ions outside

      c. an equal amount of positive and negative ions on either side of the membrane

      d. an equal amount of positive ions on either side of the membrane

OBJ. 6   _____ 15. If the resting membrane potential is –70 mV, a hyperpolarized membrane is:

      a. 0 mV

      b. +30 mV

      c. – 80 mV

      d. –65 mV

OBJ. 7   _____ 16. A node along the axon represents an area where there is:

      a. a layer of fat

      b. interwoven layers of myelin and protein

      c. a gap in the cell membrane

      d. an absence of myelin

OBJ. 7   _____ 17. The larger the diameter of the axon:

      a. the slower an action potential is conducted

      b. the greater the number of action potentials

      c. the faster an action potential will be conducted

      d. the less effect it will have on action potential conduction

OBJ. 7   _____ 18. The two most important factors that determine the rate of action potential conduction are:

      a. the number of neurons and the length of their axons

      b. the strength of the stimulus and the rate at which the stimulus is applied

      c. the presence or absence of a myelin sheath and the diameter of the axon

      d. a, b, and c are correct

OBJ. 8   _____ 19. At an electrical synapse, the presynaptic and postsynaptic membranes are locked together at:

      a. gap junctions

      b. synaptic vesicles

      c. myelinated axons

      d. neuromuscular junctions

OBJ. 8   _____ 20. Exocytosis and the release of acetylcholine into the synaptic cleft is triggered by:

      a. calcium ions leaving the cytoplasm

      b. calcium ions flooding into the axoplasm

      c. reabsorption of calcium into the endoplasmic reticulum

      d. active transport of calcium into synaptic vesicles

Level
-1-

OBJ. 9   _____ 21. Inhibitory or hyperpolarizing CNS neurotransmitters include:

      a. acetylcholine and norepinephrine

      b. dopamine and serotonin

      c. glutamate and aspartate

      d. substance P and endorphins

OBJ. 9   _____ 22. An excitatory postsynaptic potential (EPSP) is:

      a. an action potential complying with the all-or-none principle

      b. a result of a stimulus strong enough to produce threshold

      c. the same as a nerve impulse along an axon

      d. a depolarization produced by the arrival of neurotransmitter

OBJ. 9   _____ 23. An inhibitory postsynaptic potential (IPSP) is a:

      a. depolarization produced by the effect of a neurotransmitter

      b. transient hyperpolarization of the postsynaptic membrane

      c. repolarization produced by the addition of multiple stimuli

      d. reflection of the activation of an opposing transmembrane potential

OBJ. 10   _____ 24. The most important determinants of neural activity are:

      a. EPSP-IPSP interactions

      b. the presence or absence of myelin on the axon of the neuron

      c. the type and number of stimuli

      d. the strength of the stimuli and the rate at which the stimuli are applied

OBJ. 10   _____ 25. The reason(s) that active neurons need ATP is to support:

      a. the synthesis, release, and recycling of neurotransmitter molecules

      b. the recovery from action potentials

      c. the movement of materials to and from the soma via axoplasmic flow

      d. a, b, and c are correct

OBJ. 11   _____ 26. The integration of all the excitatory and inhibitory stimuli affecting a neuron at any given moment is the:

      a. neurotransmitter

      b. neuromodulator

      c. transmembrane potential

      d. a, b, c are correct

OBJ. 11   _____ 27. Compounds that alter the rate of neurotransmitter release or the response of a postsynaptic neuron to specific neurotransmitters are called:

   a. neuromodulators
   b. extracellular chemicals
   c. curares
   d. biogenic amines

OBJ. 12   _____ 28. Sensory neurons are responsible for carrying impulses:

   a. to the CNS
   b. away from the CNS
   c. to the PNS
   d. from the CNS to the PNS

OBJ. 12   _____ 29. Interneurons, or associated neurons, differ from sensory and motor neurons because of their:

   a. structural characteristics
   b. inability to generate action potentials
   c. exclusive location in the brain and spinal cord
   d. functional capabilities

OBJ. 12   _____ 30. Efferent pathways consist of axons that carry impulses:

   a. toward the CNS
   b. from the PNS to the CNS
   c. away from the CNS
   d. to the spinal cord and into the brain

[L1]   Completion:

Using the terms below, complete the following statements.

cholinergic	spatial summation	microglia
proprioceptors	temporal summation	saltatory
afferent	autonomic nervous system	threshold
collaterals	neuromodulators	adrenergic
postsynaptic potentials	electrochemical gradient	divergence

OBJ. 1   31. The visceral motor system that provides automatic, involuntary regulation of smooth and cardiac muscle and glandular secretions is the _____.

OBJ. 2   32. In times of infection or injury the type of neuroglia that will increase in numbers is _____.

OBJ. 3   33. The "branches" that enable a single neuron to communicate with several other cells are called _____.

OBJ. 4   34. Sensory information is brought to the CNS by means of the _____ fibers.

OBJ. 5   35. The sum of all the chemical and electrical forces active across the cell membrane is known as the _____.

OBJ. 6   36. An action potential occurs only if the membrane is lowered to the level known as _____.

OBJ. 7   37. The process that conducts impulses along an axon at a high rate of speed is called _____ conduction.

OBJ. 8   38. Chemical synapses that release the neurotransmitter acetylcholine are known as _____ synapses.

OBJ. 8   39. Chemical synapses that release the neurotransmitter norepinephrine are known as _____ synapses.

OBJ. 9   40. Compounds that influence the postsynaptic cells' response to a neuro-transmitter are called _____.

OBJ. 10   41. Addition of stimuli occurring in rapid succession at a single synapse is called _____.

OBJ. 10   42. Addition of stimuli arriving at different locations of the nerve cell membrane is called _____.

OBJ. 11   43. Excitatory and inhibitory stimuli are integrated through interactions between _____.

OBJ. 12   44. Sensory neurons that monitor the position of skeletal muscles and joints are called _____.

OBJ. 12   45. The spread of nerve impulses from one neuron to several neurons is called _____.

[L1] Matching:

Match the terms in column B with the terms in column A. Use letters for answers in the spaces provided.

PART I

		COLUMN A		COLUMN B
OBJ. 1	____	46. somatic nervous system	A.	astrocytes
OBJ. 1	____	47. autonomic nervous system	B.	–70 mV
			C.	interoceptors
OBJ. 2	____	48. neuroglia	D.	transmit action
OBJ. 2	____	49. maintain blood-brain barriers		potentials
			E.	+30 mV
OBJ. 3	____	50. axons	F.	involuntary control
OBJ. 4	____	51. visceral sensory neurons	G.	supporting brain cells
OBJ. 4	____	52. somatic sensory neurons	H.	voluntary control
OBJ. 5	____	53. sodium channel inactivation	I.	exteroceptors
OBJ. 5	____	54. resting membrane potential (neuron)		

PART II

		COLUMN A		COLUMN B
OBJ. 6	____	55. potassium ion movement	J.	gap synapsis
OBJ. 7	____	56. unmyelinated axons	K.	cAMP
OBJ. 7	____	57. nodes of Ranvier	L.	continuous conduction
OBJ. 8	____	58. electrical synapses	M.	serotonin
OBJ. 9	____	59. CNS neurotransmitter	N.	hyperpolarization
OBJ. 10	____	60. second messenger	O.	saltatory conduction
OBJ. 11	____	61. parallel processing	P.	simultaneous responses
OBJ. 12	____	62. IPSP	Q.	repolarization

Level
-1-

[L1]   Drawing/Illustration Labeling:

Identify each numbered structure by labeling the following figures:

OBJ. 3
OBJ. 4

FIGURE 12.1   Structure and Classification of Neurons

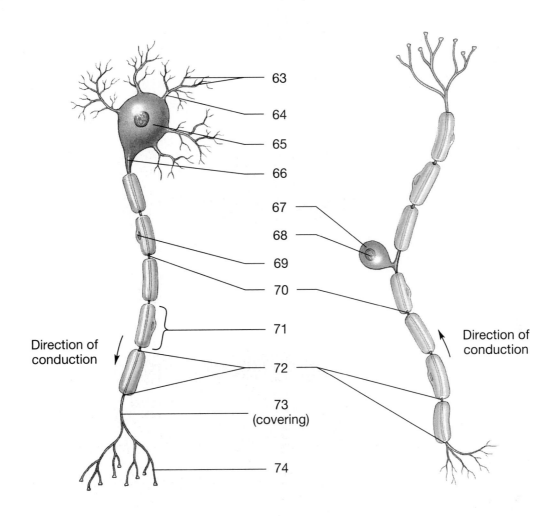

(75) _____ ← BASED ON FUNCTION → (76) _____
(type of neuron)   (type of neuron)

63 _____   70 _____

64 _____   71 _____

65 _____   72 _____

66 _____   73 _____

67 _____   74 _____

68 _____   75 _____

69 _____   76 _____

OBJ. 4   ***FIGURE 12.2***   Neuron Classification – (Based on Structure)
(Identify the types of neurons)

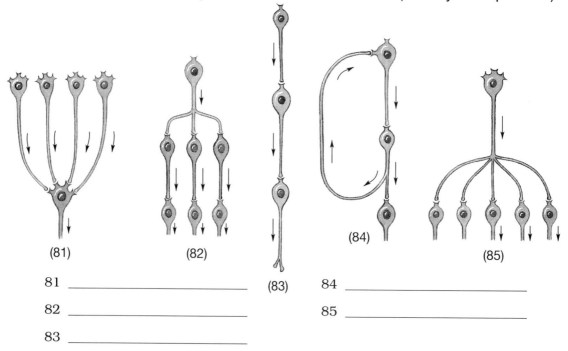

77 _____   78 _____   79 _____   80 _____

OBJ. 12   ***FIGURE 12.3***   Organization of Neuronal Pools (Identify each process)

(81)   (82)   (83)   (84)   (85)

81 _____   84 _____

82 _____   85 _____

83 _____

Level
–1–   **When you have successfully completed the exercises in L1 proceed to L2.**

☐ LEVEL 2   CONCEPT SYNTHESIS

Concept Map I:

Using the following terms, fill in the circled, numbered, blank spaces to complete the concept map. Follow the numbers that comply with the organization of the map.

Glial cells
Surround peripheral ganglia
Central Nervous System

Transmit Nerve Impulses
Schwann cells

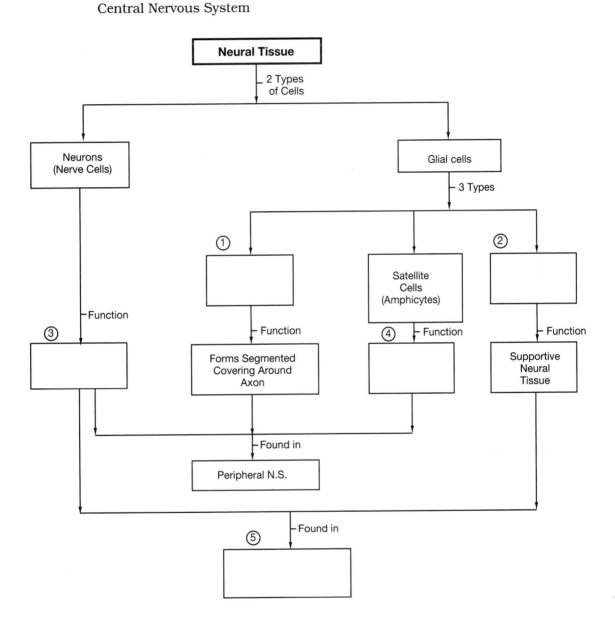

Concept Map II:

Using the following terms, fill in the circled, numbered, blank spaces to complete the concept map. Follow the numbers that comply with the organization of the map.

Brain
Afferent division
Sympathetic N.S.
Motor System

Peripheral nervous system
Smooth muscle
Somatic Nervous System

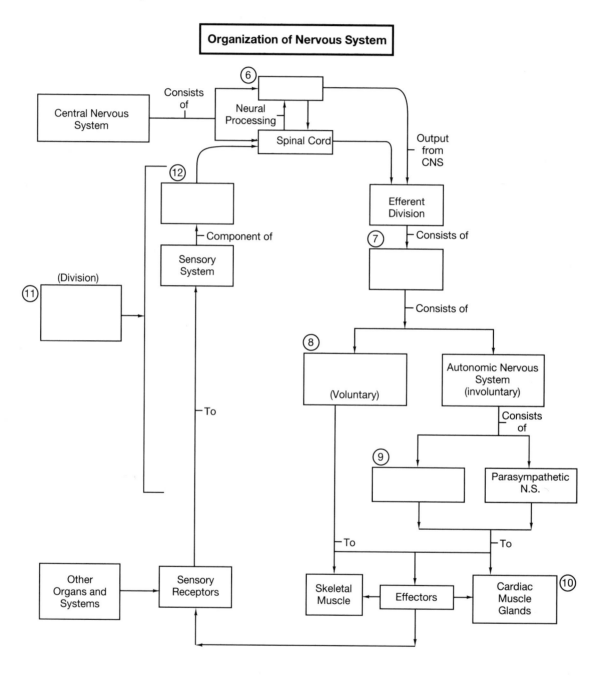

[L2]   Multiple Choice:

Place the letter corresponding to the correct answer in the space provided.

_____ 13. When sensory information is relayed from one processing center to another in the brain the pattern is called:

         a. convergence

         b. divergence

         c. serial processing

         d. reverberation

_____ 14. Interneurons are responsible for:

         a. carrying instructions from the CNS to peripheral effectors

         b. delivery of information to the CNS

         c. collecting information from the external or internal environment

         d. analysis of sensory inputs and coordination of motor outputs

_____ 15. Sensory (ascending) pathways distribute information:

         a. from peripheral receptors to processing centers in the brain

         b. from processing centers in the brain to peripheral receptors

         c. from motor pathways to interneurons in the CNS

         d. the central nervous system to the peripheral nervous system

_____ 16. The type of cells that surround the nerve cell bodies in peripheral ganglia are:

         a. Schwann cells

         b. satellite cells

         c. microglia

         d. oligodendrocytes

_____ 17. Schwann cells are glial cells responsible for:

         a. producing a complete neurilemma around peripheral axons

         b. secretion of cerebrospinal fluid

         c. phagocytic activities in the neural tissue of the PNS

         d. surrounding nerve cell bodies in peripheral ganglia

_____ 18. When a barrier prevents the movement of opposite charges toward one another a (an):

         a. action potential occurs

         b. current is produced

         c. potential difference exists

         d. generation potential is produced

_____ 19. In the formula, Current = Voltage/Resistance, i.e., I = V/R, the

    a. current is directly proportional to resistance

    b. current is inversely proportional to voltage and directly proportional to resistance

    c. current is inversely proportional to voltage

    d. current is directly proportional to voltage

_____ 20. The all-or-nothing principle states:

    a. When a stimulus is applied it triggers an action potential in the membrane.

    b. A given stimulus either triggers a typical action potential or does not produce one at all.

    c. A hyperpolarized membrane always results in the production of an action potential.

    d. Action potentials occur in all neurons if a stimulus is applied that lowers the membrane potential.

_____ 21. During the relative refractory period a larger-than-normal depolarizing stimulus can:

    a. bring the membrane to threshold and initiate a second action potential

    b. cause the membrane to hyperpolarize

    c. inhibit the production of an action potential

    d. cause a membrane to reject a response to further stimulation

_____ 22. Saltatory conduction conducts impulses along an axon:

    a. two to three times slower than the continuous conduction

    b. five to seven times faster than continuous conduction

    c. at a rate determined by the strength of the stimulus

    d. at a velocity determined by the rate at which the stimulus is applied

_____ 23. In type C fibers action potentials are conducted at speeds of approximately:

    a. 2 mph

    b. 40 mph

    c. 150 mph

    d. 500 mph

_____ 24. The larger the diameter of the axon, the:

    a. slower the rate of transmission

    b. greater the resistance

    c. faster the rate of transmission

    d. the size of the axon doesn't affect rate of transmission or resistance

_____ 25. *Facilitation* in the neuron's transmembrane potential toward threshold refers to:

    a. any shift that makes the cell more sensitive to further stimulation

    b. repolarization produced by the addition of multiple stimuli

    c. transient hyperpolarization of a postsynaptic membrane

    d. a, b, and c are correct

_____ 26. Sensory neurons that provide information about the external environment through the sense of sight, smell, hearing, and touch are called:

    a. proprioceptors

    b. exteroceptors

    c. enviroceptors

    d. interoceptors

_____ 27. The main functional difference between the autonomic nervous system and the somatic nervous system is that the activities of the ANS are:

    a. primarily voluntary controlled

    b. primarily involuntary or under "automatic" control

    c. involved with affecting skeletal muscle activity

    d. involved with carrying impulses to the CNS

_____ 28. Motor (descending) pathways begin at CNS centers concerned with motor control and end at:

    a. the cerebral cortex in the brain

    b. the cerebrum for conscious control

    c. the skeletal muscles they control

    d. reflex arcs within the spinal cord

_____ 29. *Reverberation* in neural circuits refers to collateral axons that:

    a. involve several neuronal pools processing the same information at one time

    b. relay sensory information from one processing center to another in the brain

    c. synapse on the same postsynaptic neuron

    d. use positive feedback to stimulate presynaptic neurons

_____ 30. *Presynaptic facilitation* refers to the:

    a. calcium channels remaining open for a longer period, thus increasing the amount of neurotransmitter released

    b. reduction of the amount of neurotransmitter released due to the closing of calcium channels in the synaptic knob

    c. a larger than usual depolarizing stimulus necessary to bring the membrane potential to threshold

    d. shift in the transmembrane potential toward threshold, which makes the cell more sensitive to further stimulation

Level
=2=

[L2]   Completion:

Using the terms below, complete the following statements.

association	tracts	preganglionic fibers	current
gated	stroke	postganglionic fibers	convergence
perikaryon	voltage	hyperpolarization	nuclei
parallel	ganglia	nerve impulse	

31. The cytoplasm that surrounds a neuron's nucleus is referred to as the _____.

32. Nerve cell bodies in the PNS are clustered together in masses called _____.

33. Movement of charged objects, such as ions, is referred to as _____.

34. The potential difference that exists across a membrane or other barrier is expressed as a(n) _____.

35. Ion channels that open or close in response to specific stimuli are called _____ channels.

36. The loss of positive ions, which causes a shift in the resting potential to as much as – 80 mV, is referred to as _____.

37. An action potential traveling along an axon is called a(n) _____.

38. Interruption of blood supply to the brain by a circulatory blockage or other vascular problem is called a(n) _____.

39. Axons extending from the CNS to a ganglion are called _____.

40. Axons connecting the ganglionic cells with peripheral effectors are known as _____.

41. Neurons that may be situated between sensory and motor neurons are called _____ neurons.

42. Several neurons synapsing on the same postsynaptic neuron are called _____.

43. Several neuronal pools processing the same information at one time are called _____ processing.

44. Collections of nerve cell bodies in the CNS are termed _____.

45. The axonal bundles that make up the white matter of the CNS are called _____.

[L2]   Short Essay:

Briefly answer the following questions in the spaces provided below.

46. What are the four (4) major functions of the nervous system?

47. What are the major components of the central nervous system and the peripheral nervous system?

48. What four types of glial cells are found in the central nervous system?

49. Functionally, what is the major difference between neurons and neuroglia?

50. Using a generalized model, list and describe the four (4) steps that describe an action potential.

51. What role do the nodes of Ranvier play in the conduction of an action potential?

52. What functional mechanism distinguishes an adrenergic synapse from a cholinergic synapse?

53. What is the difference between an EPSP and an IPSP and how does each type affect the generation of an action potential?

54. How are neurons categorized functionally and how does each group function?

55. What is the difference between *divergence* and *convergence*?

When you have successfully completed the exercises in L2 proceed to L3.

Level
=2=

☐ LEVEL 3   CRITICAL THINKING/APPLICATION

Using principles and concepts learned about the nervous system, answer the following questions. Write your answers on a separate sheet of paper.

1. How does starting each day with a few cups of coffee and a cigarette affect a person's behavior and why?

2. Guillain-Barré syndrome is a degeneration of the myelin sheath that ultimately may result in paralysis. What is the relationship between degeneration of the myelin sheath and muscular paralysis?

3. Even though microglia are found in the CNS, why might these specialized cells be considered a part of the body's immune system?

4. Substantiate the following statement concerning neurotransmitter function: "The effect on the postsynaptic membrane depends on the characteristics of the receptor, not on the nature of the neurotransmitter."

5. What physiological mechanisms operate to induce threshold when a single stimulus is not strong enough to initiate an action potential?

13

The Spinal Cord and Spinal Nerves

■ Overview

The central nervous system (CNS) consists of the brain and the spinal cord. Both of these parts of the CNS are covered with meninges, both are bathed in cerebrospinal fluid, and both areas contain gray and white matter. Even though there are structural and functional similarities, the brain and the spinal cord show significant independent structural and functional differences.

Chapter 13 focuses on the structure and function of the spinal cord and the spinal nerves. Emphasis is placed on the importance of the spinal cord as a communication link between the brain and the peripheral nervous system (PNS), and as an integrating center that can be independently involved with somatic reflex activity.

The questions and exercises for this chapter will help you to identify the integrating sites and the conduction pathways that make up reflex mechanisms necessary to maintain homeostasis throughout the body.

❑ LEVEL 1   REVIEW OF CHAPTER OBJECTIVES

1. Discuss the structure and functions of the spinal cord.
2. Describe the three meningeal layers that surround the central nervous system.
3. Explain the roles of white matter and gray matter in processing and relaying sensory and motor information.
4. Describe the major components of a spinal nerve.
5. Relate the distribution pattern of spinal nerves to the regions they innervate.
6. Describe the process of a neural reflex.
7. Classify the different types of reflexes and explain the functions of each.
8. Distinguish between the types of motor responses produced by various reflexes.
9. Explain how reflexes combine to produce complex behaviors.
10. Explain how higher centers control and modify reflex responses.

[L1]   Multiple Choice:

Place the letter corresponding to the correct answer in the space provided.

OBJ. 1   _____   1. The spinal cord is part of the:

      a. peripheral nervous system

      b. somatic nervous system

      c. autonomic nervous system

      d. central nervous system

OBJ. 1   _____   2. The identifiable areas of the spinal cord that are based on the regions they serve include:

      a. cervical, thoracic, lumbar, sacral

      b. pia mater, dura mater, arachnoid mater

      c. axillary, radial, median, ulnar

      d. cranial, visceral, autonomic, spinal

OBJ. 1   _____   3. The cervical enlargement of the spinal cord supplies nerves to the:

      a. shoulder girdle and arms

      b. pelvis and legs

      c. thorax and abdomen

      d. back and lumbar region

OBJ. 2   _____   4. If cerebrospinal fluid was withdrawn during a spinal tap, a needle would be inserted into the:

      a. pia mater

      b. subdural space

      c. subarachnoid space

      d. epidural space

OBJ. 2   _____   5. The meninx that is firmly bound to neural tissue and deep to the other meninges is the:

      a. pia mater

      b. arachnoid membrane

      c. dura mater

      d. epidural space

OBJ. 3   _____   6. The white matter of the spinal cord contains:

      a. cell bodies of neurons and glial cells

      b. somatic and visceral sensory nuclei

      c. large numbers of myelinated and unmyelinated axons

      d. sensory and motor nuclei

OBJ. 3   _____   7. The area of the spinal cord that surrounds the central canal and is dominated by the cell bodies of neurons and glial cells is the:

      a. white matter

      b. gray matter

      c. ascending tracts

      d. descending tracts

OBJ. 3   _____   8. The posterior gray horns of the spinal cord contain:

      a. somatic and visceral sensory nuclei

      b. somatic and visceral motor nuclei

      c. ascending and descending tracts

      d. anterior and posterior columns

OBJ. 4   _____   9. The delicate connective tissue fibers that surround individual axons of spinal nerves comprise a layer called the:

      a. perineurium

      b. epineurium

      c. endoneurium

      d. commissures

OBJ. 5   _____   10. The branches of the cervical plexus innervate the muscles of the:

      a. shoulder girdle and arm

      b. pelvic girdle and leg

      c. neck and extend into the thoracic cavity to control the diaphragm

      d. back and lumbar region

OBJ. 5   _____   11. The brachial plexus innervates the:

      a. neck and shoulder girdle

      b. shoulder girdle and arm

      c. neck and arm

      d. thorax and arm

OBJ. 6   _____   12. The final step involved in a neural reflex is:

      a. information processing

      b. the activation of a motor neuron

      c. a response by an effector

      d. the activation of a sensory neuron

OBJ. 6   _____   13. The goals of information processing during a neural reflex are the selection of:

      a. appropriate sensor selections and specific motor responses

      b. an appropriate motor response and the activation of specific motor neurons

      c. appropriate sensory selections and activation of specific sensory neurons

      d. appropriate motor selections and specific sensory responses

OBJ. 6   _____   14. In the reflex arc, information processing is performed by the:

      a. activation of a sensory neuron

      b. activation of a receptor

      c. release of the neurotransmitter by the synaptic terminal

      d. motor neuron that controls peripheral effectors

Level
-1-

OBJ. 7   _____ 15. The basic motor patterns of innate reflexes are:

     a. learned responses

     b. genetically programmed

     c. a result of environmental conditioning

     d. usually unpredictable and extremely complex

OBJ. 7   _____ 16. A professional skier making a rapid, automatic adjustment in body position while racing is an example of a (an):

     a. innate reflex

     b. cranial reflex

     c. patellar reflex

     d. acquired reflex

OBJ. 8   _____ 17. When a sensory neuron synapses directly on a motor neuron, which itself serves as the processing center, the reflex is called a (an):

     a. polysynaptic reflex

     b. innate reflex

     c. monosynaptic reflex

     d. acquired reflex

OBJ. 8   _____ 18. The sensory receptors in the stretch reflex are the:

     a. muscle spindles

     b. Golgi tendon organs

     c. Pacinian corpuscles

     d. Meissner's corpuscles

OBJ. 9   _____ 19. When one set of motor neurons is stimulated, those controlling antagonistic muscles are inhibited. This statement illustrates the principle of:

     a. contralateral reflex

     b. reciprocal inhibition

     c. crossed extensor reflex

     d. ipsilateral reflex

OBJ. 9   _____ 20. All polysynaptic reflexes share the same basic characteristic(s), which include:

     a. they involve pools of neurons

     b. they are intersegmental in distribution

     c. they involve reciprocal innervation

     d. a, b, and c are correct

OBJ. 10   _____ 21. As descending inhibitory synapses develop:

     a. the Babinski response disappears

     b. there is a positive Babinski response

     c. there will be a noticeable fanning of the toes in adults

     d. there will be a decrease in the facilitation of spinal reflexes

Level
-1-

OBJ. 10   _____ 22. The highest level of motor control involves a series of interactions that occur:

> a. as monosynaptic reflexes that are rapid but stereo-typed
> b. in centers in the brain that can modulate or build upon reflexive motor patterns
> c. in the descending pathways that provide facilitation and inhibition
> d. in the gamma efferents to the intrafusal fibers

[L1]   Completion:

Using the terms below, complete the following statements.

conus medullaris	neural reflexes	reinforcement
receptor	nuclei	epineurium
flexor reflex	cranial reflexes	somatic reflexes
Babinski sign	dermatome	filum terminale
columns	innate reflexes	crossed extensor reflex
acquired reflexes	dorsal ramus	

OBJ. 1   23. The terminal portion of the spinal cord is called the _____.

OBJ. 2   24. The supportive fibrous strand of the pia mater is the _____.

OBJ. 3   25. The cell bodies of neurons in the gray matter of the spinal cord form groups called _____.

OBJ. 4   26. The white matter of the spinal cord is divided into regions called _____.

OBJ. 4   27. The outermost layer of a spinal nerve is called the _____.

OBJ. 5   28. The branch of each spinal nerve that provides sensory and motor innervation to the skin and muscles of the back is the _____.

OBJ. 5   29. The area where each spinal nerve monitors a specific region of the body surface is known as a(n) _____.

OBJ. 6   30. Automatic motor responses, triggered by specific stimuli, are called _____.

OBJ. 6   31. A specialized cell that monitors conditions in the body or the external environment is called a(n) _____.

OBJ. 7   32. Connections that form between neurons during development produce _____.

OBJ. 7   33. Complex, learned motor patterns are called _____.

OBJ. 8   34. Reflexes processed in the brain are called _____.

OBJ. 8   35. Reflexes that control activities of the muscular system are called _____.

OBJ. 9   36. A motor response that occurs on the side opposite the stimulus is referred to as a(n) _____.

OBJ. 9   37. The withdrawal reflex affecting the muscles of a limb is a(n) _____.

OBJ. 10   38. Elevated facilitation leading to an enhancement of spinal reflexes is called _____.

OBJ. 10   39. Stroking an infant's foot on the side of the sole produces a fanning of the toes known as the _____.

[L1] Matching:

Match the terms in column B with the terms in column A. Use letters for answers in the spaces provided.

PART I

		COLUMN A		COLUMN B
OBJ. 1	_____ 40.	dorsal roots	A.	specialized membranes
OBJ. 1	_____ 41.	ventral roots	B.	outermost layer of spinal nerve
OBJ. 2	_____ 42.	spinal meninges	C.	spinal nerves $C_1 - C_5$
OBJ. 3	_____ 43.	white matter	D.	spinal nerves $C_5 - T_1$
OBJ. 4	_____ 44.	epineurium	E.	ascending, descending tracts
OBJ. 5	_____ 45.	cervical plexus	F.	contains axons of motor neurons
OBJ. 5	_____ 46.	brachial plexus	G.	sensory information to spinal cord

PART II

		COLUMN A		COLUMN B
OBJ. 6	_____ 47.	neural "wiring"	H.	monosynaptic reflex
OBJ. 6	_____ 48.	peripheral effector	I.	controls activities of muscular system
OBJ. 7	_____ 49.	somatic reflexes	J.	muscle or gland cells
OBJ. 8	_____ 50.	stretch reflex	K.	reflex arc
OBJ. 8	_____ 51.	tendon reflex	L.	Golgi tendon organ
OBJ. 9	_____ 52.	crossed extensor reflex	M.	provides facilitation, inhibition
OBJ. 9	_____ 53.	same-side reflex	N.	contralateral response
OBJ. 10	_____ 54.	descending pathways	O.	ipsilateral response

[L1] Drawing/Illustration Labeling:

Identify each numbered structure by labeling the following figures:

OBJ. 1 **FIGURE 13.1** Organization of the Spinal Cord

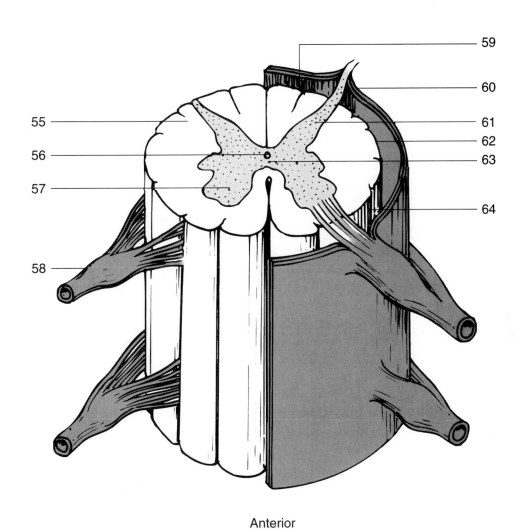

Anterior

55 _____	60 _____	
56 _____	61 _____	
57 _____	62 _____	
58 _____	63 _____	
59 _____	64 _____	

Level
-1-

OBJ. 6   **FIGURE 13.2**   Structure of the Reflex Arc

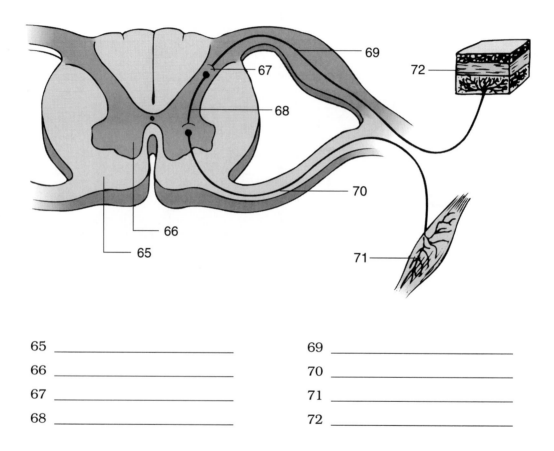

65 _____   69 _____

66 _____   70 _____

67 _____   71 _____

68 _____   72 _____

❑ LEVEL 2   CONCEPT SYNTHESIS

Concept Map I:

Using the following terms, fill in the circled, numbered, blank spaces to complete the concept map. Follow the numbers that comply with the organization of the map.

Viscera
Motor commands
Gray matter
Brain

Glial cells
Posterior gray horns
Peripheral effectors
Ascending tract

Sensory information
Anterior white columns
Somatic motor neurons

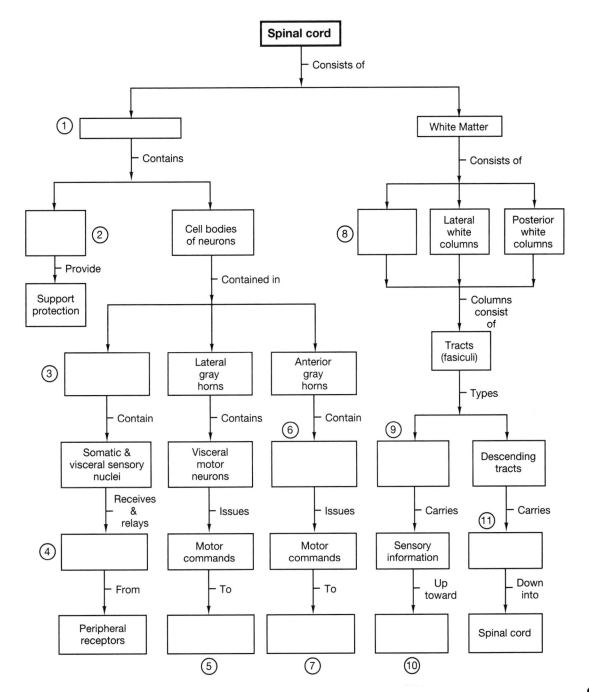

Concept Map II:

Using the following terms, fill in the circled, numbered, blank spaces to complete the concept map. Follow the numbers that comply with the organization of the map.

Stability, support   Diffusion medium   Pia mater

Shock absorber   Subarachnoid space   Dura mater

Blood vessels   Lymphatic fluid

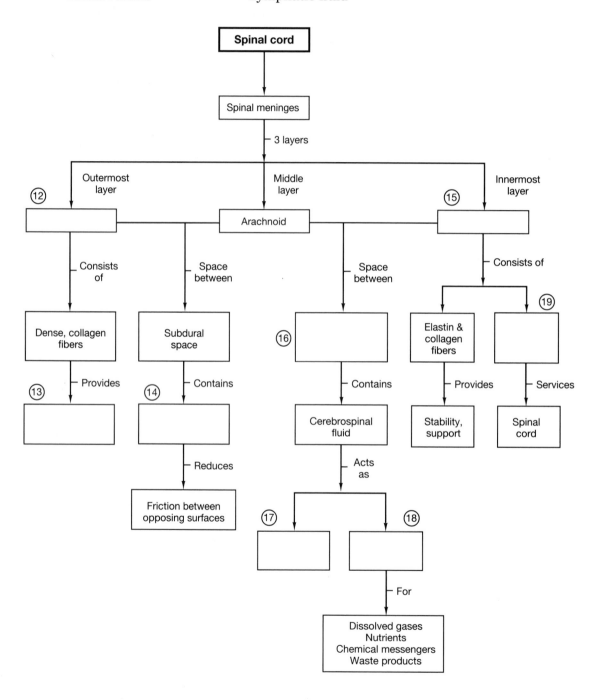

[L2]   Body Trek:

Using the terms below, complete the trek through a typical body neural processing system. Your trek begins at any one of three receptor areas in the peripheral nervous system. Assume that the consecutive numbers represent directions that eventually get you to the finish line. As you trek through the system you will confirm your route and location by identifying each numbered space. Place your answers in the spaces provided below the map. When the trek is over and all the information from the trek is completed you will have a record of how neural processing occurs in the body. Good luck, happy trekking!

motor	exteroceptors	peripheral sensory ganglion
glands	efferent fibers (motor)	skeletal muscles
fat cells	afferent fibers (sensory)	somatic motor neurons
sensory	proprioceptors	preganglionic efferent fibers
interoceptors	smooth muscle	peripheral motor ganglion
interneurons	postganglionic efferent fibers	visceral motor neurons

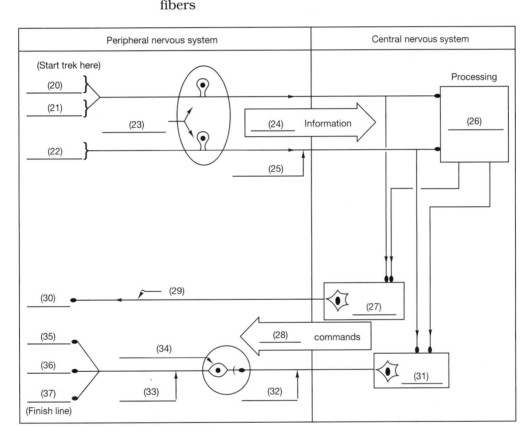

Place your answers in the spaces provided below:

20	_____	29	_____
21	_____	30	_____
22	_____	31	_____
23	_____	32	_____
24	_____	33	_____
25	_____	34	_____
26	_____	35	_____
27	_____	36	_____
28	_____	37	_____

Level
=2=

[L2]   Multiple Choice:

Place the letter corresponding to the correct answer in the space provided.

_____ 38. The designation C_3 relative to the spinal cord refers to the:

a. third cranial nerve

b. third coccygeal ligament

c. third cervical segment

d. third cerebrospinal layer

_____ 39. The specialized membranes that provide physical stability and protection as shock absorbers are the:

a. conus medullaris

b. spinal meninges

c. filum terminale

d. denticulate ligaments

_____ 40. Spinal nerves are classified as mixed nerves because they contain:

a. both sensory and motor fibers

b. both dorsal and ventral roots

c. white matter and gray matter

d. ascending and descending pathways

_____ 41. The three (3) meningeal layers of the spinal cord include:

a. spinal meninges, cranial meninges, conus medullaris

b. cauda equina, conus medullaris, filum terminale

c. white matter, gray matter, central canal

d. dura mater, pia mater, arachnoid mater

_____ 42. The *cauda equina* is a complex of the spinal cord that includes:

a. the enlargements of the cervical and thoracic regions of the cord

b. the posterior and anterior median sulci

c. the coccygeal and denticulate ligaments

d. the filum terminale and the ventral and dorsal roots caudal to the conus medullaris

_____ 43. The epidural space is an area that contains:

a. blood vessels and sensory and motor fibers

b. loose connective tissue, blood vessels, adipose tissue

c. a delicate network of collagen and elastic fibers

d. the spinal fluid

_____ 44. In the spinal cord the cerebrospinal fluid is found within the:

a. central canal and subarachnoid space

b. subarachnoid and epidural spaces

c. central canal and epidural space

d. subdural and epidural spaces

Level
=2=

_____ 45. The meninx in contact with the brain and the spinal cord is the:

      a. dura mater

      b. arachnoid mater

      c. pia mater

      d. subdural space

_____ 46. The division between the left and right sides of the spinal cord is:

      a. the anterior median fissure and the posterior median sulcus

      b. the anterior gray horns and the posterior white columns

      c. the white matter and the gray matter

      d. the anterior and posterior white columns

_____ 47. The axons in the white matter of the spinal cord that carry sensory information up toward the brain are organized into:

      a. descending tracts

      b. anterior white columns

      c. ascending tracts

      d. posterior white columns

_____ 48. In reference to the vertebral column, C_2 refers to the cervical nerve that:

      a. is between vertebrae C_2 and C_3

      b. precedes vertebra C_2

      c. follows vertebra C_2

      d. precedes vertebra C_3

_____ 49. The white ramus is the branch of a spinal nerve that consists of:

      a. myelinated postganglionic axons

      b. unmyelinated preganglionic axons

      c. unmyelinated postganglionic axons

      d. myelinated preganglionic axons

_____ 50. Unmyelinated fibers that innervate glands and smooth muscles in the body wall or limbs form the:

      a. autonomic ganglion

      b. gray ramus

      c. white ramus

      d. ramus communicans

_____ 51. The lumbo-sacral plexus that supplies the pelvic girdle and the leg includes spinal nerves:

      a. $T_{12} - L_4$

      b. $L_4 - S_4$

      c. $T_1 - T_{12}$

      d. $T_{12} - S_4$

Level
=2=

_____ 52. Pain receptors are literally the:

      a. dendrites of sensory neurons

      b. axons of motor neurons

      c. specialized cells that respond to a limited range of stimuli

      d. activators of motor neurons

_____ 53. Reflexive removal of a hand from a hot stove and blinking when the eyelashes are touched are examples of:

      a. acquired reflexes

      b. cranial reflexes

      c. innate reflexes

      d. synaptic reflexes

_____ 54. The most complicated responses are produced by polysynaptic responses because:

      a. the interneurons can control several different muscle groups

      b. a delay between stimulus and response is minimized

      c. the postsynaptic motor neuron serves as the processing center

      d. there is an absence of interneurons, which minimizes delay

_____ 55. The duration of synaptic delay is:

      a. determined by the length of the interneuron

      b. proportional to the amount of neurotransmitter substance released

      c. determined by the number of sensory neurons

      d. proportional to the number of synapses involved

_____ 56. The activity occurring in the Golgi tendon organ involves:

      a. a cooperative effort by many different muscle groups

      b. the stretching of the collagen fibers and stimulation of the sensory neuron

      c. many complex polysynaptic reflexes

      d. excitatory or inhibitory postsynaptic potentials

_____ 57. Examples of somatic reflexes include:

      a. pupillary, ciliospinal, salivary, micturition reflexes

      b. contralateral, ipsilateral, pupillary, defecation reflexes

      c. stretch, crossed extensor, corneal, gag reflexes

      d. micturition, Hering-Breuer, defecation, carotid sinus reflex

_____ 58. Examples of autonomic reflexes include:

      a. pupillary, salivary, ciliospinal reflexes

      b. patellar, Achilles, abdominal reflex

      c. plantar, corneal, gag reflex

      d. stretch, crossed extensor, corneal reflex

Level
=2=

_____ 59. In an adult, CNS injury is suspected if:

        a. there is a plantar reflex

        b. there is an increase in facilitating synapses

        c. there is a positive Babinski reflex

        d. there is a negative Babinski reflex

[L2]   Completion:

Using the terms below, complete the following statements. Place the answers in the spaces provided.

brachial plexus	dorsal ramus	nerve plexus
dura mater	gamma efferents	reflexes
motor nuclei	cauda equina	descending tracts
gray ramus	gray commissures	myelography
pia mater	spinal meninges	perineurium
sensory nuclei		

60. The connective tissue partition that separates adjacent bundles of nerve fibers in a peripheral nerve is referred to as the _____.

61. Axons convey motor commands down into the spinal cord by way of _____.

62. The outermost covering of the spinal cord is the _____.

63. The specialized protective membrane that surrounds the spinal cord is the _____.

64. The blood vessels servicing the spinal cord are found in the _____.

65. The nerves that collectively radiate from the conus medullaris are known as the _____.

66. The cell bodies of neurons that receive and relay information from peripheral receptors are called _____.

67. The cell bodies of neurons that issue commands to peripheral effectors are called _____.

68. The introduction of radiopaque dyes into the cerebrospinal fluid of the subarachnoid space involves _____.

69. The area of the spinal cord where axons cross from one side of the cord to the other is the _____.

70. Unmyelinated, postganglionic fibers that innervate glands and smooth muscles in the body wall or limbs form the _____.

71. Spinal nerve sensory and motor innervation to the skin and muscles of the back is provided by the _____.

72. A complex, interwoven network of nerves is called a _____.

73. The radial, median, and ulnar nerves would be found in the _____.

74. Programmed, automatic, involuntary motor responses and motor patterns are called _____.

75. Contraction of the myofibrils within intrafusal muscle occurs when impulses arrive over _____.

[L2] Short Essay:

Briefly answer the following questions in the spaces provided below.

76. What three membranes make up the spinal meninges?

77. Why are spinal nerves classified as *mixed* nerves?

78. What structural components make the white matter different from the gray matter in the spinal cord?

79. What is a dermatome and why are dermatomes clinically important?

80. List the five (5) steps involved in a neural reflex.

81. What four (4) criteria are used to classify reflexes?

82. Why can polysynaptic reflexes produce more complicated responses than monosynaptic reflexes?

83. What is reciprocal inhibition?

84. What is the difference between a contralateral reflex arc and an ipsilateral reflex arc?

85. What are the clinical implications of a positive Babinski reflex in an adult?

When you have successfully completed the exercises in L2 proceed to L3.

☐ LEVEL 3   CRITICAL THINKING/APPLICATION

Using principles and concepts learned about the nervous system, answer the following questions. Write your answers on a separate sheet of paper.

1. What structural features make it possible to do a spinal tap without damaging the spinal cord?
2. What is spinal meningitis, what causes the condition, and how does it ultimately affect the victim?
3. What relationship does the spinal cord have to the development of multiple sclerosis?
4. Why are reflexes such as the patellar reflex an important part of a routine physical examination?
5. How might damage to the following areas of the vertebral column affect the body?

   a. damage to the 4th or 5th vertebrae
   b. damage to C_3 to C_5
   c. damage to the thoracic vertebrae
   d. damage to the lumbar vertebrae

Level
≡**3**≡

14

The Brain and the Cranial Nerves

■ Overview

In Chapter 13 we ventured into the central nervous system (CNS) via the spinal cord. Chapter 14 considers the other division of the CNS, the brain. This large, complex organ is located in the cranial cavity completely surrounded by cerebrospinal fluid (CSF), the cranial meninges, and the bony structures of the cranium.

The development of the brain is one of the primary factors that have been used to classify us as "human." The ability to think and reason separates us from all other forms of life.

The brain consists of billions of neurons organized into hundreds of neuronal pools with extensive interconnections that provide great versatility and variability. The same basic principles of neural activity and information processing that occur in the spinal cord also apply to the brain; however, because of the increased amount of neural tissue, there is an expanded versatility and complexity in response to neural stimulation and neural function.

Included in Chapter 14 are exercises and test questions that focus on the major regions and structures of the brain and their relationships with the cranial nerves. The concept maps of the brain will be particularly helpful in making the brain come "alive" structurally.

❏ LEVEL 1   REVIEW OF CHAPTER OBJECTIVES

1. Name the major regions of the brain and describe their functions.
2. Name the ventricles of the brain and describe their locations and the connections between them.
3. Explain how the brain is protected and supported.
4. Discuss the formation, circulation, and functions of the cerebrospinal fluid.
5. Locate the motor, sensory, and association areas of the cerebral cortex and discuss their functions.
6. Identify important structures within each region of the brain and explain their functions.
7. Identify the cranial nerves and relate each pair of nerves to its principal destinations and functions.
8. Discuss important cranial reflexes.

[L1]   Multiple Choice:

Place the letter corresponding to the correct answer in the space provided.

OBJ. 1   _____   1. The major region of the brain responsible for conscious thought processes, sensations, intellectual functions, memory, and complex motor patterns is the:

    a. cerebellum
    b. medulla
    c. pons
    d. cerebrum

OBJ. 1   _____   2. The region of the brain that adjusts voluntary and involuntary motor activities on the basis of sensory information and stored memories of previous movements is the:

    a. cerebrum
    b. cerebellum
    c. medulla
    d. diencephalon

OBJ. 1   _____   3. The brain stem consists of:

    a. diencephalon, mesencephalon, pons, medulla oblongata
    b. cerebrum, cerebellum, medulla, pons
    c. thalamus, hypothalamus, cerebellum, medulla
    d. diencephalon, spinal cord, cerebellum, medulla

OBJ. 2   _____   4. The lateral ventricles of the cerebral hemisphere communicate with the third ventricle of the diencephalon through the:

    a. septum pellucidum
    b. aqueduct of Sylvius
    c. mesencephalic aqueduct
    d. interventricular foramen

OBJ. 2   _____   5. The slender canal that connects the third ventricle with the fourth ventricle is the:

    a. aqueduct of Sylvius
    b. foramen of Munro
    c. septum pellucidum
    d. diencephalic chamber

OBJ. 3   _____   6. Mechanical protection for the brain is provided by:

    a. the bones of the skull
    b. the cranial meninges
    c. the cerebrospinal fluid
    d. the dura mater

OBJ. 3   _____   7. The cranial meninges offer protection to the brain by:

    a. protecting the brain from extremes of temperature
    b. affording mechanical protection to the brain tissue
    c. providing a barrier against invading pathogenic organisms
    d. acting as shock absorbers that prevent contact with surrounding bones

Level
-1-

OBJ. 3

_____ 8. The brain is protected from shocks generated during locomotion by the:

    a. bones of the cranium

    b. the cerebrospinal fluid

    c. the spinal meninges

    d. intervertebral discs of the vertebral column and the muscles and vertebrae of the neck

OBJ. 4

_____ 9. Through a combination of active and passive transport mechanisms, ependymal cells secrete cerebrospinal fluid at a rate of approximately:

    a. 100 ml/day

    b. 500 ml/day

    c. 1200 ml/day

    d. 1500 ml/day

OBJ. 4

_____ 10. Monitoring the composition of the CSF and removal of waste products from the CSF occurs in the:

    a. subarachnoid space

    b. choroid plexus

    c. lateral and medial apertures

    d. superior saggital sinus

OBJ. 4

_____ 11. Excess cerebrospinal fluid is returned to venous circulation by:

    a. diffusion across the arachnoid villi

    b. active transport across the choroid plexus

    c. diffusion through the lateral and medial apertures

    d. passage through the subarachnoid space

OBJ. 5

_____ 12. The neurons in the _primary sensory cortex_ receive somatic sensory information from:

    a. commissural fibers in the white matter

    b. touch, pressure, pain, taste, and temperature receptors

    c. visual and auditory receptors in the eyes and ears

    d. receptors in muscle spindles and Golgi tendon organs

OBJ. 5

_____ 13. The neurons of the _primary motor cortex_ are responsible for directing:

    a. visual and auditory responses

    b. responses to taste and temperature

    c. voluntary movements

    d. involuntary movements

OBJ. 5

_____ 14. The _somatic motor association_ area is responsible for the:

    a. ability to hear, see, and smell

    b. coordination of learned motor responses

    c. coordination of all reflex activity throughout the body

    d. assimilation of neural responses to tactile stimulation

Level
-1-

OBJ. 6   _____ 15. All communication between the brain and the spinal cord involves tracts that ascend or descend through the:

        a. cerebellum
        b. diencephalon
        c. medulla
        d. thalamus

OBJ. 6   _____ 16. The thalamus contains:

        a. centers involved with emotions and hormone production
        b. relay and processing centers for sensory information
        c. centers for involuntary somatic motor responses
        d. nuclei involved with visceral motor control

OBJ. 6   _____ 17. Abstract intellectual functions are integrated and performed in the:

        a. prefrontal cortex of the frontal lobe
        b. somatic motor association area
        c. primary sensory cortex of the precentral gyrus
        d. globus pallidus

OBJ. 6   _____ 18. The hypothalamus contains centers involved with:

        a. voluntary somatic motor responses
        b. somatic and visceral motor control
        c. maintenance of consciousness
        d. emotions, autonomic function, and hormone production

OBJ. 7   _____ 19. The *special sensory* cranial nerves include the:

        a. oculomotor, trochlear, and abducens
        b. spinal accessory, hypoglossal, and glossopharyngeal
        c. olfactory, optic, and vestibulocochlear
        d. vagus, trigeminal, and facial

OBJ. 7   _____ 20. Cranial nerves III, IV, VI, and XI, which provide motor control, are:

        a. trigeminal, facial, glossopharyngeal, vagus
        b. oculomotor, trochlear, abducens, spinal accessory
        c. olfactory, optic, vestibulocochlear, hypoglossal
        d. oculomotor, hypoglossal, optic, olfactory

OBJ. 7   _____ 21. The cranial nerves that carry sensory information and involuntary motor commands are:

        a. I, II, III, IV
        b. II, IV, VI, VIII
        c. V, VI, VIII, XII
        d. V, VII, IX, X

OBJ. 8   _____ 22. Cranial reflexes are reflex arcs that involve the:

        a. sensory and motor fibers of cranial nerves
        b. sensory and motor fibers of spinal nerves
        c. sensory and motor fibers of the cerebellum
        d. sensory and motor fibers in the medulla oblongata

Level
–1–

OBJ. 8

_____ 23. The corneal reflex produces:

    a. closure of the eye due to neural stimulation

    b. rapid blinking of the affected eye following contact to the cornea

    c. eye movement in the opposite direction of the stimulus

    d. constriction of the pupil of the eye

OBJ. 8

_____ 24. A sudden sound behind and to the right, which causes a rotation of the eyes, head, and neck to that side, is a response to the:

    a. vestibulo-ocular reflex

    b. consensual light reflex

    c. auditory reflex

    d. tympanic reflex

OBJ. 8

_____ 25. The reflex often used to test the sensory function of the trigeminal nerve is the:

    a. tympanic reflex

    b. auditory reflex

    c. vestibulo-ocular reflex

    d. corneal reflex

[L1]   Completion:

Using the terms below, complete the following statements.

falx cerebri	tentorium cerebelli	epithalamus
amygdaloid body	neural cortex	postcentral gyrus
hypoglossal	spinal accessory	aqueduct of Sylvius
pyramidal cells	third ventricle	vestibulo-ocular
shunt	corpora quadrigemina	choroid plexus

OBJ. 1

26. The part of the diencephalon that does not contain neural tissue is the _____.

OBJ. 1

27. The layer of gray matter found on the surface of the cerebrum is the _____.

OBJ. 2

28. The diencephalic chamber is called the _____.

OBJ. 2

29. Instead of a ventricle, the mesencephalon has a slender canal known as the _____.

OBJ. 3

30. The fold of dura mater projecting between the cerebral hemispheres in the midsagittal plane that provides stabilization and support for the brain is the _____.

OBJ. 3

31. The dural sheet that separates and protects the cerebellar hemispheres from those of the cerebrum is the _____.

OBJ. 4

32. The site of cerebrospinal fluid production is the _____.

OBJ. 4

33. A bypass that drains excess cerebrospinal fluid and reduces intercranial pressure is called a(n) _____.

OBJ. 5

34. The primary sensory cortex forms the surface of the _____.

OBJ. 5

35. The neurons in the primary motor cortex are called _____.

OBJ. 6

36. The cerebral contribution to the limbic system is the _____.

Level
-1-

OBJ. 6   37. The roof or tectum of the mesencephalon contains two pairs of sensory nuclei known collectively as the _____.

OBJ. 7   38. The cranial nerve that controls the tongue muscles is N XII, the _____.

OBJ. 7   39. The cranial nerve that directs voluntary motor control over large superficial muscles of the back is the _____.

OBJ. 8   40. Rotational movement of the head with eye movement in the opposite direction illustrates the _____ reflex.

[L1]   Matching:

Match the terms in column B with the terms in column A. Use letters for answers in the spaces provided.

PART I

		COLUMN A		COLUMN B
OBJ. 1	_____ 41.	tectum	A.	voluntary motor control
OBJ. 1	_____ 42.	epithalamus	B.	CSF production
OBJ. 2	_____ 43.	ventricles	C.	diencephalic roof
OBJ. 2	_____ 44.	falx cerebri	D.	short association fibers
OBJ. 3	_____ 45.	arachnoid	E.	cerebrospinal fluid circulation
OBJ. 4	_____ 46.	choroid plexus	F.	fold of dura mater
OBJ. 5	_____ 47.	pyramidal system	G.	brain cover
OBJ 5	_____ 48.	arcuate	H.	roof of mesencephalon

PART II

		COLUMN A		COLUMN B
OBJ 6	_____ 49.	cerebellum	I.	optic
OBJ 6	_____ 50.	medulla oblongata	J.	olives
OBJ 6	_____ 51.	limbic system	K.	vagus
OBJ 6	_____ 52.	lentiform nucleus	L.	amygdaloid body
OBJ 7	_____ 53.	N X	M.	tympanic reflex
OBJ 7	_____ 54.	N II	N.	putamen and globus pallidus
OBJ 8	_____ 55.	N VII, N VIII	O.	coordinates motor activities

[L1]   Drawing/Illustration Labeling:

Identify each numbered structure by labeling the following figures:

OBJ. 1
OBJ. 6   **FIGURE 14.1**   Lateral View of the Human Brain

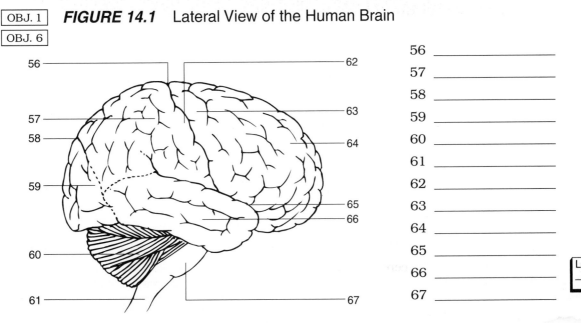

56 _____

57 _____

58 _____

59 _____

60 _____

61 _____

62 _____

63 _____

64 _____

65 _____

66 _____

67 _____

Level
–1–

OBJ. 1 *FIGURE 14.2*   Sagittal View of the Human Brain
OBJ. 6

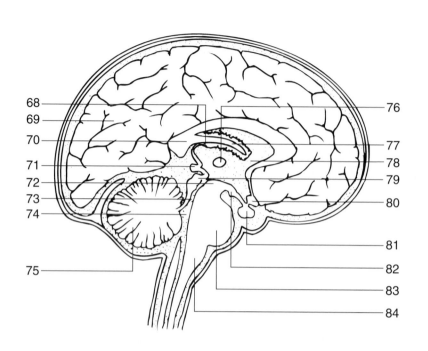

68 _____
69 _____
70 _____
71 _____
72 _____
73 _____
74 _____
75 _____
76 _____
77 _____
78 _____
79 _____
80 _____
81 _____
82 _____
83 _____
84 _____

OBJ. 1 *FIGURE 14.3*   Origins of Cranial Nerves—Inferior View of the Human Brain-Partial
OBJ. 6

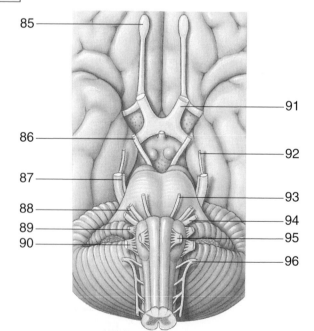

85 _____
86 _____
87 _____
88 _____
89 _____
90 _____
91 _____
92 _____
93 _____
94 _____
95 _____
96 _____

Level
-1-

When you have successfully completed the exercises in L1 proceed to L2.

☐ LEVEL 2 CONCEPT SYNTHESIS

[L2] Concept Maps:

In order to organize and study the brain, a series of concept maps will be used. All the information to complete the maps will be found from the selections that follow. Using the terms below, fill in the circled, numbered spaces to complete the concept maps. Follow the numbers which comply with the organization of the maps.

Concept Map I — Major Regions of the Brain

2 Cerebellar hemispheres	Pons	Hypothalamus
Corpora quadrigemina	Diencephalon	Medulla oblongata

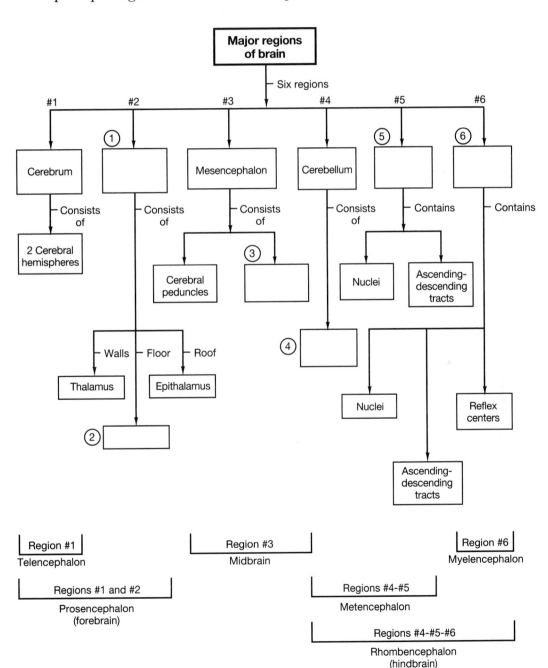

Concept Map II — Region I: Cerebrum

Occipital   Projections   Somesthetic cortex
Nuclei   Temporal   Commissural
Cerebral cortex

```
                          ┌──────────────┐
                          │  Region 1    │
                          └──────┬───────┘
                                 │                          ┌──────────────────────┐
                          ┌──────▼───────┐   Functions   ┌─▶│  Conscious thought   │
                          │   Cerebrum   │───────────────┤  └──────────────────────┘
                          └──────┬───────┘               │  ┌──────────────────────┐
                                 │ Divided               ├─▶│  Sensory perception  │
                                 │ into                  │  └──────────────────────┘
                          ┌──────▼───────┐               │  ┌──────────────────────┐
                          │ 2 Cerebral   │               └─▶│    Motor control     │
                          │ hemispheres  │                  └──────────────────────┘
                          └──────┬───────┘
                                 │ Consists of
                                 │ 5 lobes
```

Frontal	⑦	Parietal	⑨	Insula
Motor area thoughts & emotions speech	Sensory integration short-term memory audition		Vision	Island of cortex speech-hearing?

⑧

Cerebral tissues

Located between

Gray matter	White matter	Gray matter
Found in	Fibers tracts / or (Myelinated)	Basal ganglia

⑩   ⑪   ⑫   ⑬

Contains

Unmyelinated neurons	Association

| Motor-cortex to brain & spinal cord | Sensory- from spinal cord to brain | Connects areas of cerebral cortex within same hemisphere | Connects R&L sides of brain ex. corpus callosum |

Level
=2=

Concept Map III — Region II: Diencephalon

Autonomic centers
Thalamus
Ventral

Geniculates
Cerebrospinal fluid
Anterior

Preoptic area
Pineal gland

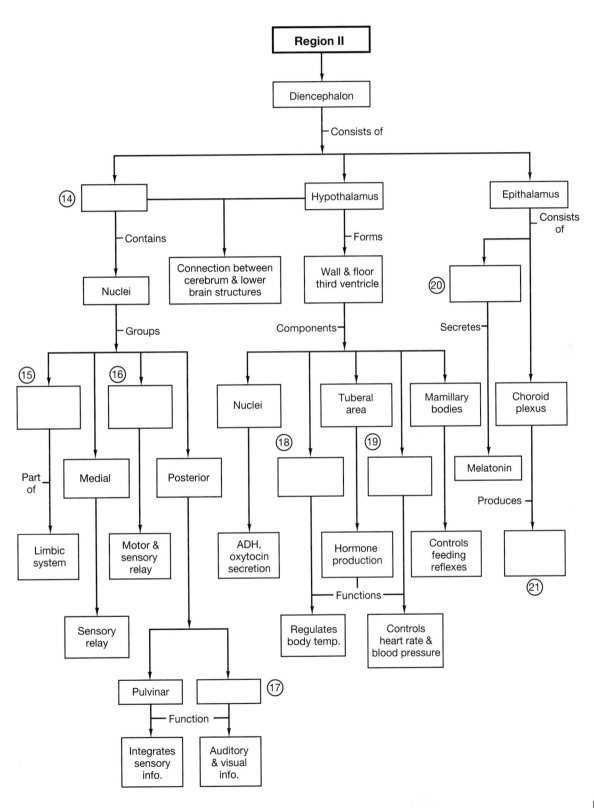

Concept Map IV — Region III: Mesencephalon

Corpora quadrigemina   Red nucleus   Gray matter
Cerebral peduncles   Substantia nigra   Inferior colliculi

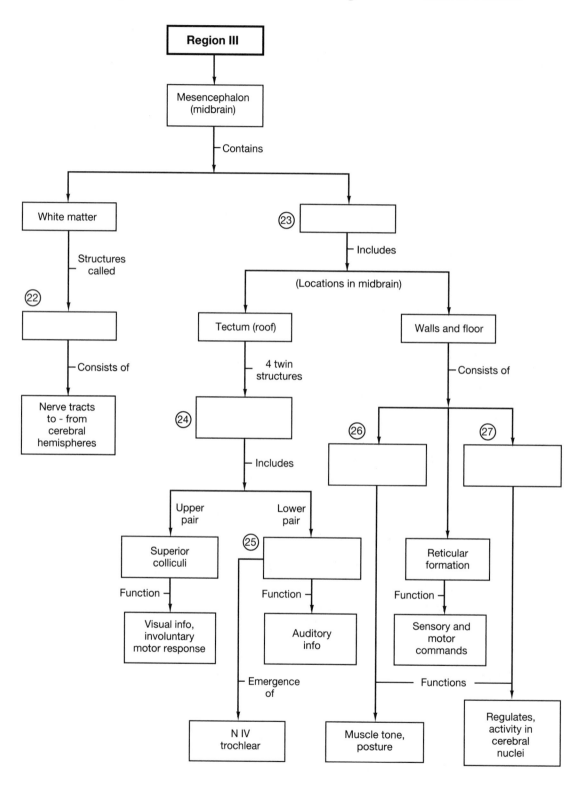

Concept Map V — Region IV: Cerebellum

Purkinje cells   Arbor vitae   Gray matter
Vermis   Cerebellar nuclei

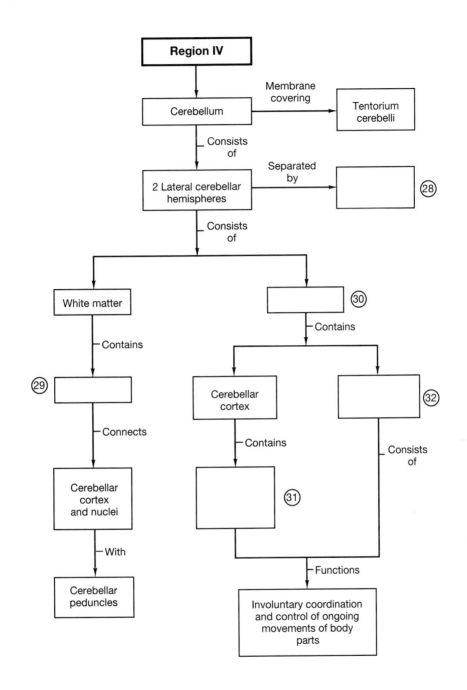

Concept Map VI — Region V: Pons

Inferior   Transverse fibers   Superior
Apneustic   White matter   Respiratory centers

```
                        ┌─────────────┐
                        │  Region V   │
                        └─────────────┘
                               │
                        ┌─────────────┐
                        │    Pons     │
                        └─────────────┘
                               │
                          ─ Consists of
```

- Gray matter

  ─ Consists of
  - Nuclei
    - Associated with
      - N V-Trigeminal
        N VI-Abducens
        N VII-Facial
        N VIII-Auditory
  - (33)
    - Contains
      - Reticular formations
        - 2 Centers
          - (34)
          - Pneumotaxic
            - Function
              - Modify activity of respiratory rhythmicity center in medulla

- (35)

  ─ Consists of
  - Cerebellar peduncles
    - Consists of
      - (36)
        - Links
          - Cerebellum
            - To
              - Mesencephalon diencephalon cerebrum
      - Middle
        - Contains
          - Transverse fibers
            - Carries info between
              - Cerebellum and pons
      - (37)
        - Links
          - Cerebellum
            - To
              - Medulla & spinal cord
  - (38)
    - Interconnects
      - Cerebellar hemispheres

Concept Map VII — Region VI: Medulla Oblongata

Cardiac
Brain and spinal cord
Distribution of blood flow

Cuneatus
Olivary
Respiratory rhythmicity
  center

Reflex centers
White matter

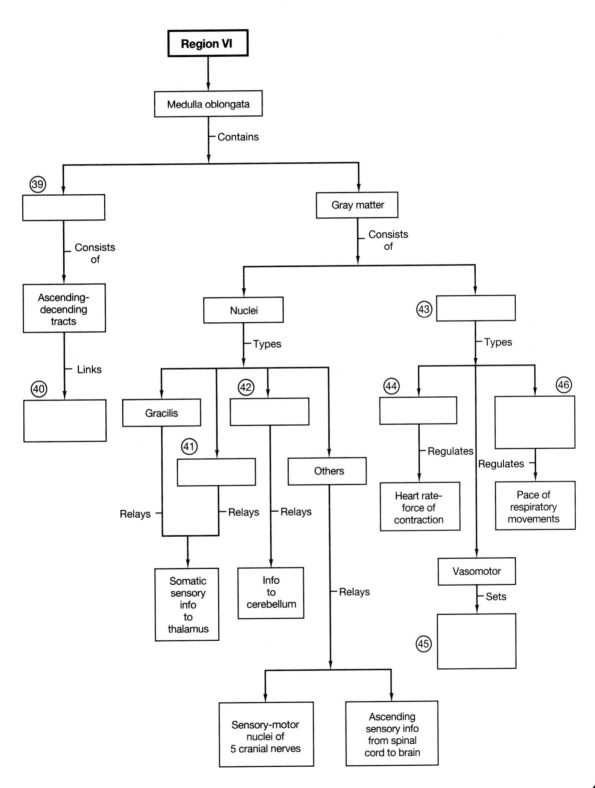

[L2] Multiple Choice:

Place the letter corresponding to the correct answer in the space provided.

_____ 47. The versatility of the brain to respond to stimuli is greater than that of the spinal cord because of:

        a. the location of the brain in the cranial vault

        b. the size of the brain and the number of myelinated neurons

        c. the fast-paced processing centers located in the brain

        d. the number of neurons and complex interconnections between the neurons.

_____ 48. The mesencephalon, or midbrain, processes:

        a. visual and auditory information and generates involuntary motor responses

        b. emotions, sensations, memory, and visceral motor control

        c. voluntary motor activities and sensory information

        d. sensory information from the cerebellum to the cerebrum

_____ 49. An individual with a damaged _visual association area_:

        a. is incapable of receiving somatic sensory information

        b. cannot read because he or she is blind

        c. is red-green color blind and experiences glaucoma

        d. can see letters clearly, but cannot recognize or interpret them

_____ 50. The three major groups of axons that comprise the central white matter are:

        a. motor, sensory, and interneurons

        b. unipolar, bipolar, and multipolar

        c. reticular, caudate, and arcuate fibers

        d. association, commissural, and projection fibers

_____ 51. The three masses of gray matter that lie between the bulging surface of the insula and the lateral wall of the diencephalon are the:

        a. amygdaloid, lentiform, and caudate

        b. claustrum, putamen, and globus pallidus

        c. formix, gyrus, and corpus striatum

        d. cingulate, reticular, and hippocampus

_____ 52. The functions of the limbic system involve:

        a. unconscious processing of visual and auditory information

        b. adjusting muscle tone and body position to prepare for a voluntary movement

        c. emotional states and related behavioral drives

        d. controlling reflex movements and processing of conscious thought

Level
=2=

_____ 53. The mamillary bodies in the floor of the hypothalamus contain motor nuclei that control the reflex movements involved with:

    a. chewing, licking, and swallowing

    b. touch, pressure, and pain

    c. eye movements and auditory responses

    d. taste and temperature responses

_____ 54. Hypothalamic or thalamic stimulation that depresses reticular formation activity in the brain stem results in:

    a. heightened alertness and a generalized excitement

    b. emotions of fear, rage, and pain

    c. sexual arousal and pleasure

    d. generalized lethargy or actual sleep

_____ 55. The pineal gland, an endocrine structure that secretes the hormone melatonin, is found in the:

    a. hypothalamus

    b. thalamus

    c. epithalamus

    d. brain stem

_____ 56. The pulvinar nuclei of the thalamus integrate:

    a. visual information from the eyes and auditory signals from the ears

    b. sensory information for projection to the association areas of the cerebral cortex

    c. sensory data concerning touch, pressure, and pain

    d. sensory information for relay to the frontal lobes

_____ 57. The part(s) of the diencephalon responsible for coordination of activities of the central nervous system and the endocrine system is (are) the:

    a. thalamus

    b. hypothalamus

    c. epithalamus

    d. a, b, and c are correct

_____ 58. The hypothalamus produces and secretes the hormones:

    a. ADH and oxytocin

    b. adrenalin and growth hormone

    c. estrogen and testosterone

    d. prolactin and FSH

_____ 59. The preoptic area of the hypothalamus controls the:

    a. visual responses of the eye

    b. dilation and constriction of the iris in the eye

    c. emotions of rage and aggression

    d. physiological responses to changes in body temperature

_____ 60. The nerve fiber bundles on the ventrolateral surfaces of the mesencephalon are the:

    a. corpora quadrigemina

    b. substantia nigra

    c. cerebral peduncles

    d. inferior and superior colliculi

_____ 61. The effortless serve of a tennis player is a result of establishing:

    a. cerebral motor responses

    b. cerebellar motor patterns

    c. sensory patterns within the medulla

    d. motor patterns in the pons

_____ 62. Huge, highly branched Purkinje cells are found in the:

    a. cerebellar cortex

    b. pons

    c. medulla oblongata

    d. cerebral cortex

_____ 63. The centers in the pons that modify the activity of the respiratory rhythmicity center in the medulla oblongata are the:

    a. vasomotor and cardiac centers

    b. ascending and descending tracts

    c. apneustic and pneumotaxic centers

    d. nucleus gracilis and nucleus cuneatus

_____ 64. The cardiovascular centers and the respiratory rhythmicity center are found in the:

    a. cerebrum

    b. cerebellum

    c. pons

    d. medulla oblongata

_____ 65. The large veins found between the inner and outer layers of the dura mater are called the:

    a. dural venules

    b. dural sinuses

    c. dural falx

    d. vasosinuses

_____ 66. Neural tissue in the CNS is isolated from the general circulation by the:

    a. blood-brain barrier

    b. falx cerebri

    c. corpus striatum

    d. anterior commissures

_____ 67. The CSF reaches the subarachnoid space via:

    a. the subdural sinuses

    b. choroid plexus

    c. three holes in the fourth ventricle

    d. superior sagittal sinus

Level
=2=

_____ 68. The blood-brain barrier remains intact throughout the CNS, except in:

     a. the pons and the medulla

     b. the thalamus and the mesencephalon

     c. portions of the cerebrum and the cerebellum

     d. portions of the hypothalamus and the choroid plexus

_____ 69. The ventricles in the brain form hollow chambers that serve as passageways for the circulation of:

     a. blood

     b. cerebrospinal fluid

     c. interstitial fluid

     d. lymph

_____ 70. The central white matter of the cerebrum is found:

     a. in the superficial layer of the neural cortex

     b. beneath the neural cortex and around the cerebral nuclei

     c. in the deep cerebral nuclei and the neural cortex

     d. in the cerebral cortex and in the cerebral nuclei

_____ 71. The series of elevated ridges that increase the surface area of the cerebral hemispheres and the number of neurons in the cortical area are called:

     a. sulci

     b. fissures

     c. gyri

     d. a, b, and c are correct

[L2]   Completion:

Using the terms below, complete the following statements.

fissures	hypothalamus	corpus striatum
drives	hippocampus	fornix
thalamus	pituitary gland	spinal cord
commissural	aqueduct of Sylvius	third ventricle
sulci	arcuate	extrapyramidal system

72. The walls of the diencephalon are formed by the _____.

73. The primary link between the nervous and endocrine system is the _____.

74. The shortest association fibers in the central white matter are called _____ fibers.

75. The fibers that permit communication between the two cerebral hemispheres are called _____ fibers.

76. The caudate and lentiform nuclei are collectively referred to as the _____.

77. The part of the limbic system that appears to be important in learning and storage of long-term memory is the _____.

78. The tract of white matter that connects the hippocampus with the hypothalamus is the _____.

79. Unfocused "impressions" originating in the hypothalamus are called _____.

80. Because the cerebrum has a pair of lateral ventricles, the diencephalic chamber is called the _____.

81. Instead of a ventricle, the mesencephalon has a slender canal known as the _____.

Level
=2=

82. In the caudal half of the medulla the fourth ventricle narrows and becomes continuous with the central canal of the _____.

83. The shallow depressions that separate the cortical surface of the cerebral hemispheres are called _____.

84. Deep grooves separating the cortical surface of the cerebral hemispheres are called _____.

85. The pathway that controls muscle tone and coordinates learned movement patterns and other somatic motor activities is the _____.

86. The component of the brain that integrates with the endocrine system is the _____.

[L2]   Review of Cranial Nerves:

On the following chart, fill in the blanks to complete the names of the cranial nerves and their functions. If the function is motor, use *M*; if sensory, use *S*; if both motor and sensory, use *B*. Questions 93–98 refer to the functions.

Number	Name	Motor	Sensory	Both	#
I	(87)		S		
II	Optic				(93)
III	(88)	M			
IV	Trochlear				(94)
V	(89)			B	
VI	Abducens				(95)
VII	(90)			B	
VIII	Vestibulocochlear				(96)
IX	(91)			B	
X	Vagus				(97)
XI	(92)	M			
XII	Hypoglossal				(98)

[L2]   Short Essay:

Briefly answer the following questions in the spaces provided below.

99. Why is the brain more versatile than the spinal cord when responding to stimuli?

100. List the six (6) major regions in the adult brain.

101. What does the term "higher centers" refer to in the brain?

102. Where is the limbic system and what are its functions?

103. What are the hypothalamic contributions to the endocrine system?

104. What are the two (2) primary functions of the cerebellum?

105. What are the four (4) primary structural components of the pons?

106. What is the primary function of the blood-brain barrier?

107. What are the three (3) major vital functions of the cerebrospinal fluid?

108. What cranial nerves are classified as "special sensory"?

109. Why are cranial reflexes clinically important?

When you have successfully completed the exercises in L2 proceed to L3.

☐ LEVEL 3   CRITICAL THINKING/APPLICATION

Using principles and concepts learned about the brain and cranial nerves, answer the following questions. Write your answers on a separate sheet of paper.

1. How does a blockage of cerebrospinal fluid (CSF) in an arachnoid granulation cause irreversible brain damage?

2. During the action in a football game one of the players was struck with a blow to the head, knocking him unconscious. The first medics to arrive at his side immediately subjected him to smelling salts. Why?

3. After taking a walk you return home and immediately feel the urge to drink water because you are thirsty. What part of the brain is involved in the urge to drink because of thirst?

4. Ever since the mid-1980s an increasing number of young people have developed Parkinson's disease. The reason has been linked to a "street drug" that had a contaminant that destroyed neurons in the substantia nigra of the mesencephalon. What clinical explanation substantiates the relationship between this street drug and the development of Parkinson's disease?

5. Suppose you are participating in a contest where you are blindfolded and numerous objects are placed in your hands to identify. The person in the contest who identifies the most objects while blindfolded receives a prize. What neural activities and areas in the CNS are necessary in order for you to name the objects correctly and win the prize?

Level
3

15

Integrative Functions

■ Overview

You probably have heard the statement that "No man is an island, no man can stand alone," implying that man is a societal creature, depending on and interacting with other people in the world in which one lives. So it is with the parts of the human nervous system.

The many diverse and complex physical and mental activities require that all component parts of the nervous system function in an integrated and coordinated way if homeostasis is to be maintained within the human body.

Chapters 13 and 14 considered the structure and functions of the spinal cord and the brain. Chapter 15 focuses on how communication processes take place via pathways, nerve tracts, and nuclei that relay sensory and motor information from the spinal cord to the higher centers in the brain. The following exercises are designed to test your understanding of patterns and principles of the brain's sensory, motor, and higher-order functions.

☐ LEVEL 1 REVIEW OF CHAPTER OBJECTIVES

1. Identify the principal sensory and motor pathways.
2. Compare the components, processes, and functions of the pyramidal and extra-pyramidal systems.
3. Explain how we can distinguish between sensations that originate in different areas of the body.
4. Describe the levels of information processing involved in motor control.
5. Discuss how the brain integrates sensory information and coordinates responses.
6. Explain how memories are created, stored, and recalled.
7. Distinguish between the levels of consciousness and unconsciousness, and identify the characteristics of brain activity associated with the different levels of sleep.
8. Describe the alterations in brain function produced by administered drugs.
9. Summarize the effects of aging on the nervous system.

[L1] Multiple Choice:

Place the letter corresponding to the correct answer in the space provided.

OBJ. 1 _____ 1. The three major somatic sensory pathways include the:

    a. first-order, second-order, and third-order pathways

    b. nuclear, cerebellum, and thalamic pathways

    c. posterior column, spinothalamic, and spinocerebellar pathways

    d. anterior, posterior, and lateral pathways

OBJ. 1 _____ 2. The axons of the posterior column ascend within the:

    a. posterior and lateral spinothalamic tracts

    b. fasciculus gracilis and fasciculus cuneatus

    c. posterior and anterior spinocerebellar tracts

    d. a, b, and c are correct

OBJ. 1 _____ 3. The spinothalamic pathway consists of:

    a. lateral and anterior tracts

    b. anterior and posterior tracts

    c. gracilis and cuneatus nuclei

    d. proprioceptors and interneurons

OBJ. 1 _____ 4. The somatic nervous system issues somatic motor commands that direct the:

    a. activities of the autonomic nervous system

    b. contractions of smooth and cardiac muscles

    c. activities of glands and fat cells

    d. contractions of skeletal muscles

OBJ. 1 _____ 5. The autonomic nervous system issues motor commands that control:

    a. the somatic nervous system

    b. smooth and cardiac muscles, glands, and fat cells

    c. contractions of skeletal muscles

    d. voluntary activities

OBJ. 2 _____ 6. The pyramidal system consists of:

    a. rubrospinal and reticulospinal tracts

    b. vestibulospinal and tectospinal tracts

    c. corticobulbar and corticospinal tracts

    d. sensory and motor neurons

OBJ. 2 _____ 7. The extrapyramidal system consists of:

    a. gracilis and cuneatus nuclei

    b. pyramidal, corticobulbar, and corticospinal tracts

    c. sensory and motor fibers

    d. vestibulospinal, tectospinal, rubrospinal, reticulospinal tracts

OBJ. 2   _____   8. The motor commands carried by the tectospinal tracts are triggered by:

a. visual and auditory stimuli

b. stimulation of the red nucleus

c. the vestibular nuclei

d. a, b, and c are correct

OBJ. 2   _____   9. Actions resulting from stimulation in the rubrospinal tract result in:

a. regulation of reflex and autonomic activity

b. involuntary regulation of posture and muscle tone

c. regulation of posture and balance

d. regulation of eye, head, neck, and arm position

OBJ. 3   _____   10. Proprioceptive data from peripheral structures, visual information from the eyes, and equilibrium-related sensations are processed and integrated in the:

a. cerebral cortex

b. cerebral nuclei

c. cerebellum

d. red nucleus

OBJ. 3   _____   11. The integrative activities performed by neurons in the cerebellar cortex and cerebellar nuclei are essential to the:

a. involuntary regulation of posture and muscle tone

b. involuntary regulation of autonomic functions

c. voluntary control of smooth and cardiac muscle

d. precise control of voluntary and involuntary movements

OBJ. 4   _____   12. The primary structural component that delays the relaying of information from one step to another as information is processed is the presence of:

a. receptors

b. synapses

c. effectors

d. neurons

OBJ. 4   _____   13. As one moves from the medulla to the cerebral cortex, the motor patterns become:

a. less complex and less variable

b. less complex and more variable

c. increasingly complex and variable

d. increasingly complex and less variable

OBJ. 4   _____   14. Motor patterns associated with eating and reproduction are controlled by the:

a. cerebral nuclei

b. cerebellum

c. hypothalamus

d. midbrain

OBJ. 4 _____ 15. Motor patterns associated with movements in response to sudden visual or auditory stimuli are controlled by the:

       a. midbrain

       b. cerebellum

       c. cerebral nuclei

       d. thalamus

OBJ. 5 _____ 16. The primary characteristic of higher-order functions is that they are performed by the:

       a. cerebral cortex

       b. thalamus

       c. medulla

       d. cerebellum

OBJ. 5 _____ 17. The general interpretation area present only in the left cerebral hemisphere is:

       a. Broca's area

       b. the premotor cortex

       c. the precentral gyrus

       d. Wernicke's area

OBJ. 5 _____ 18. The speech center regulating patterns of breathing and vocalization needed for normal speech is:

       a. Wernicke's area

       b. Broca's area

       c. the prefrontal cortex

       d. the cerebral nuclei

OBJ. 6 _____ 19. Long-term memories that seem to be part of consciousness, such as your name, are referred to as:

       a. reflexive memories

       b. tertiary memories

       c. secondary memories

       d. primary memories

OBJ. 6 _____ 20. The two components of the limbic system that are essential to memory consolidation are the:

       a. cingulate gyrus and the reticular formations

       b. mamillary bodies and fornix

       c. parahippocampal gyrus and the pineal body

       d. amygdaloid body and the hippocampus

OBJ. 7 _____ 21. When the entire body relaxes and activity at the cerebral cortex is at a minimum, the level of sleep is called:

       a. deep sleep

       b. slow-wave sleep

       c. non-REM sleep

       d. a, b, and c are correct

Level
-1-

OBJ. 8   _____ 22. Compounds that inhibit serotonin production or block its action cause:

        a. exhilaration

        b. severe depression and anxiety

        c. motor problems of Parkinson's disease

        d. schizophrenia

OBJ. 8   _____ 23. Excessive production of dopamine may be associated with pronounced disturbances of:

        a. mood, thought patterns, and behavior

        b. performing voluntary and involuntary movements

        c. intellectual abilities

        d. motor problems of Parkinson's disease

OBJ. 9   _____ 24. A reduction in brain size and weight associated with aging results primarily from:

        a. a decrease in blood flow to the brain

        b. a change in the synaptic organization of the brain

        c. a decrease in the volume of the cerebral cortex

        d. an accumulation of a fibrillar protein

OBJ. 9   _____ 25. Abnormal intracellular and extracellular deposits in CNS neurons associated with the aging process include:

        a. lipofuscin

        b. amyloid

        c. neurofibrillary tangles

        d. a, b, and c are correct

[L1]   Completion:

Using the terms below, complete the following statements.

reflexive	sensation	special senses
pyramids	arteriosclerosis	pyramidal system
declarative	depression	general senses
REM	cerebral nuclei	Parkinson's disease
cerebellum	Alzheimer's disease	reticular activating system
first-order	acetylcholine	

OBJ. 1   26. The information that arrives in the form of an action potential in a sensory fiber is called a _____.

OBJ. 1   27. The sensory neuron that delivers the sensations to the CNS is referred to as a _____ neuron.

OBJ. 2   28. Voluntary control of skeletal muscles is provided by the _____.

OBJ. 2   29. The corticospinal tracts are visible along the ventral surface of the medulla as a pair of thick bands known as the _____.

OBJ. 3   30. The processing centers whose functions blur the distinctions between the "conscious" and "unconscious" motor control are the _____.

OBJ. 4   31. Cerebral nuclei stimulate motor neurons by releasing _____.

OBJ. 4   32. Motor patterns associated with learned movement patterns are controlled by the _____.

OBJ. 5 33. Touch, pressure, vibration, pain, and temperature are examples of _____.

OBJ. 5 34. Sight, smell, taste, and hearing are examples of _____.

OBJ. 6 35. Memories that can be voluntarily retrieved and verbally expressed are called _____ memories.

OBJ. 6 36. Memories that are not voluntarily accessible are called _____.

OBJ. 7 37. The poorly defined network formation that extends from the hypothalamus to the medulla is the _____.

OBJ. 7 38. Active dreaming accompanied by alterations in blood pressure and respiratory rates constitutes sleep known as _____.

OBJ. 8 39. Inadequate dopamine production causes the motor problems of _____.

OBJ. 8 40. Drugs that depress the release of norepinephrine cause _____.

OBJ. 9 41. A progressive disorder characterized by the loss of higher cerebral functions is _____.

OBJ. 9 42. As aging occurs, fatty deposits accumulate in the blood vessels reducing the rate of flow in the arteries and resulting in a condition known as _____.

[L1] Matching:

Match the terms in column B with the terms in column A. Use letters for answers in the spaces provided.

PART I

		COLUMN A		COLUMN B
OBJ. 1	____ 43.	interneuron	A.	speech center
OBJ. 1	____ 44.	cross-over	B.	voluntary motor control
OBJ. 2	____ 45.	pyramidal system	C.	reticular formation
OBJ. 2	____ 46.	extrapyramidal system	D.	indicates presence of brain tumor
OBJ. 3	____ 47.	rubospinal tract	E.	second-order neuron
OBJ. 3	____ 48.	reticulospinal tract	F.	involuntary motor control
OBJ. 4	____ 49.	delta waves	G.	red nucleus
OBJ. 4	____ 50.	theta waves	H.	decussate
OBJ. 5	____ 51.	broca's area	I.	low-frequency waves

PART II

		COLUMN A		COLUMN B
OBJ. 6	____ 52.	fact memory	J.	learned motor behavior
OBJ. 6	____ 53.	skill memory	K.	neurotransmitter
OBJ. 7	____ 54.	norepinephrine	L.	promotes deep sleep
OBJ. 7	____ 55.	serotonin	M.	sedative
OBJ. 8	____ 56.	acetylcholine	N.	awake, alert state
OBJ. 8	____ 57.	valium	O.	color of a stop sign
OBJ. 9	____ 58.	lipofusion	P.	amyloid
OBJ. 9	____ 59.	plaques	Q.	granular pigment

Level
—1—

[L1]   Drawing/Illustration Labeling:

Identify each numbered structure by labeling the following figures:

OBJ. 1   **FIGURE 15.1**   Ascending and Descending Tracts of the Spinal Cord

Select your answers from the following choices:

Posterior spinocerebellar   Anterior corticospinal
Tectospinal   Anterior spinocerebellar
Lateral corticospinal   Vestibulospinal
Anterior spinothalamic   Rubrospinal
Fasciculus gracilis   Lateral spinothalamic
Reticulospinal   Fasciculus cuneatus

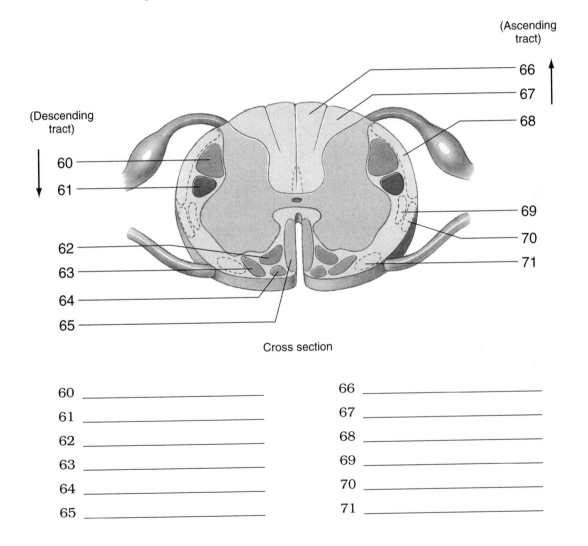

Cross section

60 _____	66 _____
61 _____	67 _____
62 _____	68 _____
63 _____	69 _____
64 _____	70 _____
65 _____	71 _____

□ LEVEL 2   CONCEPT SYNTHESIS

Concept Map I:

Using the following terms, fill in the circled, numbered, blank spaces to complete the concept map. Follow the numbers that comply with the organization of the map.

Spinothalamic
Fasciculus cuneatus
Posterior column
Medulla (nucleus gracilis)

Posterior tract
Thalamus (ventral nuclei)
Proprioceptors

Anterior tract
Interneurons
Cerebellum

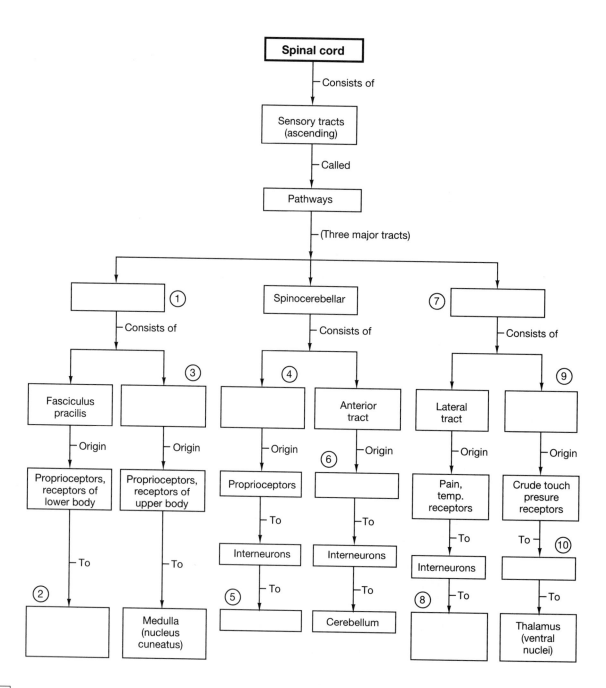

Concept Map II:

Using the following terms, fill in the circled, numbered, blank spaces to complete the concept map. Follow the numbers that comply with the organization of the map.

Primary motor cortex
Lateral corticospinal tracts
Balance, muscle tone
Pyramidal tracts

Tectum (midbrain)
Reticular formation
  (brain stem)
Vestibulospinal tract

Rubrospinal tract
Posture, muscle tone
Voluntary motor control
  of skeletal muscles

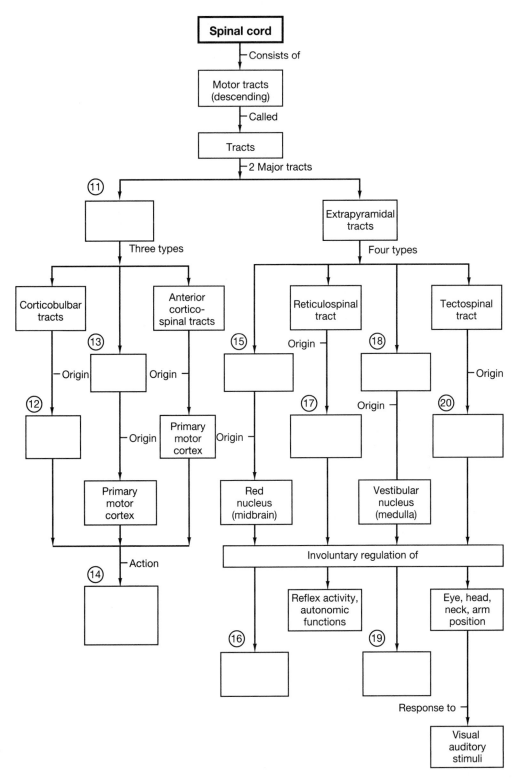

Concept Map III:

Using the following terms, fill in the circled, numbered, blank spaces to complete the concept map. Follow the numbers that comply with the organization of the map.

Auditory information   Equilibrium sensations
Cerebrum   Superior Colliculi
Reticular formation   Thalamus
Sensory information   Red Nucleus

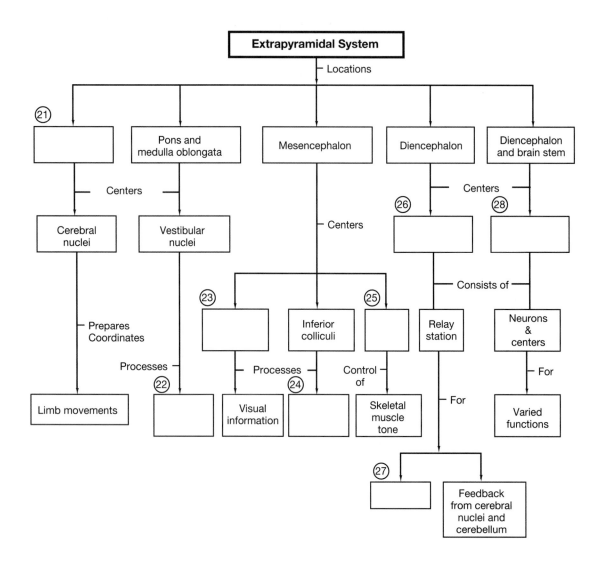

Body Trek:

Using the terms below, fill in the blanks to complete the trek through the spinal cord on the spinothalamic tract.

cerebral cortex	synapse	pons
primary	somesthetic	receptor
medulla	posterior horn	midbrain
tertiary	diencephalon	secondary
commissural	pyramidal	thalamus
dorsal root ganglion		

Robo is activated and programmed to trek along an ascending tract in the spinal cord while "carrying" an impulse that is a result of a stimulus applied to the left hand of a subject. The robot is injected into the spinal canal and immediately begins the trek by proceeding to the synaptic area of the primary neuron in the posterior horn of the spinal cord.

The stimulus is applied at a (29)_____ area on the left hand of the subject. The impulse is transmitted on the primary neuron into the (30)_____ and relayed into the (31)_____ of the spinal cord. Robo "picks up" the impulse at the synapse between the (32)_____ neuron and the (33)_____ neuron. The impulse is "carried" from the left side of the spinal cord to the right side across the (34)_____ fibers in the anterior of the gray and white matter. Continuing on this postsynaptic pathway Robo carries the impulse "uphill" as it ascends through the entrance to the brain, the (35)_____ , on into the (36)_____ and into the (37)_____, finally arriving in the (38)_____ where the robot rushes the impulse into the (39)_____ , an important relay "station." After a brief pause at a (40)_____ , Robo proceeds farther "uphill" on a (41)_____ neuron to the (42)_____ center of the (43)_____ where the determination is made that the stimulus caused PAIN! (Ouch!) Robo is anxious to get out of the "pain area" because the "pain" interferes with its electronic abilities to operate normally. With its extended ultra-micro-arm it grasps onto an upper motor neuron and travels "downhill" until it can complete its trek on a lower motor neuron of the (44)_____ system into the spinal cord where it proceeds to the spinal canal for its exit during the next spinal tap.

[L2]   Multiple Choice:

Place the letter corresponding to the correct answer in the space provided.

_____ 45. If a tract name begins with *spino-* it must start in the:

 a. brain and end in the PNS, bearing motor commands

 b. spinal cord and end in the brain, bearing motor commands

 c. spinal cord and end in the PNS, carrying sensory information

 d. spinal cord and end in the brain, carrying sensory information

_____ 46. If the name of a tract ends in *-spinal* its axons must:

 a. start in the higher centers and end in the spinal cord, bearing motor commands

 b. start in the higher centers and end in the spinal cord, bearing sensory information

 c. start in the spinal cord and end in the brain, carrying sensory information

 d. start in the spinal cord and end in the PNS, carrying motor commands

Level
=2=

_____ 47. The posterior column pathway receives sensations associated with:

     a. pain and temperature

     b. crude touch and pressure

     c. highly localized fine touch, pressure, vibration, and position

     d. a, b, and c are correct

_____ 48. In the thalamus, data arriving over the posterior column pathway are integrated, sorted, and projected to the:

     a. cerebellum

     b. spinal cord

     c. primary sensory cortex

     d. peripheral nervous system (PNS)

_____ 49. If a sensation arrives at the wrong part of the sensory cortex we will:

     a. experience pain in the posterior column pathway

     b. reach an improper conclusion about the source of the stimulus

     c. be incapable of experiencing pain or pressure

     d. lose all capability of receiving and sending information

_____ 50. If the central cortex were damaged or the projection fibers cut, a person would be able to detect light touch but would be unable to:

     a. determine its source

     b. determine the magnitude of the stimulus

     c. determine the amount of pressure

     d. a, b, and c are correct

_____ 51. The spinothalamic pathway relays impulses associated with:

     a. "fine" touch, pressure, vibration, and proprioception

     b. "crude" sensations of touch, pressure, pain, and temperature

     c. the position of muscles, tendons, and joints

     d. proprioceptive information and vibrations

_____ 52. The spinocerebellar pathway includes:

     a. lateral and anterior tracts

     b. gracilis and cuneatus nuclei

     c. lateral and posterior tracts

     d. anterior and posterior tracts

_____ 53. The spinocerebellar pathway carries information concerning the:

     a. pressure on each side of the body

     b. sensations that cause referred and phantom limb pain

     c. position of muscles, tendons, and joints to the cerebellum

     d. sensations of touch, pain, and temperature

Level
=2=

_____ 54. Somatic motor pathways always involve a (an):

      a. upper and lower motor neuron

      b. ganglionic and preganglionic neuron

      c. sensory and motor fibers

      d. anterior and lateral nuclei

_____ 55. Voluntary and involuntary somatic motor commands issued by the brain reach peripheral targets by traveling over the:

      a. sensory and motor fibers

      b. pyramidal and extrapyramidal systems

      c. ganglionic and preganglionic fibers

      d. spinothalamic tracts

_____ 56. The primary goal of the vestibular nuclei is the:

      a. monitoring of muscle tone throughout the body

      b. triggering of visual and auditory stimuli

      c. maintenance of posture and balance

      d. control of involuntary eye movements

_____ 57. The reticulospinal tract is involved with regulation of:

      a. involuntary reflex activity and autonomic functions

      b. involuntary regulation of balance and posture

      c. balance and muscle tone

      d. voluntary motor control of skeletal muscles

_____ 58. An individual whose primary motor cortex has been destroyed retains the ability to walk and maintain balance but the movements:

      a. are restricted and result in partial paralysis

      b. are under involuntary control and are poorly executed

      c. are characteristic involuntary motor commands

      d. lack precision and are awkward and poorly controlled

_____ 59. Because the background activity of cerebral nuclei is inhibition, the cerebral nuclei:

      a. provide a pattern and rhythm for movement

      b. initiate specific movements

      c. exert direct control over CNS motor neurons

      d. initiate a generalized state of muscular contraction

_____ 60. When someone touches a hot stove, the rapid, automatic preprogrammed response that preserves homeostasis is provided by the:

      a. cerebral cortex

      b. primary sensory cortex

      c. spinal reflex

      d. cerebellum

_____ 61. At the highest level of processing, the complex, variable, and voluntary motor patterns are dictated by the:

      a. cerebral cortex

      b. frontal lobe

      c. cerebellum

      d. diencephalon

Level
=2=

_____ 62. The brain grows in size and complexity until at least:

      a. age 1

      b. age 4

      c. age 12

      d. age 18

_____ 63. The brain waves typical of individuals under stress or in states of psychological tension are:

      a. alpha waves

      b. beta waves

      c. theta waves

      d. delta waves

_____ 64. The brain waves found in the brains of normal resting adults when their eyes are closed are:

      a. beta waves

      b. theta waves

      c. delta waves

      d. alpha waves

_____ 65. An individual who can understand language and knows how to respond, but lacks the motor control necessary to produce the right combinations of sounds, has:

      a. global aphasia

      b. developmental dyslexia

      c. major motor aphasia

      d. disconnection syndrome

_____ 66. If connections between the prefrontal cortex and other brain regions are severed:

      a. vocalizations for normal speech are impossible

      b. tensions, frustrations, and anxieties are removed

      c. letters are often reversed or written in wrong order

      d. the individual would experience hallucinations and severe pain

_____ 67. Hemispheric specialization does not mean that the two hemispheres are independent but that:

      a. there is a genetic basis for the distribution of functions

      b. the left hemisphere is categorized in left-handed individuals

      c. 90 percent of the population have an enlarged right hemisphere at birth

      d. specific centers have evolved to process information gathered by the system as a whole

_____ 68. Two neurotransmitters that appear to be involved in the regulation of asleep-awake cycles are:

      a. serotonin and norepinephrine

      b. acetylcholine and dopamine

      c. histamine and GABA

      d. endorphins and enkephalins

_____ 69. An EEG pattern during REM sleep resembles the types of waves typical of awake adults known as:

  a. alpha waves

  b. delta waves

  c. theta waves

  d. beta waves

[L2]   Completion:

Using the terms below, complete the following statements.

cerebellum	anencephaly	global aphasia
aphasia	dopamine	ascending
GABA	medial lemniscus	homunculus
arousal	cerebral nuclei	electroencephalogram
dyslexia	thalamus	descending

70. The vestibulospinal tracts in the spinal cord consist of _____ fibers.

71. The spinothalamic tracts in the spinal cord consist of _____ fibers.

72. Cerebral nuclei inhibit motor neurons by releasing _____.

73. A rare condition in which the brain fails to develop at levels above the mesencephalon or lower diencephalon is called _____.

74. Motor patterns associated with walking and body positioning are controlled by _____.

75. One of the most common methods involved in monitoring the electrical activity of the brain is the _____.

76. Extensive damage to the general interpretive area or to the associated sensory tracts results in _____.

77. A disorder affecting the ability to speak or read is _____.

78. A disorder affecting the comprehension and use of words is referred to as _____.

79. The tract that relays information from the nuclei gracilis and cuneatus to the thalamus is the _____.

80. One of the functions of the reticular formation is _____.

81. Parkinson's disease is associated with inadequate production of the neurotransmitter _____.

82. The destination of the impulses that travel along the spinocerebellar pathways is the _____.

83. The destination of the impulses that travel along the spinothalamic pathways is the _____.

84. A sensory map created by electricity stimulating the cortical surface is called a sensory _____.

[L2]   Short Essay:

Briefly answer the following questions in the spaces provided below.

85. What are the three major somatic sensory pathways in the body?

Level
=2=

86. What motor pathways make up the pyramidal system?

87. What motor pathways make up the extrapyramidal system?

88. What is the primary functional difference between the pyramidal and extrapyramidal system?

89. What are the two (2) primary functions of the cerebellum?

90. What are the primary anatomical factors that contribute to the maturation of the cerebral and cerebellar cortex?

91. What behavior characteristics produce (a) alpha waves, (b) beta waves, (c) theta waves, and (d) delta waves?

92. What are the four (4) primary characteristics of higher-order functions?

93. What names are given for the cortical areas where general interpretation and speech are located?

94. What functions serve to identify the left cerebral hemisphere as the *categorical* hemisphere and what functions serve to identify the right cerebral hemisphere as the *representational* hemisphere?

95. What three (3) cellular mechanisms are thought to be involved in memory formation and storage?

96. List four (4) of the anatomical changes in the nervous system that are commonly associated with the aging process.

When you have successfully completed the exercises in L2 proceed to L3.

☐ LEVEL 3   CRITICAL THINKING/APPLICATION

Using principles and concepts learned about the nervous system, answer the following questions. Write your answers on a separate sheet of paper.

1. A friend of yours complains of pain radiating down through the left arm. After having an EKG it was discovered that there were signs of a mild heart attack. Why would the pain occur in the arm instead of the cardiac area?

2. As a result of an accident, axons in the descending motor tract are severed. This causes a reappearance of the Babinski reflex and the disappearance of the abdominal reflex. Why does the Babinski reflex reappear and the abdominal reflex disappear?

3. Cerebral nuclei do not exert direct control over CNS motor neurons; however, they do adjust the motor commands issued in other processing centers. What three major pathways do they use to accomplish this?

4. An individual whose primary motor cortex has been destroyed retains the ability to walk, maintain balance, and perform other voluntary and involuntary movements. Even though the movements lack precision and may be awkard and poorly controlled, why is the ability to walk and maintain balance possible?

5. The proud parents of a newborn take the infant home and experience all the joy of watching the baby suckle, stretch, yawn, cry, kick, track movements with the eyes, and stick its fingers in its mouth. Suddenly, after three months, the baby dies for no *apparent* reasons. What do you suspect and how do you explain the suspicious cause of death?

6. You are assigned the task of making a diagnosis from an EEG that reveals the presence of very large amplitude, low-frequency delta waves. What possibilities would you cite that are associated with these abnormal wave patterns?

7. Your parents have agreed to let your grandmother live in their home because recently she has exhibited some "abnormal" behaviors usually associated with Alzheimer's disease. What behaviors would you look for to confirm the possibility of Alzheimer's?

Level
=**3**=

16

Autonomic Nervous System

■ Overview

Can you imagine what life would be like if you were responsible for coordinating the involuntary activities that occur in the cardiovascular, respiratory, digestive, excretory, and reproductive systems? Your entire life would be consumed with monitoring and giving instructions to maintain homeostasis in the body. The autonomic nervous system (ANS), a division of the peripheral nervous system, does this involuntary work in the body with meticulous efficiency.

The ANS consists of efferent motor fibers that innervate smooth and cardiac muscle as well as glands. The controlling centers of the ANS are found in the brain, whereas the nerve fibers of the ANS, which are subdivided into the sympathetic and parasympathetic fibers, belong to the peripheral nervous system. The sympathetic nervous system originates in the thoracic and lumbar regions of the spinal cord, whereas the parasympathetic nervous system originates in the brain and sacral region of the spinal cord. Both divisions terminate on the smooth muscle, cardiac muscle, or glandular tissue of numerous organs.

The primary focus of Chapter 16 is on the structural and functional properties of the sympathetic and parasympathetic divisions of the ANS. Integration and control of autonomic functions are included to show how the ANS coordinates involuntary activities throughout the body to make homeostatic adjustments necessary for survival.

☐ LEVEL 1 REVIEW OF CHAPTER OBJECTIVES

1. Compare the autonomic nervous system with the other divisions of the nervous system.
2. Contrast the structures and functions of the sympathetic and parasympathetic divisions of the autonomic nervous system.
3. Describe the mechanisms of neurotransmitter release in the autonomic nervous system.
4. Compare the effects of the various autonomic neurotransmitters on target organs and tissues.
5. Discuss the relationship between the divisions of the autonomic nervous system and the significance of dual innervation.
6. Explain the importance of autonomic tone.
7. Describe the hierarchy of interacting levels of control in the autonomic nervous system.
8. Describe the effects of mimetic drugs on autonomic nervous system activities.

[L1]   Multiple Choice:

Place the letter corresponding to the correct answer in the space provided.

OBJ. 1   _____   1. The autonomic nervous system is a subdivision of the:

>   a. central nervous system
>   b. somatic nervous system
>   c. peripheral nervous system
>   d. visceral nervous system

OBJ. 1   _____   2. The lower motor neurons of the somatic nervous system (SNS) exert direct control over skeletal muscles. By contrast, in the ANS there is:

>   a. voluntary and involuntary control of skeletal muscles
>   b. always voluntary control of skeletal muscles
>   c. indirect voluntary control of skeletal muscles
>   d. a synapse interposed between the CNS and the peripheral effector

OBJ. 2   _____   3. Sympathetic (first-order) neurons originate between:

>   a. T_1 and T_2 of the spinal cord
>   b. T_1 and L_2 of the spinal cord
>   c. L_1 and L_4 of the spinal cord
>   d. S_2 and S_4 of the spinal cord

OBJ. 2   _____   4. The nerve bundle that carries preganglionic fibers to a nearby sympathetic chain ganglion is the:

>   a. collateral ganglion
>   b. autonomic nerve
>   c. gray ramus
>   d. white ramus

OBJ. 2   _____   5. The important function(s) of the postganglionic fibers that enter the thoracic cavity in autonomic nerves include:

>   a. accelerating the heart rate
>   b. increasing the force of cardiac contractions
>   c. dilating the respiratory passageways
>   d. a, b, and c are correct

OBJ. 2   _____   6. Preganglionic (first-order) neurons in the parasympathetic division of the ANS originate in the:

>   a. peripheral ganglia adjacent to the target organ
>   b. thoracolumbar area of the spinal cord
>   c. walls of the target organ
>   d. brain stem and sacral segments of the spinal cord

OBJ. 2   _____   7. Since second-order neurons in the parasympathetic division are all located in the same ganglion, the effects of parasympathetic stimulation are:

        a. more diversified and less localized than those of the sympathetic division.

        b. less diversified but less localized than those of the sympathetic division

        c. more specific and localized than those of the sympathetic division

        d. more diversified and more localized than those of the sympathetic division

OBJ. 2   _____   8. The parasympathetic division of the ANS includes visceral motor nuclei associated with cranial nerves:

        a. I, II, III, IV

        b. III, VII, IX, X

        c. IV, V, VI, VIII

        d. V, VI, VIII, XII

OBJ. 2   _____   9. In the parasympathetic division, second-order neurons originate in:

        a. intramural ganglia or ganglia associated with the target organs

        b. collateral ganglia or chain ganglia

        c. the adrenal medulla and collateral ganglia

        d. spinal segments or cranial nerves

OBJ. 3   _____   10. The effect of modified neurons in the sympathetic division which secrete neurotransmitters is that they:

        a. do not last as long as those produced by direct sympathetic stimulation

        b. are limited to peripheral tissues and CNS activity

        c. serve to dilate blood vessels and elevate blood pressure

        d. last longer than those produced by direct sympathetic stimulation

OBJ. 3   _____   11. At their synapses with ganglionic neurons, all ganglionic neurons in the sympathetic division release:

        a. epinephrine

        b. norepinephrine

        c. acetylcholine

        d. dopamine

OBJ. 3   _____   12. At neuroeffector junctions, typical sympathetic postganglionic fibers release:

        a. epinephrine

        b. norepinephrine

        c. acetylcholine

        d. dopamine

Level
-1-

OBJ. 3   _____ 13. At synapses and neuroeffector junctions, all preganglionic and postganglionic fibers in the parasympathetic division release:

        a. epinephrine

        b. norepinephrine

        c. acetylcholine

        d. a, b, and c are correct

OBJ. 4   _____ 14. When specific neurotransmitters are released by postganglionic fibers, stimulation or inhibition of activity depends on:

        a. the target organ that is affected

        b. the arrangement of the postganglionic fibers

        c. the rate at which the neurotransmitter is released to the receptor

        d. the response of the membrane receptor to the presence of the neurotransmitter

OBJ. 4   _____ 15. The sympathetic division can change tissue and organ activities by:

        a. releasing norepinephrine at peripheral synapses

        b. distribution of norepinephrine by the bloodstream

        c. distribution of epinephrine by the bloodstream

        d. a, b, and c are correct

OBJ. 4   _____ 16. Cholinergic postganglionic sympathetic fibers which innervate the sweat glands of the skin and the blood vessels of the skeletal muscles are stimulated during exercise to:

        a. constrict the blood vessels and inhibit sweat gland secretion

        b. keep the body cool and provide oxygen and nutrients to active skeletal muscles

        c. increase the smooth muscle activity in the digestive tract for better digestion

        d. decrease the body temperature and decrease the pH in the blood

OBJ. 5   _____ 17. When vital organs receive dual innervation, the result is usually:

        a. a stimulatory effect within the organ

        b. an inhibitory effect within the organ

        c. sympathetic-parasympathetic opposition

        d. sympathetic-parasympathetic inhibitory effect

OBJ. 5   _____ 18. In the parasympathetic division postganglionic fibers release transmitter substance to:

        a. target organs

        b. target cells

        c. postganglionic fibers

        d. special receptor surfaces

OBJ. 5   _____ 19. Autonomic tone exists in the heart because:

    a. ACh (acetylcholine) released by the parasympathetic division decreases the heart rate, and norepinephrine (NE) released by the sympathetic division accelerates the heart rate

    b. ACh released by the parasympathetic division accelerates the heart rate, and NE released by the sympathetic division decreases the heart rate

    c. NE released by the parasympathetic division accelerates the heart rate, and ACh released by the sympathetic division decreases the heart rate

    d. NE released by the sympathetic division and ACh released by the parasympathetic division accelerate the heart rate

OBJ. 6   _____ 20. In the absence of stimuli, autonomic tone is important to autonomic motor neurons because:

    a. it can increase their activity on demand

    b. it can decrease their activity to avoid overstimulation

    c. it shows a resting level of spontaneous activity

    d. its activity is determined by the degree to which they are stimulated

OBJ. 7   _____ 21. The lowest level of integration in the ANS consists of:

    a. centers in the medulla that control visceral functions

    b. regulatory centers in the posterior and lateral hypothalamus

    c. regulatory centers in the midbrain that control the viscera

    d. lower motor neurons that participate in cranial and spinal visceral reflexes

OBJ. 7   _____ 22. Both coordination and regulation of sympathetic function generally occur in centers in the:

    a. cerebral cortex

    b. medulla and spinal cord

    c. posterior and lateral hypothalamus

    d. cerebellum

OBJ. 7   _____ 23. Control of parasympathetic coordination and regulation generally occurs in centers in the:

    a. pons and the medulla oblongata

    b. anterior and medial hypothalamus

    c. cerebellum and the cerebral cortex

    d. medulla oblongata and spinal cord

OBJ. 8   _____ 24. Drugs that reduce the effects of autonomic stimulation by keeping the neurotransmitter from affecting the postsynaptic membranes are known as:

    a. reducing agents

    b. blocking agents

    c. receptor agents

    d. autonomic agents

Level
–1–

[L1]   Completion:

Using the terms below, complete the following statements.

visceral reflexes	acetylcholine	"fight or flight"
involuntary	sympathomimetic	excitatory
norepinephrine	opposing	epinephrine
limbic	autonomic tone	"rest and repose"
synapse		

OBJ. 1   25. The point at which information is transmitted from one excitable cell to another is the _____.

OBJ. 1   26. The lower motor neurons of the SNS exert voluntary control over skeletal muscles, while in the ANS the control of smooth and cardiac muscle is _____.

OBJ. 2   27. Because the sympathetic division of the ANS stimulates tissue metabolism and increases alertness, it is called the _____ subdivision.

OBJ. 2   28. Because the parasympathetic division of the ANS conserves energy and promotes sedentary activity, it is known as the _____ subdivision.

OBJ. 3   29. At their synaptic terminals, cholinergic, preganglionic autonomic fibers release _____.

OBJ. 3   30. The two neurotransmitters released into circulation by sympathetic stimulation of the adrenal medulla are norepinephrine and _____.

OBJ. 4   31. Preganglionic fibers that release acetylcholine are always _____.

OBJ. 4   32. The neurotransmitter that is released by sympathetic postganglionic fibers and that causes constriction of most peripheral arteries is _____.

OBJ. 5   33. Sympathetic-parasympathetic innervation in an organ produces action that is described as _____.

OBJ. 6   34. The release of small amounts of acetylcholine and norepinephrine on a continual basis in an organ innervated by sympathetic and parasympathetic fibers is referred to as _____.

OBJ. 7   35. The simplest functional units in the ANS are _____.

OBJ. 7   36. The system (area) in the brain that would most likely exert an influence on autonomic control if an emotional condition is present is the _____ system.

OBJ. 8   37. Drugs that selectively target B receptors causing an increase in heart rate and blood pressure are _____.

[L1]   Matching:

Match the terms in column B with the terms in column A. Use letters for answers in the spaces provided.

	COLUMN A		COLUMN B
OBJ. 1	____ 38. first-order neurons	A.	nicotinic, muscarinic
OBJ. 1	____ 39. second-order neurons	B.	defecation and urination
OBJ. 2	____ 40. thoracolumbar axons and ganglia	C.	preganglionic neurons
OBJ. 2	____ 41. craniosacral axons and ganglia	D.	parasympathetic division
OBJ. 3	____ 42. cholinergic receptors	E.	autonomic tone
OBJ. 3	____ 43. adrenergic receptors	F.	constricts blood vessels
OBJ. 4	____ 44. acetylcholine	G.	postganglionic neurons
OBJ. 4	____ 45. norepinephrine	H.	pupillary dilation
OBJ. 5	____ 46. sympathetic division	I.	stimulates metabolism
OBJ. 6	____ 47. sympathetic-parasympathetic opposition	J.	sympathetic division
		K.	alpha and beta
OBJ. 7	____ 48. parasympathetic reflex	L.	decrease heart rate
OBJ. 7	____ 49. sympathetic reflex	M.	dilates blood vessels
OBJ. 8	____ 50. beta-blocker		

[L1] Drawing/Illustration Labeling:

Identify each numbered structure by labeling the following figures:

OBJ. 2   *FIGURE 16.1*   Preganglionic and Postganglionic Cell Bodies
of Sympathetic Neurons

Using the terms below, identify and label the numbered structures of the sympathetic system.

inferior mesenteric ganglion	postganglionic neuron	symphatic
preganglionic axon	superior mesenteric ganglion	T_1
splanchnic nerve	spinal cord	L_1
sympathetic chain ganglion	celiac ganglion	
autonomic nerves	superior cervical ganglion	

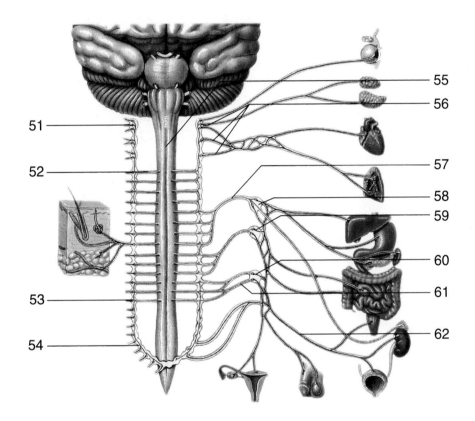

51
52
53
54

55
56
57
58
59
60
61
62

63 ———————————— (Division of ANS)

51 _____   58 _____

52 _____   59 _____

53 _____   60 _____

54 _____   61 _____

55 _____   62 _____

56 _____   63 _____

57 _____

Level
-1-

OBJ. 2 **FIGURE 16.2** Preganglionic and Postganglionic Cell Bodies
of Parasympathetic Neurons

Using the terms below, identify and label the numbered structures of the parasympathetic system.

ciliary ganglion	postganglionic axon	midbrain
preganglionic axon	pterygopalatine ganglion	medulla
otic ganglion	submandibular ganglion	N IX
N III	S_2	N X
N VII	S_4	

64 _____	71 _____
65 _____	72 _____
66 _____	73 _____
67 _____	74 _____
68 _____	75 _____
69 _____	76 _____
70 _____	77 _____

Level
-1-
When you have successfully completed the exercises in L1 proceed to L2.

☐ LEVEL 2   CONCEPT SYNTHESIS

Concept Map I:

Using the following terms, fill in the circled, numbered, blank spaces to complete the concept map. Follow the numbers that comply with the organization of the map.

Ganglia outside CNS   Motor neurons   Sympathetic
Smooth muscle   Postganglionic   First-order neurons

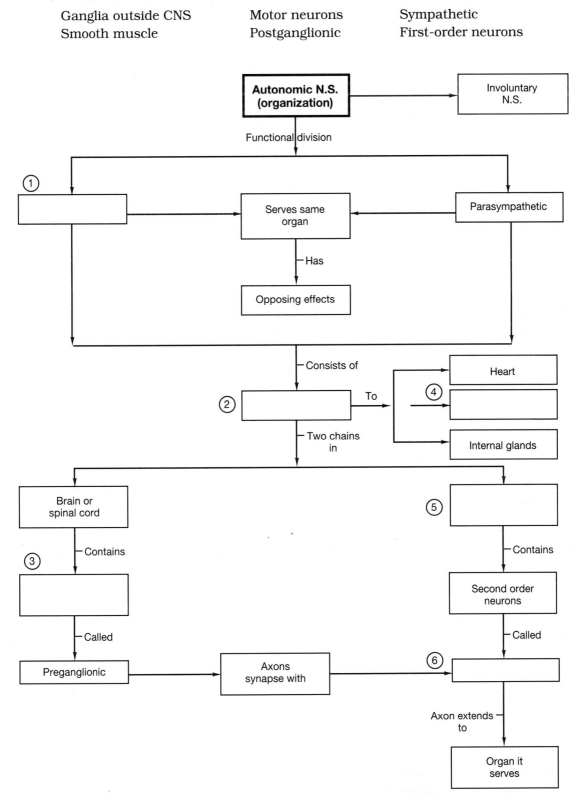

Concept Map II:

Using the following terms, fill in the circled, numbered, blank spaces to complete the concept map. Follow the numbers that comply with the organization of the map.

Adrenal medulla (paired)
Second-order neurons (postganglionic)
Thoracolumbar
Sympathetic chain of ganglia (paired)

Spinal segments $T_1 - L_2$
Visceral effectors
General circulation

Concept Map III:

Using the following terms, fill in the circled, numbered, blank spaces to complete the concept map. Follow the numbers that comply with the organization of the map.

Lower abdominopelvic cavity
Nasal, tear, salivary glands
Craniosacral
Segments $S_2 - S_4$

Otic ganglia
Intramural ganglia
Ciliary ganglion

N X
N VII
Brain stem

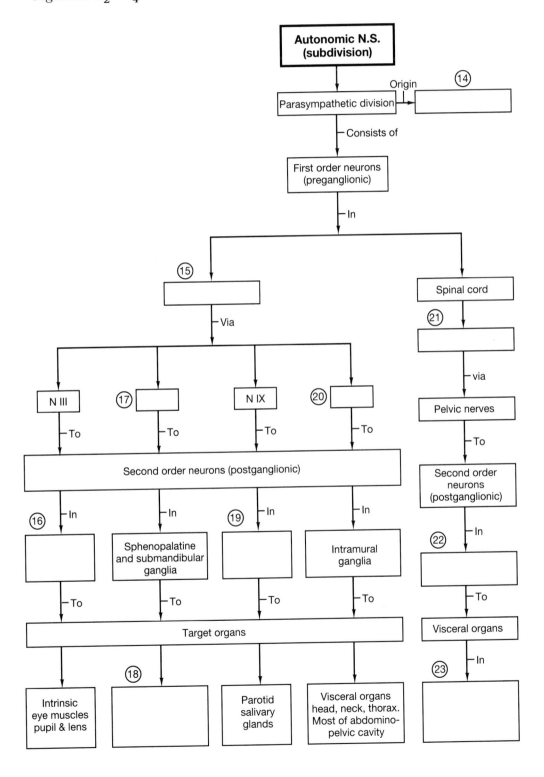

Concept Map IV:

Using the following terms, fill in the circled, numbered, blank spaces to complete the concept map. Follow the numbers which comply with the organization of the concept map.

Respiratory

Pons

Thalamus

Sympathetic visceral reflexes

Parasympathetic visceral reflexes

Spinal cord $T_1 - L_2$

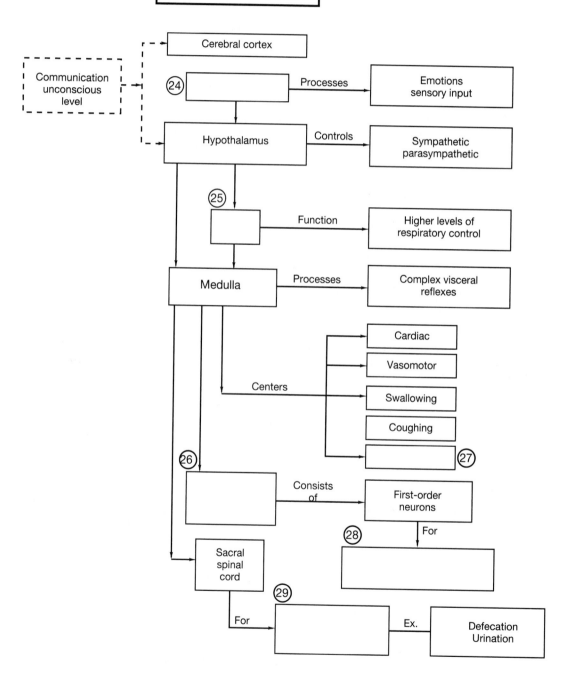

Body Trek:

Using the terms below, fill in the blanks to complete the trek through the spinal cord on the spinothalamic tract.

synapse   target organ   postganglionic
epinephrine   axon   sympathetic chain ganglion
circulation   collateral   white ramus
splanchnic   spinal nerve   abdominopelvic
gray ramus   gray matter

Robo is activated and equipped with an electrical detection device for a trek through the sympathetic nervous system. The robot will monitor the neuronal pathways of action potentials as they are transmitted through the sympathetic division. After the robot is injected into the spinal canal, the current of the CSF will move it to the T_1 position of the spinal column. From this location the electronic tracking of the first action potential will take place.

Robo's electrical sensor "picks up" the action potential (A.P.) in the cell body of the sympathetic preganglionic neuron located in the lateral horn of the (30) _____ of the spinal cord. The A.P. is carried away from the cell body by way of the (31) _____, which passes through the ventral root of the (32) _____ located at T_1. At this point it projects to the (33) _____ on either side of the vertebral column. The tracking device follows the A.P. in the axon through a short connection between the spinal nerve and the sympathetic chain ganglia called the (34) _____ communican. After entering the sympathetic chain ganglia, the preganglionic axon synapses with the (35) _____ neuron. Robo detects the A.P. or the postganglionic neuron as it passes through the (36) _____ communicans and then enters the spinal nerve or projects through small nerves to the (37) _____. Robo descends to the T_5 level of the spinal cord to "track" a second A.P., which travels a different route in the sympathetic division. Robo "picks up" the A.P. on an axon as it exits the chain ganglia. Without a delay at a synapse, it continues on to the (38) _____ nerve to a (39) _____ ganglion. There it will be delayed because of a (40) _____ preceding the postganglionic neuron, which eventually innervates the viscera of the (41) _____ cavity.

Robo's trek continues as the micro-robot descends to the T_9 level where it gets in position to monitor the A.P. in a preganglionic axon, which extends to the adrenal medulla. In the adrenal medulla, the A.P. will continue on a postsynaptic ganglionic neuron that secretes the neurotransmitters norepinephrine and (42) _____ into the general (43) _____.

From its location at T_9 Robo treks a short distance to the spinal canal where it will be removed during the next spinal tap and return to Mission Control.

[L2]   Multiple Choice:

Place the letter corresponding to the correct answer in the space provided.

_____ 44. The axon of a ganglionic neuron is called a postganglionic fiber because:

        a. it carries impulses away from the ganglion

        b. it carries impulses to the target organ

        c. it carries impulses toward the ganglion

        d. a, b, and c are correct

_____ 45. The effect(s) produced by sympathetic postganglionic fibers in spinal nerves include(s):

        a. stimulation of secretion by sweat glands

        b. acceleration of blood flow to skeletal muscles

        c. dilation of the pupils and focusing of the eyes

        d. a, b, and c are correct

Level =2=

_____ 46. The summary effects of the collateral ganglia include:

      a. accelerating the heart rate and dilation of respiratory passageways

      b. redirection of blood flow and energy use by visceral organs and release of stored energy

      c. dilation of the pupils and focusing of the eyes

      d. acceleration of blood flow and stimulation of energy production

_____ 47. Stimulation of alpha-1 receptors on a target cell triggers a:

      a. depolarization that has an inhibitory effect

      b. hyperpolarization that results in a refractory period

      c. depolarization that has an excitatory effect

      d. repolarization that has an inhibitory effect

_____ 48. The major structural difference between sympathetic preganglionic and ganglionic fibers is that:

      a. preganglionic fibers are short and postganglionic fibers are long

      b. preganglionic fibers are long and postganglionic fibers are short

      c. preganglionic fibers are close to target organs and postganglionic fibers are close to the spinal cord

      d. preganglionic fibers innervate target organs while postganglionic fibers originate from cranial nerves

_____ 49. The parasympathetic division innervates areas serviced by the cranial nerves and innervates:

      a. general circulation

      b. collateral ganglia

      c. the adrenal medulla

      d. organs in the thoracic and abdominopelvic cavities

_____ 50. The effects of parasympathetic stimulation are usually:

      a. brief in duration and restricted to specific organs and sites

      b. long in duration and diverse in distribution

      c. brief in duration and diverse in distribution

      d. long in duration and restricted to specific organs and sites

_____ 51. The functions of the parasympathetic division center on:

      a. accelerating the heart rate and the force of contraction

      b. dilation of the respiratory passageways

      c. relaxation, food processing, and energy absorption

      d. a, b, and c are correct

_____ 52. During a crisis, the event necessary for the individual to cope with stressful and potentially dangerous situations is called:

      a. splanchnic innervation

      b. sympathetic activation

      c. parasympathetic activation

      d. the effector response

Level =2=

_____ 53. The two classes of sympathetic receptors include:

      a. alpha and beta receptors

      b. alpha-1 and alpha-2 receptors

      c. beta-1 and beta-2 receptors

      d. ganglionic and preganglionic receptors

_____ 54. Intramural ganglia are components of the parasympathetic division that are located:

      a. in the nasal and salivary glands

      b. in the intrinsic eye muscles

      c. inside the tissues of visceral organs

      d. in the adrenal medulla

_____ 55. Parasympathetic preganglionic fibers of the vagus nerve entering the abdominopelvic cavity join the:

      a. cardiac plexus

      b. pulmonary plexus

      c. hypogastric plexus

      d. celiac plexus

_____ 56. Sensory nerves deliver information to the CNS along:

      a. spinal nerves

      b. cranial nerves

      c. autonomic nerves that innervate peripheral effectors

      d. a, b, and c are correct

_____ 57. Of the following selections, the one that includes only sympathetic reflexes is:

      a. swallowing, coughing, sneezing, vomiting reflexes

      b. baroreceptor, vasomotor, pupillary dilation, sexual reflexes

      c. defecation, urination, digestion, secretion reflexes

      d. light, consenual light, sexual arousal, baroreceptor

_____ 58. In general, the primary result(s) of treatment with beta blockers is (are):

      a. reducing the strain on the heart and lowering peripheral blood pressure

      b. decreasing the heart rate

      c. decreasing the force of contraction of the heart

      d. a, b, and c are correct

Level
=2=

[L2]   Completion:

Using the terms below, complete the following statements.

hypothalamus	collateral	white ramus
postganglionic	gray ramus	norepinephrine
acetylcholine	blocking agents	splanchnic
adrenal medulla		

59. The fibers that innervate peripheral organs are called _____ fibers.

60. Adrenergic postganglionic sympathetic terminals release the neurotransmitter _____.

61. Modified sympathetic ganglia containing second-order neurons are located in the _____.

62. Prevertebral ganglia found anterior to the vertebral centra are called _____ ganglia.

63. The nerve bundle that carries sympathetic postganglionic fibers is known as the _____.

64. The nerve bundle containing the myelinated preganglionic axons of sympathetic motor neurons en route to the sympathetic chain or collateral ganglion is the _____.

65. In the dorsal wall of the abdominal cavity, preganglionic fibers that innervate the collateral ganglia form the _____ nerves.

66. Sympathetic activation is controlled by sympathetic centers in the _____.

67. Muscarinic and nicotine receptors are stimulated by the neurotransmitter _____.

68. Drugs that reduce the effects of autonomic stimulation by keeping the neurotransmitter from affecting the postsynaptic membranes are known as _____.

[L2]   Short Essay:

Briefly answer the following questions in the spaces provided below.

69. Why is the sympathetic division of the ANS called the "fight or flight" system?

70. Why is the parasympathetic division of the ANS known as the "rest and repose" system?

71. What three patterns of neurotransmission are of primary importance in the autonomic nervous system?

Level
=2=

72. What are the three major components of the sympathetic division?

73. What two general patterns summarize the effects of the functions of the collateral ganglia?

74. What are the two major components of the parasympathetic division of the ANS?

75. In what four cranial nerves do parasympathetic preganglionic fibers travel when leaving the brain?

76. Why is the parasympathetic division sometimes referred to as the anabolic system?

77. List four (4) adrenergic receptors found in the human body.

78. Name two (2) cholinergic receptors found in the human body.

79. What effect does dual innervation have on autonomic control throughout the body?

80. Why is autonomic tone an important aspect of ANS function?

81. (a) What parasympathetic reflexes would distention of the rectum and urinary bladder initiate?

(b) What sympathetic reflexes would low light levels and changes in blood pressure in major arteries initiate?

When you have successfully completed the exercises in L2 proceed to L3.

☐ LEVEL 3   CRITICAL THINKING/APPLICATION

Using principles and concepts learned about the autonomic nervous system, answer the following questions. Write your answers on a separate sheet of paper.

1. You have probably heard stories about "superhuman" feats that have been performed during times of crisis. What autonomic mechanisms are involved in producing a sudden, intensive physical activity that, under ordinary circumstances, would be impossible?

2. A question on your next anatomy and physiology exam reads:

"How do alpha-2 receptors provide negative feedback that controls the amount of norepinephrine at synapses?" How will you answer the question? (See my answer if you need help!)

3. Recent surveys show that 30 to 35 percent of the American adult population is involved in some type of exercise program. What contributions does sympathetic activation make to help the body adjust to the changes that occur during exercise and still maintain homeostasis?

4. Perhaps you know someone who has high blood pressure. What clinical explanation would you offer to express how beta-blockers are useful in controlling high blood pressure?

5. A student in a speech class who is a health fanatic decides to make a speech about cigarette smoking and its adverse effects on the health of the body. The first statement made by the student is, "Cigarette smoking causes high blood pressure." The smokers in the class want to know how and why. How should the health fanatic, who has already taken anatomy and physiology, answer the smokers?

Level
3

CHAPTER

17

Sensory Function

■ Overview

Do we really see with our eyes, hear with our ears, smell with our nose, and taste with our tongue? Ask the ordinary layperson and the answer will probably be yes. Ask the student of anatomy and physiology and the most likely response will be "not really." Our awareness of the world within and around us is aroused by sensory receptors that react to stimuli within the body or to stimuli in the environment outside the body. The previous chapters on the nervous system described the mechanisms by which the body perceives stimuli as neural events and not necessarily as environmental realities.

Sensory receptors receive stimuli and neurons transmit action potentials to the CNS where the sensations are processed, resulting in motor responses that serve to maintain homeostasis.

Chapter 17 considers the *general senses* of temperature, pain, touch, pressure, vibration, and proprioception and the *special senses* of smell, taste, balance, hearing, and vision. The activities in this chapter will reinforce your understanding of the structure and function of general sensory receptors and specialized receptor cells that are structurally more complex than those of the general senses.

☐ LEVEL 1   REVIEW OF CHAPTER OBJECTIVES

1. Distinguish between the general and special senses.
2. Explain why receptors respond to specific stimuli and how the organization of a receptor affects its sensitivity.
3. Identify the receptors for the general senses and describe how they function.
4. Describe the sensory organ of smell and trace the olfactory pathways to their destinations in the brain.
5. Describe the sensory organ of taste and trace the gustatory pathways to their destinations in the brain.
6. Identify the accessory structures of the eye and explain their functions.
7. Describe the internal structures of the eye and explain their functions.
8. Explain how we are able to distinguish colors and perceive depth.

9. Explain how light is converted into nerve impulses and trace the visual pathways to their destinations in the brain.

10. Describe the structures of the outer and middle ear and explain how they function.

11. Describe the parts of the inner ear and their roles in the process of hearing.

12. Trace the pathways for the sensations of equilibrium and hearing to their respective destinations in the brain.

[L1]   Multiple Choice:

Place the letter corresponding to the correct answer in the space provided.

OBJ. 1   _____   1. The term *general senses* refers to sensations of:

   a. smell, taste, balance, hearing, and vision
   b. pain, smell, pressure, balance, and vision
   c. temperature, pain, touch, pressure, vibration, and proprioception
   d. touch, taste, balance, vibration, and hearing

OBJ. 1   _____   2. The special senses refer to:

   a. balance, taste, smell, hearing, and vision
   b. temperature, pain, taste, touch, and hearing
   c. touch, pressure, vibration, and proprioception
   d. proprioception, smell, touch, and taste

OBJ. 2   _____   3. When sensations of slow or burning and aching pain are carried on Type C fibers, the individual becomes:

   a. aware of the pain, and can specifically locate the affected area
   b. aware of the pain, but has only a general idea of the affected area
   c. aware of the pain, but has no idea about the affected area
   d. aware of the pain, but perceives it in an area different from the affected area

OBJ. 2   _____   4. We cannot hear high-frequency sounds that dolphins respond to nor detect scents that excite bloodhounds because:

   a. our CNS is incapable of interpreting these stimuli
   b. our receptors have characteristic ranges of sensitivity
   c. these stimuli cause abnormal receptor function
   d. the stimuli produce inappropriate stimulation with no basis of fact

OBJ. 2   _____   5. The transduction process converts a real stimulus into a neural event that:

   a. must be carried by the PNS to the spinal cord
   b. must be peripherally regulated
   c. must be interpreted by the CNS
   d. must affect numerous receptors for proper interpretation

OBJ. 3 _____ 6. The receptors for the general senses are the:

     a. axons of the sensory neurons

     b. dendrites of sensory neurons

     c. cell bodies of the sensory neurons

     d. a, b, and c are correct

OBJ. 3 _____ 7. An injection or deep cut causes sensations of fast or prickling pain carried on:

     a. unmyelinated Type A fibers

     b. myelinated Type C fibers

     c. myelinated Type B fibers

     d. myelinated Type A fibers

OBJ. 3 _____ 8. The three classes of mechanoreceptors are:

     a. Meissner's corpuscles, Pacinian corpuscles, Merkel's discs

     b. fine touch, crude touch, and pressure receptors

     c. tactile, baroreceptors, proprioceptors

     d. slow-adapting, fast-adapting, and central adapting receptors

OBJ. 4 _____ 9. The first step in olfactory reception occurs on the surface of the:

     a. olfactory cilia

     b. columnar cells

     c. basal cells

     d. olfactory glands

OBJ. 4 _____ 10. The CNS interprets smell on the basis of the particular pattern of:

     a. cortical arrangement

     b. neuronal replacement

     c. receptor activity

     d. sensory impressions

OBJ. 5 _____ 11. The three different types of papillae on the human tongue are:

     a. kinocilia, cupula, and maculae

     b. sweet, sour, and bitter

     c. cuniculate, pennate, and circumvate

     d. filiform, fungiform, and circumvallate

OBJ. 5 _____ 12. Taste buds are monitored by cranial nerves:

     a. VII, IX, X

     b. IV, V, VI

     c. I, II, III

     d. VIII, XI, XII

OBJ. 5 _____ 13. After synapsing in the thalamus, gustatory information is projected to the appropriate portion of the:

     a. medulla

     b. medial lemniscus

     c. primary sensory cortex

     d. VII, IX, and X cranial nerves

Level 1

OBJ. 6 _____ 14. A lipid-rich product that helps to keep the eyelids from sticking together is produced by the:

      a. gland of Zeis

      b. meibomian gland

      c. lacrimal glands

      d. conjunctiva

OBJ. 6 _____ 15. The fibrous tunic, the outermost layer covering the eye, consists of the:

      a. iris and choroid

      b. pupil and ciliary body

      c. sclera and cornea

      d. lacrimal sac and orbital fat

OBJ. 6 _____ 16. The vascular tunic consists of three distinct structures that include:

      a. sclera, cornea, iris

      b. choroid, pupil, lacrimal sac

      c. retina, cornea, iris

      d. iris, ciliary body, choroid

OBJ. 7 _____ 17. The function of the vitreous body in the eye is to:

      a. provide a fluid cushion for protection of the eye

      b. serve as a route for nutrient and waste transport

      c. stabilize the shape of the eye and give physical support to the retina

      d. serve as a medium for cleansing the inner eye

OBJ. 7 _____ 18. The primary function of the lens of the eye is to:

      a. absorb light after it passes through the retina

      b. biochemically interact with the photoreceptors of the retina

      c. focus the visual image on retinal receptors

      d. integrate visual information for the retina

OBJ. 7 _____ 19. When looking directly at an object, its image falls upon the portion of the retina called the:

      a. fovea centralis

      b. choroid layer

      c. sclera

      d. focal point

OBJ. 8 _____ 20. When photons of all wavelengths stimulate both rods and cones, the eye perceives:

      a. "black" objects

      b. all the colors of the visible light spectrum

      c. either "red" or "blue" light

      d. "white" light

Level
-1-

OBJ. 9   _____ 21. Axons converge on the optic disc, penetrate the wall of the eye, and proceed toward the:

　　　　　　a. retina at the posterior part of the eye
　　　　　　b. diencephalon as the optic nerve (N II)
　　　　　　c. retinal processing areas below the choroid coat
　　　　　　d. cerebral cortex area of the parietal lobes

OBJ. 9   _____ 22. The sensation of vision arises from the integration of information arriving at the:

　　　　　　a. lateral geniculate of the left side
　　　　　　b. visual cortex of the cerebrum
　　　　　　c. lateral geniculate of the right side
　　　　　　d. reflex centers in the brain stem

OBJ. 9   _____ 23. Axons that leave the olfactory bulbs in the cerebrum travel along the olfactory tract (N I) to reach the:

　　　　　　a. olfactory cortex
　　　　　　b. hypothalamus
　　　　　　c. portions of the limbic system
　　　　　　d. a, b, and c are correct

OBJ. 10   _____ 24. The bony labyrinth of the ear is subdivided into the:

　　　　　　a. auditory meatus, auditory canal, ceruminous glands
　　　　　　b. saccule, utricle, vestibule
　　　　　　c. vestibule, semicircular canals, and the cochlea
　　　　　　d. ampulla, crista, cupula

OBJ. 10   _____ 25. The dividing line between the external ear and the middle ear is the:

　　　　　　a. pharyngotympanic tube
　　　　　　b. tympanic membrane
　　　　　　c. sacculus
　　　　　　d. utriculus

OBJ. 10   _____ 26. The auditory ossicles of the middle ear include the:

　　　　　　a. sacculus, utriculus, ampulla
　　　　　　b. vestibule, cochlea, organ of Corti
　　　　　　c. malleus, stapes, incus
　　　　　　d. otoliths, maculae, otoconia

OBJ. 11   _____ 27. The structure in the cochlea of the inner ear that provides information to the CNS is the:

　　　　　　a. scala tympani
　　　　　　b. organ of Corti
　　　　　　c. tectorial membrane
　　　　　　d. basilar membrane

OBJ. 11 _____ 28. The receptors that provide the sensation of hearing are located in the:

    a. vestibule
    b. ampulla
    c. tympanic membrane
    d. cochlea

OBJ. 12 _____ 29. The senses of equilibrium and hearing are provided by receptors in the:

    a. external ear
    b. middle ear
    c. inner ear
    d. a, b, and c are correct

OBJ. 12 _____ 30. Ascending auditory sensations synapse in the thalamus and then are delivered by projection fibers to the:

    a. auditory cortex of the parietal lobe
    b. auditory cortex of the temporal lobe
    c. auditory cortex of the occipital lobe
    d. auditory cortex of the frontal lobe

[L1] Completion:

Using the terms below, complete the following statements.

rods	endolymph	occipital	olfactory
sensitivity	mechanoreceptors	somatosensory	pupil
cerebral cortex	sclera	saccule, utricle	midbrain
thermoreceptors	taste buds	round window	cones

OBJ. 1 31. The ultimate destination of sensations from the general senses is the _____ cortex.

OBJ. 1 32. The information provided by receptors of the special senses is distributed to specific areas of the _____.

OBJ. 2 33. The reason we cannot hear extreme high frequency sounds is that our receptors have characteristic ranges of _____.

OBJ. 3 34. The receptors that respond to extremes of temperature are called _____.

OBJ. 3 35. Receptors that are sensitive to stimuli that distort their cell membranes are _____.

OBJ. 4 36. The only type of sensory information that reaches the cerebral cortex without synapsing in the thalamus is _____ stimuli.

OBJ. 5 37. Gustatory receptors are clustered in individual _____.

OBJ. 6 38. Most of the ocular surface of the eye is covered by the _____.

OBJ. 6 39. The opening surrounded by the iris is called the _____.

OBJ. 7 40. The photoreceptors that enable us to see in dimly lit rooms, at twilight, or in pale moonlight are the _____.

OBJ. 8 41. The photoreceptors that account for the perception of color are the _____.

Level
-1-

OBJ. 9   42. Visual information is integrated in the cortical area of the _____ lobe.

OBJ. 10   43. The chambers and canals in the inner ear contain a fluid called _____.

OBJ. 10   44. The thin membranous partition that separates the perilymph of the cochlear chambers from the air spaces of the middle ear is the _____.

OBJ. 11   45. The receptors in the inner ear that provide sensations of gravity and linear acceleration are located in the _____ and the _____.

OBJ. 12   46. The reflex processing center responsible for changing the position of the head in response to a sudden loud noise is located in the inferior colliculus of the _____.

[L1] Matching:

Match the terms in column B with the terms in column A. Use letters for answers in the spaces provided.

PART I   **COLUMN A**   **COLUMN B**

OBJ. 1   ____ 47. nose, eye, ear   A. general senses
OBJ. 1   ____ 48. somatosensory cortex   B. fast-adapting receptors, urination
OBJ. 2   ____ 49. phasic receptors
OBJ. 2   ____ 50. tonic receptors   C. smell
OBJ. 3   ____ 51. nociceptors   D. sense organs
OBJ. 3   ____ 52. chemoreceptors   E. taste
OBJ. 4   ____ 53. olfaction   F. sensation of pain
OBJ. 5   ____ 54. gustation   G. slow adapting receptors
                                          H. response to specific molecules

PART II   **COLUMN A**   **COLUMN B**

OBJ. 6   ____ 55. eyelids   I. tears
OBJ. 6   ____ 56. lacrimal glands   J. visual pigment
OBJ. 7   ____ 57. sharp vision   K. fovea centralis
OBJ. 8   ____ 58. rhodopsin   L. equilibrium
OBJ. 9   ____ 59. suprachiasmatic nucleus   M. inner ear
                                          N. palpebral
OBJ. 10   ____ 60. ceruminous glands   O. ear wax
OBJ. 11   ____ 61. membranous labyrinth   P. hypothalamus
OBJ. 12   ____ 62. utricle, saccule

[L1] Drawing/Illustration Labeling:

Identify each numbered structure by labeling the following figures:

OBJ. 3   *FIGURE 17.1*   Sensory Receptors of the Skin

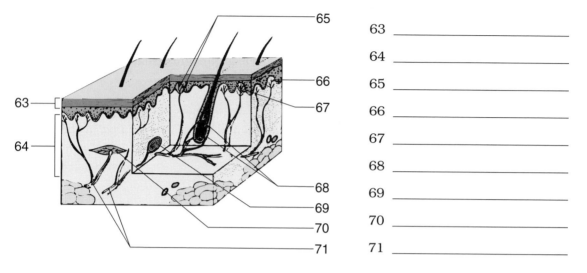

63	_____
64	_____
65	_____
66	_____
67	_____
68	_____
69	_____
70	_____
71	_____

OBJ. 6
OBJ. 7   *FIGURE 17.2*   Sectional Anatomy of the Eye

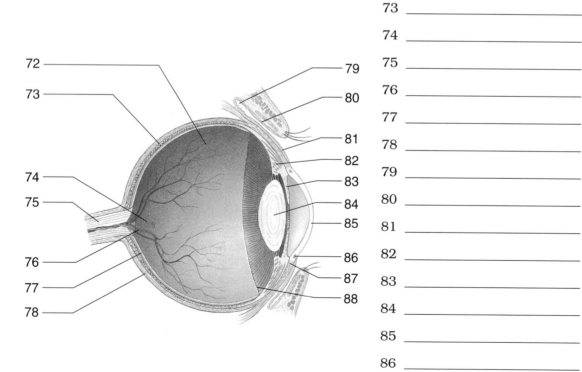

72	_____
73	_____
74	_____
75	_____
76	_____
77	_____
78	_____
79	_____
80	_____
81	_____
82	_____
83	_____
84	_____
85	_____
86	_____
87	_____
88	_____

Level
-1-

OBJ. 10
OBJ. 11

FIGURE 17.3   Anatomy of the Ear: External, Middle, and Inner Ears

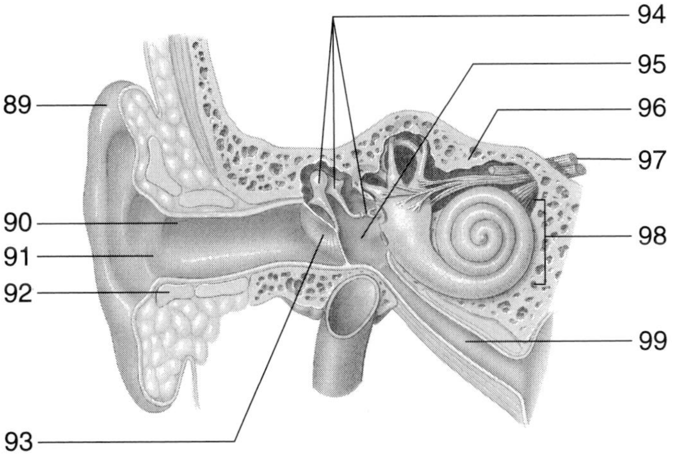

89 _____

90 _____

91 _____

92 _____

93 _____

94 _____

95 _____

96 _____

97 _____

98 _____

99 _____

OBJ. 10
OBJ. 11

FIGURE 17.4   Anatomy of the Ear—(Bony Labyrinth)

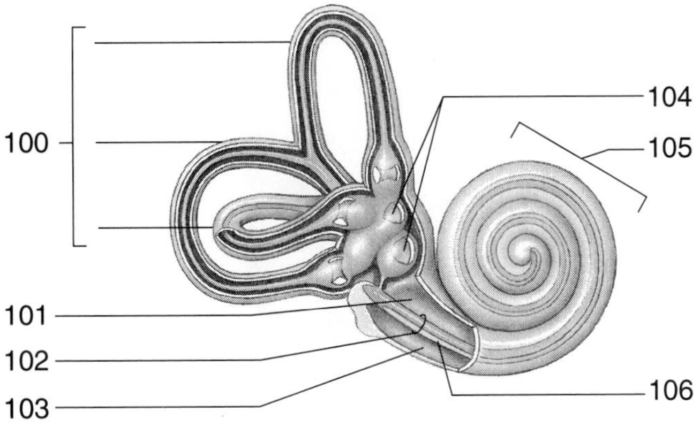

100 _____

101 _____

102 _____

103 _____

104 _____

105 _____

106 _____

OBJ. 11

FIGURE 17.5   Gross Anatomy of the Cochlea: Details Visible in Section

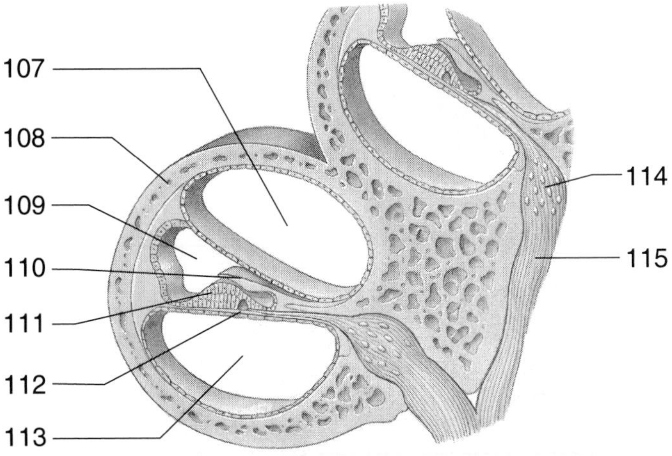

107 _____

108 _____

109 _____

110 _____

111 _____

112 _____

113 _____

114 _____

115 _____

OBJ. 11   **FIGURE 17.6**   Three-Dimensional Structure of Tectorial Membrane and Hair Complex of the Organ of Corti

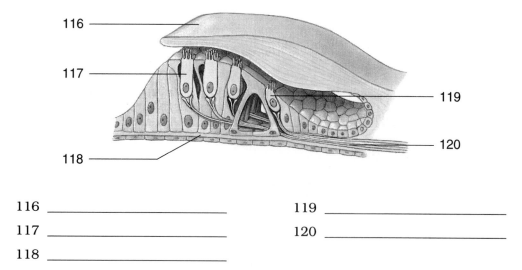

116 _____   119 _____

117 _____   120 _____

118 _____

OBJ. 4   **FIGURE 17.7**   Gustatory Pathways (Identify taste areas on tongue and cranial nerves to medulla)

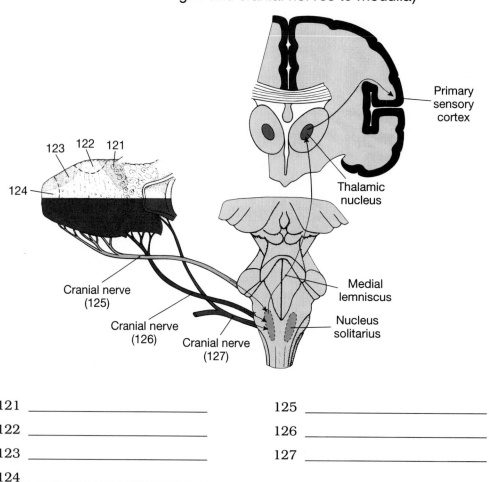

121 _____   125 _____

122 _____   126 _____

123 _____   127 _____

124 _____

Level
–1–   **When you have successfully completed the exercises in L1 proceed to L2.**

☐ LEVEL 2   CONCEPT SYNTHESIS

Concept Map I:

Using the following terms, fill in the circled, numbered, blank spaces to complete the concept map. Follow the numbers that comply with the organization of the map.

Pacinian corpuscles	Merkel's discs	Baroreceptors
Proprioception	Tactile	Pressure
Thermoreceptors	Muscle spindles	Dendritic processes
Aortic sinus	Pain	

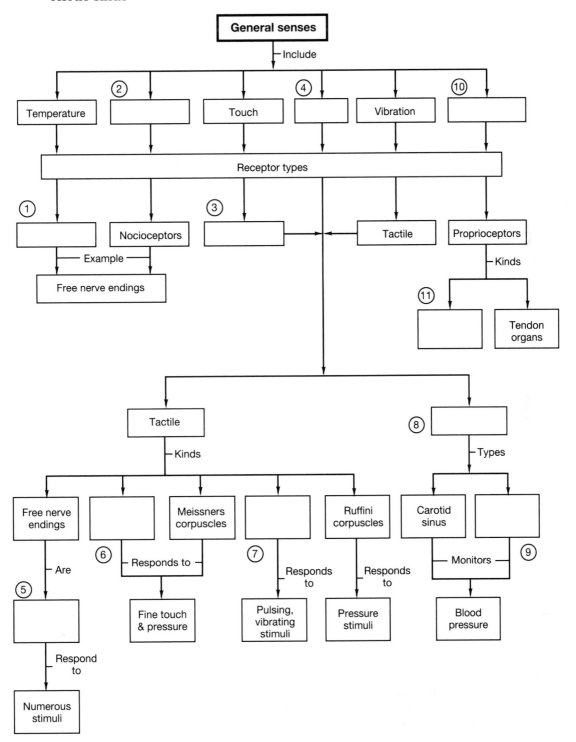

Concept Map II:

Using the following terms, fill in the circled, numbered, blank spaces to complete the concept map. Follow the numbers that comply with the organization of the map.

Retina   Audition   Rods and cones   Olfaction
Smell   Ears   Taste buds   Tongue
Balance and hearing

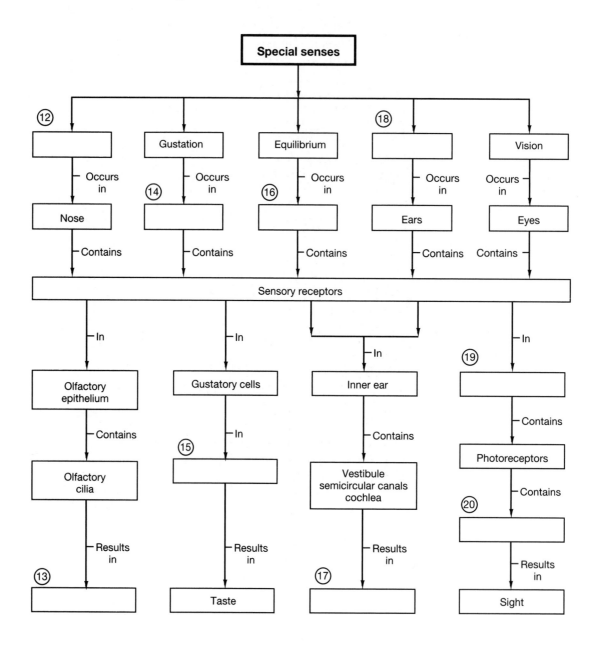

Concept Map III:

Using the following terms, fill in the circled, numbered, blank spaces to complete the concept map. Follow the numbers that comply with the organization of the map.

Round window
Oval window
Middle ear
Vestibulocochlear nerve (VIII)

Hair cells of organ of Corti
Incus
Endolymph in cochlear duct
External acoustic meatus

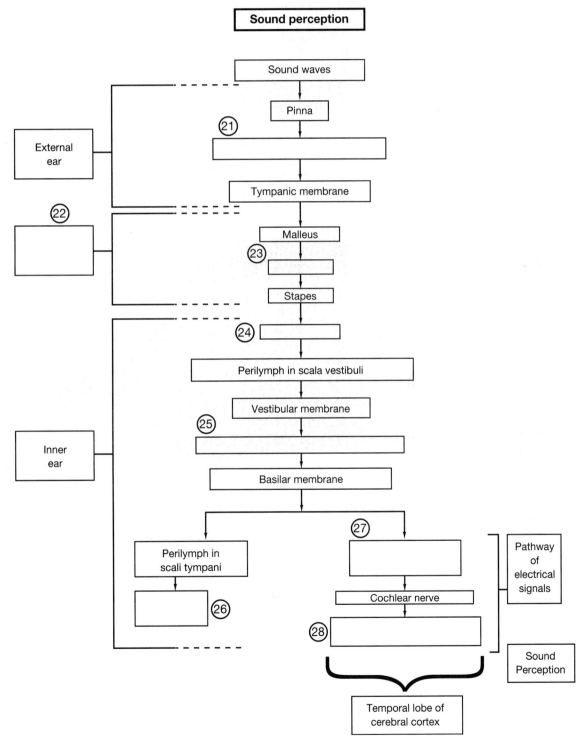

Body Trek:

Using the terms below, fill in the blanks to complete the trek through the outer, middle, and inner ear.

pharyngotympanic	oval window	stereocilia
endolymph	scala vestibuli	scala tympani
incus	organ of Corti	round window
pinna	cochlear	middle ear
tympanic membrane	malleus	external ear
ceruminous	vestibular duct	nasopharynx
ossicles	tectorial	basilar membrane
external auditory meatus	stapes	

Robo's programming for this trek involves following sound waves through the outer and middle ear until the sound (mechanical) waves are converted into nerve (electrical) impulses in the inner ear.

Robo's trek begins as sound waves are funnelled by the (29) _____ into the (30) _____ , the entrance to the external auditory canal. The robot's progress is slowed in this area because of the presence of a "waxy" material secreted by (31) _____ glands along the canal. All of a sudden Robo is thrust against the (32) _____ , or eardrum, causing a tremendous vibration that pushes the robot from the (33) _____ into the (34) _____ , or tympanic cavity. This compartment is an air-filled space containing three bones, the auditory (35) _____. One of the bones looks like a "hammer," the (36) _____ , while another one looks like an "anvil," the (37) _____, and the third has an appearance like a "stirrup," or (38) _____. Robo's trek is momentarily halted because of an opening that leads into an elongated channel, the (39) _____ tube, which allows for communication between the tympanic cavity and the (40) _____. After trekking around the opening Robo is "waved" through an ovoid "pane-like" covering, the (41) _____ , which is waving back and forth and creating vibrations that set up pressure waves in a clear fluid, the (42) _____ of the (43) _____ in the inner ear. The robot is propelled by the pressure waves as they are propagated through the perilymph of the (44) _____ and (45) _____. These pressure waves distort the (46) _____ on their way to the (47) _____ of the tympanic duct. As Robo treks into the duct, it senses a distorted membrane forcing hair cells of the (48) _____ toward or away from the (49) _____ membrane. This movement moves Robo along, and leads to displacement of (50) _____ and stimulation of sensory neurons of the (51) _____ nerve.

Mission Control orders Robo to use the programmed exit route by picking up a pressure wave as it "rolls" back toward the entrance to the vestibular duct. Robo's handlers are happy to retrieve their robot, which is covered with wax and needs recharging because of a run-down battery.

[L2] Multiple Choice:

Place the letter corresponding to the correct answer in the space provided.

_____ 52. Our knowledge of the environment is limited to those characteristics that stimulate our:

a. eyes and ears

b. peripheral receptors

c. nose and tongue

d. central nervous system

_____ 53. The labeled line activity that provides information about the strength, duration, variation, and movement of the stimulus is called:

     a. sensory coding

     b. tonic reception

     c. phasic reception

     d. peripheral reception

_____ 54. Sensory neurons that are always active are called:

     a. phasic receptors

     b. peripheral receptors

     c. tonic receptors

     d. mechanoreceptors

_____ 55. Inactive receptors that become active for a short time when there is a change in conditions they are monitoring are called:

     a. tonic receptors

     b. phasic receptors

     c. nociceptors

     d. chemoreceptors

_____ 56. The system that helps to focus attention and either heightens or reduces awareness of arriving sensations is the:

     a. reticular activating system in the midbrain

     b. limbic system in the cerebrum

     c. endocrine system

     d. autonomic nervous system

_____ 57. Of the following selections, the one that includes only tactile receptors is:

     a. free nerve endings, carotid and aortic bodies, and Merkel's discs

     b. Ruffini's and Meissner's corpuscles, carotid and aortic bodies

     c. root hair plexus, free nerve endings, muscle spindles, tendon organs

     d. Merkel's discs, Meissner's, Pacinian, and Ruffini's corpuscles

_____ 58. The receptors that provide information which plays a major role in regulating cardiac function and adjusting blood flow to vital tissues are:

     a. mechanoreceptors

     b. baroreceptors

     c. tactile receptors

     d. a, b, and c are correct

_____ 59. The position of joints, the tension in tendons and ligaments, and the state of muscular contractions are monitored by:

     a. proprioceptors

     b. baroreceptors

     c. mechanoreceptors

     d. tactile receptors

_____ 60. A taste receptor sensitive to dissolved chemicals but insensitive to pressure illustrates the concept of:

      a. transduction

      b. receptor specificity

      c. receptor potential

      d. phasic reception

_____ 61. When a neuron responds to an excitatory or inhibitory neurotransmitter, it functions as a:

      a. mechanoreceptor

      b. tactile receptor

      c. chemoreceptor

      d. proprioceptor

_____ 62. The chemoreceptive neurons that monitor the oxygen concentration of the arterial blood are found within the:

      a. carotid and aortic sinus

      b. supporting and stem cells

      c. filiform and circumvallate papillae

      d. carotid and aortic bodies

_____ 63. Considerable convergence along the olfactory pathway prevents sensations from reaching the cerebral cortex because of:

      a. excessive mucins in the nasal passageways

      b. inhibition at intervening synapses

      c. dissolved chemicals interacting with the receptor surface

      d. cyclic-AMP opening sodium channels in the membrane

_____ 64. There are no known structural differences between warm and cold thermoreceptors because they are all:

      a. mechanically regulated ion channels

      b. affected by excitatory or inhibitory neurotransmitters

      c. free nerve endings

      d. affected by the extremes of temperature

_____ 65. The only known example of neuronal replacement in the human adult is the:

      a. olfactory receptor population

      b. gustatory receptor population

      c. optic receptor population

      d. vestibuloreceptor population

_____ 66. During the focusing process, when light travels from the air into the relatively dense cornea:

      a. the sclera assumes an obvious color

      b. the light path is bent

      c. pupillary reflexes are triggered

      d. reflexive adjustments occur in both pupils

_____ 67. Exposure to bright light produces a:

      a. rapid reflexive increase in pupillary diameter

      b. slow reflexive increase in pupillary diameter

      c. very slow reflexive decrease in pupillary diameter

      d. rapid reflexive decrease in pupillary diameter

_____ 68. The color of the eye is determined by:

      a. light reflecting through the cornea onto the retina

      b. the thickness of the iris and the number and distribution of pigment cells

      c. the reflection of light from the aqueous humor

      d. the number and distribution of pigment cells in the lens

_____ 69. In rating visual acuity, a person whose vision is rated 20/15 is better than normal since this person can:

      a. read letters at 20 feet that are only discernible by the normal eye at 15 feet from the chart

      b. read letters at 15 feet that are only discernible by the normal eye at 20 feet from the chart

      c. read letters at 15 or 20 feet that normal individuals cannot read at all

      d. read letters without the aid of a lens at 35 feet from the chart

_____ 70. After an intense exposure to light, a "ghost" image remains on the retina because a photoreceptor cannot respond to stimulation until:

      a. the rate of neurotransmitter release declines at the receptor membrane

      b. the receptor membrane channels close

      c. opsin activates transducin

      d. its rhodopsin molecules have been regenerated

_____ 71. The role that vitamin A plays in the eye is:

      a. the visual pigment *retinal* is synthesized from vitamin A

      b. the protein part of rhodopsin is synthesized from vitamin A

      c. it acts as a coenzyme for the activation of transducin

      d. it activates opsin for rhodopsin-based photoreception

_____ 72. The reason everything appears to be black and white when we enter dimly lighted surroundings is:

      a. rods and cones are stimulated

      b. only cones are stimulated

      c. only rods are stimulated

      d. only the blue cones are stimulated

_____ 73. When one or more classes of cones are nonfunctional, the result is:

      a. a blind spot occurs

      b. color blindness

      c. image inversion

      d. the appearance of "ghost" images

_____ 74. The most detailed information about the visual image is provided by the:

    a. cones

    b. rods

    c. optic disc

    d. rods and cones

_____ 75. The region of the retina called the "blind spot" is an area that structurally comprises the:

    a. choroid coat

    b. suprachiasmatic nucleus

    c. optic disc

    d. visual cortex

_____ 76. The partial cross-over that occurs at the optic chiasm ensures that the visual cortex receives:

    a. an image that is inverted

    b. an image that is reversed before reaching the cortex

    c. different images from the right and left eyes

    d. a composite picture of the entire visual field

[L2]   Completion:

Using the terms below, complete the following statements.

retina	accommodation	afferent fiber
receptive field	receptor potential	transduction
referred pain	pupil	primary sensory cortex
hyperopia	adaptation	cataract
Hertz	sensory receptor	sensation
labeled line	myopia	aqueous humor
ampulla	nystagmus	central adaptation

77. A specialized cell that monitors conditions in the body or the external environment is a _____.

78. The information passed to the CNS is called a _____.

79. The area monitored by a single receptor cell is its _____.

80. The process of translating a stimulus into an action potential is called _____.

81. The change in the transmembrane potential that accompanies receptor stimulation is called a _____.

82. An action potential travels to the CNS over an _____.

83. The neural link between receptor and cortical neuron is called a _____.

84. A reduction in sensitivity in the presence of a constant stimulus is _____.

85. The inhibition of nuclei along a sensory pathway usually involves _____.

86. Pain sensations from visceral organs that are often perceived as originating in more superficial regions innervated by the same spinal nerves are known as _____.

87. Painful stimuli can be "ignored" because the impulses do not reach the _____.

88. Sensory receptors in the semicircular canals in the inner ear are located in the _____.

89. The condition in which short, jerky eye movements appear after damage to the brain stem or inner ear is referred to as _____.

90. The frequency of sound is measured in units called _____.

91. When the muscles of the iris contract, they change the diameter of the central opening referred to as the _____.

92. The posterior chamber of the eye contains a fluid called _____.

93. The visual receptors and associated neurons in the eye are contained in the _____.

94. When the lens of the eye loses its transparency, the abnormal lens is known as a _____.

95. When the lens of the eye becomes rounder to focus on the image of a nearby object on the retina, the mechanism is called _____.

96. If a person sees objects at close range under normal conditions, the individual is said to be "nearsighted," a condition formally termed _____.

97. If a person sees distant objects more clearly under normal conditions, the individual is said to be "farsighted," a condition formally termed _____.

[L2]   Short Essay:

Briefly answer the following questions in the spaces provided below.

98. What sensations are included as *general senses*?

99. What sensations are included as *special senses*?

100. What three (3) major steps are involved in the process of translating a stimulus into an action potential (transduction)?

101. What four (4) kinds of *general sense* receptors are found throughout the body and to what kind(s) of stimuli do they respond?

102. What is the functional difference between a *baroreceptor* and a *proprioceptor*?

Level
=2=

103. Trace an olfactory sensation from the time it leaves the olfactory bulb until it reaches its final destinations in the higher centers of the brain.

104. What are the four (4) *primary taste sensations*?

105. What sensations are provided by the *saccule* and *utricle* in the vestibule of the inner ear?

106. What are the three (3) primary functions of the *fibrous tunic* which consists of the *sclera* and the *cornea*?

107. What are the three (3) primary functions of the *vascular tunic* which consists of the iris, the ciliary body, and the choroid?

108. What are the primary functions of the *neural tunic*, which consists of an outer pigment layer and an inner retina that contains the visual receptors and associated neurons?

109. When referring to the eye, what is the purpose of *accommodation* and how does it work?

110. Why are cataracts common in the elderly?

When you have successfully completed the exercises in L2 proceed to L3.

□ LEVEL 3   CRITICAL THINKING/APPLICATION

Using principles and concepts learned about sensory functions, answer the following questions. Write your answers on a separate sheet of paper.

1. The perfume industry expends considerable effort to develop odors that trigger sexual responses. How might this effort be anatomically related to the processes involved in olfaction?

2. Grandmother and Granddad are all "decked out" for a night of dining and dancing. Grandma has doused herself with her finest perfume and Grandpa has soaked his face with aftershave. The pleasant scents are overbearing to others; however, the old folks can barely smell the aromas. Why?

3. Suppose you were small enough to wander around in the middle ear. What features in this region would make you think that you might be in a blacksmith's shop or perhaps a shop involved with equestrian enterprises?

4. We have all heard the statement "Eat carrots because they are good for your eyes." What is the relationship between eating carrots and the "health" of the eye?

5. A bright flash of light temporarily blinds normal, well-nourished eyes. As a result, a "ghost" image remains on the retina. Why?

18

The Endocrine System

■ Overview

The nervous system and the endocrine system are the metabolic control systems in the body. Together they monitor and adjust the physiological activities throughout the body to maintain homeostasis. The effects of nervous system regulation are usually rapid and short term, whereas the effects of endocrine regulation are ongoing and long term.

Endocrine cells are glandular secretory cells that release chemicals, called hormones, into the bloodstream for distribution to target tissues throughout the body. The influence of these chemical "messengers" results in facilitating processes that include growth and development, sexual maturation and reproduction, and the maintenance of homeostasis within other systems.

Chapter 18 consists of exercises that will test your knowledge of the endocrine organs and their functions. The integrative and complementary activities of the nervous and endocrine systems are presented to show the coordinated effort necessary to regulate adequately physiological activities in the body.

☐ LEVEL 1   REVIEW OF CHAPTER OBJECTIVES

1. Compare the endocrine and nervous systems.
2. Compare the cellular components of the endocrine system with those of other tissues and systems.
3. Compare the major chemical classes of hormones.
4. Explain the general mechanisms of hormonal action.
5. Describe how endocrine organs are controlled.
6. Describe the location, hormones, and functions of the following endocrine glands and tissues: pituitary, thyroid, parathyroids, thymus, adrenals, kidneys, heart, pancreas, ovaries, testes, and pineal gland.
7. Discuss the results of abnormal hormone production.
8. Explain how hormones interact to produce coordinated physiological responses.
9. Identify the hormones that are especially important to normal growth and discuss their roles.
10. Define the general adaptation syndrome and compare homeostatic responses with stress responses.
11. Describe the effects that hormones have on behavior.

[L1]   Multiple Choice:

Place the letter corresponding to the correct answer in the space provided.

OBJ. 1   _____   1. Response patterns in the endocrine system are particularly effective in:

     a. regulating ongoing metabolic processes

     b. rapid short-term specific responses

     c. the release of chemical neurotransmitters

     d. crisis management

OBJ. 1   _____   2. An example of a functional similarity between the nervous system and the endocrine system is:

     a. both systems secrete hormones into the bloodstream

     b. the cells of the endocrine and nervous systems are functionally the same

     c. compounds used as hormones by the endocrine system and may also function as neurotransmitters inside the CNS

     d. both produce very specific responses to environmental stimuli

OBJ. 2   _____   3. Neurons communicate with one another and with effectors by:

     a. releasing chemicals into the bloodstream

     b. interacting with a hormone and a receptor complex

     c. receptor binding at the cell membrane

     d. releasing chemical neurotransmitters

OBJ. 2   _____   4. The release of hormones by endocrine cells alters the:

     a. rate at which chemical neurotransmitters are released.

     b. metabolic activities of many tissues and organs simultaneously

     c. very specific responses to environmental stimuli

     d. anatomical boundary between the nervous and endocrine systems

OBJ. 3   _____   5. The protein hormone _prolactin_ is involved with the:

     a. melanin synthesis

     b. production of milk

     c. production of testosterone

     d. labor contractions and milk ejection

OBJ. 3   _____   6. The amino acid derivative hormone _epinephrine_ is responsible for:

     a. increased cardiac activity

     b. glycogen breakdown

     c. release of lipids by adipose tissue

     d. a, b, and c are correct

OBJ. 4   _____   7. The most notable effect of ADH produced in the neurohypophysis of the pituitary gland is to:

     a. increase the amount of water lost at the kidneys

     b. decrease the amount of water lost at the kidneys

     c. stimulate the contraction of uterine muscles

     d. increase or decrease calcium ion concentrations in body fluids

OBJ. 4   _____   8. Stimulation of contractile cells in mammary tissue and uterine muscles in the female is initiated by secretion of:

    a. oxytocin from the posterior pituitary

    b. melatonin from the pineal gland

    c. oxytocin from the adenohypophysis

    d. melatonin from the neurohypophysis

OBJ. 4   _____   9. Hormones alter cellular operations by changing the:

    a. cell membrane permeability properties

    b. identities, activities, quantities, or properties of important enzymes

    c. arrangement of the molecular complex of the cell membrane

    d. rate at which hormones affect the target organ cells

OBJ. 4   _____   10. Catecholamines and peptide hormones affect target organ cells by:

    a. binding to receptors in the cytoplasm

    b. binding to target receptors in the nucleus

    c. enzymatic reactions that occur in the ribosomes

    d. second messengers released when receptor binding occurs at the membrane surface

OBJ. 4   _____   11. Steroid hormones affect target organ cells by:

    a. target receptors in peripheral tissues

    b. releasing second messengers at cell membrane receptors

    c. binding to receptors in the cell membrane

    d. binding to target receptors in the nucleus

OBJ. 5   _____   12. Endocrine cells responding directly to changes in the composition of the extracellular fluid is an example of:

    a. a neural reflex

    b. a reflex arc

    c. an endocrine reflex

    d. a hypothalamic control mechanism

OBJ. 5   _____   13. The hypothalamus is a major coordinating and control center because:

    a. it contains autonomic centers and acts as an endocrine organ

    b. it initiates endocrine and neural reflexes

    c. it stimulates appropriate responses by peripheral target cells

    d. it stimulates responses to restore homeostasis

OBJ. 6   _____   14. The glycoprotein hormone _FSH_ in the male is responsible for:

    a. maturation of germinative cells in the gonads

    b. production of interstitial cells in the male

    c. sperm formation and testosterone secretion

    d. the male doesn't secrete FSH

Level
-1-

OBJ. 6   _____ 15. The inability to tolerate stress due to underproduction of gluco-
                       corticoids results in a syndrome known as:

                       a. Cushing's disease
                       b. Addison's disease
                       c. myxedema
                       d. eunuchoidism

OBJ. 6   _____ 16. The *posterior pituitary* secretes:

                       a. growth hormone and prolactin
                       b. oxytocin and antidiuretic hormone
                       c. thyroid stimulating and adrenocorticotrophic hormone
                       d. melanocyte-stimulating and luteinizing hormone

OBJ. 6   _____ 17. The *gonadotropic* hormones secreted by the *anterior pituitary* are:

                       a. TSH, ACTH, ADH
                       b. PRL, GH, MSH
                       c. FSH, LH, ICSH
                       d. ADH, MSH, FSH

OBJ. 6   _____ 18. The hormone synthesized from molecules of the neurotrans-
                       mitter serotonin in the *pineal gland* is:

                       a. melanocyte-stimulating hormone
                       b. pinealtonin
                       c. oxytocin
                       d. melatonin

OBJ. 7   _____ 19. The overproduction syndrome aldosteronism results in:

                       a. high blood potassium concentrations
                       b. low blood volume
                       c. dependence on lipids for energy
                       d. increased body weight due to water retention

OBJ. 8   _____ 20. When a cell receives instructions from two different hormones at
                       the same time the results may be:

                       a. antagonistic or synergistic
                       b. permissive
                       c. integrative
                       d. a, b, and c are correct

OBJ. 9   _____ 21. The hormones that are of primary importance to normal growth
                       include:

                       a. TSH, ACTH, insulin, parathormone, and LH
                       b. prolactin, insulin, growth hormone, ADH, and MSH
                       c. growth hormone, prolactin, thymosin, androgens, and
                          insulin
                       d. growth hormone, thyroid hormones, insulin, para-
                          thormone, and gonadal hormones

Level
-1-

OBJ. 9 _____ 22. The reason insulin is important to normal growth is that it promotes:

    a. changes in skeletal proportions and calcium deposition in the body

    b. passage of glucose and amino acids across cell membranes

    c. muscular development via protein synthesis

    d. a, b, and c are correct

OBJ. 10 _____ 23. Despite the variety of potential stresses, the human body responds to each one:

    a. in a variety of ways to make the adjustments necessary for homeostasis

    b. with the same basic pattern of hormonal and physiological adjustments

    c. by suppressing the stress using psychological conditioning

    d. by reducing the stress via relaxation exercises

OBJ. 10 _____ 24. The body's response to stress constitutes the general adaptation syndrome (GAS), which consists of:

    a. alarm

    b. resistance

    c. exhaustion

    d. a, b, and c are correct

OBJ. 11 _____ 25. In an adult, changes in the mixture of hormones reaching the CNS have significant effects on:

    a. intellectual capabilities

    b. memory and learning

    c. emotional status

    d. a, b, and c are correct

[L1] Completion:

Using the terms below, complete the following statements.

gigantism	cytoplasm	neurotransmitters
aggressive	exhaustive	general adaptation syndrome
G-protein	parathyroids	diabetes mellitus
reflex	catecholamines	adrenal gland
testosterone	hypothalamus	thymus

OBJ. 1   26. Many compounds used as hormones by the endocrine system also function inside the CNS as _____.

OBJ. 2   27. An example of a gland in the body that functions as a neural tissue and an endocrine tissue is the _____.

OBJ. 3   28. Epinephrine, norepinephrine, and dopamine are structurally similar and belong to a class of compounds sometimes called _____.

OBJ. 4   29. The hormone that stimulates the production of enzymes and proteins in skeletal muscle fibers, causing an increase in muscle size and strength is _____.

OBJ. 5   30. The link between the first messenger and the second messenger is a _____.

OBJ. 5 31. The control of calcium ion concentrations by parathormone or calcitonin is an example of an endocrine _____.

OBJ. 5 32. The most complex endocrine responses are directed by the _____.

OBJ. 6 33. The glands that are primarily responsible for increasing calcium ion concentrations in body fluids are the _____.

OBJ. 6 34. The gland that affects maturation and functional competence of the immune system is the _____.

OBJ. 7 35. The underproduction syndrome of insulin results in the condition called _____.

OBJ. 8 36. Steroid hormones bind to receptors in the _____.

OBJ. 9 37. Oversecretion of the growth hormone (GH) leads to a disorder called _____.

OBJ. 10 38. The collapse of vital systems due to stress represents the _____ phase of the GAS syndrome.

OBJ. 10 39. Patterns of hormonal and physiological adjustments to stress constitute the _____.

OBJ. 11 40. In precocious puberty, sexual hormones are produced at an inappropriate time causing affected children to become _____.

[L1] Matching:

Match the terms in column B with the terms in column A. Use letters for answers in the spaces provided.

PART I

	COLUMN A	COLUMN B
OBJ. 1 ____	41. neurotransmitters	A. catecholamine
OBJ. 1 ____	42. release of hormones	B. peptide hormone
OBJ. 2 ____	43. endocrine cells	C. pituitary
OBJ. 2 ____	44. neural cells	D. nervous system
OBJ. 3 ____	45. thyroid stimulating hormone	E. stimulates RBC production
		F. regulates blood glucose concentration
OBJ. 3 ____	46. epinephrine	G. endocrine system
OBJ. 4 ____	47. erythropoietin	H. action potentials
OBJ. 4 ____	48. glucagon	I. release of chemicals
OBJ. 5 ____	49. "master gland"	J. produces releasing–inhibiting hormones
OBJ. 5 ____	50. hypothalamus	

PART II

	COLUMN A	COLUMN B
OBJ. 6 ____	51. thyroid C cells	K. overproduction of T_4
OBJ. 6 ____	52. adrenal medulla	L. dwarfism
OBJ. 7 ____	53. cretinism	M. abnormal skeletal development
OBJ. 7 ____	54. Graves' disease	N. calcitonin
OBJ. 8 ____	55. enzyme activator or inhibitor	O. underproduction of T_4
OBJ. 8 ____	56. hormone	P. infertility
OBJ. 9 ____	57. undersecretion of GH	Q. second messenger
OBJ. 9 ____	58. thyroxine in elderly	R. immediate response to crisis
OBJ. 10 ____	59. alarm phase of GAS	S. secretion of epinephrine
OBJ. 11 ____	60. excessive Androgens	T. first messenger

Labeling:

Identify each numbered structure by labeling the following figures:

OBJ. 6   **FIGURE 18.1**   The Endocrine System

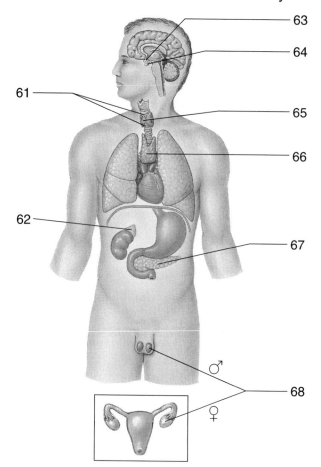

61 _____

62 _____

63 _____

64 _____

65 _____

66 _____

67 _____

68 _____

OBJ. 3   **FIGURE 18.2**   Structural Classification of Hormones

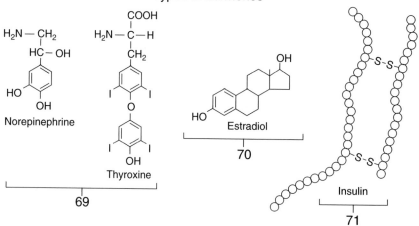

69 _____   71 _____   70 _____

Level
–1–

When you have successfully completed the exercises in L1 proceed to L2.

☐ LEVEL 2 CONCEPT SYNTHESIS

Concept Map I:

Using the following terms, fill in the circled, numbered, blank spaces to complete the concept map. Follow the numbers that comply with the organization of the map.

Male/female gonads   Epinephrine   Pineal
Hormones   Bloodstream   Testosterone
Heart   Parathyroids   Peptide hormones
Pituitary

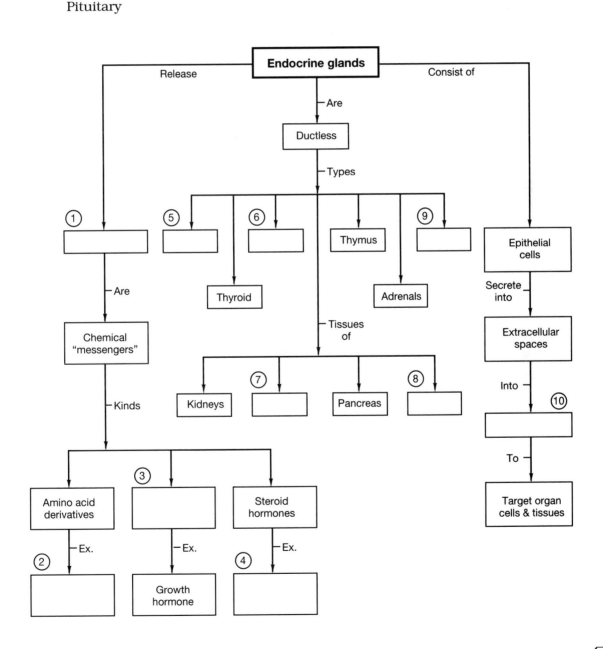

Concept Map II:

Using the following terms, fill in the circled, numbered, blank spaces to complete the concept map. Follow the numbers that comply with the organization of the map.

Pars nervosa   Thyroid   Adenohypophysis
Mammary glands   Melanocytes   Posterior pituitary
Sella turcica   Kidneys   Adrenal cortex
Oxytocin   Gonadotropic hormones   Pars distalis

Note to student: The following concept map provides an example of how each endocrine gland can be mapped. You are encouraged to draw concept maps for each gland in the endocrine system.

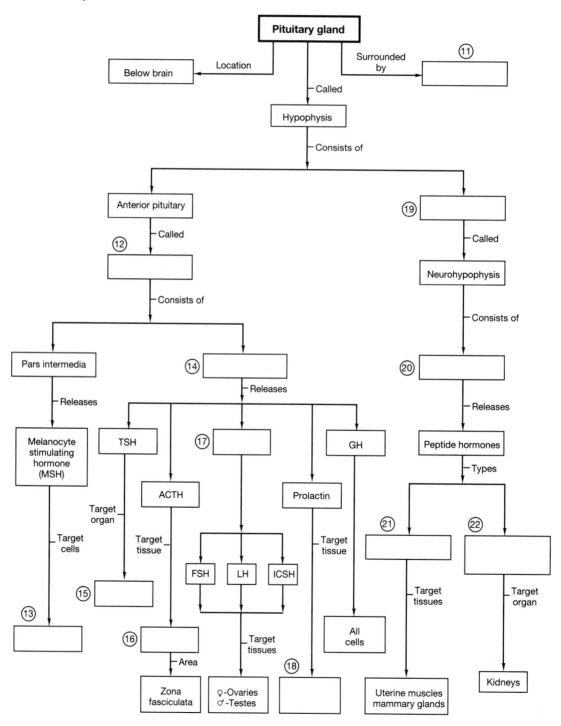

Level =2=

Concept Map III:

Using the following terms, fill in the circled, numbered, blank spaces to complete the concept map. Follow the numbers that comply with the organization of the map.

Homeostasis   Ion channel opening
Contraction   Target cells
Hormones   Cellular communication

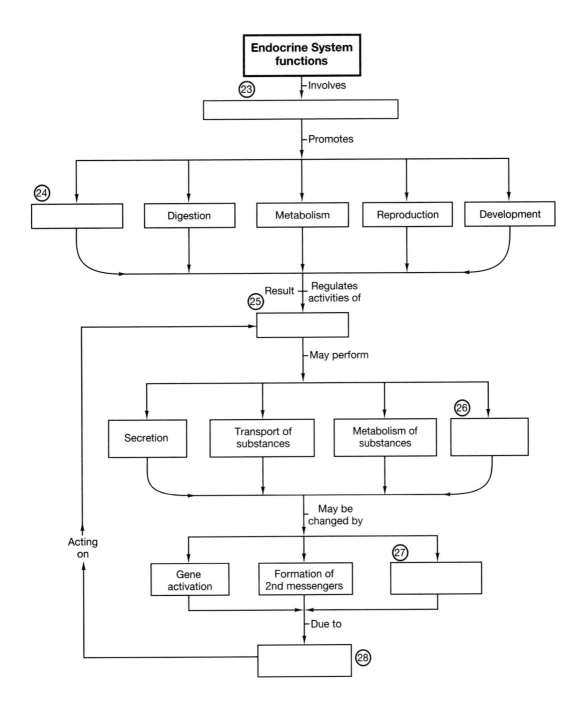

Body Trek:

Using the terms below, fill in the blanks to complete a trek through the endocrine system

progesterone	erythropoietin	adrenal	sella turcica
kidneys	heart	estrogen	ovaries
thymus	pituitary	testosterone	insulin
infundibulum	melatonin	pineal	pancreas
testes	parathyroid	RBC	thyroid

The body trek to "visit" the glands of the endocrine system will be taken by using an imaginary pathway through the body. Specific directions will be used to locate and identify the glands comprising the endocrine system.

The trek begins in the roof of the thalamus where the (29) _____ gland synthesizes the hormone (30) _____. Continuing in an anterior-inferior direction, the trek down the stalk of the hypothalamus, called the (31) _____, leads to the base of the brain, the location of the (32) _____ gland, which is housed in a bony encasement, the (33) _____. This pea-sized "master" gland secretes approximately nine tropic hormones that exert their effects on target organs, tissues, and cells through-out the body. Trekking southward in the body, the next "landmarks" are located in the neck region. Situated on the anterior surface of the trachea is the (34) _____ gland with four small (35) _____ glands embedded in its posterior surface. Just below and posterior to the sternum is the (36) _____ , a gland that is primarily concerned with the functional competence of the immune system.

Moving slightly downward and away from the midline of the thorax, the production of *atrial natriuretic hormone* is taking place in the tissues of the (37) _____. Trekking inferiorly and moving to a position behind the stomach, the (38) _____ can be seen secreting glucagon and (39) _____ , which are used to control blood glucose levels. Leaving the stomach region and progressing to the superior borders of the kidneys, the (40) _____ glands, which secrete hormones that affect the kidneys and most cells throughout the body, are identified. While in the area the much larger (41) _____ are sighted. Their endocrine tissues release (42) _____ which stimulates the production of (43) _____ by the bone marrow.

The imaginary trek within a female ends with a visit to the (44) _____, located near the lateral wall of the pelvic cavity. Here (45) _____, which supports follicle maturation, and (46) _____, which prepares the uterus for implantation, are released. The trek ends in the male by a visit to the scrotal sac, the location of the (47) _____ , which release (48) _____, the hormone that supports functional maturation of sperm and promotes development of male characteristics in the body.

[L2] Multiple Choice:

Place the letter corresponding to the correct answer in the space provided.

_____ 49. The hormone secreted in large amounts by the thyroid gland is:

    a. triiodothyronine

    b. thyroxine

    c. thyroglobulin

    d. calcitonin

_____ 50. The specialized cells in the parathyroid glands that secrete parathormone are:

    a. chief cells

    b. C cells

    c. follicular epithelial cells

    d. cells of the pars distalis

Level
=2=

_____ 51. Corticosteroids are hormones that are produced and secreted by the:

        a. medulla of the adrenal glands

        b. cortex of the cerebral hemispheres

        c. somatosensory cortex of the cerebrum

        d. cortex of the adrenal glands

_____ 52. The adrenal medulla produces and releases the hormones:

        a. mineralocorticoids and glucocorticoids

        b. epinephrine and norepinephrine

        c. androgens and estrogens

        d. aldosterone and cortisone

_____ 53. The endocrine functions of the kidney and the heart include the production and secretion of the hormones:

        a. renin and angiotensinogen

        b. insulin and glucagon

        c. erythropoietin and atrial natriuretic peptide

        d. epinephrine and norepinephrine

_____ 54. The endocrine tissues of the pancreas secrete the hormones:

        a. insulin, glycogen, and pancreatin

        b. insulin, renin, and angiotensin

        c. melatonin, pancreatin, and insulin

        d. somatostatin, glucagon, and insulin

_____ 55. The reproductive system of the male produces the endocrine hormones:

        a. FSH and ICSH

        b. inhibin and testosterone

        c. LH and ICSH

        d. a, b, and c are correct

_____ 56. The endocrine tissues of the reproductive system of the female produce:

        a. estrogens, inhibin, and progesterone

        b. androgens and estrogens

        c. FSH and LH

        d. a, b, and c are correct

_____ 57. The overall effect of the hormones secreted by the thyroid gland is:

        a. to increase calcium ion concentrations in body fluids

        b. maturation and functional competence of the immune system

        c. an increase in cellular rates of metabolism and oxygen consumption

        d. increased reabsorption of sodium ions and water from urine

_____ 58. The parathyroid glands secrete hormones that:

        a. affect the circulating concentration of calcium ions

        b. affect the functional competence of the immune system

        c. affect blood pressure and other cardiac activity

        d. affect the release of amino acids from skeletal muscles

Level
=2=

_____ 59. The zona fasciculata of the adrenal cortex produces steroid hormones that are known as glucocorticoids because of their:

        a. anti-inflammatory activities

        b. effects on glucose metabolism

        c. effect on the electrolyte composition of body fluids

        d. effect on the production of glucagon in the pancreas

_____ 60. One of the major effects of the ANP produced by the heart is that it:

        a. assists in the production of red blood cells

        b. constricts peripheral blood vessels

        c. promotes water retention at the kidneys

        d. depresses thirst

_____ 61. Glucagon and insulin secreted by the islets of Langerhans regulate:

        a. blood calcium levels

        b. blood glucose concentrations

        c. protein synthesis in muscle cells

        d. blood glucose utilization by the brain

_____ 62. Testosterone produced in the interstitial cells of the testes is responsible for:

        a. promoting the production of functional sperm

        b. determining male secondary sexual characteristics

        c. affecting metabolic operations throughout the body

        d. a, b, and c are correct

_____ 63. The estrogens produced in follicle cells surrounding the eggs in the ovary are responsible for:

        a. supporting the maturation of eggs and stimulating the growth of the uterine lining

        b. preparing the uterus for the arrival of the developing embryo

        c. accelerating the movement of the fertilized egg along the oviduct

        d. enlargement of the mammary glands

_____ 64. The calorigenic effect of thyroid hormones enables us to:

        a. obtain iodine molecules from surrounding interstitial tissues

        b. regulate calcium ion concentrations in body fluids

        c. control blood volume and blood pressure

        d. adapt to cold temperatures

_____ 65. The three different groups of hormones based on their chemical structures are:

        a. catecholamines, prehormones, and prohormones

        b. amino acid derivatives, peptide hormones, and steroids

        c. intracellular, extracellular, and G proteins

        d. peptides, dipeptides, and polypeptides

Level
=2=

_____ 66. All of the hormones secreted by the hypothalamus, pituitary gland, heart, kidneys, thymus, digestive tract, and pancreas are:

        a. steroid hormones

        b. amino acid derivatives

        c. peptide hormones

        d. catecholamines

_____ 67. The binding of a peptide hormone to its receptor starts a biochemical chain of events that changes the pattern of:

        a. enzymatic activity within the cell

        b. calcium ion release or entry into the cell

        c. diffusion through the lipid portion of the cell membrane

        d. an RNA transcription in the nucleus

_____ 68. The hypothalamus has a profound effect on endocrine functions through the secretion of:

        a. neurotransmitters involved in reflex activity

        b. anterior and posterior pituitary hormones

        c. releasing and inhibiting hormones

        d. a, b, and c are correct

_____ 69. A pituitary endocrine cell that is stimulated by a releasing hormone is usually:

        a. also stimulated by the peripheral hormone it controls

        b. not affected by the peripheral hormone it controls

        c. stimulated or inhibited by the peripheral hormone it controls

        d. inhibited by the peripheral hormone it controls

_____ 70. Even though a thyroid hormone is a small molecule, it cannot escape by diffusing through the follicular walls of the thyroid because:

        a. the thyroid hormone is attached to a large protein

        b. the openings in the follicular cells are too small

        c. thyroid hormones are released at a low background level

        d. a, b, and c are correct

_____ 71. Aldosterone targets kidney cells and causes the:

        a. loss of sodium ions and water, increasing fluid losses in the urine

        b. retention of sodium ions and water, reducing fluid losses in the urine

        c. production and release of angiotensen II by the kidneys

        d. secretion of the enzyme renin by the kidneys

_____ 72. The most dramatic functional change that occurs in the endocrine system due to aging is:

        a. a decrease in blood and tissue concentrations of ADH and TSH

        b. an overall decrease in circulating hormone levels

        c. a decline in the concentration of reproductive hormones

        d. a, b, and c are correct

Level
=2=

_____ 73. The two basic categories of endocrine disorders are:

a. abnormal hormone production or abnormal cellular sensitivity
b. the underproduction and overproduction syndrome
c. muscular weakness and the inability to tolerate stress
d. sterility and cessation of ovulation

[L2]   Completion:

Using the terms below, complete the following statements.

reticularis	prehormones	target cells
parathormone	portal	cyclic-AMP
somatomedins	nucleus	fenestrated
hypothalamus	calmodulin	tyrosine
glucocorticoids	epinephrine	aldosterone

74. The interface between the neural and endocrine systems is the _____.

75. The peripheral cells that are modified by the activities of a hormone are referred to as _____.

76. Thyroid hormones target receptors in the _____.

77. The hormone that promotes the absorption of calcium salts for bone deposition is _____.

78. Catecholamines and thyroid hormones are synthesized from molecules of the amino acid _____.

79. Structurally distinct peptide chains produced at the rough endoplasmic reticulum (RER) are called _____.

80. Many hormones produce their effects by increasing intracellular concentrations of _____.

81. An intracellular protein that acts as an intermediary for oxytocin and several regulatory hormones secreted by the hypothalamus is _____.

82. Liver cells respond to the presence of growth hormone by releasing hormones called _____.

83. Capillaries that have a "swiss cheese" appearance and are found only where relatively large molecules enter or leave the circulatory system are described as _____.

84. Blood vessels that link two capillary networks are called _____ vessels.

85. The principal mineralocorticoid produced by the human adrenal cortex is _____.

86. The zona of the adrenal cortex involved in the secretion of androgens is the _____.

87. The dominant hormone of the alarm phase of the general adaptation syndrome is _____.

88. The dominant hormones of the resistance phase of the GAS are _____.

[L2]   Short Essay:

Briefly answer the following questions in the spaces provided below.

89. List the three groups of hormones on the basis of chemical structure.

90. What are the four (4) possible events set in motion by activation of a G protein, each of which leads to the appearance of a second messenger in the cytoplasm?

91. What three mechanisms does the hypothalamus use to regulate the activities of the nervous and endocrine systems?

92. What hypothalamic control mechanisms are used to effect endocrine activity in the anterior pituitary?

93. How does the calorigenic effect of thyroid hormones help us to adapt to cold temperatures?

94. How does the kidney hormone erythropoietin cause an increase in blood pressure?

95. How do you explain the relationship between the pineal gland and circadian rhythms?

96. What are the possible results when a cell receives instructions from two different hormones at the same time?

When you have successfully completed the exercises in L2 proceed to L3.

☐ LEVEL 3   CRITICAL THINKING/APPLICATION

Using principles and concepts learned about the endocrine system, answer the following questions. Write your answers on a separate sheet of paper.

1. Negative feedback is an important mechanism for maintaining homeostasis in the body. Using the criteria below, create a model that shows how negative feedback regulates hormone production.

   Criteria:

   Gland X secretes hormone 1.

   Gland Y secretes hormone 2.

   Organ Z contains gland Y.

2. (a) As a diagnostic clinician, what endocrine malfunctions and conditions would you associate with the following symptoms in patients A, B, and C?

   Patient A: Marked increase in daily fluid intake, 12–15 liters of urine per day, glycosuria, lethargic.

   Patient B: Blood pressure 100/60 mm Hg, pulse 50 beats/min, abnormal weight gain, dry skin and hair, always tired, body temperature 95°F, feelings of weakness.

   Patient C: Blood pressure 160/100 mm Hg, glucose 130 mg %, muscle weakness, poor wound healing, thin arms, legs, and skin, red cheeks.

   (b) What glands are associated with each disorder?

3. An anatomy and physiology peer of yours makes a claim stating that the hypothalamus is exclusively a part of the CNS. What argument(s) would you make to substantiate that the hypothalamus is an interface between the nervous system and the endocrine system?

4. Topical steroid creams are used effectively to control irritating allergic responses or superficial rashes. Why are they ineffective and dangerous to use in the treatment of open wounds?

5. The recommended daily allowance (RDA) for iodine in adult males and females is 150 mg/day. Even though only a small quantity is needed each day, what symptoms are apparent if the intake of this mineral is insufficient or there is a complete lack of it in the diet?

6. If you have participated in athletic competitions, you have probably heard the statement, "I could feel the adrenalin flowing." Why does this feeling occur and how is it manifested in athletic performance?

7. Even though there has been a great deal of negative publicity about the use of anabolic steroids, many athletes continue to use them as a means of increasing their muscle mass and improving their endurance and performance. What are the dangers and health risks associated with taking anabolic steroids?

Level
≡**3**≡

19

Blood

■ Overview

Fact: Every cell of the body needs a continuous supply of water, nutrients, energy, and oxygen.

Ten and one-half pints of blood, making up 7 percent of the body's total weight, circulate constantly through the body of the average adult, bringing to each cell the oxygen, nutrients, and chemical substances necessary for its proper functioning, and, at the same time, removing waste products. This life-sustaining fluid, along with the heart and a network of blood vessels, comprises the cardiovascular system, which, acting in concert with other systems, plays an important role in the maintenance of homeostasis. This specialized fluid connective tissue consists of two basic parts: the formed elements and cell fragments, and the fluid plasma in which they are carried.

Functions of the blood, including transportation, regulation, protection, and coagulation, serve to maintain the integrity of cells in the human body.

Chapter 19 provides an opportunity for you to identify and study the components of blood and review the functional roles the components play in order for cellular needs to be met throughout the body. Blood types, hemostasis, hematopoiesis, and blood abnormalities are also considered.

❒ LEVEL 1   REVIEW OF CHAPTER OBJECTIVES

1. Describe the important components and major functions of blood.
2. Identify locations of the body used for blood collection and list the basic physical characteristics of the blood samples drawn from these locations.
3. Discuss the composition and functions of plasma.
4. Describe the origin and production of the formed elements in blood.
5. List the characteristics and functions of red blood cells.
6. Describe the structure of hemoglobin and indicate its functions.
7. Describe the recycling system for aged or damaged or abnormal erythrocytes.
8. Define erythropoiesis, identify the stages involved in erythrocyte maturation, and describe the homeostatic regulation of erythrocyte production.
9. List examples of important blood tests and cite the normal values for each test.

10. Explain the importance of blood typing on the basis of ABO and Rh incompatibilities.

11. Categorize the various white blood cells on the basis of their structure and functions and discuss the factors that regulate production of each class.

12. Describe the structure, function, and production of platelets.

13. Discuss mechanisms that control blood loss after an injury and describe the reaction sequences responsible for blood clotting.

[L1]   Multiple choice:

Place the letter corresponding to the correct answer in the space provided.

OBJ. 1 _____ 1. The formed elements of the blood consist of:

    a. antibodies, metalloproteins, and lipoproteins

    b. red and white blood cells, and platelets

    c. albumins, globulins, and fibrinogen

    d. electrolytes, nutrients, and organic wastes

OBJ. 1 _____ 2. Loose connective tissue and cartilage contain a network of insoluble fibers, whereas plasma, a fluid connective tissue, contains:

    a. dissolved proteins

    b. a network of collagen and elastic fibers

    c. elastic fibers only

    d. collagen fibers only

OBJ. 1 _____ 3. Blood transports dissolved gases, bringing oxygen from the lungs to the tissues and carrying:

    a. carbon dioxide from the lungs to the tissues

    b. carbon dioxide from one peripheral cell to another

    c. carbon dioxide from the interstitial fluid to the cell

    d. carbon dioxide from the tissues to the lungs

OBJ. 1 _____ 4. The "patrol agents" in the blood that defend the body against toxins and pathogens are:

    a. hormones and enzymes

    b. albumins and globulins

    c. white blood cells and antibodies

    d. red blood cells and platelets

OBJ. 2 _____ 5. When checking the efficiency of gas exchange at the lungs, blood may be required via:

    a. arterial puncture

    b. venepuncture

    c. puncturing the tip of a finger

    d. a, b, or c could be used

OBJ. 2 _____ 6. Blood temperature is roughly _____ °C, and the blood pH averages _____.

    a. 0 °C; 6.8

    b. 32 °C; 7.0

    c. 38 °C; 7.4

    d. 98 °C; 7.8

Level
-1-

OBJ. 3 _____ 7. Which one of the following statements is correct?

    a. Plasma contributes approximately 92 percent of the volume of whole blood, and H_2O accounts for 55 percent of the plasma volume.

    b. Plasma contributes approximately 55 percent of the volume of whole blood, and H_2O accounts for 92 percent of the plasma volume.

    c. H_2O accounts for 99 percent of the volume of the plasma, and plasma contributes approximately 45 percent of the volume of whole blood.

    d. H_2O accounts for 45 percent of the volume of the plasma, and plasma contributes approximately 99 percent of the volume of whole blood

OBJ. 3 _____ 8. The three primary classes of plasma proteins are:

    a. antibodies, metalloproteins, lipoproteins

    b. serum, fibrin, fibrinogen

    c. albumins, globulins, fibrinogen

    d. heme, porphyrin, globin

OBJ. 3 _____ 9. In addition to water and proteins, the plasma consists of:

    a. erythrocytes, leukocytes, and platelets

    b. electrolytes, nutrients, and organic wastes

    c. albumins, globulins, and fibrinogen

    d. a, b, and c are correct

OBJ. 4 _____ 10. Formed elements in the blood are produced through the process of:

    a. hemolysis

    b. hemopoiesis

    c. diapedesis

    d. erythrocytosis

OBJ. 4 _____ 11. The stem cells that produce all of the blood cells are called:

    a. erythroblasts

    b. rouleaux

    c. hemocytoblasts

    d. plasma cells

OBJ. 5 _____ 12. Circulating mature RBCs lack:

    a. mitochondria

    b. ribosomes

    c. nuclei

    d. a, b, and c are correct

OBJ. 5 _____ 13. The *primary* function of a mature red blood cell is:

    a. transport of respiratory gases

    b. delivery of enzymes to target tissues

    c. defense against toxins and pathogens

    d. a, b, and c are correct

Level
–1–

OBJ. 6   _____ 14. The part of the hemoglobin molecule that *directly* interacts with oxygen is:

      a. globin
      b. the porphyrin
      c. the iron ion
      d. the sodium ion

OBJ. 6   _____ 15. Iron is necessary in the diet because it is involved with:

      a. the prevention of sickle cell anemia
      b. hemoglobin production
      c. prevention of hematuria
      d. a, b, and c are correct

OBJ. 7   _____ 16. The iron extracted from heme molecules during hemoglobin recycling is stored in the protein-iron complexes:

      a. transferrin and porphyrin
      b. biliverdin and bilirubin
      c. urobilin and stercobilin
      d. ferritin and hemosiderin

OBJ. 7   _____ 17. During RBC recycling each heme unit is stripped of its iron and converted to:

      a. biliverdin
      b. urobilin
      c. transferrin
      d. ferritin

OBJ. 8   _____ 18. The primary site of erythropoiesis in the adult is the:

      a. bone marrow
      b. kidney
      c. liver
      d. heart

OBJ. 8   _____ 19. Erythropoietin appears in the plasma when peripheral tissues, especially the kidneys, are exposed to:

      a. extremes of temperature
      b. high urine volumes
      c. excessive amounts of radiation
      d. low oxygen concentrations

OBJ. 9   _____ 20. The RBC test(s) that may be used to determine if a person is anemic is (are):

      a. hematocrit
      b. hemoglobin
      c. RBC count
      d. a, b, and/or c may be used

OBJ. 9   _____ 21. The normal concentration of hemoglobin in blood is:

      a. 4–8 g/dl
      b. 9–11 g/dl
      c. 12–18 g/dl
      d. 19–21 g/dl

Level
–1–

OBJ. 10   _____ 22. Agglutinogens are contained (on, in) the _____ , while the agglutinins are found (on, in) the _____ .

        a. plasma; cell membrane of RBC

        b. nucleus of the RBC; mitochondria

        c. cell membrane of RBC; plasma

        d. mitochondria; nucleus of the RBC

OBJ. 10   _____ 23. If you have type A blood, your plasma holds circulating _____ that will attack _____ erythrocytes.

        a. anti-B agglutinins; Type B

        b. anti-A agglutinins; Type A

        c. anti-A agglutinogens; Type A

        d. anti-A agglutinins; Type B

OBJ. 10   _____ 24. A person with type O blood contains:

        a. anti-A and anti-B agglutinins

        b. anti-O agglutinins

        c. anti-A and anti-B agglutinogens

        d. type O blood lacks agglutinins altogether

OBJ. 11   _____ 25. The two types of *agranular* leukocytes found in the blood are:

        a. neutrophils, eosinophils

        b. leukocytes, lymphocytes

        c. monocytes, lymphocytes

        d. neutrophils, monocytes

OBJ. 11   _____ 26. Based on their staining characteristics, the types of *granular* leukocytes found in the blood are:

        a. lymphocytes, monocytes, erythrocytes

        b neutrophils, monocytes, lymphocytes

        c. eosinophils, basophils, lymphocytes

        d. neutrophils, eosinophils, basophils

OBJ. 11   _____ 27. The number of eosinophils increases dramatically during:

        a. an allergic reaction or a parasitic infection

        b. an injury to a tissue or a bacterial infection

        c. tissue degeneration or cellular deterioration

        d. a, b, and c are correct

OBJ. 12   _____ 28. *Megakaryocytes* are specialized cells of the bone marrow responsible for:

        a. specific immune responses

        b. engulfing invading bacteria

        c. formation of platelets

        d. production of scar tissue in an injured area

OBJ. 12   _____ 29. The rate of megakaryocyte activity and platelet formation is regulated by:

        a. thrombopoietin

        b. interleukin -6

        c. multi–CSF

        d. a, b, and c are correct

Level
-1-

OBJ. 13   _____ 30. *Basophils* are specialized in that they:

a. contain microphages that engulf invading bacteria

b. contain histamine that exaggerates the inflammation response at the injury site

c. are enthusiastic phagocytes, often attempting to engulf items as large or larger than themselves

d. produce and secrete antibodies that attack cells or proteins in distant portions of the body

OBJ. 13   _____ 31. The process of hemostasis includes *five* phases. The correct order of the phases as they occur after injury is as follows:

a. vascular, coagulation, platelet, clot retraction, clot destruction

b. coagulation, vascular, platelet, clot destruction, clot retraction

c. platelet, vascular, coagulation, clot retraction, clot destruction

d. vascular, platelet, coagulation, clot retraction, clot destruction

OBJ. 13   _____ 32. The *extrinsic* pathway involved in blood clotting involves the release of:

a. platelet factors and platelet thromboplastin

b. Ca^{++} and clotting factors VIII, IX, X, XI, XIII

c. tissue factors and tissue thromboplastin

d. prothrombin and fibrinogen

OBJ. 13   _____ 33. The "common pathway" in blood clotting involves the following events in sequential order as follows:

a. tissue factors $\rightarrow$ Ca^{++} $\rightarrow$ plasminogen $\rightarrow$ plasmin

b. prothrombin $\rightarrow$ thrombin $\rightarrow$ fibrinogen $\rightarrow$ fibrin

c. platelet factors $\rightarrow$ Ca^{++} $\rightarrow$ fibrinogen $\rightarrow$ fibrin

d. vascular $\rightarrow$ platelet $\rightarrow$ coagulation $\rightarrow$ destruction

[L1]   Completion:

Using the terms below, complete the following statements.

vascular	diapedesis	platelets
hemoglobin	hemopoiesis	rouleaux
lymphocytes	viscosity	formed elements
plasma	agglutinins	serum
erythroblasts	Coumadin	lymphopoiesis
agglutinogens	vitamin B12	fixed macrophages
venepuncture	hematocrit	monocyte
transferrin		

OBJ. 1   34. The ground substance of the blood is the _____.

OBJ. 1   35. The blood cells and cell fragments suspended in the ground substance are referred to as _____.

OBJ. 2   36. The most common clinical procedure for collecting blood for blood tests is the _____.

OBJ. 3   37. Interactions between the dissolved proteins and the surrounding water molecules determine the blood's _____.

OBJ. 3   38. When the clotting proteins are removed from the plasma, the remaining fluid is the _____.

OBJ. 4   39. The process of blood cell formation is called _____.

OBJ. 5   40. The flattened shape of RBCs enables them to form stacks called _____.

OBJ. 6   41. The part of the red blood cell responsible for the cell's ability to transport oxygen and carbon dioxide is _____.

OBJ. 7   42. During recycling, when iron is extracted from the heme molecules and released into the bloodstream, it binds to the plasma protein _____.

OBJ. 8   43. For erythropoiesis to proceed normally, the myeloid tissues must receive adequate amounts of amino acids, iron, and _____.

OBJ. 8   44. *Very* immature RBCs that actively synthesize hemoglobin are the _____.

OBJ. 9   45. The blood test used to determine the percentage of formed elements in whole blood is the _____.

OBJ. 10   46. Antigens on the surfaces of RBCs whose presence and structure are genetically determined are called _____.

OBJ. 10   47. Immunoglobulins in plasma that react with antigens on the surfaces of foreign red blood cells when donor and recipient differ in blood type are called _____.

OBJ. 11   48. T cells and B cells are representative cell populations of WBCs identified as _____.

OBJ. 11   49. Immobile monocytes found in many connective tissues are called _____.

OBJ. 11   50. The process of WBCs migrating across the endothelial lining of a capillary, squeezing between adjacent endothelial cells, is called _____.

OBJ. 11   51. The production of lymphocytes from stem cells is called _____.

Level -1-

OBJ. 12   52. After a clot has formed, the clot shrinks due to the action of actin and myosin filaments contained in _____.

OBJ. 13   53. The leukocyte that would be involved in producing scar tissue that "walls off" an injured area is a _____.

OBJ. 13   54. The anticoagulant that inactivates thrombin is _____.

OBJ. 13   55. The first phase of homeostasis involves a period of local vasoconstriction called the _____ phase.

[L1]   Matching:

Match the terms in column B with the terms in column A. Use letters for answers in the spaces provided.

PART I   **COLUMN A**   **COLUMN B**

OBJ. 1   _____   56. whole blood   A. hematocytoblasts
OBJ. 1   _____   57. pH regulation   B. heme
OBJ. 2   _____   58. median cubital vein   C. buffer in blood
OBJ. 3   _____   59. antibodies   D. hormone
OBJ. 3   _____   60. globulin   E. hemoglobin
OBJ. 4   _____   61. myeloid tissue   F. jaundice
OBJ. 4   _____   62. stem cells   G. bone marrow
OBJ. 5   _____   63. red blood cells   H. immunoglobulin
OBJ. 6   _____   64. porphyrin   I. transport protein
OBJ. 7   _____   65. bilirubin   J. plasma and formed elements
OBJ. 8   _____   66. erythropoietin   K. venepuncture

PART II   **COLUMN A**   **COLUMN B**

OBJ. 8   _____   67. reticulocytes   L. specific immunity
OBJ. 9   _____   68. excessive RBCs   M. depresses clotting
OBJ. 10   _____   69. agglutinogens   N. A, B, and D
OBJ. 10   _____   70. cross-reaction   O. neutrophils, eosinophils
OBJ. 11   _____   71. microphages   P. syneresis
OBJ. 11   _____   72. neutrophils   Q. polymorphonuclear leukocytes
OBJ. 11   _____   73. lymphocytes   R. immature RBCs
OBJ. 11   _____   74. monocyte   S. agglutination
OBJ. 12   _____   75. platelets   T. macrophage
OBJ. 13   _____   76. coumadin   U. erythrocytosis
OBJ. 13   _____   77. clot retraction   V. cell fragments

Level
-1-

[L1]   Drawing/Illustration Labeling:

Identify each numbered structure by labeling the following figures:

OBJ. 11   **FIGURE 19.1**   Agranular Leukocytes

79 _____

78 _____

80 _____

OBJ. 11   **FIGURE 19.2**   Granular Leukocytes

82 _____

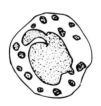

81 _____

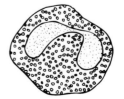

83 _____

Level
–1–

OBJ. 5   *FIGURE 19.3*

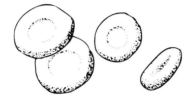

84 _____

☐ LEVEL 2   CONCEPT SYNTHESIS

Concept Map I:

Using the following terms, fill in the circled, numbered, blank spaces to complete the concept map. Follow the numbers that comply with the organization of the map.

Neutrophils	Leukocytes	Solutes	Plasma
Albumins	Monocytes	Gamma	

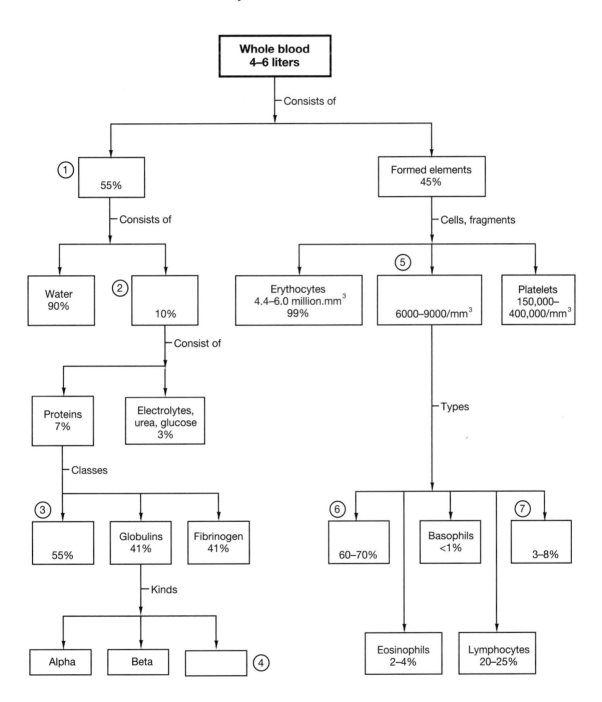

Concept Map II:

Using the following terms, fill in the circled, numbered, blank spaces to complete the concept map. Follow the numbers that comply with the organization of the map.

Clot dissolves   Plasminogen   Forms plug

Platelet phase   Plasmin   Clot retraction

Forms blood clot   Vascular spasm

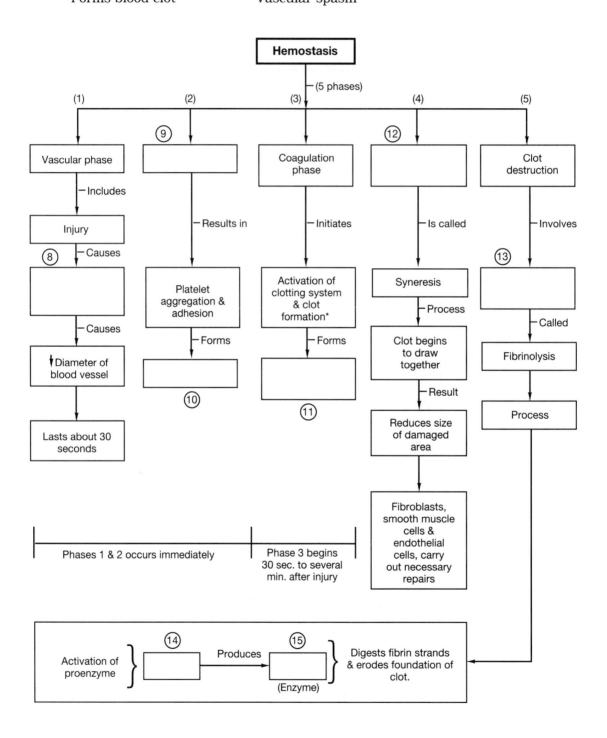

Concept Map III:

Using the following terms, fill in the circled, numbered, blank spaces to complete the concept map. Follow the numbers that comply with the organization of the map.

Sickle cell anemia   Iron deficiency   Polycythemia
Abnormally large   Polycythemia vera   Hemorrhagic anemia
    number of RBCs   Thalassemia

Erythrocyte disorders

├─ Abnormal numbers ─ 4 types
│
│   ⑯ [____]   Erythrocytosis   ⑱ [____]   Anemia
│
│   Result ├─ Elevated hematocrit; normal blood volume
│   ⑰ [____]
│   Result ├─ ↑In hematopoiesis. Affects all cell elements of blood
│
└─ Abnormal structure ─ 2 types
    ㉑ [____]   ㉒ [____]
    Result ├─ Change in A:A. Sequence in one of globin chains of heme molecule
    Result ├─ Genetic inability to produce adequate amount of one of four globin chains

↓Hematocrit; abnormal small # of blood cells   Low hemo. conc.; normal hematocrit   Tissues become O₂ starved due to circulatory blockage — Result

Types of anemia

⑲ [____]   Aplastic anemia   ⑳ [____]   Pernicious anemia

Result ├─ Severe bleeding
├─ ↓ Marrow prod. of RBC: # of reticulocytes low
Result ├─ RCB↓ in synthesis of hemoglobin
├─ RBC maturation ceases due to inadequate supply of vit B₁₂

Concept Map IV:

Using the following terms, fill in the circled, numbered, blank spaces to complete the concept map. Follow the numbers that comply with the organization of the map.

Factor VII & Ca⁺⁺   Intrinsic pathway

Platelets   Inactive Factor XI

Thrombin   Fibrinogen

Active Factor IX   Tissue damage

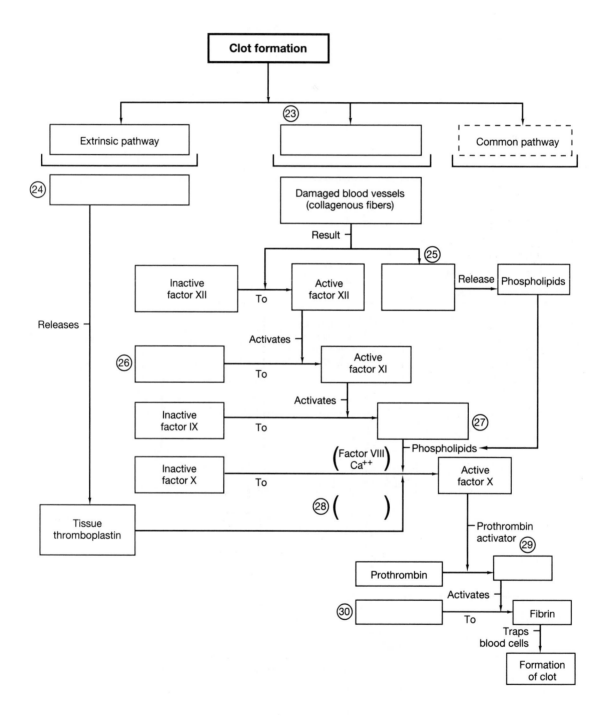

Body Trek:

Using the terms below, fill in the blanks to complete a trek into the blood vessel lining of a superficial vein.

plasminogen	coagulation	platelet	fibrinogen
vascular	fibrin	clot	clot retraction
endothelium	thrombocytes	platelet plug	fibrinolysis
RBCs	vascular spasm	plasmin	

Robo's programmed assignment for this trek is to monitor and collect information related to the clotting process in a superficial vein that has been incised and damaged because of an injury. The micro-robot is located just inside the damaged area of the blood vessel lining, the (31) _____. There is an immediate "feeling" of being crushed owing to a (32) _____ that decreases the diameter of the vessel at the site of the injury. Not only is Robo in danger of being crushed, but the turbulent and sticky "atmospheric" conditions threaten the robot's sensory devices. The turbulence increases as fragments with rough edges, the (33) _____, flow into the damaged area, where they immediately adhere to the exposed surfaces of the interior lining of the vessel wall. As more fragments arrive, they begin to stick to each other, forming a (34) _____. The vessel constriction, fragment aggregation, and adhesion comprise the (35) _____ phase and the (36) _____ phase of the clotting response. Thirty seconds to several minutes later, the damaged site is a mecca of chemical activity as the (37) _____ phase begins. Robo's "chemical sensors" are relaying messages to mission control about a series of complex, sequential steps involving the conversion of circulating (38) _____ into the insoluble protein (39) _____. The robot is maneuvered from the immediate area to make room for additional (40) _____ and fragments that are arriving and becoming trapped within the fibrous tangle, forming a (41) _____ that seals off the damaged portion of the vessel. The injury site still has torn edges, initiating a robotic response that facilitates Robo's exit to the outside of the vessel, where the final stages of the clotting response will be monitored. The trek to the outside is "tight" due to the (42) _____ process, which involves the actions of fibroblasts, smooth muscle cells, and endothelial cells in the area to carry out the necessary repairs. As the repairs proceed, the robot's "chemical sensors" are again active owing to the process of (43) _____, or clot destruction. The proenzyme (44) _____ activates the enzyme (45) _____, which begins digesting the fibrin strands and ending the formation of the clot.

Mission accomplished, Robo is returned to Mission Control for reprogramming and recharging in preparation for its next trek.

[L2] Multiple Choice:

Place the letter corresponding to the correct answer in the space provided.

_____ 46. Because the concentration of dissolved gases is different between the plasma and the tissue fluid:

a. O_2 will tend to diffuse from the plasma to the interstitial fluid, and CO_2 will tend to diffuse in the opposite direction

b. CO_2 will tend to diffuse from the plasma to the interstitial fluid, and O_2 will tend to diffuse in the opposite direction

c. both O_2 and CO_2 will tend to diffuse from the plasma to the interstitial fluid

d. none of the above statements is true

_____ 47. If blood comprises 7 percent of the body weight in kilograms, how many liters of blood would there be in an individual who weighs 85 kg?

    a. 5.25
    b. 5.95
    c. 6.25
    d. 6.95

_____ 48. The reason liver disorders can alter the composition and functional properties of the blood is because:

    a. the liver synthesizes immunoglobulins and protein hormones
    b. the proteins synthesized by the liver are filtered out of the blood by the kidneys
    c. the liver is the primary source of plasma proteins
    d. the liver serves as a filter for plasma proteins and pathogens

_____ 49. When plasma O_2 concentrations are falling, the rising plasma CO_2 binds to the _____ portion of the hemoglobin molecule.

    a. globin
    b. heme
    c. porphyrin
    d. none of these

_____ 50. The fate of iron after hemoglobin breakdown occurs as follows:

    a. iron $\rightarrow$ bloodstream $\rightarrow$ transferrin $\rightarrow$ new hemoglobin
    b. iron $\rightarrow$ bilirubin $\rightarrow$ liver $\rightarrow$ bile $\rightarrow$ excreted
    c. iron $\rightarrow$ bloodstream $\rightarrow$ transferrin $\rightarrow$ (stored) ferritin and hemosiderin
    d. all of the above

_____ 51. Signs of iron deficiency anemia include:

    a. hematocrit declines; hemoglobin content and O_2-carrying capacity reduced
    b. hematocrit increases; hemoglobin decreases; O_2-carrying capacity decreases
    c. hematocrit increases; hemoglobin increases; O_2-carrying capacity increases
    d. hematocrit decreases; hemoglobin decreases; O_2-carrying capacity increases

_____ 52. On the average, one microliter of blood contains _____ erythrocytes.

    a. 260 million
    b. 10^{13}
    c. 5.2 million
    d. 25 million

_____ 53. In a mature RBC, energy is obtained exclusively by anaerobic respiration, i.e., the breakdown of glucose in the cytoplasm.

    a. true
    b. false

Level
=2=

_____ 54. Protein synthesis in a mature RBC occurs primarily in:

    a. ribosomes

    b. mitochondria

    c. nucleus

    d. none of these

_____ 55. If agglutinogen "B" meets with agglutinin "anti-A," the result would be:

    a. a cross-reaction

    b. no agglutination would occur

    c. the patient would be comatose

    d. the patient would die

_____ 56. The precursor of all blood cells in the human body is the:

    a. myeloblast

    b. hemocytoblast

    c. megakaryocyte

    d. none of these

_____ 57. _Reticulocytes_ are nucleated immature cells that develop into mature:

    a. lymphocytes

    b. platelets

    c. leukocytes

    d. erythrocytes

_____ 58. The blood cells that may originate in the thymus, spleen, and lymph nodes as well as in the bone marrow are the:

    a. erythrocytes

    b. lymphocytes

    c. leukocytes

    d. monocytes

_____ 59. Rh-negative blood indicates:

    a. the presence of the Rh agglutinogen

    b. the presence of agglutinogen A

    c. the absence of the Rh agglutinogen

    d. the absence of agglutinogen B

_____ 60. In hemolytic disease of the newborn:

    a. the newborn's agglutinins cross the placental barrier and cause the newborn's RBCs to degenerate

    b. the mother's agglutinins cross the placental barrier and destroy fetal red blood cells

    c. the mother's agglutinogens destroy her own RBC, causing deoxygenation of the newborn

    d. all of the above occur

_____ 61. _Circulating_ leukocytes represent a small fraction of the total population, since most WBCs are found in peripheral tissues.

    a. true

    b. false

Level
=2=

_____ 62. A typical microliter of blood contains _____ leukocytes.

        a. 1,000–2,000

        b. 3,000–5,000

        c. 6,000–9,000

        d. 10,000–12,000

_____ 63. Under "normal" conditions, neutrophils comprise _____ of the circulatory white blood cells.

        a. 10–20 percent

        b. 20–40 percent

        c. 50–70 percent

        d. 85–95 percent

_____ 64. A noticeable feature of a leukemia is:

        a. presence of abnormal granulocytes

        b. decline of RBC and platelet formations

        c. presence of abnormal neutrophils, eosinophils, basophils

        d. all of the above are noticeable

_____ 65. Platelets are unique formed elements of the blood because they:

        a. initiate the immune response and destroy bacteria

        b. are cytoplasmic enzyme packets rather than individual cells

        c. can reproduce because they contain a nucleus

        d. none of the above apply to platelets

_____ 66. Clot destruction involves a process that begins with:

        a. activation of the proenzyme fibrinogen which initiates the production of fibrin

        b. activation of prothrombin, which initiates the production of thrombin

        c. activation of Ca^{++} to produce tissue plasmin

        d. activation of the proenzyme plasminogen, which initiates the production of plasmin

_____ 67. The major effect of a vitamin K deficiency in the body is that it leads to:

        a. a breakdown of the common pathway, inactivating the clotting system

        b. the body becomes insensitive to situations that would necessitate the clotting mechanism

        c. an overactive clotting system, which might necessitate thinning of the blood

        d. all of the above are correct

_____ 68. A vitamin B_{12} deficiency results in the type of anemia known as:

        a. pernicious

        b. aplastic

        c. hemorrhagic

        d. sickle cell

Level
=2=

92. What is the difference between an *embolus* and a *thrombus*?

93. What two primary effects does erythropoietin have on the process of erythropoiesis?

When you have successfully completed the exercises in L2 proceed to L3.

☐ LEVEL 3   CRITICAL THINKING/APPLICATION

Using principles and concepts learned about the blood, answer the following questions. Write your answers on a separate sheet of paper.

1. A man weighs 154 pounds and his hematocrit is 45 percent. The weight of blood is approximately 7 percent of his total body weight.

    a. What is his total blood volume in liters?

    b. What is his total cell volume in liters?

    c. What is his total plasma volume in liters?

    d. What percent of his blood constitutes the plasma volume?

    e. What percent of his blood comprises the formed elements?

    Note: 1 kilogram = 2.2 pounds

2. Considering the following condition, what effect would each have on an individual's hematocrit (percentage of erythrocytes to the total blood volume)?

    a. an increase in viscosity of the blood

    b. an increased dehydration in the body

    c. a decrease in the number of RBCs

    d. a decrease in the diameter of a blood vessel

3. A 70-year-old woman receives a laboratory report containing the following information:

    WBC: 15,000/mm^3 platelets: 600,000/mm^3

    RBC: 10.5 million/mm^3 hematocrit: 75%

    total blood volume: 8 liters

    Her clinical symptoms included: increased viscosity of blood, clogging of capillaries, and sluggish flow of blood through vessels.

    What condition is associated with these clinical signs and symptoms, and what usually occurs with this type of abnormality?

4. A person with type B blood has been involved in an accident and excessive bleeding necessitates a blood transfusion. Due to an error by a careless laboratory technician, the person is given type A blood. Explain what will happen.

5. Suppose you are a young woman who has Rh-negative blood. You are intent on marrying a young man who has Rh-positive blood. During your first pregnancy your gynecologist suggests that you be given RhoGAM injections. What is the purpose of the injections and how does the RhoGAM affect the prospective mother's immune system?

6. What type of anemia would you associate with the following results of RBC tests? What causes this type of anemia?

      Hematocrit: 35   hemoglobin: 9.0 g/dl reticulocyte count: 0.2%

      MGV: 103.6   MCHC: 37

7. Your doctor has recommended that you take a vitamin supplement of your choice. You look for a supplement that lists all the vitamins; however, you cannot find one that has vitamin K.

    a. Why?

    b. What are some dietary sources of vitamin K?

    c. What is the function of vitamin K in the body?

Level
3

20

The Heart

■ Overview

The heart is one of nature's most efficient and durable pumps. Throughout life, it beats 60 to 80 times per minute, supplying oxygen and other essential nutrients to every cell in the body and removing waste for elimination from the lungs or the kidneys. Every day approximately 2100 gallons of blood is pumped through a vast network of about 60,000 miles of blood vessels. The "double pump" design of this muscular organ facilitates the continuous flow of blood through a closed system that consists of a pulmonary circuit to the lungs and a systemic circuit serving all other regions of the body. A system of valves in the heart maintains the "one-way" direction of blood flow through the heart. Nutrient and oxygen supplies are delivered to the heart's tissues by coronary arteries that encircle the heart like a crown. Normal cardiac rhythm is maintained by the heart's electrical system, centered primarily in a group of specialized pacemaker cells located in a sinus node in the right atrium.

Chapter 20 examines the structural features of the heart, cardiac physiology, cardiac dynamics, and the heart and the cardiovascular system.

☐ LEVEL 1   REVIEW OF CHAPTER OBJECTIVES

1. Describe the location and general features of the heart.
2. Describe the structure of the pericardium and explain its functions.
3. Trace the flow of blood through the heart, identifying the major blood vessels, chambers, and heart valves.
4. Identify the layers of the heart wall.
5. Describe the vascular supply and innervation of the heart.
6. Describe the events of an action potential in cardiac muscle and explain the importance of calcium ions to the contractile process.
7. Discuss the differences between nodal cells and conducting fibers and describe the components and functions of the conducting system of the heart.
8. Identify the electrical events associated with a normal electrocardiogram.
9. Explain the events of the cardiac cycle, including atrial and ventricular systole and diastole, and relate the heart sounds to specific events in this cycle.

10. Define stroke volume and cardiac output and describe the factors that influence these values.

11. Explain how adjustments in stroke volume and cardiac output are coordinated at different levels of activity.

12. Describe the effects of autonomic activity on heart function.

13. Describe the effects of hormones, drugs, temperature, and changes in ion concentrations on the heart.

[L1]   Multiple Choice:

Place the letter corresponding to the correct answer in the space provided.

OBJ. 1 _____ 1. The "double pump" function of the heart includes the right side, which serves as the _____ circuit pump, while the left side serves as the _____ pump.

    a. systemic; pulmonary

    b. pulmonary; hepatic portal

    c. hepatic portal; cardiac

    d. pulmonary; systemic

OBJ. 1 _____ 2. The major difference between the left and right ventricles relative to their role in heart function is:

    a. the L.V. pumps blood through the short, low-resistance pulmonary circuit

    b. the R.V. pumps blood through the low-resistance systemic circulation

    c. the L.V. pumps blood through the high-resistance systemic circulation

    d. The R.V. pumps blood through the short, high-resistance pulmonary circuit

OBJ. 1 _____ 3. The average *maximum* pressure developed in the right ventricle is about:

    a. 15–28 mm Hg

    b. 50-60 mm Hg

    c. 67–78 mm Hg

    d. 80–120 mm Hg

OBJ. 2 _____ 4. The visceral pericardium, or epicardium, covers the:

    a. inner surface of the heart

    b. outer surface of the heart

    c. vessels in the mediastinum

    d. endothelial lining of the heart

OBJ. 2 _____ 5. The valves of the heart are covered by a squamous epithelium, the:

    a. endocardium

    b. epicardium

    c. visceral pericardium

    d. parietal pericardium

Level -1-

OBJ. 3 _____ 6. Atrioventricular valves prevent backflow of blood into the _____; semilunar valves prevent backflow into the _____.

    a. atria; ventricles

    b. lungs; systemic circulation

    c. ventricles; atria

    d. capillaries; lungs

OBJ. 3 _____ 7. Blood flows from the left atrium into the left ventricle through the _____ valve.

    a. bicuspid

    b. L. atrioventricular

    c. mitral

    d. a, b, and c are correct

OBJ. 3 _____ 8. When deoxygenated blood leaves the right ventricle through a semilunar valve, it is forced into the:

    a. pulmonary veins

    b. aortic arch

    c. pulmonary arteries

    d. lung capillaries

OBJ. 3 _____ 9. Blood from systemic circulation is returned to the right atrium by the:

    a. superior and inferior vena cava

    b. pulmonary veins

    c. pulmonary arteries

    d. brachiocephalic veins

OBJ. 3 _____ 10. Oxygenated blood from the systemic arteries flows into:

    a. peripheral tissue capillaries

    b. systemic veins

    c. the right atrium

    d. the left atrium

OBJ. 3 _____ 11. The lung capillaries receive deoxygenated blood from the:

    a. pulmonary veins

    b. pulmonary arteries

    c. aorta

    d. superior and inferior vena cava

OBJ. 4 _____ 12. One of the important differences between skeletal muscle tissue and cardiac muscle tissue is that cardiac muscle tissue is:

    a. striated voluntary muscle

    b. multinucleated

    c. comprised of unusually large cells

    d. striated involuntary muscle

Level
-1-

OBJ. 4

_____ 13. Cardiac muscle tissue:

   a. will not contract unless stimulated by nerves

   b. does not require nerve activity to stimulate a contraction

   c. is under voluntary control

   d. a, b, and c are correct

OBJ. 5

_____ 14. Blood from coronary circulation is returned to the right atrium of the heart via:

   a. anastomoses

   b. the circumflex branch

   c. coronary sinus

   d. the anterior interventricular branch

OBJ. 5

_____ 15. The right coronary artery supplies blood to:

   a. the right atrium

   b. portions of the conducting system of the heart

   c. portions of the right and left ventricles

   d. a, b, and c are correct

OBJ. 6

_____ 16. The correct sequential path of a normal action potential in the heart is:

   a. SA node → AV bundle → AV node → Purkinje fibers

   b. AV node → SA node → AV bundle → bundle of His

   c. SA node → AV node → bundle of His → bundle branches → Purkinje fibers

   d. SA node → AV node → bundle branches → AV bundle → Purkinje fibers

OBJ. 6

_____ 17. An above-normal increase in K^+ concentrations in the blood would cause:

   a. the heart to contract spastically

   b. an increase in membrane potentials from the SA node

   c. the heart to dilate and become flaccid and weak

   d. action potentials to increase in intensity until the heart fails

OBJ. 7

_____ 18. The sinoatrial node acts as the pacemaker of the heart because these cells are:

   a. located in the wall of the left atrium

   b. the only cells in the heart that can conduct an impulse

   c. the only cells in the heart innervated by the autonomic nervous system

   d. the ones that depolarize and reach threshold first

OBJ. 7

_____ 19. After the AV node is depolarized and the impulse spreads through the atria, there is a slight delay before the impulse spreads to the ventricles. The reason for this delay is to allow:

   a. the atria to finish contracting

   b. the ventricles to repolarize

   c. a greater venous return

   d. nothing; there is no reason for the delay

Level
–1–

OBJ. 8   _____ 20. If each heart muscle cell contracted at its own individual rate, the condition would resemble:

   a. heart flutter
   b. bradycardia
   c. fibrillation
   d. myocardial infarction

OBJ. 8   _____ 21. The P wave of a normal electrocardiogram indicates:

   a. atrial repolarization
   b. atrial depolarization
   c. ventricular repolarization
   d. ventricular depolarization

OBJ. 8   _____ 22. The QRS complex of the ECG appears as the:

   a. atria depolarize
   b. atria repolarize
   c. ventricles repolarize
   d. ventricles depolarize

OBJ. 8   _____ 23. ECGs are useful in detecting and diagnosing abnormal patterns of cardiac activity called:

   a. myocardial infarctions
   b. cardiac arrhythmias
   c. excitation-contraction coupling
   d. autorhythmicity

OBJ. 8   _____ 24. Ultrasound analysis producing images that follow the details of cardiac contractions is called:

   a. electrocardiography
   b. echocardiography
   c. arteriography
   d. angiography

OBJ. 9   _____ 25. The "lubb-dubb" sounds of the heart have practical clinical value because they provide information concerning:

   a. the strength of ventricular contraction
   b. the strength of the pulse
   c. the efficiency of heart valves
   d. the relative time the heart spends in systole and diastole

OBJ. 9   _____ 26. When a chamber of the heart fills with blood and prepares for the start of the next cardiac cycle the heart is in:

   a. systole
   b. ventricular ejection
   c. diastole
   d. isovolumetric contraction

OBJ. 9 _____ 27. At the start of atrial systole, the ventricles are filled to around:

    a. 10 percent of capacity

    b. 30 percent of capacity

    c. 50 percent of capacity

    d. 70 percent of capacity

OBJ. 10 _____ 28. The amount of blood ejected by the left ventricle per minute is the:

    a. stroke volume

    b. cardiac output

    c. end-diastolic volume

    d. end-systolic volume

OBJ. 10 _____ 29. The amount of blood pumped out of each ventricle during a single beat is the:

    a. stroke volume

    b. EDV

    c. cardiac output

    d. ESV

OBJ. 11 _____ 30. Starling's law of the heart states:

    a. stroke volume is equal to cardiac output

    b. venous return is inversely proportional to cardiac output

    c. cardiac output is equal to venous return

    d. a, b, and c are correct

OBJ. 11 _____ 31. Physicians are interested in cardiac output because it provides a useful indication of:

    a. ventricular efficiency over time

    b. valvular malfunctions

    c. atrial efficiency in respect to time

    d. the amount of blood ejected by each ventricle

OBJ. 12 _____ 32. The factor that prevents tetanization of cardiac muscle is:

    a. the cardiac muscle fibers are specialized for contraction

    b. the refractory period of cardiac muscle is relatively brief

    c. the action potential of the muscle triggers a single contraction

    d. the muscle cell completes its contraction before the membrane can respond to an additional stimulus

OBJ. 12 _____ 33. The cardioacceleratory center in the medulla is responsible for the activation of:

    a. baroreceptors

    b. parasympathetic neurons

    c. sympathetic neurons

    d. chemoreceptors

Level -1-

OBJ. 13   _____ 34. Caffeine and nicotine cause:

    a. a decrease in the heart rate

    b. a depressing effect on excitable membranes in the nervous system

    c. an increase in the heart rate

    d. do not have an effect on the herat rate

OBJ. 13   _____ 35. An elevated body temperature:

    a. depresses the heart rate, but increases the contractile force

    b. accelerates the heart rate and increases the contractile force

    c. depresses the heart rate and contractile force

    d. does not affect the heart rate

[L1]   Completion:

Using the terms below, complete the following statements.

autonomic nervous system	repolarization	hyperpolarization
intercalated discs	atria	left ventricle
Bainbridge reflex	nodal cells	stroke volume
pulmonary	auscultation	chemoreceptors
electrocardiogram	carbon dioxide	fibrous skeleton
pulmonary veins	coronary sinus	automaticity
pericardium	myocardium	

OBJ. 1   36. The chambers in the heart with thin walls that are highly distensible are the _____.

OBJ. 1   37. The internal connective tissue network of the heart is called the _____.

OBJ. 1   38. Each cardiac muscle fiber contacts several others at specialized sites known as _____.

OBJ. 2   39. The serous membrane lining the pericardial cavity is called the _____.

OBJ. 3   40. The right side of the heart contains blood with an abundance of the gas _____.

OBJ. 3   41. When blood leaves the left atrium, it is forced through an atrioventricular valve into the _____.

OBJ. 3   42. Oxygenated blood is returned to the left atrium via the _____.

OBJ. 3   43. The only arteries in the body that carry deoxygenated blood are the _____ arteries.

OBJ. 4   44. The muscular wall of the heart which forms both atria and ventricles is the _____.

OBJ. 5   45. Blood returns to the heart from coronary circulation via the _____.

OBJ. 6   46. When slow potassium channels begin opening in cardiac muscle, the result is a period of _____.

OBJ. 7   47. The individual units responsible for establishing the rate of cardiac contraction are the _____.

OBJ. 7

48. The action of cardiac muscle tissue contracting on its own in the absence of neural stimulation is called _____.

OBJ. 8

49. The recording and evaluating of electrical events that occur in the heart constitutes a(n) _____.

OBJ. 9

50. Listening to sounds in the chest to determine the condition of the heart and the lungs is called _____.

OBJ. 10

51. An increase in heart rate in response to stretching the right atrial wall is the _____.

OBJ. 10

52. Blood CO_2, pH, and O_2 levels are monitored by _____.

OBJ. 11

53. The amount of blood ejected by a ventricle during a single beat is the _____.

OBJ. 12

54. The most important control of heart rate is the effect of the _____.

OBJ. 13

55. An increase in extracellular potassium concentraiton leads to membrane _____.

[L1]   Matching:

Match the terms in column B with the terms in column A. Use letters for answers in the spaces provided.

PART I

			COLUMN A	COLUMN B
OBJ. 1	____	56.	fossa ovalis	A. blood to systemic arteries
OBJ. 1	____	57.	cardiocytes	B. semilunar
OBJ. 2	____	58.	serous membrane	C. muscular wall of the heart
OBJ. 3	____	59.	tricuspid valve	D. sodium ion permeability
OBJ. 3	____	60.	aortic valve	E. atrial septum structure
OBJ. 3	____	61.	brachiocephalic artery	F. right atrioventricular
OBJ. 3	____	62.	aorta	G. interconnections between blood vessels
OBJ. 4	____	63.	myocardium	H. cardiac muscle fibers
OBJ. 5	____	64.	anastomoses	I. pericardium
OBJ. 6	____	65.	rapid depolarization	J. blood to the arm

PART II

			COLUMN A	COLUMN B
OBJ. 7	____	66.	cardiac pacemaker	K. Bainbridge reflex
OBJ. 7	____	67.	bundle of His	L. AV conducting fibers
OBJ. 8	____	68.	P wave	M. monitor blood pressure
OBJ. 9	____	69.	"lubb" sound	N. decreasing heart rate
OBJ. 9	____	70.	"dubb" sound	O. SA node
OBJ. 10	____	71.	atrial reflex	P. AV valves close; semilunar valves open
OBJ. 10	____	72.	baroreceptors	Q. increases heart rate
OBJ. 11	____	73.	cardiac output	R. semilunar valve closes
OBJ. 12	____	74.	sympathetic activation	S. atrial depolarization
OBJ. 13	____	75.	increasing extracellular K^+	T. SV X HR
OBJ. 13	____	76.	epinephrine	U. release of norepinephrine

Level
–1–

[L1]   Drawing/Illustration Labeling:

Identify each numbered structure by labeling the following figures:

OBJ. 1   **FIGURE 20.1**   Anatomy of the Heart (ventral view)

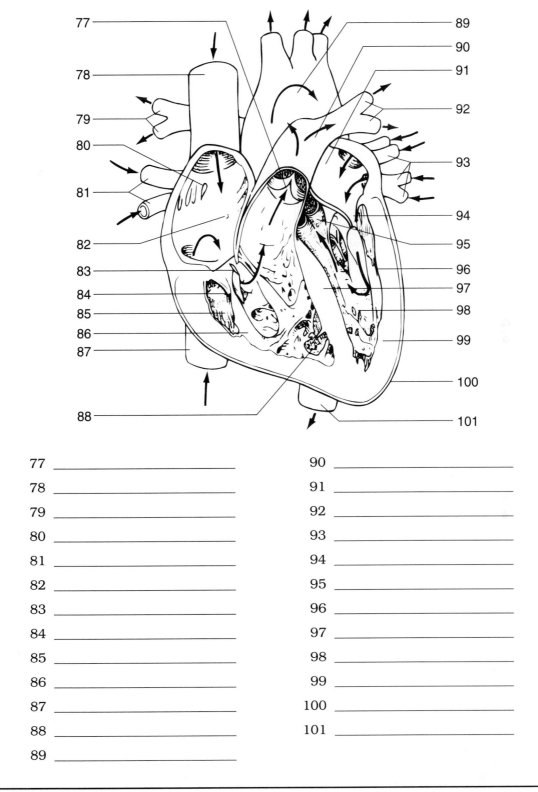

77 _____	90 _____
78 _____	91 _____
79 _____	92 _____
80 _____	93 _____
81 _____	94 _____
82 _____	95 _____
83 _____	96 _____
84 _____	97 _____
85 _____	98 _____
86 _____	99 _____
87 _____	100 _____
88 _____	101 _____
89 _____	

When you have successfully completed the exercises in L1 proceed to L2.

Level
–1–

□ LEVEL 2   CONCEPT SYNTHESIS

Concept Map I:

Using the following terms, fill in the circled, numbered, blank spaces to complete the concept map. Follow the numbers that comply with the organization of the map.

Epicardium	Tricuspid	Two atria
Pacemaker cells	Blood from atria	Deoxygenated blood
Endocardium	Two semilunar	Oxygenated blood
Aortic		

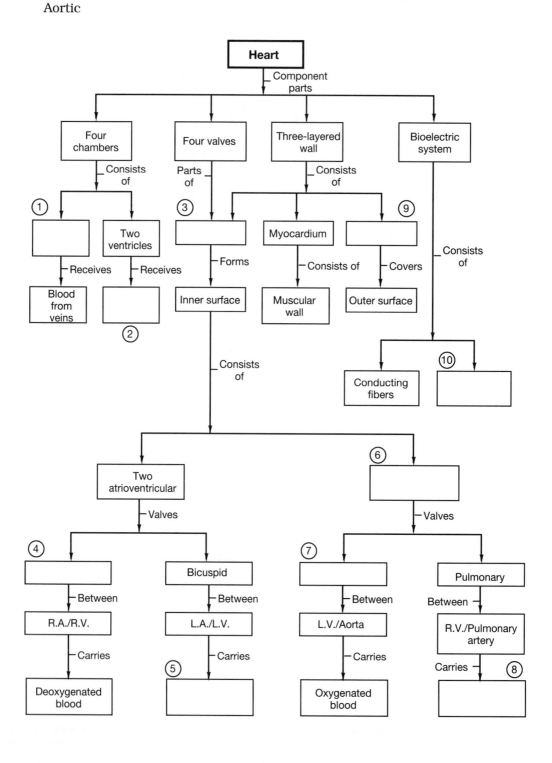

Concept Map II:

Using the following terms, fill in the circled, numbered, blank spaces to complete the concept map. Follow the numbers that comply with the organization of the map.

Decreasing parasympathetic stimulation to heart
Increasing end-diastolic ventricular volume
Increasing cardiac output
Increasing atrial pressure
Increasing stroke volume
Decreasing intrathoracic pressure

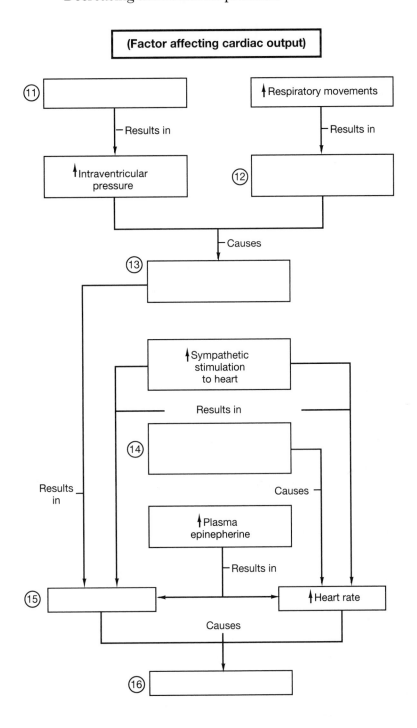

Concept Map III:

Using the following terms, fill in the circled, numbered, blank spaces to complete the concept map. Follow the numbers that comply with the organization of the map.

Increasing blood pressure, decreasing blood pressure

Increasing epinephrine, norepinephrine

Increasing heart rate

Decreasing CO_2

Increasing CO_2

Increasing acetylcholine

Carotid and aortic bodies

Decreasing heart rate

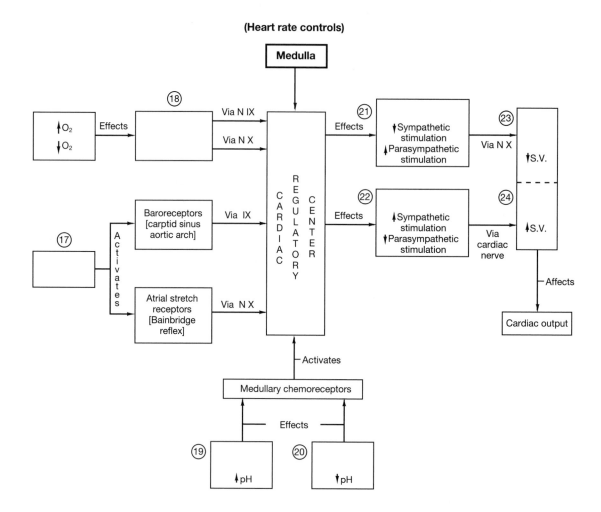

(Heart rate controls)

Body Trek:

Using the terms below, fill in the blanks to complete a trek through the heart and peripheral blood vessels.

aortic semi-lunar valve
pulmonary semi-lunar valve
inferior vena cava
left atrium
tricuspid valve

L. common carotid artery
systemic arteries
bicuspid valve
pulmonary arteries
superior vena cava

pulmonary veins
right atrium
left ventricle
systemic veins
aorta
right ventricle

Robo's instructions: On this trek you will use the "heart map" to follow a drop of blood as it goes through the heart and peripheral blood vessels. Identify each structure at the sequential, circled, numbered locations — 25 through 40.

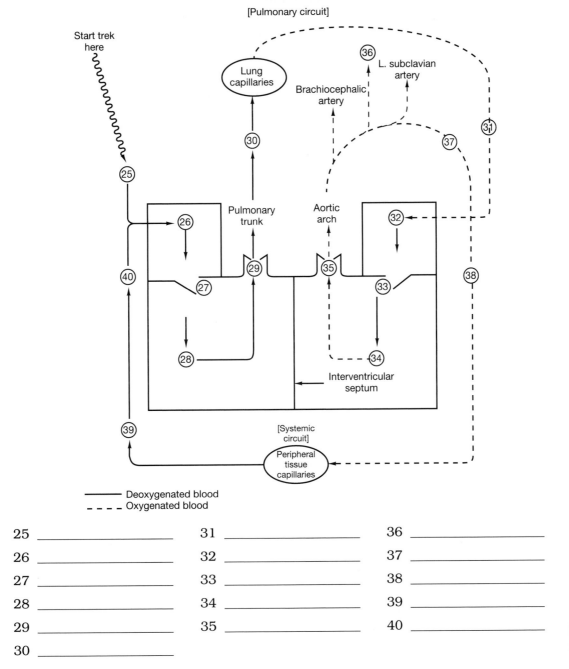

**Ventral view of heart–
path of circulation–"Heart Map"**

25 _____

26 _____

27 _____

28 _____

29 _____

30 _____

31 _____

32 _____

33 _____

34 _____

35 _____

36 _____

37 _____

38 _____

39 _____

40 _____

Level
=2=

[L2] Multiple Choice:

Place the letter corresponding to the correct answer in the space provided.

_____ 41. Assuming anatomic position, the *best* way to describe the *specific* location of the heart in the body is:

      a. within the mediastinum of the thorax

      b. in the region of the fifth intercostal space

      c. just behind the lungs

      d. in the center of the chest

_____ 42. The function of the chordae tendinae is to:

      a. anchor the semilunar valve flaps and prevent backward flow of blood into the ventricles

      b. anchor the AV valve flaps and prevent backflow of blood into the atria

      c. anchor the bicuspid valve flaps and prevent backflow of blood into the ventricle

      d. anchor the aortic valve flaps and prevent backflow into the ventricles

_____ 43. Which one of the following would not show up on an electrocardiogram?

      a. abnormal heart block

      b. murmurs

      c. heart block

      d. bundle branch block

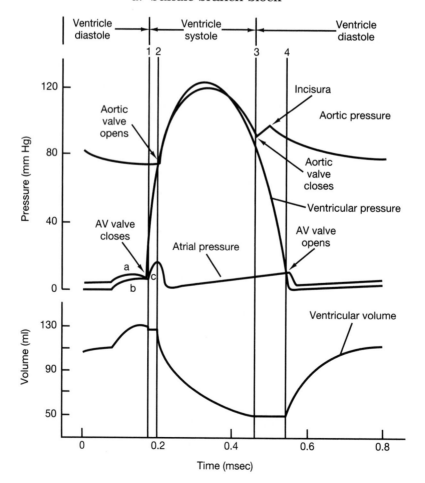

[To answer questions 44 – 51 refer to the graph on page 374, which records the events of the cardiac cycle.]

_____ 44. During ventricular diastole, when the pressure in the left ventricle rises above that in the left atrium:

      a. the left AV valve closes

      b. the left AV valve opens

      c. the aortic valve closes

      d. all the valves close

_____ 45. During ventricular systole, the blood volume in the atria is _____, and the volume in the ventricle is _____.

      a. decreasing; increasing

      b. increasing; decreasing

      c. increasing; increasing

      d. decreasing; decreasing

_____ 46. During _most_ of ventricular diastole:

      a. the pressure in the L. atrium is slightly lower than the pressure in the L. ventricle

      b. the pressure in the L. ventricle reaches 120 mm Hg while the pressure in the L. atrium reaches 30 mm Hg

      c. the pressure in the L. ventricle is slightly lower than the pressure in the L. atrium

      d. the pressures are the same in the L. ventricle and the L. atrium

_____ 47. When the pressure within the L. ventricle becomes greater than the pressure within the aorta:

      a. the pulmonary semilunar valve is forced open

      b. the aortic semilunar valve closes

      c. the pulmonary semilunar valve closes

      d. the aortic semilunar valve is forced open

_____ 48. The volume of blood in the L. ventricle is at its lowest when:

      a. the ventricular pressure is 40 mm of Hg

      b. the AV valve opens

      c. the atrial pressure is 30 mm of Hg

      d. the AV valve closes

_____ 49. The incisura on the graph produces a brief rise in the aortic pressure. The rise in pressure is due to:

      a. the closure of the semilunar valve

      b. a decrease in the ventricular pressure

      c. the opening of the AV valve

      d. an increase in atrial pressure

_____ 50. During isovolumetric systole, pressure is highest in the:

      a. pulmonary veins

      b. left atrium

      c. left ventricle

      d. aorta

Level =2=

_____ 51. Blood pressure in the large systemic arteries is greatest during:

      a. isovolumetric diastole

      b. atrial systole

      c. isovolumetric systole

      d. systolic ejection

_____ 52. Decreased parasympathetic (vagus) stimulation to the heart results in a situation known as:

      a. bradycardia

      b. tachycardia

      c. stenosis

      d. carditis

_____ 53. Serious arrhythmias that reduce the pumping efficiency of the heart may indicate:

      a. damage to the myocardium

      b. injury to the SA and AV nodes

      c. variations in the ionic composition of the extracellular fluids

      d. a, b, and c are correct

_____ 54. During exercise the _most important_ control mechanism to increase cardiac output is:

      a. increased body temperature

      b. increased end-systolic volume

      c. increased sympathetic activity to the ventricles

      d. increased epinephrine from the adrenal medulla

_____ 55. The diastole phase of the blood pressure relative to the left ventricle indicates that:

      a. both the atria are filling

      b. the ventricles are "resting"

      c. the AV valves are closed

      d. presystolic pressures are high

_____ 56. Which of the following _does not_ control the movement of blood through the heart?

      a. opening and closing of the valves

      b. contraction of the myocardium

      c. size of the atria and ventricles

      d. relaxation of the myocardium

_____ 57. Valvular malfunction in the heart:

      a. causes an increase in the amount of blood pumped out of each ventricle

      b. interferes with ventricular contraction

      c. increases the cardiac output

      d. interferes with movement of blood through the heart

Level =2=

_____ 58. If the bicuspid valve is defective and valvular regurgitation occurs, the end result is:

       a. an insufficient amount of blood is available to be moved into the aorta and systemic circulation

       b. an overstretching of the L. ventricle causing a decrease in the force of contraction

       c. blood gushes into the L. ventricle causing excessive pressure

       d. a, b, and c are correct

_____ 59. Valvular stenosis refers to:

       a. the inability of a valve to seal tightly

       b. the growth of scar tissue on the valve

       c. excessive narrowing of a valve

       d. an abnormal increase of blood volume

_____ 60. One of the major results of congestive heart failure is:

       a. the heart, veins, and capillaries are excessively restricted

       b. an abnormal increase in blood volume and interstitial fluid

       c. the valves in the heart cease to function

       d. the heart begins to contract spastically

[L2]   Completion:

Using the terms below, complete the following statements.

angina pectoris	endocardium	auricle
cardiac reserve	anastomoses	systemic
pectinate muscles	myocardium	nodal
trabeculae carneae	bradycardia	diuretics
ventricular fibrillation	tachycardia	infarct

61. Interconnections between vessels are referred to as _____.

62. An improperly functioning bicuspid (mitral) valve would affect _____ circulation.

63. The comb-like muscle ridges in the atria are called _____.

64. The pitted-ridged muscle bundles in the ventricular walls are called _____.

65. The inner surface of the heart that is continuous with the endothelium of the attached blood vessels is the _____.

66. The muscular wall of the heart that contains cardiac muscle tissue, associated connective tissue, blood vessels, and nerves is the _____.

67. The term used to indicate a heart rate that is slower than normal is _____.

68. The term used to indicate a heart rate that is faster than the normal rate is _____.

69. The spontaneous depolarization in the heart takes place in specialized _____ cells.

70. The heart condition that poses an immediate threat to life is _____.

71. The difference between resting and maximum cardiac output is the _____.

72. The most common treatment for congestive heart failure is to administer _____.

73. The expanded ear-like extension of the atrium is called a(n) _____.

74. Temporary heart insufficiency and ischemia causing severe chest pain is referred to as _____.

75. Tissue degeneration of the heart due to coronary circulation blockage creates a nonfunctional area known as a(n) _____.

[L2]   Short Essay:

Briefly answer the following questions in the spaces provided below.

76. If the heart beat rate (HR) is 80 beats per minute and the stroke volume (SV) is 75 m*l*, what is the cardiac output (CO) (*l*/min)?

77. If a normal cardiac output is 5.0 *l*/min and the maximum CO is 10.0 *l*/min., what is the percent increase in CO above resting?

78. If the heart rate increases to 250 beats/min, what conditions would occur relative to the following factors? (Use arrows to indicate an increase or decrease.)

    _____ CO, _____ SV, _____ length of diastole, _____ ventricular filling

79. If the end-systolic volume (ESV) is 60 m*l* and the end-diastolic volume (EDV) is 140 m*l*, what is the stroke volume (SV)?

80. If the cardiac output (CO) is 5 *l*/min. and the heart rate (HR) is 100 beats/min, what is the stroke volume (SV)?

81. What is the difference between the visceral pericardium and the parietal pericardium?

82. What is the purpose of the chordae tendinae, papillary muscles, trabeculae carneae, and pectinate muscles? Where are they located?

83. What three (3) distinct layers comprise the tissues of the heart wall?

84. What are the seven (7) important functions of fibrous skeleton of the heart?

85. Because of specialized sites known as intercalated discs, cardiac muscle is a functional syncytium. What does this statement mean?

86. Beginning with the SA node, trace the pathway of an action potential through the conducting network of the heart. (Use arrows to indicate direction.)

87. What two common clinical problems result from abnormal pacemaker function? Explain what each term means.

88. What three important factors have a direct effect on the heart rate and the force of contraction?

☐ LEVEL 3   CRITICAL THINKING/APPLICATION

Using principles and concepts learned about the heart, answer the following questions. Write your answers on a separate sheet of paper.

1. What is the thoracoabdominal pump and how does it affect stroke volume, cardiac output, and heart rate?

2. Unlike the situation in skeletal muscle, cardiac muscle contraction is an active process, but relaxation is entirely passive. Why?

3. After a vigorous tennis match, Ted complains of chest pains. He was advised to see his doctor, who immediately ordered an ECG. An evaluation of the ECG showed a slight irregular wave pattern. The physician ordered a PET scan, which showed an obstruction due to an embolus (clot) in a branch of a coronary artery. What is the relationship between chest pains and the possibility of a heart attack?

4. If the oxygen debt in skeletal muscle tissue is not paid back, the muscle loses its ability to contract efficiently owing to the presence of lactic acid and a decrease in the O_2 supply to the muscle. Why is the inability to develop a significant oxygen debt consistent with the function of the heart?

5. A friend of yours is depressed over a lifestyle that includes excessive smoking, lack of sleep, and excessive drinking of coffee and alcohol. You tell him that these habits are harmful to the heart. He decides to have an ECG. What cardiac arrythmias are associated with his lifestyle, and what abnormalities would show up on the ECG?

6. You are responsible for conducting a clinical evaluation of a patient who has previously been diagnosed as having a heart murmur. While auscultating the patient you detect a rushing sound immediately before the first heart sound. What is your diagnosis and what is causing the abnormal sound?

7. Your anatomy and physiology instructor is lecturing on the importance of coronary circulation to the overall functional efficiency of the heart. Part of this efficiency is due to the presence of arterial anastomoses. What are arterial anastomoses and why are they important in coronary circulation?

8. The right and left atria of the heart look alike and perform similar functional demands. The right and left ventricles are very different, structurally and functionally. Why are the atrial similarities and the differences in the ventricles significant in the roles they play in the functional activities of the heart?

C H A P T E R

21

Blood Vessels and Circulation

■ Overview

Blood vessels form a closed system of ducts that transport blood and allow exchange of gases, nutrients, and wastes between the blood and the body cells. The general plan of the cardiovascular system includes the heart, arteries that carry blood away from the heart, capillaries that allow exchange of substances between blood and the cells, and veins that carry blood back to the heart. The pulmonary circuit carries deoxygenated blood from the right side of the heart to the lung capillaries where the blood is oxygenated. The systemic circuit carries the oxygenated blood from the left side of the heart to the systemic capillaries in all parts of the body. The cyclic nature of the system results in deoxygenated blood returning to the right atrium to continue the cycle of circulation.

Blood vessels in the muscles, the skin, the cerebral circulation, and the hepatic portal circulation are specifically adapted to serve the functions of organs and tissues in these important special areas of cardiovascular activity.

Chapter 21 describes the structure and functions of the blood vessels, the dynamics of the circulatory process, cardiovascular regulation, and patterns of cardiovascular response.

❑ LEVEL 1   REVIEW OF CHAPTER OBJECTIVES

1. Distinguish between the types of blood vessels on the basis of their structure and function.
2. Describe how and where fluid and dissolved materials enter and leave the circulatory system.
3. Explain the mechanisms that regulate blood flow through arteries, capillaries, and veins.
4. Describe the factors that influence blood pressure and how blood pressure is regulated.
5. Discuss the mechanisms and various pressures involved in the movement of fluids between capillaries and interstitial spaces.
6. Describe how central and local control mechanisms interact to regulate blood flow and pressure in tissues.
7. Identify the principal blood vessels and the functional characteristics of the *(special circulation)* to the brain, heart and lungs.
8. Explain how the activities of the cardiac, vasomotor, and respiratory centers are coordinated to control the blood flow through the tissues.

9. Explain how the circulatory system responds to the demands of exercise, hemorrhaging and shock.

10. Identify the major arteries and veins and the areas they serve.

11. Discuss the effects of aging on the cardiovascular system.

[L1]   Multiple Choice:

Place the letter corresponding to the correct answer in the space provided.

OBJ. 1   _____ 1. The layer of vascular tissue that consists of an endothelial lining and an underlying layer of connective tissue dominated by elastic fibers is the:

       a. tunica interna
       b. tunica media
       c. tunica externa
       d. tunica adventitia

OBJ. 1   _____ 2. Smooth muscle fibers in arteries and veins are found in the:

       a. endothelial lining
       b. tunica externa
       c. tunica interna
       d. tunica media

OBJ. 1   _____ 3. One of the major characteristics of the arteries supplying peripheral tissues is that they are:

       a. elastic
       b. muscular
       c. rigid
       d. a, b, and c are correct

OBJ. 2   _____ 4. The only blood vessels whose walls permit exchange between the blood and the surrounding interstitial fluids are:

       a. arterioles
       b. venules
       c. capillaries
       d. a, b, and c are correct

OBJ. 2   _____ 5. The *primary* route for substances entering or leaving a continuous capillary is:

       a. diffusion through gaps between adjacent endothelial cells
       b. crossing the endothelia of fenestrated capillaries
       c. bulk transport of moving vesicles
       d. active transport or secretion across the capillary wall

OBJ. 2   _____ 6. The unidirectional flow of blood in venules and medium-sized veins is maintained by:

       a. the muscular walls of the veins
       b. pressure from the left ventricle
       c. arterial pressure
       d. the presence of valves

Level
–1–

OBJ. 3   _____ 7. The "specialized" arteries that are able to tolerate the pressure shock produced each time ventricular systole occurs and blood leaves the heart are:

         a. muscular arteries
         b. elastic arteries
         c. arterioles
         d. fenestrated arteries

OBJ. 3   _____ 8. Of the following blood vessels, the greatest resistance to blood flow occurs in the:

         a. veins
         b. capillaries
         c. venules
         d. arterioles

OBJ. 3   _____ 9. If the systolic pressure is 120 mm Hg and the diastolic pressure is 90 mm Hg, the mean arterial pressure (MAP) is:

         a. 30 mm Hg
         b. 210 mm Hg
         c. 100 mm Hg
         d. 80 mm Hg

OBJ. 3   _____ 10. The distinctive sounds of Korotkoff heard when taking the blood pressure are produced by:

         a. turbulences as blood flows past the constricted portion of the artery
         b. the contraction and relaxation of the ventricles
         c. the opening and closing of the atrioventricular valves
         d. a, b, and c are correct

OBJ. 3   _____ 11. When taking a blood pressure, the first sound picked up by the stethoscope as blood pulses through the artery is the:

         a. mean arterial pressure
         b. pulse pressure
         c. diastolic pressure
         d. peak systolic pressure

OBJ. 4   _____ 12. The most important determinant of vascular resistance is:

         a. a combination of neural and hormonal mechanisms
         b. differences in the length of the blood vessels
         c. friction between the blood and the vessel walls
         d. the diameter of the arterioles

OBJ. 4   _____ 13. Venous pressure is produced by:

         a. the skeletal muscle pump
         b. increasing sympathetic activity to the veins
         c. increasing respiratory movements
         d. a, b, and c are correct

OBJ. 4 _____ 14. From the following selections, choose the answer that correctly identifies all the factors which would increase blood pressure. (Note: CO = cardiac output; SV = stroke volume; VR = venous return; PR = peripheral resistance; BV = blood volume.)

    a. increasing CO, increasing SV, decreasing VR, decreasing PR, increasing BV

    b. increasing CO, increasing SV, increasing VR, increasing PR, increasing BV

    c. increasing CO, increasing SV, decreasing VR, increasing PR, decreasing BV

    d. increasing CO, decreasing SV, increasing VR, decreasing PR, increasing BV

OBJ. 5 _____ 15. The two major factors affecting blood flow rates are:

    a. diameter and length of blood vessels

    b. pressure and resistance

    c. neural and hormonal control mechanisms

    d. turbulence and viscosity

OBJ. 5 _____ 16. The formula $F = P/R$ means: (Note: F = flow; P = pressure; R = resistance.)

    a. increasing P, decreasing R, increasing F

    b. decreasing P, increasing R, increasing F

    c. increasing P, increasing R, decreasing F

    d. decreasing P, decreasing R, increasing F

OBJ. 6 _____ 17. Atrial natriuretic peptide (ANP) reduces blood volume and pressure by:

    a. blocking release of ADH

    b. stimulating peripheral vasodilation

    c. increased water loss by kidneys

    d. a, b, and c are correct

OBJ. 7 _____ 18. The four large blood vessels, two from each lung, that empty into the left atrium, completing the pulmonary circuit, are the:

    a. venae cavae

    b. pulmonary arteries

    c. pulmonary veins

    d. subclavian veins

OBJ. 7 _____ 19. The blood vessels that provide blood to capillary networks that surround the alveoli in the lungs are:

    a. pulmonary arterioles

    b. fenestrated capillaries

    c. left and right pulmonary arteries

    d. left and right pulmonary veins

OBJ. 8 _____ 20. The central regulation of cardiac output primarily involves the activities of the:

    a. somatic nervous system

    b. autonomic nervous system

    c. central nervous system

    d. a, b, and c are correct

Level
–1–

OBJ. 8   _____ 21. An increase in cardiac output normally occurs during:

        a. widespread sympathetic stimulation

        b. widespread parasympathetic stimulation

        c. the process of vasomotion

        d. stimulation of the vasomotor center

OBJ. 8   _____ 22. Stimulation of the vasomotor center in the medulla causes _____, and inhibition of the vasomotor center causes _____.

        a. vasodilation; vasoconstriction

        b. increasing diameter of arteriole; decreasing diameter of arteriole

        c. hyperemia; ischemia

        d. vasoconstriction; vasodilation

OBJ. 8   _____ 23. Hormonal regulation by vasopressin, epinephrine, angiotensin II, and norepinephrine results in:

        a. increasing peripheral vasodilation

        b. decreasing peripheral vasoconstriction

        c. increasing peripheral vasoconstriction

        d. a, b, and c are correct

OBJ. 9   _____ 24. The three primary interrelated changes that occur as exercise begins are:

        a. decreasing vasodilation, increasing venous return, increasing cardiac output

        b. increasing vasodilation, decreasing venous return, increasing cardiac output

        c. increasing vasodilation, increasing venous return, increasing cardiac output

        d. decreasing vasodilation, decreasing venous return, decreasing cardiac output

OBJ. 9   _____ 25. The only area of the body where the blood supply is unaffected while exercising at maximum levels is the:

        a. hepatic portal circulation

        b. pulmonary circulation

        c. brain

        d. peripheral circulation

OBJ. 10   _____ 26. The three elastic arteries that originate along the aortic arch and deliver blood to the head, neck, shoulders, and arms are the:

        a. axillary, R. common carotid, right subclavian

        b. R. dorsoscapular, R. thoracic, R. vertebral

        c. R. axillary, R. brachial, L. internal carotid

        d. brachiocephalic, L. common carotid, left subclavian

OBJ. 10   _____ 27. The large blood vessel that collects most of the venous blood from organs below the diaphragm is the:

        a. superior vena cava

        b. inferior vena cava

        c. hepatic portal vein

        d. superior mesenteric vein

OBJ. 10 _____ 28. The three blood vessels that provide blood to all of the digestive organs in the abdominopelvic cavity are the:

   a. thoracic aorta, abdominal aorta, superior phrenic artery
   b. intercostal, esophageal, and bronchial arteries
   c. celiac artery and the superior and inferior mesenteric arteries
   d. suprarenal, renal, and lumbar arteries

OBJ. 11 _____ 29. The primary effect of a decrease in the hematocrit of elderly individuals is:

   a. thrombus formation in the blood vessels
   b. a lowering of the oxygen-carrying capacity of the blood
   c. a reduction in the maximum cardiac output
   d. damage to ventricular cardiac muscle fibers

OBJ. 11 _____ 30. The primary cause of varicose veins is:

   a. improper diet
   b. aging
   c. swelling due to edema in the veins
   d. inefficient venous valves

[L1]   Completion:

Using the terms below, complete the following statements.

hydrostatic pressure   aortic arch   sphygmomanometer
autoregulation   vasomotion   circulatory pressure
thoracoabdominal pump   shock   total peripheral resistance
pulse pressure   vasoconstriction   hepatic portal system
arterioles   central ischemic   precapillary sphincter
arteriosclerosis   venules   venous thrombosis
osmotic pressure   fenestrated

OBJ. 1   31. The smallest vessels of the arterial system are the _____.

OBJ. 1   32. Blood flowing out of the capillary complex first enters small _____.

OBJ. 2   33. Capillaries that have an incomplete endothelial lining are _____ capillaries.

OBJ. 3   34. The entrance to each capillary is guarded by a band of smooth muscle, the _____.

OBJ. 4   35. The difference between the systolic and diastolic pressures is the _____.

OBJ. 4   36. The instrument used to determine blood pressure is called a _____.

OBJ. 4   37. The pressure difference between the base of the ascending aorta and the entrance to the right atrium is the _____.

OBJ. 4   38. For circulation to occur, the circulatory pressure must be greater than the _____.

OBJ. 5   39. The blood flow within any one capillary occurs in a series of pulses called _____.

OBJ. 5   40. The force that pushes water molecules *out* of solution is _____.

Level
-1-

OBJ. 5   41. A force that pulls water *into* a solution is _____.

OBJ. 6   42. The regulation of blood flow at the tissue level is called _____.

OBJ. 7   43. Blood leaving the capillaries supplied by the celiac, superior and inferior mesenteric arteries flows into the _____.

OBJ. 8   44. Stimulation of the vasomotor center in the medulla causes _____.

OBJ. 8   45. A rise in the respiratory rate accelerates venous return through the action of the _____.

OBJ. 9   46. An acute circulatory crisis marked by low blood pressure and inadequate peripheral blood flow is referred to as _____.

OBJ. 9   47. The reflex causing circulation to be reduced to an absolute minimum is called the _____ response.

OBJ. 10 48. The part of the vascular system that connects the ascending aorta with the caudally directed descending aorta is the _____.

OBJ. 11 49. The aging process may cause peripheral veins to become constricted or blocked by formation of a stationary blood clot resulting in a condition called _____.

OBJ. 11 50. Most of the age-related changes in the circulatory system are related to _____.

[L1]   Matching:

Match the terms in column B with the terms in column A. Use letters for answers in the spaces provided.

PART I

			COLUMN A		COLUMN B
OBJ. 1		51.	tunica externa	A.	changes in blood pressure
OBJ. 1		52.	vasa vasorum	B.	preferred channel
OBJ. 2		53.	sinusoids	C.	venous return
OBJ. 3		54.	metarteriole	D.	minimum blood pressure
OBJ. 3		55.	thoracoabdominal pump	E.	fenestrated capillaries
				F.	connected by tight junction
OBJ. 4		56.	systole		
OBJ. 4		57.	diastole	G.	"vessels of vessels"
OBJ. 5		58.	continuous capillaries	H.	hydrostatic force = osmotic force
OBJ. 5		59.	dynamic center		
OBJ. 6		60.	baroreceptors	I.	connective tissue sheath
				J.	"peak" blood pressure

PART II

			COLUMN A		COLUMN B
OBJ. 6		61.	mean arterial pressure	K.	vasomotor center
OBJ. 6		62.	atrial natriuretic peptide	L.	increasing venous return
OBJ. 7		63.	circle of Willis	M.	received blood from arm
OBJ. 8		64.	medulla	N.	blood clot in lungs
OBJ. 9		65.	atrial reflex	O.	bracheocephalic artery
OBJ. 9		66.	rise in body temperature	P.	weakened vascular wall
				Q.	vasodilation
OBJ. 10		67.	innominate artery	R.	increased blood flow to skin
OBJ. 10		68.	axillary vein	S.	single value for blood pressure
OBJ. 11		69.	pulmonary embolism		
OBJ. 11		70.	aneurysm	T.	brain circulation

[L1] Drawing/Illustration Labeling:

Identify each numbered structure by labeling the following figures:

OBJ. 10 | *FIGURE 21.1* The Arterial System

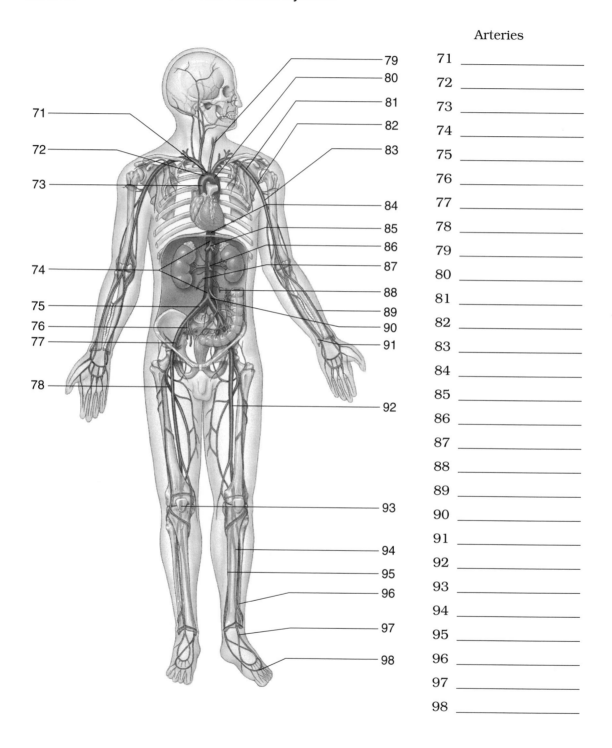

Arteries

71 _____

72 _____

73 _____

74 _____

75 _____

76 _____

77 _____

78 _____

79 _____

80 _____

81 _____

82 _____

83 _____

84 _____

85 _____

86 _____

87 _____

88 _____

89 _____

90 _____

91 _____

92 _____

93 _____

94 _____

95 _____

96 _____

97 _____

98 _____

OBJ. 10   *FIGURE 21.2*   The Venous System

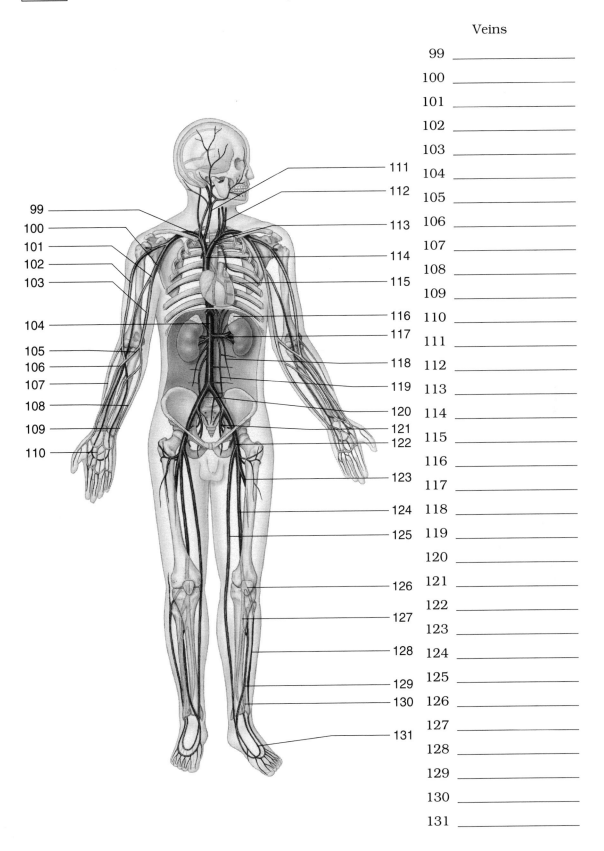

Veins

99 _____

100 _____

101 _____

102 _____

103 _____

104 _____

105 _____

106 _____

107 _____

108 _____

109 _____

110 _____

111 _____

112 _____

113 _____

114 _____

115 _____

116 _____

117 _____

118 _____

119 _____

120 _____

121 _____

122 _____

123 _____

124 _____

125 _____

126 _____

127 _____

128 _____

129 _____

130 _____

131 _____

Level
–1–

OBJ. 10   *FIGURE 21.3*   Major Arteries of the Head and Neck

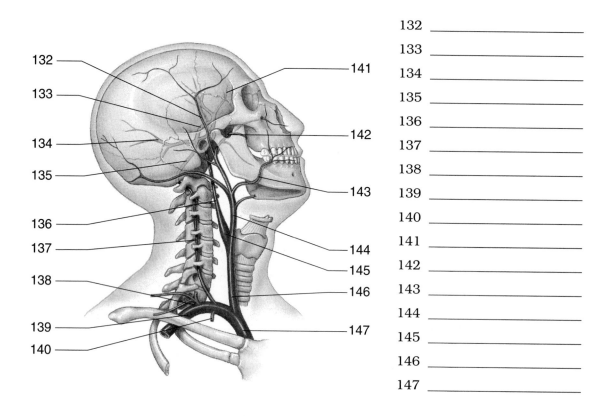

132 _____
133 _____
134 _____
135 _____
136 _____
137 _____
138 _____
139 _____
140 _____
141 _____
142 _____
143 _____
144 _____
145 _____
146 _____
147 _____

OBJ. 10   *FIGURE 21.4*   Major Veins Draining the Head and Neck

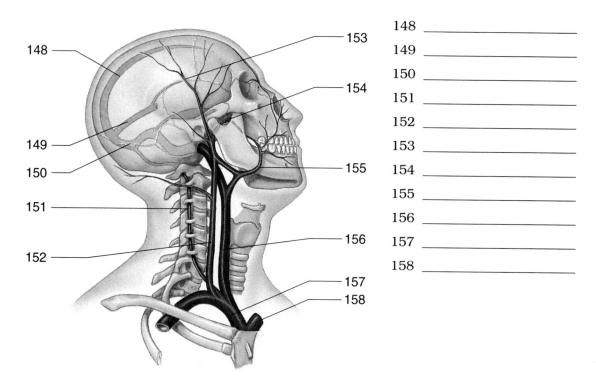

148 _____
149 _____
150 _____
151 _____
152 _____
153 _____
154 _____
155 _____
156 _____
157 _____
158 _____

Level
–1–

When you have successfully completed the exercises in L1 proceed to L2.

☐ LEVEL 2   CONCEPT SYNTHESIS

Concept Map I:

Using the following terms, fill in the circled, numbered, blank spaces to complete the concept map. Follow the numbers that comply with the organization of the map.

Pulmonary arteries   Veins and venules   Systemic circuit
Arteries and arterioles   Pulmonary veins

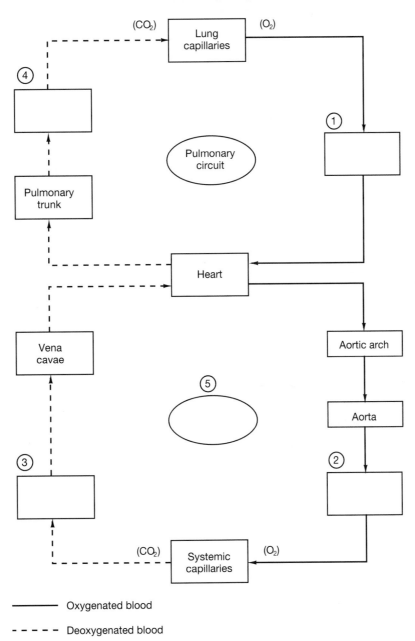

The cardiovascular system

Concept Map II:

Using the following terms, fill in the circled, numbered, blank spaces to complete the concept map. Follow the numbers that comply with the organization of the map.

Erythropoietin   Increasing plasma volume   Adrenal cortex
ADH (vasopressin)   Epinephrine, norepinephrine   Kidneys
Increasing fluid   Increasing blood pressure   Atrial natriuretic factor

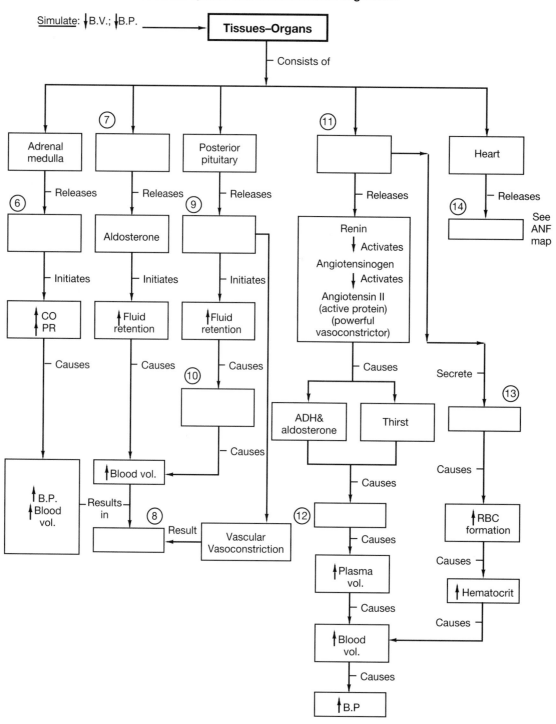

Endocrine system and cardiovascular regulation

Concept Map III:

Using the following terms, fill in the circled, numbered, blank spaces to complete the concept map. Follow the numbers that comply with the organization of the map.

Vasoconstriction   Vasomotor center   Hyperemia
Reactive   Epinephrine   Vasodilation
ADH (vasopressin)

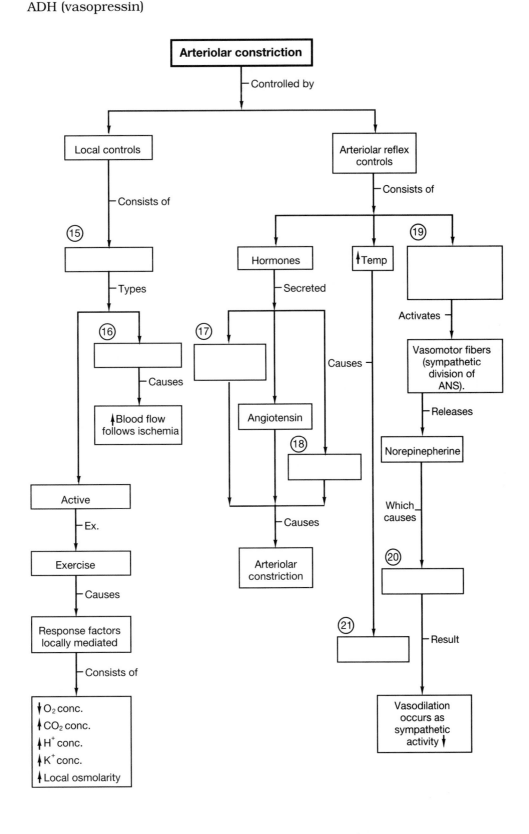

Concept Map IV:

Using the following terms, fill in the circled, numbered, blank spaces to complete the concept map. Follow the numbers that comply with the organization of the map.

Decreasing blood pressure   Increasing H_2O loss by kidneys

Decreasing H_2O intake   Increasing blood flow (l/min.)

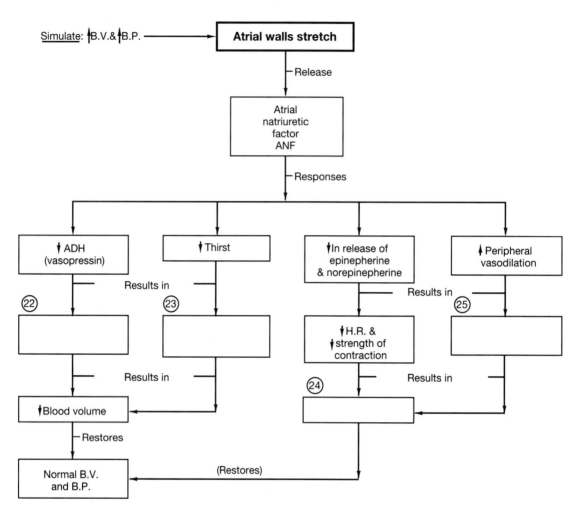

Concept Map V:

Using the following terms, fill in the circled, numbered, blank spaces to complete the concept map. Follow the numbers that comply with the organization of the map.

L. Subclavian artery

L. common iliac artery

Ascending aorta

Celiac trunk

Superior mesenteric artery

Thoracic aorta

R. Gonadal artery

Brachiocephalic artery

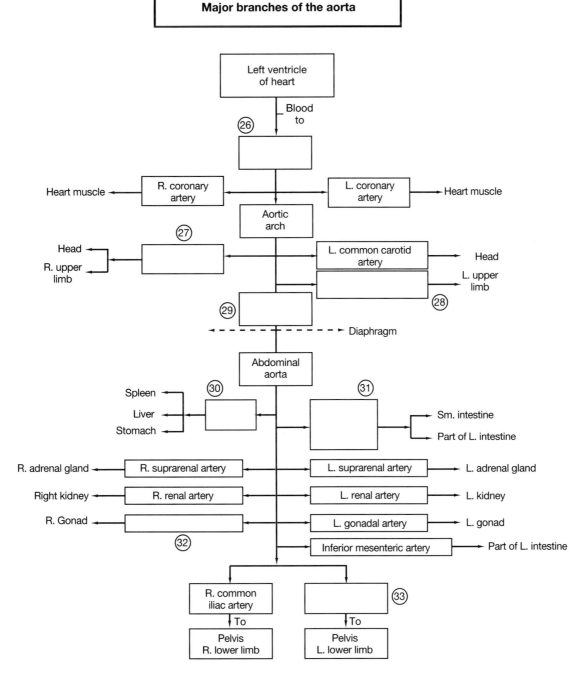

Concept Map VI:

Using the following terms, fill in the circled, numbered, blank spaces to complete the concept map. Follow the numbers that comply with the organization of the map. (Note the direction of the arrows on this map.)

Azygous vein   L. common iliac vein   L. renal vein
L. hepatic veins   superior vena cava   R. suprarenal vein

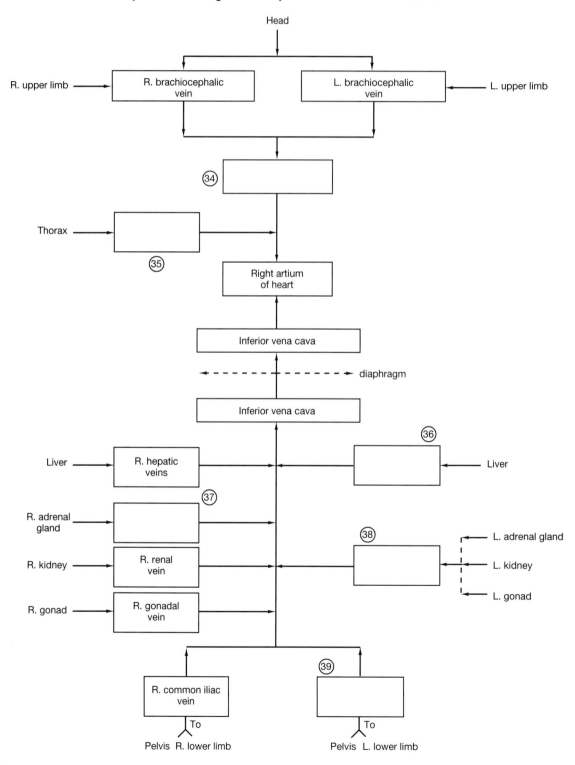

Major veins draining into the superior and inferior venae cavae

BODY TREK:

Using the terms below, fill in the blanks to complete a trek through the major arteries of the body.

lumbar	abdominal aorta	superior phrenic
intercostal	L. subclavian	common iliacs
brachiocephalic	aortic valve	inferior mesenteric
suprarenal	L. common carotid	celiac
renal	mediastinum	superior mesenteric
descending aorta	gonadal	inferior phrenic
thoracic aorta	aortic arch	

Robo has been "charged" and "programmed" to take an "aortic trek." The objective of the mission is to identify the regions and major arteries that branch off the aorta from its superior position in the thorax to its inferior location in the abdominopelvic region.

The micro-robot is catheterized directly into the left ventricle of the heart, via a technique devised exclusively for introducing high-tech devices directly into the heart. A ventricular contraction "swooshes" the robot through the (40) _____ into the (41) _____, which curves across the superior surface of the heart, connecting the ascending aorta with the caudally directed (42) _____. The first artery branching off the aortic arch, the (43) _____ serves as a passageway for blood to arteries that serve the arms, neck, and head. As the robot continues its ascent, the (44) _____ and the (45) _____ artery arise separately from the aortic arch, both supplying arterial blood to vessels that provide blood to regions that include the neck, head, shoulders, and arms. All of a sudden the robot is directed downward to begin its descent into the (46) _____ located within the (47) _____ and providing blood to the (48) _____ arteries that carry blood to the spinal cord and the body wall. Proceeding caudally, near the diaphragm, a pair of (49) _____ arteries deliver blood to the muscular diaphragm that separates the thoracic and abdominopelvic cavities. Leaving the thoracic aorta Robo continues trekking along the same circulatory pathway into the (50) _____, which descends posteriorly to the peritoneal cavity. The robot's "thermosensors" detect record high temperatures in the area, probably due to the cushion of adipose tissue surrounding the aorta in this region.

Just below the diaphragm, the (51) _____ arteries, which carry blood to the diaphragm, precede the (52) _____ artery, which supplies blood to arterial branches that circulate blood to the stomach, liver, spleen, and pancreas. Trekking downward, the (53) _____ arteries serving the adrenal glands and the (54) _____ arteries supplying the kidneys arise along the posteriolateral surface of the abdominal aorta behind the peritoneal lining. In the same region the (55) _____ arteries arise on the anterior surface of the abdominal aorta and branch into the connective tissue of the mesenteries. A pair of (56) _____ arteries supplying blood to the testicles in the male and the ovaries in the female are located slightly superior to the (57) _____ artery, which branches into the connective tissues of the mesenteries supplying blood to the last third of the large intestine, colon, and rectum.

On the final phase of the aortic trek, prior to the bifurcation of the aorta, small (58) _____ arteries begin on the posterior surface of the aorta and supply the spinal cord and the abdominal wall. Near the level of L4 the abdominal aorta divides to form a pair of muscular arteries, the (59) _____ , which are the major blood suppliers to the pelvis and the legs.

As Robo completes the programmed mission, the message relayed to Mission Control reads, "Mission accomplished, please remove." Mission Control instructs the robot to continue its descent into the femoral artery where an arterial removal technique for high-tech gadgetry will facilitate Robo's exit and return to Mission Control for the next assignment.

Level
=2=

[L2] Multiple Choice:

Place the letter corresponding to the correct answer in the space provided.

_____ 60. In traveling from the heart to the peripheral capillaries, blood passes through:

    a. arteries, arterioles, and venules

    b. elastic arteries, muscular arteries, and arterioles

    c. veins, venules, arterioles

    d. a, b, and c are correct

_____ 61. At any given moment, the systemic circulation contains about _____ of the total blood volume.

    a. 10%

    b. 51%

    c. 71%

    d. 91%

_____ 62. The goal of cardiovascular regulation is:

    a. to equalize the stroke volume with the cardiac output

    b. to increase pressure and reduce resistance

    c. to equalize the blood flow with the pressure differences

    d. maintenance of adequate blood flow through peripheral tissues and organs

_____ 63. Along the length of a typical capillary, blood pressure gradually falls from about:

    a. 120 to 80 mm Hg

    b. 75 to 50 mm Hg

    c. 30 to 18 mm Hg

    d. 15 to 5 mm Hg

_____ 64. In which of the following organs would you find *fenestrated* capillaries?

    a. kidneys

    b. liver

    c. endocrine glands

    d. both a and c

_____ 65. The *average* pressure in arteries is approximately:

    a. 120 mm Hg

    b. 100 mm Hg

    c. 140 mm Hg

    d. 80 mm Hg

_____ 66. The *average* pressure in veins is approximately:

    a. 2 mm Hg

    b. 10 mm Hg

    c. 15 mm Hg

    d. 80 mm Hg

Level =2=

_____ 67. When fluid moves across capillary membranes, the pressures forcing fluid *out* of the capillaries are:

    a. hydrostatic pressure; interstitial osmotic pressure

    b. interstitial fluid pressure; osmotic pressure of plasma

    c. hydrostatic pressure; interstitial fluid pressure

    d. interstitial osmotic pressure; osmotic pressure of plasma

_____ 68. The formula for resistance, $R = Ln/r^4$, where (L) is the length of the vessel, (n) is the viscosity of the blood, and (r) is the radius of the vessel, would confirm the following correct relationship:

    a. increasing radius, increasing friction, increasing flow

    b. decreasing radius, decreasing friction, decreasing flow

    c. increasing radius, decreasing friction, increasing flow

    d. decreasing radius, increasing friction, decreasing flow

_____ 69. Considering cardiac output (CO), mean arterial pressure (AP), and total peripheral resistance (PR), the relationship among flow, pressure, and resistance could be expressed as:

    a. $AP = CO \times PR$

    b. $CO = AP/PR$

    c. $CO/AP = 1/PR$

    d. a, b, and c are correct

_____ 70. To increase blood flow to an adjacent capillary, the local controls that operate are:

    a. decreasing O_2, increasing CO_2, decreasing pH

    b. increasing O_2, decreasing CO_2, decreasing pH

    c. increasing O_2, increasing CO_2, increasing pH

    d. decreasing O_2, decreasing CO_2, decreasing pH

_____ 71. The adrenergic fibers innervating arterioles are _____ fibers that release _____ and cause _____.

    a. parasympathetic; norepinephrine; vasodilation

    b. sympathetic; epinephrine; vasodilation

    c. sympathetic; norepinephrine; vasoconstriction

    d. parasympathetic; epinephrine; vasoconstriction

_____ 72. Two arteries formed by the bifurcation of the brachiocephalic artery are the:

    a. aorta, subclavian

    b. common iliac, common carotid

    c. jugular, carotid

    d. common carotid, subclavian

_____ 73. The artery that serves the posterior thigh is the:

    a. deep femoral

    b. common iliac

    c. internal iliac

    d. celiac

Level
=2=

_____ 74. The _____ return blood to the heart and the _____ transport blood away from the heart.

        a. arteries; veins

        b. venules; arterioles

        c. veins; arteries

        d. arterioles; venules

_____ 75. The large vein that drains the thorax is the:

        a. superior vena cava

        b. internal jugular

        c. vertebral vein

        d. azygous vein

_____ 76. The veins that drain the head, neck, and upper extremities are the:

        a. jugulars

        b. brachiocephalics

        c. subclavian

        d. azygous

_____ 77. The veins that drain venous blood from the legs and the pelvis are:

        a. posterior tibials

        b. femorals

        c. great saphenous

        d. common iliacs

_____ 78. The vein that drains the knee region of the body is the:

        a. femoral

        b. popliteal

        c. external iliacs

        d. great saphenous

_____ 79. The large artery that serves the brain is the:

        a. internal carotid

        b. external carotid

        c. subclavian

        d. cephalic

_____ 80. The "link" between the subclavian and brachial artery is the:

        a. brachiocephalic

        b. vertebral

        c. cephalic

        d. axillary

_____ 81. The three arterial branches of the celiac trunk are the:

        a. splenic, pancreatic, mesenteric

        b. phrenic, intercostal, adrenolumbar

        c. L. gastric, splenic, hepatic

        d. brachial, ulnar, radial

Level
=2=

_____ 82. The artery that supplies most of the small intestine and the first half of the large intestine is the:

        a. suprarenal artery

        b. superior mesenteric

        c. hepatic

        d. inferior mesenteric

_____ 83. The artery that supplies the pelvic organs is the:

        a. internal iliac artery

        b. external iliac artery

        c. common iliacs

        d. femoral artery

_____ 84. The subdivision(s) of the popliteal artery is (are):

        a. great saphenous artery

        b. anterior and posterior tibial arteries

        c. peroneal artery

        d. femoral and deep femoral arteries

[L2]   Completion:

Using the terms below, complete the following statements.

radial	precapillary sphincters	aorta
mesoderm	great saphenous	circle of Willis
edema	reactive hyperemia	brachial
lumen	elastic rebound	venous return
veins	endothelium	recall of fluids

85. The anastomosis that encircles the infundibulum of the pituitary gland is the _____.

86. When the arteries absorb part of the energy provided by ventricular systole and give it back during ventricular diastole, the phenomenon is called _____.

87. When the blood volume increases at the expense of interstitial fluids, the process is referred to as _____.

88. The abnormal accumulation of interstitial fluid is called _____.

89. Rhythmic alterations in blood flow in capillary beds are controlled by _____.

90. The response in which tissue appears red in a given area is referred to as _____.

91. The lining of the lumen of blood vessels is comprised of a tissue layer called the _____.

92. The heart and blood vessels develop from the germ layer called the _____.

93. The blood vessels that contain valves are the _____.

94. Changes in thoracic pressure during breathing and the action of skeletal muscles serve to aid in _____.

95. The largest artery in the body is the _____.

96. The artery generally auscultated to determine the blood pressure in the arm is the _____ artery.

97. The artery generally used to take the pulse at the wrist is the _____ artery.

98. The longest vein in the body is the _____.

99. The central cavity through which blood flows in a blood vessel is referred to as the _____.

[L2]   Short Essay

Briefly answer the following questions in the spaces provided below.

100. What three distinct layers comprise the histological composition of typical arteries and veins?

101. List the types of blood vessels in the cardiovascular tree and briefly describe their anatomical associations (use arrows to show this relationship).

102. (a) What are sinusoids?

(b) Where are they found?

(c) Why are they important functionally?

103. Relative to gaseous exchange, what is the primary difference between the pulmonary circuit and the systemic circuit?

104. (a) What are the three (3) primary sources of peripheral resistance?

    (b) Which one can be adjusted by the nervous or endocrine system to regulate blood flow?

105. Symbolically summarize the relationship among blood pressure (BP), peripheral resistance (PR), and blood flow (F). State what the formula means.

106. Explain what is meant by: BP = 120 mm Hg/80 mm Hg

107. What is the mean arterial pressure (MAP) if the systolic pressure is 110 mm Hg and the diastolic pressure is 80 mm Hg?

108. What are the three (3) important functions of continual movement of fluid from the plasma into tissues and lymphatic vessels?

109. What are the three (3) primary factors that influence blood pressure and blood flow?

110. What three (3) major baroreceptor populations enable the cardiovascular system to respond to alterations in blood pressure?

111. What hormones are responsible for long-term and short-term regulation of cardio-vascular performance?

112. What are the clinical signs and symptoms of age-related changes in the blood?

113. How does arteriosclerosis affect blood vessels?

When you have successfully completed the exercises in L2 proceed to L3.

☐ LEVEL 3   CRITICAL THINKING/APPLICATION

Using principles and concepts learned about blood vessels and circulation, answer the following questions. Write your answers on a separate sheet of paper.

1. Trace the circulatory pathway a drop of blood would take if it begins in the L. ventricle, travels down the left arm, and returns to the R. atrium. (Use arrows to indicate direction of flow.)

2. During a clinical experience, taking a pulse is one of the primary responsibilities of the clinician. What arteries in the body serve as locations for detecting a pulse?

3. Suppose you are lying down and quickly rise to a standing position. What causes dizziness or a loss of consciousness to occur as a result of standing up quickly?

4. While pedalling an exercycle, Mary periodically takes her pulse by applying pressure to the carotid artery in the upper neck. Why might this method give her an erroneous perception of her pulse rate while exercising?

5. Recent statistics show that approximately 30 to 35 percent of the adult American population is involved in some type of exercise program. What cardiovascular changes occur and what benefits result from exercising?

C H A P T E R

22

The Lymphatic System and Immunity

■ Overview

The lymphatic system includes lymphoid organs and tissues, a network of lymphatic vessels that contain a fluid called lymph, and a dominant population of individual cells, the lymphocytes. The cells, tissues, organs, and vessels containing lymph perform three major functions:

1. The lymphoid organs and tissues serve as operating sites for the phagocytes and cells of the immune system that provide the body with protection from pathogens;
2. the lymphatic vessels containing lymph help to maintain blood volume in the cardio-vascular system and absorb fats and other substances from the digestive tract; and
3. the lymphocytes are the "defensive specialists" that protect the body from pathogenic microorganisms, foreign tissue cells, and diseased or infected cells in the body that pose a threat to the normal cell population.

Chapter 22 provides exercises focusing on topics that include the organization of the lymphatic system, the body's defense mechanisms, patterns of immune response, and interactions between the lymphatic system and other physiological systems.

☐ LEVEL 1   REVIEW OF CHAPTER OBJECTIVES

1. Identify the major components of the lymphatic system and explain their functions.
2. Discuss the importance of lymphocytes and describe their distribution in the body.
3. Describe the structure of the lymphoid tissues and organs and explain their functions.
4. List the body's nonspecific defenses and describe the components and mechanisms of each.
5. Define specific resistance and identify the forms and properties of immunity.
6. Distinguish between cell-mediated immunity and antibody-mediated (humoral) immunity and identify the cells responsible for each.
7. Discuss the different types of T cells and the role played by each in the immune response.

8. Describe the general structure of an antibody molecule and discuss the different types of antibodies present in body fluids and secretions.

9. Explain the effects of antibodies and how they are produced.

10. Discuss the primary and secondary responses to antigen exposure.

11. List the hormones of the immune system and explain their significance.

12. Describe the origin, development, activation, and regulation of resistance.

13. Explain the origin of autoimmune disorders, immunodeficiency diseases and allergies, and list important examples of each type of disorder.

14. Describe the effects of stress and aging on the immune response.

[L1]   Multiple Choice:

Place the letter corresponding to the correct answer in the space provided.

OBJ. 1   _____   1. The *major components* of the lymphatic system include:

    a. lymph nodes, lymph, lymphocytes

    b. spleen, thymus, tonsils

    c. thoracic duct, R. lymphatic duct, lymph nodes

    d. lymphatic vessels, lymph, lymphatic organs

OBJ. 1   _____   2. Lymphatic *organs* found in the lymphatic system include:

    a. thoracic duct, R. lymphatic duct, lymph nodes

    b. lymphatic vessels, tonsils, lymph nodes

    c. spleen, thymus, lymph nodes

    d. a, b, and c are correct

OBJ. 1   _____   3. The *primary* function of the lymphatic system is:

    a. transporting of nutrients and oxygen to tissues

    b. removal of carbon dioxide and waste products from tissues

    c. regulation of temperature, fluid, electrolytes, and pH balance

    d. production, maintenance, and distribution of lymphocytes

OBJ. 2   _____   4. Lymphocytes that assist in the regulation and coordination of the immune response are:

    a. plasma cells

    b. helper T and suppressor T cells

    c. B cells

    d. NK and B cells

OBJ. 2   _____   5. Normal lymphocyte populations are maintained through lympho-poiesis in the:

    a. bone marrow and lymphatic tissues

    b. lymph in the lymphatic tissues

    c. blood and the lymph

    d. spleen and liver

OBJ. 3   _____ 6. The largest collection of lymphoid tissue in the body is contained within the:

    a. adult spleen

    b. thymus gland

    c. the tonsils

    d. the lymphatic nodules

OBJ. 3   _____ 7. The reticular epithelial cells in the cortex of the thymus maintain the blood thymus barrier and secrete the hormones that:

    a. form distinctive structures known as Hassall's Corpuscles

    b. cause the T cells to leave circulation via blood vessels

    c. stimulate stem cell divisions and T cell differentation

    d. cause the T cells to enter the circulation via blood vessels

OBJ. 4   _____ 8. Of the following selections, the one that includes only nonspecific defenses is:

    a. T and B-cell activation, complement, inflammation, phagocytosis

    b. hair, skin, mucous membranes, antibodies

    c. hair, skin, complement, inflammation, phagocytosis

    d. antigens, antibodies, complement, macrophages

OBJ. 4   _____ 9. The protective categories that prevent the approach of, deny entrance to, or limit the spread of microorganisms or other environmental hazards are called:

    a. specific defenses

    b. nonspecific defenses

    c. specific immunity

    d. immunological surveillance

OBJ. 4   _____ 10. NK (natural killer) cells sensitive to the presence of abnormal cell membranes are primarily involved with:

    a. defenses against specific threats

    b. complex and time-consuming defense mechanisms

    c. phagocytic activity for defense

    d. immunological surveillance

OBJ. 5   _____ 11. The four general characteristics of specific defenses include:

    a. specificity, versatility, memory, and tolerance

    b. innate, active, acquired, passive

    c. accessibility, recognition, compatibility, immunity

    d. a, b, and c are correct

OBJ. 5   _____ 12. The two major ways that the body "carries out" the immune response are:

    a. phagocytosis and the inflammatory response

    b. immunological surveillance and fever

    c. direct attack by T cells and attack by circulating antibodies

    d. physical barriers and the complement system

OBJ. 5   _____ 13. A *specific* defense mechanism is always activated by:

        a. an antigen

        b. an antibody

        c. inflammation

        d. fever

OBJ. 6   _____ 14. The type of immunity that develops as a result of natural exposure to an antigen in the environment is:

        a. naturally acquired immunity

        b. natural innate immunity

        c. naturally acquired active immunity

        d. naturally acquired passive immunity

OBJ. 6   _____ 15. The fact that people are not subject to the same diseases as goldfish describes the presence of:

        a. active immunity

        b. passive immunity

        c. acquired immunity

        d. innate immunity

OBJ. 6   _____ 16. When an antigen appears, the immune response begins with:

        a. the presence of immunoglobulins in body fluids

        b. the release of endogenous pyrogens

        c. the activation of the complement system

        d. activation of specific T cells and B cells

OBJ. 6   _____ 17. When the immune "recognition" system malfunctions, activated B cells begin to:

        a. manufacture antibodies against other cells and tissues

        b. activate cytotoxic T killer cells

        c. secrete lymphotoxins to destroy foreign antigens

        d. recall memory T cells to initiate the proper response

OBJ. 7   _____ 18. T-cell activation leads to the formation of cytotoxic T cells and memory T cells that provide:

        a. humoral immunity

        b. cellular immunity

        c. phagocytosis and immunological surveillance

        d. stimulation of inflammation and fever

OBJ. 7   _____ 19. Before an antigen can stimulate a lymphocyte, it must first be processed by a:

        a. macrophage

        b. NK cell

        c. cytotoxic T cell

        d. neutrophil

OBJ. 8   _____ 20. An active antibody is shaped like a(n):

        a. T

        b. A

        c. Y

        d. B

Level
-1-

OBJ. 8   _____ 21. The most important antibody action in the body is:

    a. alteration in the cell membrane to increase phagocytosis

    b. to attract macrophages and neutrophils to the infected areas

    c. activation of the complement systems

    d. cell lysis and the cell membrane digestion

OBJ. 9   _____ 22. Antibodies may promote inflammation through the stimulation of:

    a. basophils and mast cells

    b. plasma cells and memory B cells

    c. suppressor T cells

    d. cytotoxic T cells

OBJ. 10   _____ 23. The antigenic determinant site is the certain portion of the antigen's exposed surface where:

    a. the foreign "body" attacks

    b. phagocytosis occurs

    c. the antibody attacks

    d. the immune surveillance system is activated

OBJ. 10   _____ 24. In order for an antigenic molecule to be a complete antigen, it must:

    a. be a large molecule

    b. be immunogenic and reactive

    c. contain a hapten and a small organic molecule

    d. be subject to antibody activity

OBJ. 11   _____ 25. The effect(s) of tumor necrosis factor (TNF) in the body is (are) to:

    a. slow tumor growth and to kill sensitive tumor cells

    b. stimulate granulocyte production

    c. increase T-cell sensitivity to interleukins

    d. a, b, and c are correct

OBJ. 11   _____ 26. The major functions of interleukins in the immune system are to:

    a. increase T-cell sensitivity to antigens exposed on macrophage membranes

    b. stimulate B-cell activity, plasma-cell formation, and antibody production

    c. enhance nonspecific defenses

    d. a, b, and c are correct

OBJ. 12   _____ 27. The ability to demonstrate an immune response upon exposure to an antigen is called:

    a. anaphylaxis

    b. passive immunity

    c. immunosuppression

    d. immunological competence

OBJ. 12   _____ 28. Fetal antibody production is uncommon because the developing fetus has:

        a. cell-mediated immunity

        b. natural passive immunity

        c. antibody mediated immunity

        d. endogenous pyrogens

OBJ. 13   _____ 29. When an immune response mistakenly targets normal body cells and tissues, the result is:

        a. immune system failure

        b. the development of an allergy

        c. depression of the inflammatory response

        d. an autoimmune disorder

OBJ. 14   _____ 30. With advancing age the immune system becomes:

        a. increasingly susceptible to viral infection

        b. increasingly susceptible to bacterial infection

        c. less effective at combatting disease

        d. a, b, and c are correct

[L1]   Completion:

Using the terms below, complete the following statements.

lymphokines	innate	immunological competence
helper T	haptens	plasma cells
diapedesis	active	lymph capillaries
lacteals	passive	cell-mediated
precipitation	monokines	immunodeficiency disease
vaccinated	neutralization	cytotoxic T cells
suppressor T	phagocytes	IgG
antibodies		

OBJ. 1   31. The special lymphatic vessels in the lining of the small intestines are the _____.

OBJ. 2   32. Lymphocytes that attack foreign cells or body cells infected by viruses are called _____ cells.

OBJ. 2   33. Plasma cells are responsible for the production and secretion of _____.

OBJ. 3   34. The lymphatic system begins in the tissues as _____.

OBJ. 4   35. Cells that represent the "first line" of cellular defense against pathogenic invasion are the _____.

OBJ. 4   36. The process during which macrophages move through adjacent endothelial cells of capillary walls is called _____.

OBJ. 5   37. An immunization where antibodies are administered to fight infection or prevent disease is _____.

OBJ. 5   38. Cytotoxic T cells are responsible for the type of immunity referred to as _____.

OBJ. 6   39. Immunity that is present at birth and has no relation to previous exposure to the pathogen involved is _____.

Level
-1-

OBJ. 6

40. Immunity that appears following exposure to an antigen as a consequence of the immune response is referred to as _____.

OBJ. 7

41. The types of cells that inhibit the responses of other T cells and B cells are called _____ cells.

OBJ. 7

42. Before B cells can respond to an antigen, they must receive a signal from _____ cells.

OBJ. 8

43. Antibodies are produced and secreted by _____.

OBJ. 9

44. When a direct attack by an antibody covers an antigen its effect is _____.

OBJ. 9

45. The formation of insoluble immune complexes is called _____.

OBJ. 10

46. Small organic molecules that are not antigens by themselves are called _____.

OBJ. 10

47. The ability to demonstrate an immune response upon exposure to an antigen is called _____.

OBJ. 11

48. Chemical messengers secreted by lymphocytes are called _____.

OBJ. 11

49. Chemical messengers released by active macrophages are called _____.

OBJ. 12

50. The only antibodies that cross the placenta from the maternal blood stream are _____ antibodies.

OBJ. 13

51. When the immune system fails to develop normally or the immune response is blocked in some way, the condition is termed _____.

OBJ. 14

52. Because of increased susceptibility of acute viral infection in the elderly, it is recommended that they be _____.

[L1]   Matching:

Match the terms in column B with the terms in column A. Use letters for answers in the spaces provided.

PART I

			COLUMN A		COLUMN B
OBJ. 1	_____	53.	Peyer's patch	A.	thymus medullary cells
OBJ. 1	_____	54.	GALT	B.	type II allergy
OBJ. 2	_____	55.	macrophages	C.	active and passive
OBJ. 2	_____	56.	microphages	D.	cellular immunity
OBJ. 3	_____	57.	Hassall's corpuscles	E.	innate or acquired
OBJ. 4	_____	58.	mast cells	F.	nonspecific immune response
OBJ. 5	_____	59.	acquired immunity		
OBJ. 5	_____	60.	specific immunity	G.	lymphatic tissue in digestive system
OBJ. 6	_____	61.	B cells		
OBJ. 6	_____	62.	passive immunity	H.	monocytes
OBJ. 6	_____	63.	antigen contact	I.	lymph nodules in small intestine
OBJ. 7	_____	64.	cytotoxic T cells		

J.   transfer of antibodies

K.   neutrophils and eosinophils

L.   humoral immunity

PART II

			COLUMN A		COLUMN B
OBJ. 7	_____	65.	T lymphocytes	M.	resistance to viral infections
OBJ. 8	_____	66.	antibody	N.	activation of T cells
OBJ. 9	_____	67.	bacteriophage	O.	type II allergy
OBJ. 9	_____	68.	opsonization	P.	activation of B cells
OBJ. 10	_____	69.	active immunity	Q.	decline in immune surveillance
OBJ. 11	_____	70.	interleukins	R.	exposure to an antigen
OBJ. 11	_____	71.	interferons	S.	cell mediated immunity
OBJ. 11	_____	72.	lymphokines	T.	coating of antibodies
OBJ. 12	_____	73.	IgG antibodies	U.	enhances nonspecific defenses
OBJ. 13	_____	74.	cytotoxic reactions	V.	agglutination
OBJ. 13	_____	75.	immune complex disorder	W.	two parallel pairs of polypeptide chains
OBJ. 14	_____	76.	cancer	X.	accompany fetail-maternal Rh incompatibility

[L1]   Drawing/Illustration Labeling:

Identify each numbered structure by labeling the following figures:

OBJ. 1   *FIGURE 22.1*   The Lymphatic System

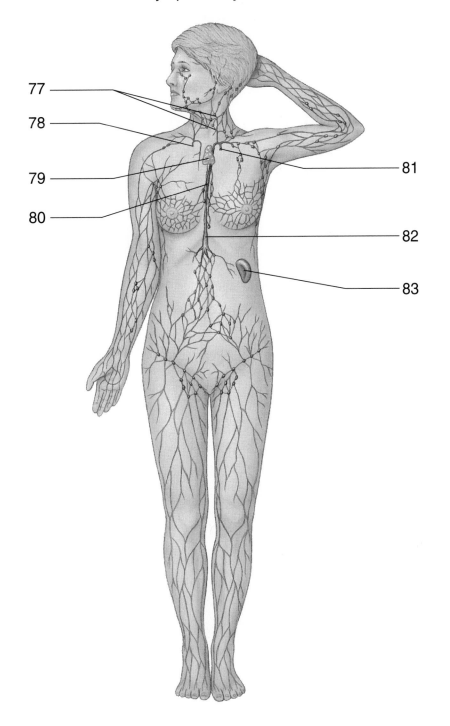

77 _____   81 _____

78 _____   82 _____

79 _____   83 _____

80 _____

OBJ. 1   *FIGURE 22.2*   The Lymphatic Ducts

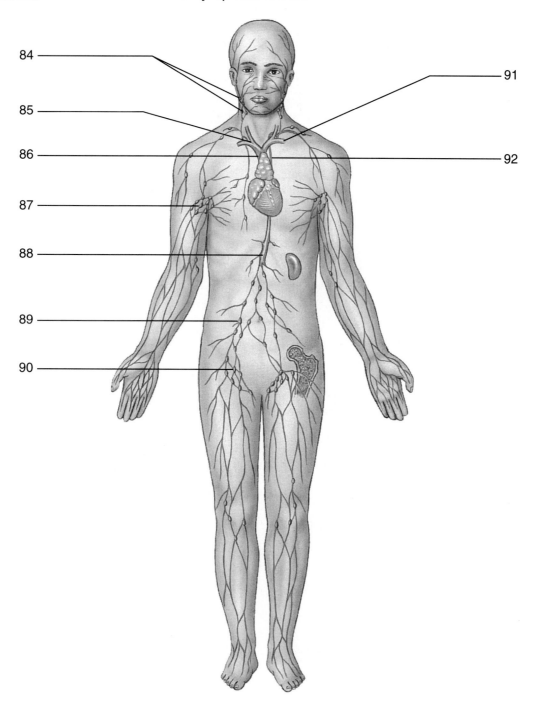

84 _____	89 _____
85 _____	90 _____
86 _____	91 _____
87 _____	92 _____
88 _____	

OBJ. 4   **FIGURE 22.3**   Nonspecific Defenses

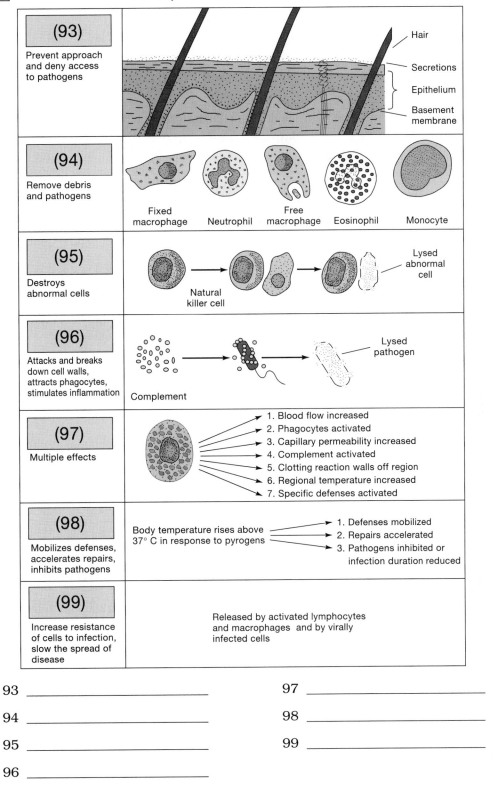

(93) Prevent approach and deny access to pathogens	Hair, Secretions, Epithelium, Basement membrane
(94) Remove debris and pathogens	Fixed macrophage, Neutrophil, Free macrophage, Eosinophil, Monocyte
(95) Destroys abnormal cells	Natural killer cell, Lysed abnormal cell
(96) Attacks and breaks down cell walls, attracts phagocytes, stimulates inflammation	Complement, Lysed pathogen
(97) Multiple effects	1. Blood flow increased 2. Phagocytes activated 3. Capillary permeability increased 4. Complement activated 5. Clotting reaction walls off region 6. Regional temperature increased 7. Specific defenses activated
(98) Mobilizes defenses, accelerates repairs, inhibits pathogens	Body temperature rises above 37° C in response to pyrogens: 1. Defenses mobilized 2. Repairs accelerated 3. Pathogens inhibited or infection duration reduced
(99) Increase resistance of cells to infection, slow the spread of disease	Released by activated lymphocytes and macrophages and by virally infected cells

93 _____   97 _____

94 _____   98 _____

95 _____   99 _____

96 _____

When you have successfully completed the exercises in L1 proceed to L2.

Level -1-

☐ LEVEL 2   CONCEPT SYNTHESIS

Concept Map I:

Using the following terms, fill in the circled, numbered, blank spaces to complete the concept map. Follow the numbers that comply with the organization of the map.

Active immunization   Inflammation   Phagocytic cells
Transfer of antibodies   Passive immunization   Specific immunity
Acquired   Active   Innate
Nonspecific immunity

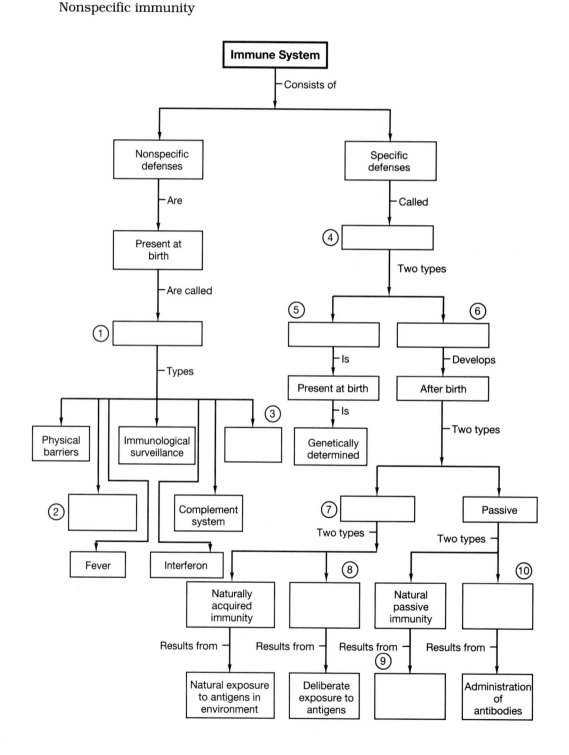

Concept Map II:

Using the following terms, fill in the circled, numbered, blank spaces to complete the concept map. Follow the numbers that comply with the organization of the map.

Increasing vascular permeability
Tissue damaged
Tissue repaired
Phagocytosis of bacteria
Bacteria not destroyed

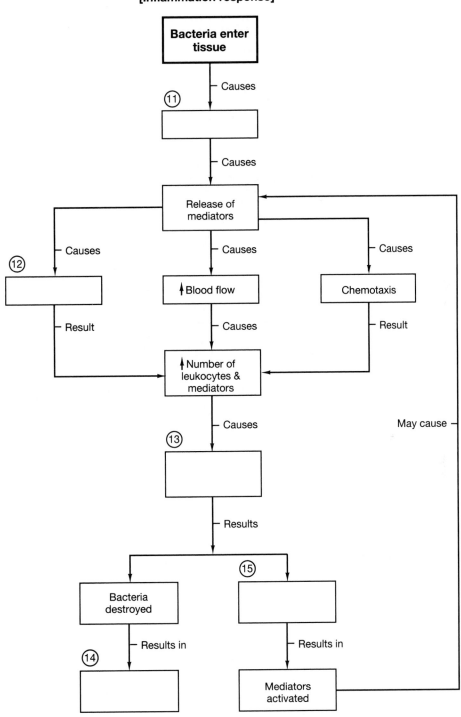

[Inflammation response]

Bacteria enter tissue

— Causes

⑪

— Causes

Release of mediators

— Causes

⑫

— Result

— Causes

↕ Blood flow

— Causes

— Causes

Chemotaxis

— Result

↕ Number of leukocytes & mediators

— Causes

⑬

— Results

Bacteria destroyed

⑮

May cause —

— Results in

⑭

— Results in

Mediators activated

Body Trek:

Using the terms below, fill in the blanks to complete a trek through the body's department of defense.

killer T cells	viruses	B cells
antibodies	helper T cells	macrophages
natural killer cells	suppressor T cells	memory T and B cells

Robo has been programmed to guide you on a trek through the body's department of defense—the immune system. You will proceed with the micro-robot from statement 1 (question 16) to statement 9 (question 24). At each site you will complete the event that is taking place by identifying the type of cells or proteins participating in the specific immune response.

The body's department of defense

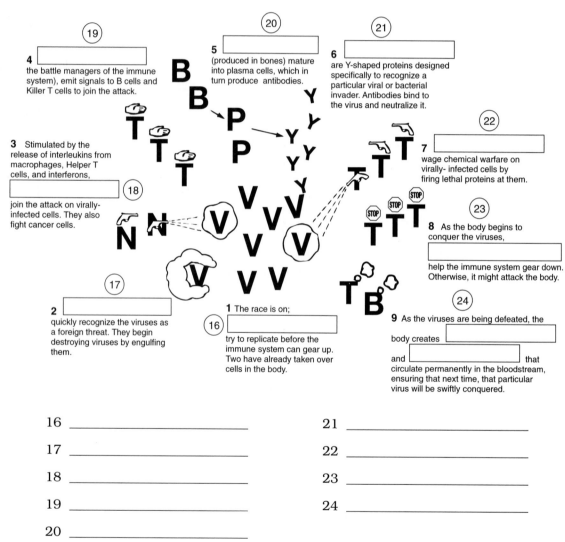

19

4 _____
the battle managers of the immune system), emit signals to B cells and Killer T cells to join the attack.

20

5 _____
(produced in bones) mature into plasma cells, which in turn produce antibodies.

21

6 _____
are Y-shaped proteins designed specifically to recognize a particular viral or bacterial invader. Antibodies bind to the virus and neutralize it.

3 Stimulated by the release of interleukins from macrophages, Helper T cells, and interferons, _____ **18**
join the attack on virally-infected cells. They also fight cancer cells.

22

7 _____
wage chemical warfare on virally- infected cells by firing lethal proteins at them.

23

8 As the body begins to conquer the viruses, _____
help the immune system gear down. Otherwise, it might attack the body.

17

2 _____
quickly recognize the viruses as a foreign threat. They begin destroying viruses by engulfing them.

1 The race is on; **16** _____
try to replicate before the immune system can gear up. Two have already taken over cells in the body.

24

9 As the viruses are being defeated, the body creates _____
and _____ that circulate permanently in the bloodstream, ensuring that next time, that particular virus will be swiftly conquered.

16 _____   21 _____

17 _____   22 _____

18 _____   23 _____

19 _____   24 _____

20 _____

[L2]   Multiple Choice:

Place the letter corresponding to the correct answer in the space provided.

_____ 25. The three different classes of lymphocytes in the blood are:

     a. cytotoxic cells, helper cells, suppressor cells

     b. T cells, B cells, NK cells

     c. plasma cells, B cells, cytotoxic cells

     d. antigens, antibodies, immunoglobulins

_____ 26. The primary effect(s) of complement activation include:

     a. destruction of target cell membranes

     b. stimulation of inflammation

     c. attraction of phagocytes and enhancement of phagocytosis

     d. a, b, and c are correct

_____ 27. Of the following selections, the one that _best_ defines the lymphatic system is that it is:

     a. an integral part of the circulatory system

     b. a one-way route from the blood to the interstitial fluid

     c. a one-way route from the interstitial fluid to the blood

     d. closely related to and a part of the circulatory system

_____ 28. Tissue fluid enters the lymphatic system via the:

     a. thoracic duct

     b. lymph capillaries

     c. lymph nodes

     d. bloodstream

_____ 29. The larger lymphatic vessels contain valves.

     a. true

     b. false

_____ 30. Systemic evidence of the inflammatory response would include:

     a. redness and swelling

     b. fever and pain

     c. production of WBC's, fever

     d. swelling and fever

_____ 31. Chemical mediators of inflammation include:

     a. histamine, kinins, prostaglandins, leukotrienes

     b. epinephrine, norepinephrine, acetylcholine, histamine

     c. kinins, opsonins, epinephrine, leukotrienes

     d. a, b, and c are correct

_____ 32. T lymphocytes comprise approximately _____ percent of circulating lymphocytes.

     a. 5 – 10

     b. 20 – 30

     c. 40 – 50

     d. 70 – 80

_____ 33. B lymphocytes differentiate into:

    a. cytotoxic and suppressor cells

    b. helper and suppressor cells

    c. memory and helper cells

    d. memory and plasma cells

_____ 34. _____ cells may activate B cells while _____ cells inhibit the activity of B cells.

    a. memory; plasma

    b. macrophages; microphages

    c. memory; cytotoxic

    d. helper T; suppressor T

_____ 35. The primary response of T-cell differentiation in cell-mediated immunity is the production of _____ cells.

    a. helper T

    b. suppressor T

    c. cytotoxic T

    d. memory

_____ 36. The vaccination of antigenic materials into the body is called:

    a. naturally acquired active immunity

    b. artificially acquired active immunity

    c. naturally acquired passive immunity

    d. artificially acquired passive immunity

_____ 37. In passive immunity _____ are induced into the body by injection.

    a. antibodies

    b. antigens

    c. T and B cells

    d. lymphocytes

_____ 38. The lymphatic function of the white pulp of the spleen is:

    a. phagocytosis of abnormal blood cell components

    b. release of splenic hormones into lymphatic vessels

    c. initiation of immune responses by B cells and T cells

    d. to degrade foreign proteins and toxins released by bacteria

_____ 39. A person with type AB blood has:

    a. anti-A and anti-B antibodies

    b. only anti-O antibodies

    c. no antigens

    d. neither anti-A nor anti-B antibodies

_____ 40. The antibodies produced and secreted by B lymphocytes are soluble proteins called:

    a. lymphokines

    b. agglutinins

    c. immunoglobulins

    d. leukotrines

Level
=2=

_____ 41. The genes found in a region called the *major histocompatibility complex* are called:

      a. immunoglobulins (IgG)

      b. human leukocyte antigens (HLAs)

      c. autoantibodies

      d. alpha and gamma interferons

_____ 42. Memory B cells do not differentiate into plasma cells unless they:

      a. are initially subjected to a specific antigen

      b. are stimulated by active immunization

      c. are exposed to the same antigen a second time

      d. are stimulated by passive immunization

_____ 43. The three-dimensional "fit" between the variable segments of the antibody molecule and the corresponding antigenic determinant site is referred to as the:

      a. immunodeficiency complex

      b. antibody-antigen complex

      c. protein-complement complex

      d. a, b, and c are correct

_____ 44. One of the primary nonspecific effects that glucocorticoids have on the immune response is:

      a. inhibition of interleukin secretion

      b. increased release of T and B cells

      c. decreased activity of cytotoxic T cells

      d. depression of the inflammatory response

[L2]   Completion:

Using the terms below, complete the following statements.

cytokines	T cells	antigen	mast
pyrogens	Kupffer cells	interferon	IgM
NK cells	properdin	helper T cells	IgG
Langerhans cells	opsonins	microglia	

45. Approximately 80 percent of circulating lymphocytes are classified as _____.

46. Fixed macrophages inside the CNS are called _____.

47. Macrophages in and around the liver sinusoids are referred to as _____.

48. Macrophages that reside within the epithelia of the skin and digestive tract are _____.

49. A negative immune response leads to tolerance of the _____.

50. Hormones of the immune system released by tissue cells to coordinate local activities are classified as _____.

51. The complement factor involved in the alternative pathway in the complement system is _____.

52. Cells that play a pivotal role in the inflammatory process are called _____ cells.

53. Circulating proteins that can reset the body's "thermostat" and cause a rise in body temperature are referred to as _____.

54. Enhanced phagocytosis is accomplished by a group of proteins called _____.

55. Immunological surveillance involves specific lymphocytes called _____.

56. The protein that appears to slow cell division and is effective in the treatment of cancer is _____.

57. Antibodies that comprise approximately 80 percent of all antibodies in the body are _____.

58. Antibodies that occur naturally in blood plasma and are used to determine an individual's blood type are _____.

59. The human immunodeficiency virus (HIV) attacks _____ cells in humans.

[L2]   Short Essay:

Briefly answer the following questions in the spaces provided below.

60. What are the three (3) primary organizational components of the lymphatic system?

61. What three (3) major functions are performed by the lymphatic system?

62. What are the three (3) different *classes* of lymphocytes found in the blood and where does each class originate?

63. What three (3) kinds of T cells comprise 80 percent of the circulating lymphocyte population? What is each type basically responsible for?

64. What is the primary function of stimulated B cells and what are they ultimately responsible for?

65. What three (3) lymphatic organs are important in the lymphatic system?

66. What are the six (6) defenses that provide the body with a defensive capability known as *nonspecific immunity?*

67. What are the primary differences between the "recognition" mechanisms of NK (natural killer) cells and T and B cells?

68. What four (4) primary effects result from complement activation?

69. What are the four (4) general characteristics of specific defenses?

70. What is the primary difference between active and passive immunity?

71. What two (2) primary mechanisms are involved in order for the body to reach its goal of the immune response?

72. What seven (7) possible processes could result when the antibody-antigen complex is found to eliminate the threat posed by the antigen?

73. What five (5) different classes of antibodies (immunoglobulins) are found in body fluids?

74. What are the four (4) primary subgroups of the lymphokines and the monokines?

75. What is an autoimmune disorder and what happens in the body when it occurs?

When you have successfully completed the exercises in L2 proceed to L3.

☐ LEVEL 3   CRITICAL THINKING/APPLICATION

Using principles and concepts learned about the lymphatic system and immunity, answer the following questions. Write your answers on a separate sheet of paper.

1. Organ transplants are common surgical procedures in a number of hospitals throughout the United States. What happens if the donor and recipient are not compatible?

2. Via a diagram, draw a schematic model to illustrate your understanding of the immune system as a whole. (*Hint: Begin with the antigen and show the interactions that result in cell-mediated and antibody-mediated immunity.*)

3. Defensins are protein molecules found in neutrophils that form peptide spears that pierce the cell membranes of pathogenic microbes. How does the role of defensin fit into the mechanisms of immunity?

4. A common cop-out for a sudden or unexplainable pathogenic condition in the body is, "He has a virus." Since viruses do not conform to the prokaryotic or eukaryotic plan of organization, how can they be classified as infectious agents?

5. Why are NK cells effective in fighting viral infections?

6. We usually associate a fever with illness or disease. In what ways may a fever be beneficial?

7. Stress of some kind is an everyday reality in the life of most human beings throughout the world. How does stress contribute to the effectiveness of the immune response?

Level
3

23

The Respiratory System

■ Overview

The human body can survive without food for several weeks, without water for several days, but if the oxygen supply is cut off for more than several minutes, the possibility of death is imminent. All cells in the body require a continuous supply of oxygen and must continuously get rid of carbon dioxide, a major waste product.

The respiratory system delivers oxygen from air to the blood and removes carbon dioxide, the gaseous waste product of cellular metabolism. It also plays an important role in regulating the pH of the body fluids.

The respiratory system consists of the lungs (which contain air sacs with moist membranes), a number of accessory structures, and a system of passageways that conduct inspired air from the atmosphere to the membranous sacs of the lungs, and expired air from the sacs in the lungs back to the atmosphere.

Chapter 23 includes exercises that examine the functional anatomy and organization of the respiratory system, respiratory physiology, the control of respiration, and respiratory interactions with other systems.

❏ LEVEL 1   REVIEW OF CHAPTER OBJECTIVES

1. Describe the primary functions of the respiratory system.
2. Explain how the delicate respiratory exchange surfaces are protected from environmental hazards.
3. Identify the organs of the upper respiratory system and describe their functions.
4. Describe the structure of the larynx and discuss its role in normal breathing and sound production.
5. Discuss the structure of the extrapulmonary airways.
6. Describe the superficial anatomy of the lungs, the structure of a pulmonary lobule, and the functional anatomy of the alveoli.
7. Describe the physical principles governing the movement of air into the lungs and the diffusion of gases into and out of the blood.
8. Describe the origins and actions of the muscles responsible for respiratory movements.
9. Differentiate between pulmonary ventilation and alveolar ventilation.

10. Describe how oxygen and carbon dioxide are transported in the blood.
11. Describe the major factors that influence the rate of respiration.
12. Identify and discuss the reflex activity and the brain centers involved in the control of respiration.

[L1] Multiple Choice:

Place the letter corresponding to the correct answer in the space provided.

OBJ. 1 _____ 1. The primary function(s) of the respiratory system is (are):

    a. to move air to and from the exchange surfaces

    b. to provide an area for gas exchange between air and circulating blood

    c. to protect respiratory surfaces from dehydration and environmental variations

    d. a, b, and c are correct

OBJ. 2 _____ 2. Pulmonary surfactant is a phospholipid secretion produced by alveolar cells to:

    a. increase the surface area of the alveoli

    b. reduce the cohesive force of H_2O molecules and lower surface tension

    c. increase the cohesive force of air molecules and raise surface tension

    d. reduce the attractive forces of O_2 molecules and increase surface tension

OBJ. 2 _____ 3. The "patrol force" of the alveolar epithelium involved with phagocytosis is comprised primarily of:

    a. alveolar NK cells

    b. alveolar cytotoxic cells

    c. alveolar macrophages

    d. alveolar plasma cells

OBJ. 3 _____ 4. The respiratory system consists of structures that:

    a. provide defense from pathogenic invasion

    b. permit vocalization and production of other sounds

    c. regulate blood volume and pressure

    d. a, b, and c are correct

OBJ. 3 _____ 5. The difference between the true and false vocal cords is that the false vocal cords:

    a. are highly elastic

    b. are involved with the production of sound

    c. play no part in sound production

    d. articulate with the corniculate cartilages

OBJ. 3 _____ 6. Structures in the trachea that prevent its collapse or over-expansion as pressures change in the respiratory system are the:

    a. O-ringed tracheal cartilages

    b. C-shaped tracheal cartilages

    c. irregular circular bones

    d. S-shaped tracheal bones

OBJ. 4 _____ 7. The entry of liquids or solid food into the respiratory passageways during swallowing is prohibited by the:

    a. glottis folding back over the epiglottis

    b. expansion of the cricoid cartilages

    c. epiglottis folding back over the glottis

    d. depression of the larynx

OBJ. 4 _____ 8. If food particles or liquids manage to touch the surfaces of the ventricular or vocal folds the:

    a. individual will choke to death

    b. coughing reflex will be triggered

    c. glottis will remain open

    d. epiglottis will not function

OBJ. 5 _____ 9. The trachea allows for the passage of large masses of food through the esophagus due to:

    a. the C-shaped tracheal cartilages

    b. the elasticity of the ringed tracheal cartilages

    c. distortion of the esophageal wall

    d. a, b, and c are correct

OBJ. 5 _____ 10. The purpose of the hilus along the medial surface of the lung at the level of the branching primary bronchi is to:

    a. mark the line of separation between the two bronchi

    b. provide access for entry to pulmonary vessels and nerves

    c. prevent foreign objects from entering the trachea

    d. a, b, and c are correct

OBJ. 6 _____ 11. Dilation and relaxation of the bronchioles are possible because of the presence of:

    a. bands of smooth muscle encircling the lumen

    b. the presence of C-shaped cartilaginous rings

    c. cuboidal epithelial cells containing cilia

    d. bands of skeletal muscles in the submucosa

OBJ. 6 _____ 12. Structural features that make the lungs highly pliable and capable of tolerating great changes in volume are:

    a. the air-filled passageways and the alveoli

    b. the bronchi, bronchioles, and the pleura

    c. the parenchyma, lobules, and the costal surfaces

    d. the elastic fibers in the trabeculae, the septa, and the pleurae

Level -1-

OBJ. 6 ____ 13. After passing through the trachea, the correct pathway a molecule of inspired air would take to reach an alveolus is:

    a. primary bronchus → secondary bronchus → bronchioles → respiratory bronchioles → terminal bronchioles → alveolus

    b. primary bronchus → bronchioles → terminal bronchioles → respiratory bronchioles → alveolus

    c. primary bronchus → secondary bronchus → bronchioles → terminal bronchioles → respiratory bronchioles → alveolus

    d. bronchi → secondary bronchus → bronchioles → terminal bronchioles → alveolus

OBJ. 6 ____ 14. The serous membrane in contact with the lung is the:

    a. visceral pleura

    b. parietal pleura

    c. pulmonary mediastinum

    d. pulmonary mesentery

OBJ. 7 ____ 15. A necessary feature for normal gas exchange in the alveoli is:

    a. for the alveoli to remain dry

    b. an increase in pressure in pulmonary circulation

    c. for fluid to move into the pulmonary capillaries

    d. an increase in lung volume

OBJ. 7 ____ 16. The movement of air into and out of the lungs is primarily dependent on:

    a. pressure differences between the air in the atmosphere and air in the lungs

    b. pressure differences between the air in the atmosphere and the anatomic dead space

    c. pressure differences between the air in the atmosphere and individual cells

    d. a, b, and c are correct

OBJ. 7 ____ 17. During inspiration there will be an increase in the volume of the thoracic cavity and a(n):

    a. decreasing lung volume, increasing intrapulmonary pressure

    b. decreasing lung volume, decreasing intrapulmonary pressure

    c. increasing lung volume, decreasing intrapulmonary pressure

    d. increasing lung volume, increasing intrapulmonary pressure

OBJ. 7 ____ 18. During expiration there is a(n):

    a. decrease in intrapulmonary pressure

    b. increase in intrapulmonary pressure

    c. increase in atmospheric pressure

    d. increase in the volume of the lungs

Level
-1-

OBJ. 7   _____ 19. If there is a P_{O_2} of 104 mm Hg and a P_{CO_2} of 40 mm Hg in the alveoli, and a P_{O_2} of 40 mm Hg and a P_{CO_2} of 45 mm Hg within the pulmonary blood, there will be a net diffusion of:

a. CO_2 into the blood from alveoli; O_2 from the blood into alveoli
b. O_2 and CO_2 into the blood from the alveoli
c. O_2 and CO_2 from the blood into the alveoli
d. O_2 into the blood from alveoli; CO_2 from the blood into the alveoli

OBJ. 7   _____ 20. Each molecule of hemoglobin has the capacity to carry _____ atoms of oxygen (O_2).

a. 6
b. 8
c. 4
d. 2

OBJ. 7   _____ 21. What percentage of total oxygen (O_2) is carried within red blood cells chemically bound to hemoglobin?

a. 5%
b. 68%
c. 98%
d. 100%

OBJ. 7   _____ 22. Factors that cause a decrease in hemoglobin saturation at a given P_{O_2} are:

a. increasing P_{O_2}, decreasing CO_2, increasing temperature
b. increasing diphosphoglycerate (DPG), increasing temperature, decreasing pH
c. decreasing DPG, increasing pH, increasing CO_2
d. decreasing temperature, decreasing CO_2, decreasing P_{O_2}

OBJ. 8   _____ 23. During expiration the diaphragm:

a. contracts and progresses inferiorly toward the stomach
b. contracts and the dome rises into the thoracic cage
c. relaxes and the dome rises into the thoracic cage
d. relaxes and progresses inferiorly toward the stomach

OBJ. 8   _____ 24. Stiffening and reduction in chest movement effectively limit the:

a. normal partial pressure of CO_2
b. respiratory minute volume
c. partial pressure of O_2
d. rate of hemoglobin saturation

OBJ. 9   _____ 25. Alveolar ventilation can be calculated by:

a. adding the tidal volume to the inspiratory reserve volume
b. adding the expiratory reserve volume and the residual volume
c. subtracting the tidal volume from the inspiratory reserve
d. subtracting the dead space from the tidal volume

Level
−1−

OBJ. 9 _____ 26. Pulmonary ventilation refers to the:

    a. gas exchange between the blood and the alveoli

    b. amount of air in the anatomic dead space

    c. movement of air in and out of the lungs

    d. movement of air in the upper respiratory tract

OBJ. 10 _____ 27. A developing fetus obtains its oxygen from the:

    a. ductus arteriosus

    b. foramen ovale

    c. ductus venosus

    d. maternal bloodstream

OBJ. 11 _____ 28. The output from baroreceptors affects the respiratory centers causing:

    a. the respiratory rate to decrease without affecting blood pressure

    b. the respiratory rate to increase with an increase in blood pressure

    c. the respiratory rate to decrease with a decrease in blood pressure

    d. the respiratory rate to increase with a decrease in blood pressure

OBJ. 11 _____ 29. Under normal conditions the greatest effect on the respiratory centers is initiated by:

    a. decreases in P_{O_2}

    b. increases and decreases in P_{O_2} and P_{CO_2}

    c. increases and decreases in P_{CO_2}

    d. increases in P_{O_2}

OBJ. 11 _____ 30. An elevated body temperature will:

    a. decrease respiration

    b. increase depth of respiration

    c. accelerate respiration

    d. not affect the respiratory rate

OBJ. 12 _____ 31. The initiation of inspiration originates with discharge of inspiratory neurons in the:

    a. diaphragm

    b. medulla

    c. pons

    d. lungs

OBJ. 12 _____ 32. Reflexes important in regulating the forced ventilations that accompany strenuous exercise are known as the:

    a. Hering-Breuer reflexes

    b. protective reflexes

    c. chemoreceptor reflexes

    d. pulmonary reflexes

Level
-1-

OBJ. 12 ____ 33. As the volume of the lungs increases, the:

    a. inspiratory center is stimulated, the expiration center inhibited

    b. the inspiratory center is inhibited, the expiratory center stimulated

    c. the inspiratory and expiratory centers are stimulated

    d. the inspiratory and expiratory centers are inhibited

OBJ. 12 ____ 34. An increase in pneumotaxic output causes:

    a. increasing rate of respiration

    b. decreasing rate of respiration

    c. increasing depth of respiration

    d. a, b, and c are correct

[L1]   Completion:

Using the terms below, complete the following statements.

chemoreceptor	glottis	carbaminohemoglobin
mucus escalator	apnea	external respiration
inspiration	vital capacity	pneumotaxic
ductus arteriosus	expiration	mechanoreceptor
internal respiration	external nares	surfactant
apneustic	Bohr effect	larynx
alveolar ventilation	respiratory bronchioles	bronchioles

OBJ. 1   35. The movement of air out of the lungs is referred to as _____.

OBJ. 1   36. The movement of air into the lungs is called _____.

OBJ. 2   37. The oily phospholipid secretion that coats the alveolar epithelium and keeps the alveoli from collapsing is called _____.

OBJ. 2   38. The mechanism in the respiratory tree that carries debris and duct particles toward the pharynx and an acid bath in the stomach is the _____.

OBJ. 3   39. Air normally enters the respiratory system via the paired _____.

OBJ. 3   40. Inspired air leaves the pharynx by passing through a narrow opening called the _____.

OBJ. 4   41. The glottis is surrounded and protected by the _____.

OBJ. 5   42. The part of the bronchial tree dominated by smooth muscle tissue is the _____.

OBJ. 6   43. Within a pulmonary lobule, the terminal bronchioles branch to form several _____.

OBJ. 7   44. The diffusion of gases between the alveoli and the circulating blood is termed _____.

OBJ. 7   45. The exchange of dissolved gases between the blood and the interstitial fluids in peripheral tissues is called _____.

OBJ. 7   46. The fetal connection between the pulmonary trunk and the aorta is the _____.

OBJ. 8   47. With increasing age, deterioration of elastic tissue reduces lung compliance, lowering the _____.

OBJ. 9   48. The primary function of pulmonary ventilation is to maintain adequate _____.

OBJ. 10   49. The effect of pH on the hemoglobin saturation curve is called the _____.

OBJ. 10   50. CO_2 molecules bound to exposed amino groups (NH_2) along the globin portions of hemoglobin molecules form _____.

OBJ. 11   51. When respiration is suspended due to exposure to toxic vapors or chemical irritants, reflex activity may result in _____.

OBJ. 12   52. Reflexes that are a response to changes in the volume of the lungs or changes in arterial blood pressure are _____ reflexes.

OBJ. 12   53. Reflexes that are a response to changes in the Po_2 and Pco_2 of the blood and cerebrospinal fluid are _____ reflexes.

OBJ. 12   54. The dorsal inspiratory center of the lower pons is called the _____ center.

OBJ. 12   55. The ventral expiratory center of the upper pons is called the _____ center.

[L1]   Matching:

Match the terms in column B with the terms in column A. Use letters for answers in the spaces provided.

PART I

	COLUMN A	COLUMN B
OBJ. 1 _____ 56.	upper respiratory tract	A. nostrils
OBJ. 2 _____ 57.	septal cells	B. lungs
OBJ. 2 _____ 58.	alveolar macrophages	C. sound production
OBJ. 3 _____ 59.	external nares	D. surfactant cells
OBJ. 4 _____ 60.	larynx	E. delivers air to lungs
OBJ. 5 _____ 61.	trachea	F. rise in arterial pCO_2
OBJ. 6 _____ 62.	lower respiratory tract	G. dust cells
OBJ. 7 _____ 63.	eupnea	H. windpipe
OBJ. 7 _____ 64.	hyperventilation	I. quiet breathing
OBJ. 7 _____ 65.	hypoventilation	J. abnormal decreasing arterial pCO_2

PART II

	COLUMN A	COLUMN B
OBJ. 8 _____ 66.	external intercostals	K. carbon dioxide
OBJ. 9 _____ 67.	pulmonary ventilation	L. interatrial connection
OBJ. 10 _____ 68.	hemoglobin	M. prevents overexpansion of lungs
OBJ. 10 _____ 69.	bicarbonate ion	
OBJ. 10 _____ 70.	foramen ovale	N. emphysema
OBJ. 11 _____ 71.	smoking	O. inhibitory effect on respiration
OBJ. 12 _____ 72.	inflation reflex	
OBJ. 12 _____ 73.	deflation reflex	P. stimulates inspiratory center
OBJ. 12 _____ 74.	apneustic center	
OBJ. 12 _____ 75.	pneumotaxic center	Q. stimulatory effect on respiration
		R. breathing
		S. oxygen transport
		T. elevation of ribs

Level -1-

[L1]   Drawing/Illustration Labeling:

Identify each numbered structure by labeling the following figures:

OBJ. 2   **FIGURE 23.1**   Upper Respiratory Tract

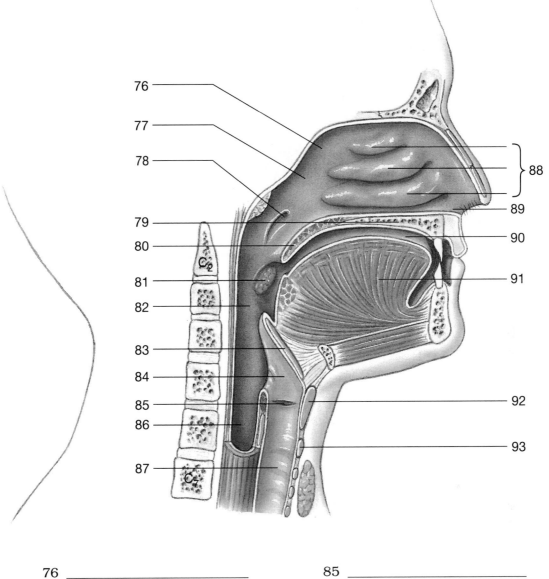

76 _____   85 _____

77 _____   86 _____

78 _____   87 _____

79 _____   88 _____

80 _____   89 _____

81 _____   90 _____

82 _____   91 _____

83 _____   92 _____

84 _____   93 _____

Level
-1-

OBJ. 2   *FIGURE 23.2*   Thorax and Lungs

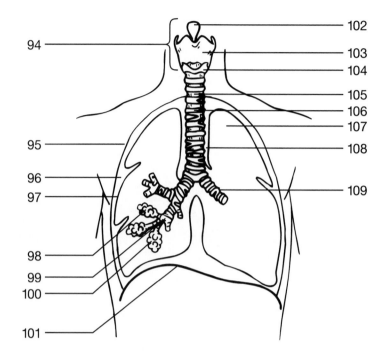

94	_____	102	_____
95	_____	103	_____
96	_____	104	_____
97	_____	105	_____
98	_____	106	_____
99	_____	107	_____
100	_____	108	_____
101	_____	109	_____

OBJ. 4   **FIGURE 23.3**   Lower Respiratory Tract

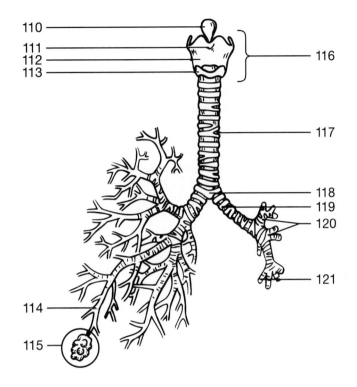

110 _____   116 _____

111 _____   117 _____

112 _____   118 _____

113 _____   119 _____

114 _____   120 _____

115 _____   121 _____

When you have successfully completed the exercises in L1 proceed to L2.

Level
-1-

☐ LEVEL 2   CONCEPT SYNTHESIS

Concept Map I:

Using the following terms, fill in the circled, numbered, blank spaces to complete the concept map. Follow the numbers that comply with the organization of the map.

Pulmonary veins   Respiratory bronchioles   Trachea
Left ventricle   Right atrium   Pharynx
Pulmonary arteries   Systemic capillaries   Alveoli
Secondary bronchi

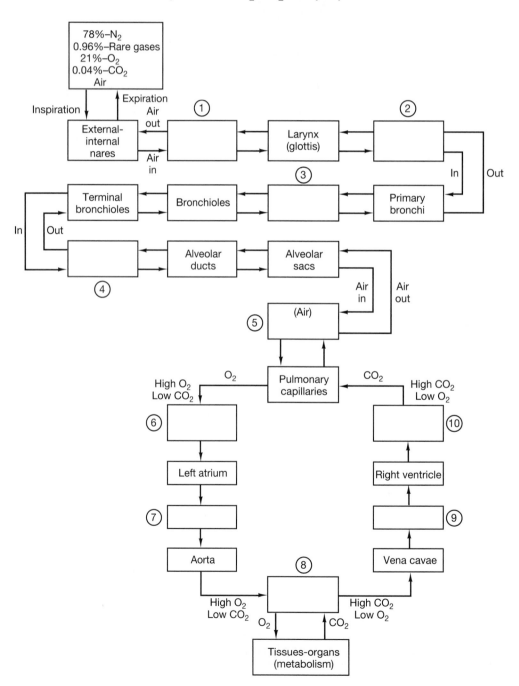

**External–internal respiration
(Air flow and O_2–CO_2 transport)**

Concept Map II:

Using the following terms, fill in the circled, numbered, blank spaces to complete the concept map. Follow the numbers that comply with the organization of the map.

Lungs	Upper respiratory tract	Paranasal sinuses
Blood	Ciliated mucous membrane	Pharynx & Larynx
Alveoli	O_2 from alveoli into blood	Speech production

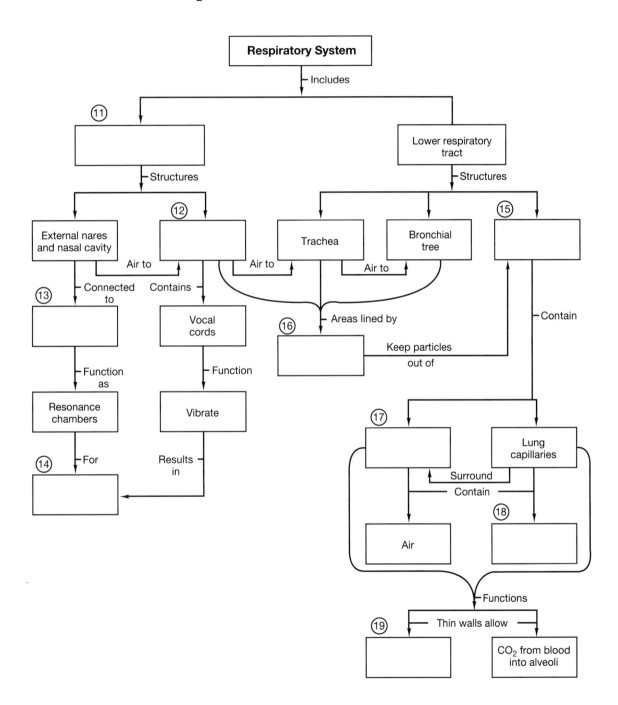

Body Trek:

Using the terms below, fill in the blanks to complete a trek through the respiratory system.

primary bronchus	nasal cavity	nasopharynx
epiglottis	smooth muscle	oral cavity
hard palate	cilia	oxygen
alveolus	mucus escalator	pharynx
carbon dioxide	larynx	nasal conchae
trachea	soft palate	bronchioles
external nares	vestibule	vocal folds
pulmonary capillaries	phonation	

Robo's trek into the upper respiratory tract is facilitated via a deep inhalation by the host. The micro-robot enters through the (20) _____ that communicate with the (21) _____. The reception area, called the (22) _____ , contains coarse hairs that trap airborne particles and prevent them from entering the respiratory tract.

As the robot treks into the "cavernous arena" its "thermosensors" are activated owing to the warm and humid conditions. From the lateral walls of the cavity, the "air-conditioning" system, consisting of superior, middle, and inferior (23) _____ , project toward the nasal septum. As air bounces off these projections it is warmed and humidified and induces turbulence, causing airborne particles to come in contact with the mucous covering the surfaces of these lobular structures. A bony (24) _____ separates the oral and nasal cavities. A fleshy (25) _____ extends behind the hard palate, marking the boundary line between the superior (26) _____ and the rest of the pharyngeal regions. As the micro-robot approaches the (27) _____ it goes into an uncontrollable skid to the bottom because of mucus secretions that coat the pharyngeal walls.

Because the host is not swallowing, the elongate, cartilaginous (28) _____ permits the passage of the robot into the (29) _____ , a region that contains the (30) _____ , which are involved with sound production called (31) _____. After passing through the "sound barrier" Robo treks into a large "flume-like" duct, the (32) _____ , which contains C-shaped cartilaginous rings that prevent collapse of the air passageway.

As the robot treks through the "wind tunnel" finger-like cellular extensions, the (33) _____ , cause a "countercurrent" that serves to move mucus toward the (34) _____. Leaving the "wind tunnel," the robot's program directs it into the right (35) _____ , a decision made by mission control. The robot's descent is complicated by increased "trek resistance" as the passageways get narrower and at times are almost impossible because of constriction caused by (36) _____ tissue in the submucosal lining of the (37) _____. After a few minor delays the trek continues into a sac-like (38) _____ , where the atmosphere is dry and gaseous exchange creates a "breeze" that flows into and out of the "see-through" spheres. Robo's chemoreceptors "pick up" the presence of (39) _____ and (40) _____ passing freely between the air inside the spheres and the blood in the (41) _____.

Robo's return trek for exit is facilitated by a "reverse program" that includes a "ride" on the (42) _____ , making the robot's ascent fast, smooth, and easy. Upon reaching the nasal cavity, removal of the little robot is immediate due to the host's need to blow his nose.

[L2]   Multiple Choice:

Place the letter corresponding to the correct answer in the space provided.

_____ 43. The paranasal sinuses include:

    a. occipital, parietal, temporal, and mandibular

    b. ethmoid, parietal, sphenoid, and temporal

    c. frontal, sphenoid, ethmoid, and maxillary

    d. mandibular, maxillary, frontal, temporal

_____ 44. A rise in arterial Pco_2 elevates cerebrospinal fluid (CSF) carbon dioxide levels and stimulates the chemoreceptive neurons of the medulla to produce:

    a. hypercapnia

    b. hyperventilation

    c. hypoventilation

    d. eupnea

_____ 45. The primary function of pulmonary ventilation is to maintain adequate:

    a. air in the anatomic dead space

    b. pulmonary surfactant

    c. vital capacity

    d. alveolar ventilation

_____ 46. The purpose of the fluid in the pleural cavity is to:

    a. provide a medium for the exchange of O_2 and CO_2

    b. reduce friction between the parietal and visceral pleura

    c. allow for the exchange of electrolytes during respiratory movements

    d. provide lubrication for diaphragmatic constriction

_____ 47. When the lungs are in the resting position, the factor that opposes their collapse is:

    a. an intrapleural pressure of 760 mm Hg

    b. an intrapulmonary pressure of 5 mm Hg

    c. an intrapleural pressure of 764 mm Hg

    d. an intrapulmonary pressure of 760 mm Hg

_____ 48. If a person is stabbed in the chest and the _thoracic wall_ is punctured but the lung is not penetrated:

    a. the lungs will collapse

    b. the intrapleural pressure becomes subatmospheric

    c. the intrapulmonary pressure becomes subatmospheric

    d. the intra-alveolar pressure becomes subatmospheric

_____ 49. The _most important_ factor determining airway resistance is:

    a. interactions between flowing gas molecules

    b. length of the airway

    c. thickness of the wall of the airway

    d. airway radius

Level
=2=

_____ 50. The sympathetic division of the ANS causes _____ of airway smooth muscle; therefore, resistance is _____.

      a. relaxation; decreased

      b. relaxation; increased

      c. constriction; increased

      d. constriction; decreased

_____ 51. The substance administered during an asthmatic attack to decrease resistance via airway dilation is:

      a. histamine

      b. epinephrine

      c. norepinephrine

      d. angiotensin

_____ 52. Decreased amounts of CO_2 concentrations in the bronchioles cause:

      a. an increase in bronchiolar dilation

      b. a decrease in bronchiolar constriction

      c. an increase in bronchiolar constriction

      d. none of these

_____ 53. The volume of alveolar air and capillary blood in the right proportion to each alveolus produces:

      a. pulmonary surfactant

      b. optimum breathing

      c. pulmonary circulation

      d. lung efficiency

_____ 54. If a person is breathing 15 times a minute and has a tidal volume of 500 ml, the _total_ minute respiratory volume is:

      a. 7500 min/ml

      b. 515 ml

      c. 7500 ml/min

      d. 5150 ml

_____ 55. The _residual volume_ is the volume of air:

      a. at the end of normal respiration if expiratory muscles are actively contracted

      b. which remains in lungs after maximal expiration

      c. which can be inspired over and above resting tidal volume

      d. inspired and expired during one breath

_____ 56. The maximum amount of air moved in and out during a single breath is:

      a. tidal volume

      b. vital capacity

      c. alveolar ventilation

      d. residual volume

_____ 57. If a person is breathing 12 times per minute and the tidal volume is 500 ml, what is the alveolar ventilation rate?

      a. 0 ml/min

      b. 5250 ml/min

      c. 6000 ml/min

      d. 4200 ml/min

_____ 58. The most effective means of increasing alveolar ventilation is:

      a. increase rapid shallow breathing

      b. breathe normally

      c. breathe slowly and deeper

      d. increase total pulmonary circulation

_____ 59. In "respiratory distress syndrome of the newborn," the administering of adrenal steroids serves to:

      a. synthesize immature surfactant-synthesizing cells

      b. increase O_2 consumption to the newborn

      c. enhance surfactant-synthesizing maturation

      d. all of these

_____ 60. The partial pressure of O_2 at sea level is:

      a. 104 mm Hg

      b. 200 mm Hg

      c. 760 mm Hg

      d. 160 mm Hg

_____ 61. As the number of the molecules of gas dissolved in a liquid increases:

      a. the pressure of the gas decreases

      b. the pressure of the gas increases

      c. the pressure of the gas is not affected

      d. none of these

_____ 62. Movement of air into and out of the lungs is accomplished by the process of _____, while all movement of gases across membranes are by _____.

      a. endocytosis; exocytosis

      b. diffusion; osmosis

      c. bulk flow; passive diffusion

      d. pinocytosis; diffusion

_____ 63. The correct sequential transport of O_2 from the tissue capillaries to O_2 consumption in cells is:

      a. lung, alveoli, plasma, erythrocytes, cells

      b. erythrocytes, plasma, interstitial fluid, cells

      c. plasma, erythrocytes, alveoli, cells

      d. erythrocytes, interstitial fluid, plasma, cells

_____ 64. It is important that free H^+ resulting from dissociation of H_2CO_3 combine with hemoglobin to reduce the possibility of:

      a. CO_2 escaping from the RBC

      b. recombining with H_2O

      c. an acidic condition within the blood

      d. maintaining a constant pH in the blood

_____ 65. In the pulmonary capillaries the bicarbonate ion is always returned to the:

      a. RBC

      b. interstitial fluid

      c. alveoli

      d. plasma

_____ 66. If you desired to control the rate of respiration voluntarily you might:

      a. sneeze

      b. engage in an emotional experience

      c. hold your breath

      d. inflict pain

_____ 67. One of the early symptoms of emphysema is:

      a. a reduced expiratory volume

      b. the lung becomes solid and airless

      c. the lungs become inflamed

      d. the vital capacity is reduced

[L2] Completion:

Using the terms below, complete the following statements.

hypercapnia	thyroid cartilage	atelectasis
hilus	intrapleural	paraenchyma
pneumothorax	phonation	epiglottis
alveolar ventilation	carina	anatomic dead space
laryngopharynx	cribiform plate	pharynx
anoxia	lamina propria	Henry's law
Dalton's law	vestibule	hypoxia

68. The layer of connective tissue between the respiratory epithelium and the underlying bones or cartilages is the _____.

69. The portion of the nasal cavity containing epithelium with coarse hairs is the _____.

70. The chamber shared by the digestive and respiratory systems is the _____.

71. The portion of the pharynx lying between the hyoid and entrance to the esophagus is the _____.

72. The elastic cartilage that prevents the entry of liquids or solid food into the respiratory passageway is the _____.

73. The "Adam's apple" is the prominent ridge on the anterior surface of the _____.

74. Sound production at the larynx is called _____.

75. The groove along the medial surface of a lung that provides access for the pulmonary vessels and nerves is called the _____.

76. The roof of the nasal cavity is formed primarily by the _____.

77. A decline in oxygen content causes affected tissues to suffer from _____.

78. The effects of cerebrovascular accidents and myocardial infarctions are the result of localized _____.

79. A collapsed lung is referred to as a(n) _____.

80. An injury to the chest wall that penetrates the parietal pleura or damages the alveoli and the visceral pleura is called a _____.

81. The pressure measured in the slender space between the parietal and visceral pleura is the _____ pressure.

82. An increase in the P_{CO_2} of arterial blood is called _____.
83. The ridge that marks the line of separation between the two bronchi is the _____.
84. The substance of each lung is referred to as the _____.
85. The volume of air in the conducting passages is known as the _____.
86. The amount of air reaching the alveoli each minute is the _____.
87. The relationship among pressure, solubility, and the number of gas molecules in solution is known as _____.
88. The principle that states, "Each of the gases that contributes to the total pressure is in proportion to its relative abundance" is known as _____.

[L2]   Short Essay:

Briefly answer the following questions in the spaces provided below.

89. What are the six (6) primary functions of the respiratory system?

90. How is the inhaled air warmed and humidified in the nasal cavities, and why is this type of environmental condition necessary?

91. What is surfactant and what is its function in the respiratory membrane?

92. What protective mechanisms comprise the respiratory defense system?

93. What is the difference among external respiration, internal respiration, and cellular respiration?

Level
=2=

94. What is the relationship between the volume and the pressure of a gas?

95. What do the terms *pneumothorax* and *atelectasis* refer to, and what is the relationship between the two terms?

96. What is vital capacity and how is it measured?

97. What is alveolar ventilation and how is it calculated?

98. What are the three (3) methods of CO_2 transport in the blood?

99. What three (3) kinds of reflexes are involved in the regulation of respiration?

100. What is the relationship between rise in arterial P_{CO_2} and hyperventilation?

Level
=2=   **When you have successfully completed the exercises in L2 proceed to L3.**

☐ LEVEL 3   CRITICAL THINKING/APPLICATION

Using principles and concepts learned about the respiratory system, answer the following questions. Write your answers on a separate sheet of paper.

1. I. M. Good decides to run outside all winter long in spite of the cold weather. How does running in cold weather affect the respiratory passageways and the lungs?

2. U. R. Hurt is a 68-year-old man with arteriosclerosis. His condition has promoted the development of a pulmonary embolism that has obstructed blood flow to the right upper lobe of the lung. Eventually he experiences heart failure. Why?

3. C. U. Later's son is the victim of a tracheal defect that resulted in the formation of complete cartilaginous rings instead of the normal C-shaped tracheal rings. After eating a meal he experiences difficulty when swallowing and feels pain in the thoracic region. Why?

4. M. I. Right has been a two-pack-a-day smoker for the last 20 years. Recently, he has experienced difficulty breathing. After a series of pulmonary tests his doctor informs Mr. Right that he has emphysema. What is the relationship between his condition and his breathing difficulties?

5. R. U. Bunting was involved in an automobile accident. Ever since the accident he has experienced a great deal of respiratory distress due to his inability to cough and sneeze. Based on these symptoms describe the extent of Mr. Bunting's injuries.

6. What effect does respiratory acidosis have on the respiratory process and what compensatory mechanisms operate to bring the body back to a homeostatic pH value?

C H A P T E R

24

The Digestive System

■ Overview

The digestive system is the "food processor" in the human body. The system works at two levels—one mechanical, the other chemical. The mechanical processes include ingestion, deglutition (swallowing), peristalsis, compaction, and defecation, while mastication (chewing) and absorption serve as mechanical and chemical mechanisms.

By the time you are 65 years old, you will have consumed more than 70,000 meals and disposed of 50 tons of food, most of which has to be converted into chemical forms that cells can utilize for their metabolic functions.

After ingestion and mastication, the food is propelled through the muscular gastrointestinal tract (GI) where it is chemically split into smaller molecular substances. The products of the digestive process are then absorbed from the GI tract into the circulating blood and transported wherever they are needed in the body or, if undigested, are eliminated from the body. As the food is processed through the GI tract, accessory organs such as the salivary glands, liver, and pancreas assist in the "breakdown" process.

Chapter 24 includes exercises organized to guide your study of the structure and function of the digestive tract and accessory organs, hormonal and nervous system influences on the digestive process, and the chemical and mechanical events that occur during absorption.

☐ LEVEL 1 REVIEW OF CHAPTER OBJECTIVES

1. Identify the organs of the digestive tract and the accessory organs of digestion.
2. List the primary functions of the digestive system.
3. Describe the histological characteristics of a representative portion of the digestive tract and relate anatomical structure to specific digestive functions.
4. Describe the processes involved in the movement of digestive materials through the gastrointestinal tract.
5. List and describe the mechanisms that regulate the activities of the digestive system.
6. Describe the anatomy and functions of the oral cavity, pharynx, and esophagus.
7. Describe the anatomy of the stomach, its histological features, and its roles in digestion and absorption.
8. Describe the anatomical and histological characteristics of the small intestine.

9. Explain the functions of the intestinal secretions and discuss the regulation of secretory activities.

10. Describe the structure and functions of the pancreas, liver, and gall bladder and explain how their activities are regulated and coordinated.

11. Discuss the regulation and coordination of gastric and intestinal movements, gastric emptying, and intestinal absorption.

12. Describe the structure of the large intestine, its movements, and the absorptive processes that take place within it.

13. Describe the processes of digestion and absorption for carbohydrates, lipids, and proteins.

14. Discuss the mechanisms and processes involved in the absorption of water, electrolytes and vitamins.

[L1]   Multiple choice:

Place the letter corresponding to the correct answer in the space provided.

OBJ. 1   _____   1. Which one of the following organs is not a part of the digestive system?

      a. liver
      b. gall bladder
      c. spleen
      d. pancreas

OBJ. 1   _____   2. Of the following selections, the *one* which contains *only* accessory structures is:

      a. pharynx, esophagus, small and large intestine
      b. oral cavity, stomach, pancreas and liver
      c. salivary glands, pancreas, liver, gall bladder
      d. tongue, teeth, stomach, small and large intestine

OBJ. 2   _____   3. The active process that occurs when materials enter the digestive tract via the mouth is:

      a. secretion
      b. ingestion
      c. absorption
      d. excretion

OBJ. 2   _____   4. The lining of the digestive tract plays a defensive role by protecting surrounding tissues against:

      a. corrosive effects of digestive acids and enzymes
      b. mechanical stresses
      c. pathogenic organisms swallowed with food
      d. a, b, and c are correct

OBJ. 3   _____   5. Sympathetic stimulation of the muscularis externa promotes:

      a. muscular inhibition and relaxation
      b. increased muscular tone and activity
      c. muscular contraction and increased excitation
      d. increased digestive and gastric motility

Level
-1-

OBJ. 3 _____ 6. The mucous-producing, unicellular glands found in the mucosal epithelium of the stomach and small and large intestine are:

a. enteroendocrine cells
b. parietal cells
c. chief cells
d. goblet cells

OBJ. 4 _____ 7. Which of the layers of the digestive tube is (are) most responsible for peristalsis along the esophagus?

a. tunica mucosa
b. circular and longitudinal layers
c. tunica submucosa
d. tunica muscularis

OBJ. 4 _____ 8. Once a bolus of food has entered the laryngopharynx, swallowing continues involuntarily due to the:

a. swallowing reflex
b. size of the bolus
c. peristaltic activity
d. a, b, and c are correct

OBJ. 4 _____ 9. Which of the following is true about peristalsis in the esophagus?

a. The rate of peristaltic waves in the esophagus is constant.
b. The peristalsis is controlled by the nervous system.
c. The peristalsis is controlled by the endocrine system.
d. Peristalsis is controlled by the nervous and the endocrine systems.

OBJ. 4 _____ 10. Swirling, mixing, and churning motions of the digestive tract provide:

a. action of acids, enzymes, and buffers
b. mechanical processing after ingestion
c. chemical breakdown of food into small fragments
d. a, b, and c are correct

OBJ. 4 _____ 11. Strong contractions of the ascending and transverse colon moving the contents of the colon toward the sigmoid colon are called:

a. defecation
b. pendular movements
c. segmentation
d. mass peristalsis

OBJ. 5 _____ 12. Accelerated secretions by the salivary glands, resulting in the production of watery saliva containing abundant enzymes, are promoted by:

a. sympathetic stimulation
b. parasympathetic stimulation
c. the gastroenteric reflex
d. excessive secretion of salivary amylase

Level
—1—

OBJ. 5   _____ 13. The hormone gastrin:

      a. is produced in response to sympathetic stimulation
      b. is secreted by the pancreatic islets
      c. increases the activity of parietal and chief cells
      d. inhibits the activity of the muscularis externa of the stomach

OBJ. 6   _____ 14. The submandibular gland produces saliva, which is:

      a. primarily a mucus secretion
      b. primarily a serous secretion
      c. both mucus and serous
      d. neither mucus nor serous

OBJ. 6   _____ 15. The three pairs of salivary glands that secrete into the oral cavity include:

      a. pharyngeal, palatoglossal, palatopharyngeal
      b. lingual, labial, frenulum
      c. uvular, ankyloglossal, hypoglossal
      d. parotid, sublingual, submandibular

OBJ. 6   _____ 16. Crushing, mashing, and grinding of food is best accomplished by the action of the:

      a. incisors
      b. bicuspids
      c. cuspids
      d. molars

OBJ. 6   _____ 17. The three phases of deglutition are:

      a. parotid, sublingual, submandibular
      b. pharyngeal, palatopharyngeal, stylopharyngeal
      c. buccal, pharyngeal, esophageal
      d. palatal, lingual, mesial

OBJ. 7   _____ 18. Stomach emptying occurs more rapidly when:

      a. there is a greater volume of stomach contents
      b. there is a greater volume of duodenal contents
      c. there is more fat in the stomach
      d. the material in the stomach is hyperosmotic to the plasma

OBJ. 7   _____ 19. The contractions of the stomach are inhibited by:

      a. secretin
      b. gastrin
      c. pepsinogen
      d. trypsin

OBJ. 7   _____ 20. Which of the following is secreted by the stomach?

      a. galactase
      b. gastrin
      c. ptyalin
      d. secretin

Level
-1-

OBJ. 8 _____ 21. The three divisions of the small intestine are:

    a. cephalic, gastric, intestinal

    b. buccal, pharyngeal, esophageal

    c. duodenum, jejunum, ileum

    d. fundus, body, pylorus

OBJ. 8 _____ 22. The myenteric plexus (Auerbach) of the intestinal tract is found:

    a. within the mucosa

    b. within the submucosa

    c. within the circular muscle layer

    d. between the circular and longitudinal muscle layers

OBJ. 9 _____ 23. An enzyme not found in pancreatic juice is:

    a. lipase

    b. amylase

    c. chymotrypsin

    d. disaccharidase

OBJ. 9 _____ 24. What happens to salivary amylase after it is swallowed?

    a. It is compacted and becomes part of the feces.

    b. It is absorbed and re-secreted by the salivary glands.

    c. It is absorbed in the duodenum and broken down into amino acids in the liver.

    d. It is digested and absorbed in the small intestine.

OBJ. 9 _____ 25. The three phases of gastric function include:

    a. parotid, sublingual, submandibular

    b. duodenal, jejunal, iliocecal

    c. cephalic, gastric, intestinal

    d. buccal, pharyngeal, esophageal

OBJ. 10 _____ 26. The functions of the gall bladder involve:

    a. contraction and synthesis

    b. contraction and absorption

    c. synthesis and absorption

    d. absorption and digestion

OBJ. 10 _____ 27. The primary function(s) of the liver is (are):

    a. metabolic regulation

    b. hematological regulation

    c. bile production

    d. a, b, and c are correct

OBJ. 10 _____ 28. The hormone that promotes the flow of bile and of pancreatic juice containing enzymes is:

    a. secretin

    b. gastrin

    c. enterogastrone

    d. cholecystokinin

Level
−1−

OBJ. 11   _____ 29. The secretion of parietal cells that facilitate absorption of vitamin B_{12} across the intestinal lining is:

        a. intrinsic factor

        b. hydrochloric acid

        c. gastrin

        d. secretin

OBJ. 11   _____ 30. The central mechanisms regulating gastric function are:

        a. direct control by the endocrine system, indirect regulation by the ANS

        b. direct control by the CNS, and indirect regulation by local hormones

        c. direct control by the CNS and the ANS

        d. direct control by local homeostatic mechanisms

OBJ. 12   _____ 31. Undigested food residues are moved through the large intestine in the following sequence:

        a. cecum, colon, and rectum

        b. colon, cecum, rectum

        c. ileum, colon, rectum

        d. duodenum, jejunum, ileum

OBJ. 12   _____ 32. The longitudinal ribbon of smooth muscle visible on the outer surfaces of the colon just beneath the serosa are the:

        a. haustrae

        b. taenia coli

        c. epiploic appendages

        d. vermiform appendix

OBJ. 13   _____ 33. The enzyme lactase, which digests lactose to glucose and galactose, is synthesized by:

        a. the stomach

        b. the pancreas

        c. epithelial cells lining the small intestine

        d. Brunner's glands

OBJ. 13   _____ 34. Hydrochloric acid in the stomach functions primarily to:

        a. facilitate protein digestion

        b. facilitate lipid digestion

        c. facilitate carbohydrate digestion

        d. hydrolyze peptide bonds

OBJ. 13   _____ 35. A molecule absorbed into the lacteals of the lymphatic system within the walls of the small intestine is:

        a. glucose

        b. fat

        c. vitamin B_{12}

        d. amino acids

Level
–1–

[L1]   Completion:

Using the terms below, complete the following statements.

gastrin	plasticity	alveolar connective tissue
vitamin B$_{12}$	cuspids	carbohydrates
dense bodies	chyme	Kupffer cells
esophagus	peristalsis	pyloric sphincter
mucins	haustrae	lamina propria
lipase	pepsin	cholecystokinin
acini	digestion	duodenum

OBJ. 1   36. The digestive tube between the pharynx and the stomach is the _____.

OBJ. 2   37. The chemical breakdown of food into small organic fragments suitable for absorption by the digestive epithelium refers to _____.

OBJ. 3   38. The primary tissue of the tunica submucosa is _____.

OBJ. 3   39. The layer of loose connective tissue that contains blood vessels, glands, and lymph nodes is the _____.

OBJ. 4   40. Because smooth muscle tissues do not contain sarcomeres, the thin filaments are attached to structures called _____.

OBJ. 4   41. The ability of smooth muscles to function over a wide range of lengths is called _____.

OBJ. 4   42. The muscular externa propels materials from one portion of the digestive tract to another by waves of muscular contractions referred to as _____.

OBJ. 5   43. Stimulation of stretch receptors in the stomach wall and chemoreceptors in the mucosa trigger the release of _____.

OBJ. 6   44. Saliva originating in the submandibular and sublingual salivary glands contains large numbers of glycoproteins called _____.

OBJ. 6   45. The conical teeth used for tearing or slashing are the _____.

OBJ. 6   46. Salivary enzymes in the oral cavity begin with the digestive process by acting on _____.

OBJ. 7   47. The agitation of ingested materials with the gastric juices secreted by the glands of the stomach produces a viscous, soupy mixture called _____.

OBJ. 8   48. The portion of the small intestine that receives chyme from the stomach and exocrine secretions from the pancreas and liver is the _____.

OBJ. 9   49. The two intestinal hormones that inhibit gastric secretion to some degree are secretin and _____.

OBJ. 10   50. The sinusoidal lining of the liver contains large numbers of phagocytic cells called _____.

OBJ. 10   51. The blind pockets lined by simple cuboidal epithelium that produce pancreatic juice are called _____.

OBJ. 11   52. The ring of muscle tissue that regulates the flow of chyme between the stomach and small intestine is the _____.

OBJ. 12   53. The series of pouches that form the wall of colon permitting considerable distension and elongation are called _____.

Level -1-   OBJ. 13   54. The pancreatic enzyme that hydrolyzes triglycerides into glycerol and fatty acids is pancreatic _____.

OBJ. 13 55. The gastric secretion secreted by stomach chief cells that digests protein into smaller peptide chains is _____.

OBJ. 14 56. A nutrient that does not have to be digested before it can be absorbed is _____.

[L1]   Matching:

Match the terms in column B with the terms in column A. Use letters for answers in the spaces provided.

PART I

			COLUMN A		COLUMN B
OBJ. 1	____	57.	stomach	A.	peptide hormone
OBJ. 2	____	58.	mechanical processing	B.	produce mucus
OBJ. 3	____	59.	lamina propria	C.	salivary amylase
OBJ. 4	____	60.	swallowing	D.	deglutition
OBJ. 4	____	61.	calmodulin	E.	gastroenteric reflex
OBJ. 5	____	62.	secretin	F.	small intestine
OBJ. 6	____	63.	chewing	G.	tearing with teeth
OBJ. 6	____	64.	parotid gland	H.	pepsinogen
OBJ. 7	____	65.	chief cells	I.	rugae
OBJ. 7	____	66.	distention of stomach	J.	loose connective tissue
OBJ. 8	____	67.	villi	K.	binding protein
OBJ. 9	____	68.	Brunner's glands	L.	mastication

PART II

			COLUMN A		COLUMN B
OBJ. 10	____	69.	liver cells	M.	vagus nerve
OBJ. 10	____	70.	cirrhosis	N.	starch digestion
OBJ. 11	____	71.	myosin light chain kinase	O.	haustra
OBJ. 11	____	72.	cephalic phase	P.	protein digestion
OBJ. 11	____	73.	gastric phase	Q.	enzyme—initiates contraction
OBJ. 11	____	74.	myenteric plexus	R.	plexus of Auerbach
OBJ. 12	____	75.	colon	S.	liver disease
OBJ. 13	____	76.	salivary amylase	T.	emulsification of fats
OBJ. 13	____	77.	trypsin	U.	hepatocytes
OBJ. 13	____	78.	bile salts	V.	rat digestion
OBJ. 13	____	79.	lipases	W.	mucosal chemoreceptors
OBJ. 14	____	80.	colonic bacteria	X.	produce vitamin K

[L1]   Drawing/Illustration Labeling:

Identify each numbered structure by labeling the following figures:

OBJ. 1   *FIGURE 24.1*   The Digestive Tract

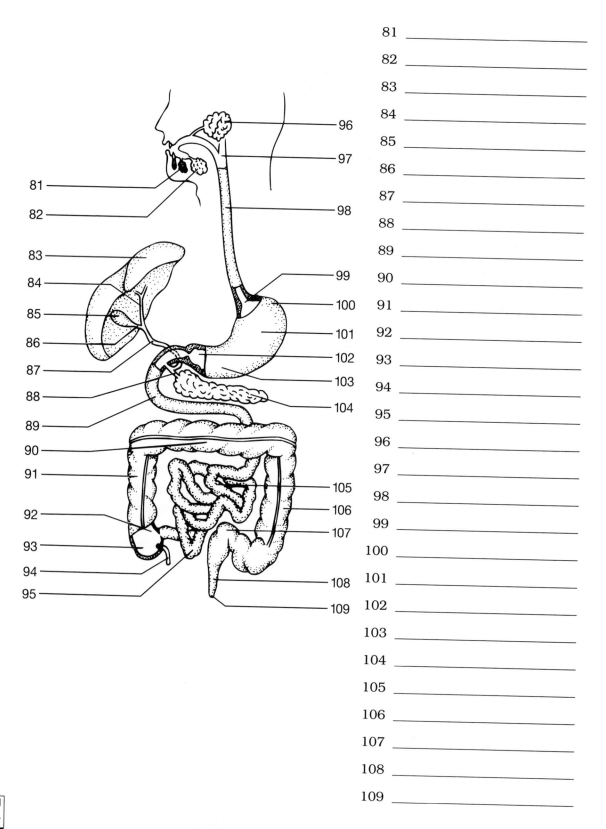

81 _____

82 _____

83 _____

84 _____

85 _____

86 _____

87 _____

88 _____

89 _____

90 _____

91 _____

92 _____

93 _____

94 _____

95 _____

96 _____

97 _____

98 _____

99 _____

100 _____

101 _____

102 _____

103 _____

104 _____

105 _____

106 _____

107 _____

108 _____

109 _____

Level
-1-

OBJ. 3   *FIGURE 24.2*   Histological Features of GI Tract Wall

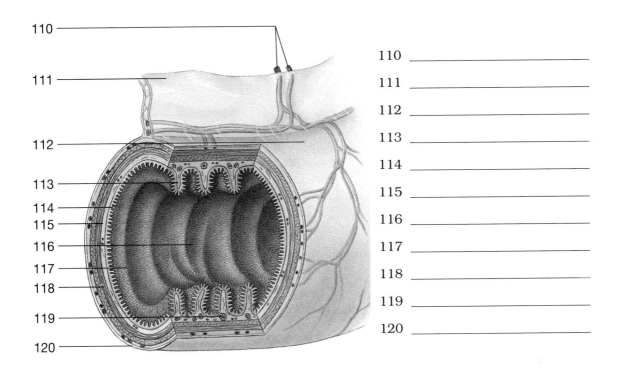

110 _____
111 _____
112 _____
113 _____
114 _____
115 _____
116 _____
117 _____
118 _____
119 _____
120 _____

OBJ. 6   *FIGURE 24.3*   Structure of a Typical Tooth

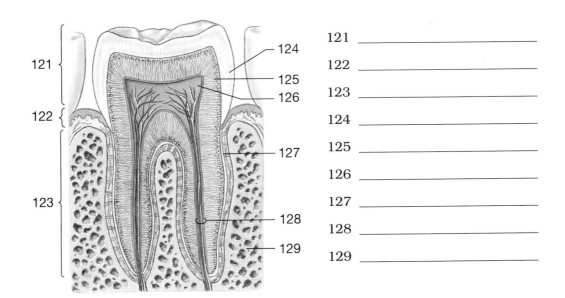

121 _____
122 _____
123 _____
124 _____
125 _____
126 _____
127 _____
128 _____
129 _____

When you have successfully completed the exercises in L1 proceed to L2.

Level
–1–

☐ LEVEL 2   CONCEPT SYNTHESIS

Concept Map I:

Using the following terms, fill in the circled, numbered, blank spaces to complete the concept map. Follow the numbers that comply with the organization of the map.

Stomach   Digestive tract movements   Hormones
Pancreas   Hydrocholoric acid   Large intestine
Amylase   Intestinal mucosa   Bile

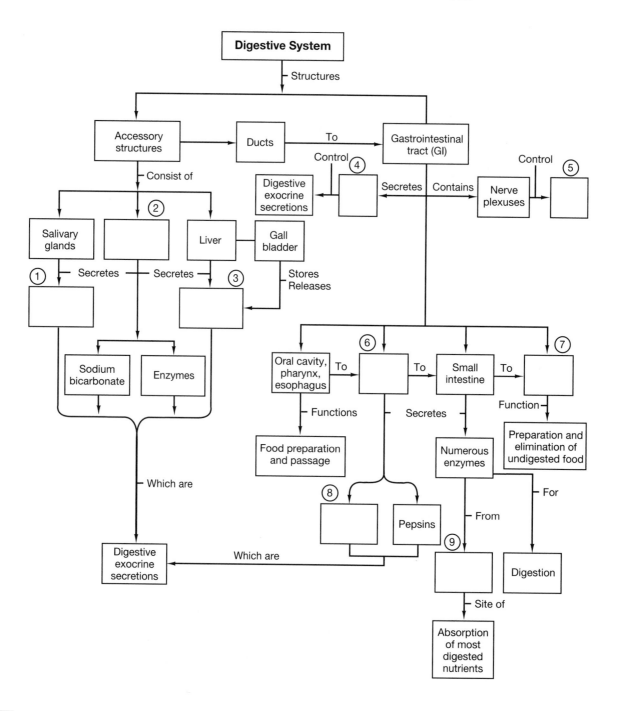

Concept Map II:

Using the following terms, fill in the circled, numbered, blank spaces to complete the concept map. Follow the numbers that comply with the organization of the map.

Simple sugars   Small intestine
Lacteal   Monoglycerides, fatty acids in micelles
Esophagus   Complex sugars and starches
Polypeptides   Disaccharides, trisaccharides
Triglycerides   Amino acids

Chemical events in digestion

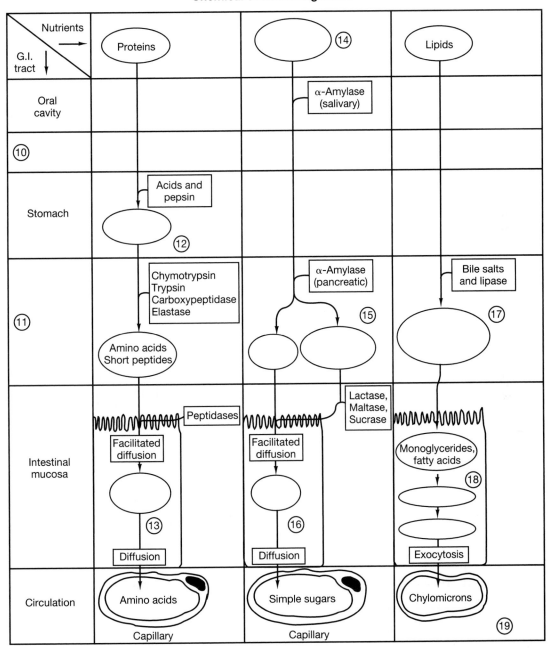

Concept Map III:

Using the following terms, fill in the circled, numbered, blank spaces to complete the concept map. Follow the numbers that comply with the organization of the map.

Enterogastric reflex

Decreasing sympathetic

Decreasing gastric emptying

Increasing stomach motility

Gastrin

Increasing sympathetic

Increasing gastric emptying

Decreasing stomach motility

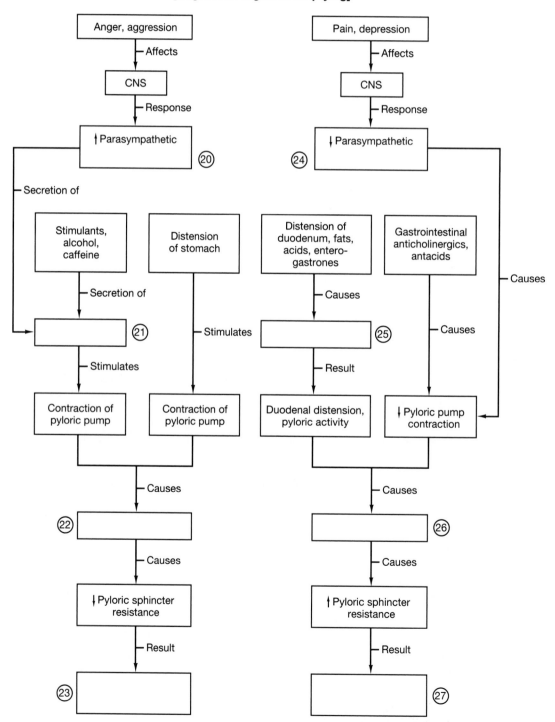

[Regulation of gastric emptying]

Body Trek:

Using the terms below, fill in the blanks to complete the trek through the digestive system.

chyme	pancreas	mastication
fauces	saliva	hydrochloric
tongue	stomach	large intestine
protein	palates	swallowing
feces	pharynx	gastric juices
teeth	lubrication	gingivae (gums)
rugae	carbohydrates	peristalsis
uvula	mucus	duodenum
pylorus	pepsin	mechanical processing
lipase	salivary amylase	gastropharyngeal sphincter
esophagus	oral cavity	

On this trek, Robo will follow a bolus of food as it passes through the GI tract. The micro-robot's metallic construction provides protection from the acids, mucins, and enzymes that are a part of the digestive process.

Robo's entry through the mouth into the (28) _____ is quick and easy. Food preparation via analysis, (29) _____ , digestion by salivary amylase, and (30) _____ occurs in this region of the processing system. Noticeable features include pink ridges, the (31) _____ that surround the base of the glistening white (32) _____ used in handling and (33) _____ of food. The roof of the cavity is formed by the hard and soft (34) _____ while the floor is comprised of the (35) _____. Three pairs of salivary glands secrete (36) _____ , which contains the enzyme (37) _____, a part of the digestive process that acts on (38) _____.

After food preparation and the start of the digestive process, the food bolus with Robo trekking along passes into the (39) _____ through the (40) _____ , a region characterized by the presence of the hanging, dangling (41) _____ supported by the posterior part of the soft palate. Additional movement is provided by powerful constrictor muscles involved in (42) _____ , causing propulsion into an elongated, flattened tube, the (43) _____. Alternating wave-like contractions called (44) _____ facilitate passage to a region that "looks" like a ring of muscle, the (45) _____, which serves as the entry "gate" into the (46) _____. In this enlarged chamber the robot's sensors and the food bolus are "agitated" because of glandular secretions called (47) _____ that produce a viscous, soupy mixture referred to as (48) _____. The walls of the chamber look "wrinkled" due to folds called (49) _____ that are covered with a slippery, slimy (50) _____ coat that offers protection from erosion and other potential damage. Robo's chemosensors detect the presence of (51) _____ acid and enzymes called (52) _____, both of which are involved in the preparation and partial digestion of (53) _____.

Following the digestive delay, the chyme-covered robot is squirted through the (54) _____ into the (55) _____ of the small intestine. The presence of fats and other partially digested food, including the chyme, creates a rather crowded condition. Some of the fat begins to "break up" owing to the action of (56) _____ , which has been released from the (57) _____ where it is secreted. Robo's descent continues into the (58) _____ where the micro-robot becomes a part of the compacted, undigestible residue that comprises the (59) _____. The remainder of the trek involves important "movements" that eventually return Robo to mission control for cleanup and re-energizing for the next adventure.

[L2]   Multiple Choice:

Place the letter corresponding to the correct answer in the space provided.

_____ 60. An error in swallowing could most likely be detected by the:

      a. larynx

      b. soft palate

      c. esophagus

      d. root of the tongue

_____ 61. Many visceral smooth muscle networks show rhythmic cycles of activity in the absence of neural stimulation due to the presence of:

      a. direct contact with motor neurons carrying impulses to the CNS

      b. an action potential generated and conducted over the sarcolemma

      c. the single motor units that contract independently of each other

      d. pacesetter cells that spontaneously depolarize and trigger contraction of entire muscular sheets

_____ 62. The reason a completely dry food bolus cannot be swallowed is:

      a. the dry food inhibits parasympathetic activity in the esophagus

      b. friction with the walls of the esophagus makes peristalsis ineffective

      c. the dry food stimulates sympathetic activity, inhibiting peristalsis

      d. a, b, and c are correct

_____ 63. The parts of the stomach include:

      a. cardia, rugae, pylorus

      b. greater omentum, lesser omentum, gastric body

      c. fundus, body, pylorus

      d. parietal cells, chief cells, gastral glands

_____ 64. The two factors that play an important part in the movement of chyme from the stomach to the small intestine are:

      a. CNS and ANS regulation

      b. release of HCl and gastric juice

      c. stomach distension and gastrin release

      d. sympathetic and parasympathetic stimulation

_____ 65. The plicae of the intestinal mucosa, which bears the intestinal villi, are structural features that provide for:

      a. increased total surface area for absorption

      b. stabilizing the mesenteries attached to the dorsal body wall

      c. gastric contractions that churn and swirl the gastric contents

      d. initiating enterogastric reflexes that accelerate the digestive process

_____ 66. The enteroendocrine cells of the intestinal crypts are responsible for producing the intestinal hormones:

      a. gastrin and pepsinogen

      b. biliverdin and bilirubin

      c. enterokinase and aminopeptidase

      d. cholecystokinin and secretin

_____ 67. Most intestinal absorption occurs in the:

      a. distal end of the duodenum

      b. body of the stomach

      c. proximal half of the duodenum

      d. proximal half of the jejunum

_____ 68. The primary function(s) of the gastrointestinal juice is (are) to:

      a. moisten the chyme

      b. assist in buffering acids

      c. dissolve digestive enzymes and products of digestion

      d. a, b, and c are correct

_____ 69. An immediate increase in the rates of glandular secretion and peristaltic activity in all segments of the small intestine are a result of the:

      a. presence of intestinal juice

      b. gastroenteric reflex

      c. gastroileal reflex

      d. action of hormonal and CNS controls

_____ 70. The primary effect of secretin is to cause a(n):

      a. decrease in duodenal submucosal secretions

      b. increase in release of bile from the gall bladder into the duodenum

      c. increase in secretion of water and buffers by the pancreas and the liver

      d. increase in gastric motility and secretory rates

_____ 71. The peptide hormone that causes the release of insulin from the pancreatic islets is:

      a. GIP

      b. CCK

      c. VIP

      d. GTP

_____ 72. The two major regions of the large intestine are the:

      a. jejunum and ileum

      b. colon and rectum

      c. haustra and taenia coli

      d. cecum and sigmoid colon

_____ 73. The muscular sphincter that guards the entrance between the ileum and the cecum is the:

> a. pyloric sphincter
>
> b. gastrointestinal sphincter
>
> c. ileocecal valve
>
> d. taenia coli

_____ 74. The contractions that force fecal material into the rectum and produce the urge to defecate are called:

> a. primary movements
>
> b. mass movements
>
> c. secondary movements
>
> d. bowel movements

_____ 75. The average composition of the fecal waste material is:

> a. 20% water, 5% bacteria, 75% indigestible remains and inorganic matter
>
> b. 45% water, 45% indigestible and inorganic matter, 10% bacteria
>
> c. 60% water, 10% bacteria, 30% indigestible and inorganic matter
>
> d. 75% water, 5% bacteria, 20% indigestible materials, inorganic matter, and epithelial remains

_____ 76. The external anal sphincter is under voluntary control.

> a. true
>
> b. false

_____ 77. The two positive feedback loops involved in the defecation reflex are:

> a. internal and external sphincter muscles
>
> b. the anorectal canal and the rectal columns
>
> c. stretch receptors in rectal walls, and the sacral parasympathetic system
>
> d. mass movements and peristaltic contractions

_____ 78. The "doorway to the liver" (porta hepatis) is a complex that includes the:

> a. hilus, bile duct, cystic duct
>
> b. bile duct, hepatic portal vein, hepatic artery
>
> c. caudate lobe, quadrate lobe, and hepatic duct
>
> d. left lobe, right lobe, and round ligament

_____ 79. Triglycerides coated with proteins create a complex known as a:

> a. micelle
>
> b. co-transport
>
> c. glycerolproteinase
>
> d. chylomicron

Level
=2=

[L2]   Completion:

Using the terms below, complete the following statements.

Brunner's glands	duodenum	lacteals
bicuspids	plexus of Meissner	ileum
mesenteries	visceral peritoneum	adventitia
segmentation	ankyloglossia	Peyer's patches
enterocrinin	incisors	alkaline tide

80. The submucosal region that contains sensory nerve cells, parasympathetic ganglia, and sympathetic postganglionic fibers is referred to as the _____.

81. The serosa that lines the inner surfaces of the body wall is called the _____.

82. Double sheets of peritoneal membrane that connect the parietal peritoneum with the visceral peritoneum are called _____.

83. The muscularis externa is surrounded by a layer of loose connective tissue called the _____.

84. Movements that churn and fragment digested materials in the small intestine are termed _____.

85. Restrictive movements of the tongue that cause abnormal eating or speaking are a result of a condition known as _____.

86. Blade-shaped teeth found at the front of the mouth used for clipping or cutting are called _____.

87. The teeth used for crushing, mashing, and grinding are the molars and the premolars called _____.

88. The sudden rush of bicarbonate ions is referred to as the _____.

89. The portion of the small intestine that receives chyme from the stomach and exocrine secretions from the pancreas and the liver is the _____.

90. The most inferior part of the small intestine is the _____.

91. Terminal lymphatics contained within each villus are called _____.

92. The submucosal glands in the duodenum that secrete mucus which provides protection and buffering action are _____.

93. Aggregations of lymphatic nodules forming lymphatic centers in the ileum are called _____.

94. The hormone that stimulates the duodenal glands when acid chyme enters the small intestine is _____.

[L2]   Short Essay:

Briefly answer the following questions in the spaces provided below.

95. What seven (7) integrated steps comprise the digestive functions?

96. What six (6) major layers show histological features reflecting specializations throughout the digestive tract? (List the layers in correct order from the inside to the outside.)

Level
=2=

97. Briefly describe the difference between a peristaltic movement of food and movement by segmentation.

98. What three (3) pairs of salivary glands secrete saliva into the oral cavity?

99. What four (4) types of teeth are found in the oral cavity and what is the function of each type?

100. Starting at the superior end of the digestive tract and proceeding inferiorly, what six (6) sphincters control movement of materials through the tract?

101. What are the similarities and differences between parietal cells and chief cells in the wall of the stomach?

102. (a) What three (3) phases are involved in the regulation of gastric function?

    (b) What regulatory mechanism(s) dominate each phase?

103. What are the three (3) most important hormones that regulate intestinal activity?

104. What are the three (3) principal functions of the large intestine?

105. What three (3) basic categories describe the functions of the liver?

106. What two (2) distinct functions are performed by the pancreas?

When you have successfully completed the exercises in L2 proceed to L3.

☐ LEVEL 3   CRITICAL THINKING/APPLICATION

Using principles and concepts learned about the digestive system, answer the following questions. Write your answers on a separate sheet of paper.

1. The study of the muscular system usually focuses on *skeletal muscle tissue*. The study of the digestive system focuses on *smooth muscle tissue* because of its essential role in mechanical processing and in moving materials along the GI tract. What are the important differences between the contraction of smooth muscle fibers and other muscle fiber types?

2. A 19-year-old male walks into a hospital emergency room and reports the following symptoms: fever—103°F, pain in the neck, headache, and swelling in front of and below the ear. What is your diagnosis and what should be done to relieve the condition and to avoid the problem in the future?

3. O. B. Good likes his four to six cups of caffeinated coffee daily, and he relieves his thirst by drinking a few alcoholic beverages during the day. He constantly complains of heartburn. What is happening in his GI tract to cause the "burning" sensation?

4. After studying anatomy and physiology for a year, I. M. Nervous complains of recurring abdominal pains, usually 2 to 3 hours after eating. Additional eating seems to relieve the pain. She finally agrees to see her doctor, who diagnoses her condition as a duodenal ulcer. Use your knowledge of anatomy and physiology to explain the diagnosis to this patient.

5. A 14-year-old girl complains of gassiness, rumbling sounds, cramps, and diarrhea after drinking a glass of milk. At first she assumes indigestion; however, every time she consumes milk or milk products the symptoms reappear. What is her problem? What strategies will be necessary for her to cope effectively with the problem?

25

Metabolism and Energetics

■ Overview

The processes of metabolism and energetics are intriguing to most students of anatomy and physiology because of the mystique associated with the "fate of absorbed nutrients." This mysterious phrase raises questions such as:

- What happens to the food we eat?
- What determines the "fate" of the food and the way it will be used?
- How are the nutrients from the food utilized by the body?

All cells require energy for anabolic processes and breakdown mechanisms involving catabolic processes. Provisions for these cellular activities are made by the food we eat; these activities provide energy, promote tissue growth and repair, and serve as regulatory mechanisms for the body's metabolic machinery.

The exercises in Chapter 25 provide for study and review of cellular metabolism; metabolic interactions including the absorptive and postabsorptive state; diet and nutrition; and bioenergetics. Much of the material you have learned in previous chapters will be of value as you "see" the integrated forces of the digestive and cardiovascular system exert a coordinated effort to initiate and advance the work of metabolism.

☐ LEVEL 1   REVIEW OF CHAPTER OBJECTIVES

1. Define metabolism and explain why cells need to synthesize new organic components.
2. Describe the basic steps in glycolysis, the TCA cycle, and the electron transport chain.
3. Summarize the energy yield of glycolysis and cellular respiration.
4. Describe the pathways involved in lipid metabolism and the mechanisms necessary for lipid transport and distribution.
5. Discuss protein metabolism and the use of proteins as an energy source.
6. Discuss nucleic acid metabolism.
7. Differentiate between the absorptive and postabsorptive metabolic states and summarize the characteristics of each.
8. Explain what constitutes a balanced diet, and why it is important.
9. Define metabolic rate and discuss the factors involved in determining an individual's BMR.
10. Discuss the homeostatic mechanisms that maintain a constant body temperature.

[L1]   Multiple Choice:

Place the letter corresponding to the correct answer in the space provided.

OBJ. 1   _____   1. The metabolic components of the body that interact to preserve homeostasis are:

    a. heart, lungs, kidneys, brain, and pancreas

    b. blood, lymph, cerebrospinal fluid, hormones, and bone marrow

    c. liver, adipose tissue, skeletal muscle, neural tissue, other peripheral tissues

    d. bones, muscles, integument, glands, and the heart

OBJ. 1   _____   2. Neurons must be provided with a reliable supply of glucose because they are:

    a. usually unable to metabolize other molecules

    b. involved primarily with transmitting nerve impulses

    c. primarily located in the brain

    d. covered with myelinated fibrous sheaths

OBJ. 1   _____   3. In resting skeletal muscles, a significant portion of the metabolic demand is met through the:

    a. catabolism of glucose

    b. catabolism of fatty acids

    c. catabolism of glycogen

    d. anabolism of ADP to ATP

OBJ. 1   _____   4. The process that breaks down organic substrates, releasing energy that can be used to synthesize ATP or other high energy compounds is:

    a. metabolism

    b. anabolism

    c. catabolism

    d. oxidation

OBJ. 2   _____   5. Acetyl-CoA cannot be used to make glucose because the:

    a. catabolism of glucose ends with the formation of Acetyl-CoA

    b. catabolic pathways in the cells are all organic substances

    c. glucose is used exclusively to provide energy for cells

    d. decarboxylation step between pyruvic acid and acetyl-CoA cannot be reversed

OBJ. 2   _____   6. In glycolysis, six carbon glucose molecules are broken down into 2 three-carbon molecules of:

    a. pyruvic acid

    b. acetyl-CoA

    c. citric acid

    d. oxaloacetic acid

OBJ. 2 _____ 7. The first step in a sequence of enzymatic reactions in the tricarboxylic acid cycle is the formation of:

    a. acetyl-CoA

    b. citric acid

    c. oxaloacetic acid

    d. pyruvic acid

OBJ. 3 _____ 8. For each glucose molecule converted to 2 pyruvates, the anaerobic reaction sequence in glycolysis provides a net gain of:

    a. 2 ATP for the cell

    b. 4 ATP for the cell

    c. 36 ATP for the cell

    d. 38 ATP for the cell

OBJ. 3 _____ 9. For each glucose molecule processed during aerobic cellular respiration the cell gains:

    a. 4 molecules of ATP

    b. 32 molecules of ATP

    c. 36 molecules of ATP

    d. 24 molecules of ATP

OBJ. 4 _____ 10. Lipids cannot provide large amounts of adenosine triphosphate (ATP) in a short period of time because:

    a. lipid reserves are difficult to mobilize

    b. lipids are insoluble and it is difficult for water-soluble enzymes to reach them

    c. most lipids are processed in mitochondria, and mitochondrial activity is limited by the availability of oxygen

    d. a, b, and c are correct

OBJ. 4 _____ 11. Although small quantities of lipids are normally stored in the liver, most of the synthesized triglycerides are bound to:

    a. glucose molecules

    b. transport proteins

    c. adipocytes

    d. hepatocytes in the liver

OBJ. 5 _____ 12. The factor(s) that make protein catabolism an impractical source of quick energy is (are):

    a. their energy yield is less than that of lipids

    b. one of the byproducts, ammonia, is a toxin that can damage cells

    c. proteins are important structural and functional cellular components

    d. a, b, and c are correct

OBJ. 5 _____ 13. The first step in amino acid catabolism is the removal of the:

    a. carboxyl group

    b. amino group

    c. keto acid

    d. hydrogen from the central carbon

Level -1-

OBJ. 6   _____ 14. All cells synthesize RNA, but DNA synthesis occurs only:

   a. in red blood cells
   b. in cells that are preparing for mitosis
   c. in lymph and cerebrospinal fluid
   d. in the bone marrow

OBJ. 6   _____ 15. The only parts of a nucleotide of RNA which can provide energy when broken down are the:

   a. sugars and phosphates
   b. sugars and purines
   c. sugars and pyrimidines
   d. purines and pyrimidines

OBJ. 7   _____ 16. After glycogen formation, if excess glucose still remains in the circulation, the hepatocytes use glucose to:

   a. provide energy
   b. repair tissues
   c. regulate metabolism
   d. synthesize triglycerides

OBJ. 7   _____ 17. When blood glucose concentrations are elevated, the glucose molecules are:

   a. catabolized for energy
   b. used to build proteins
   c. used for tissue repair
   d. a, b, and c are correct

OBJ. 7   _____ 18. Before the large vitamin B_{12} molecule can be absorbed, it must be bound to:

   a. the gastric epithelium
   b. another water-soluble vitamin
   c. vitamin C
   d. intrinsic factor

OBJ. 8   _____ 19. A balanced diet contains all of the ingredients necessary to:

   a. prevent starvation
   b. prevent life-threatening illnesses
   c. prevent deficiency diseases
   d. maintain homeostasis

OBJ. 8   _____ 20. Foods that are deficient in dietary fiber are:

   a. vegetables and fruits
   b. breads and cereals
   c. milk and meat
   d. rice and pastas

OBJ. 8   _____ 21. Foods that are low in fats, calories, and proteins are:

   a. vegetables and fruits
   b. milk and cheese
   c. meat, poultry and fish
   d. bread, cereal, and rice

OBJ. 8 _____ 22. Minerals, vitamins, and water are classified as *essential* nutrients because:

    a. they are used by the body in large quantities

    b. the body cannot synthesize the nutrients in sufficient quantities

    c. they are the major providers of calories for the body

    d. a, b, and c are correct

OBJ. 8 _____ 23. The *trace* minerals found in extremely small quantities in the body include:

    a. sodium, potassium, chloride, calcium

    b. phosphorus, magnesium, calcium, iron

    c. iron, zinc, copper, manganese

    d. phosphorus, zinc, copper, potassium

OBJ. 8 _____ 24. Hypervitaminosis involving water-soluble vitamins is relatively uncommon because:

    a. excessive amounts are stored in adipose tissue

    b. excessive amounts are readily excreted in the urine

    c. the excess amount is stored in the bones

    d. excesses are readily absorbed into skeletal muscle tissue

OBJ. 9 _____ 25. To examine the metabolic state of an individual, results may be expressed as:

    a. calories per hour

    b. calories per day

    c. calories per unit of body weight per day

    d. a, b, and c are correct

OBJ. 9 _____ 26. An individual's basal metabolic rate ideally represents:

    a. the minimum, resting energy expenditure of an awake, alert person

    b. genetic differences among ethnic groups

    c. the amounts of circulating hormone levels in the body

    d. a measurement of the daily energy expenditures for a given individual

OBJ. 10 _____ 27. The greatest amount of the daily water intake is obtained by:

    a. consumption of food

    b. drinking fluids

    c. metabolic processes

    d. decreased urination

OBJ. 10 _____ 28. The four processes involved in heat exchange with the environment are:

    a. physiological, behavioral, generational, acclimatization

    b. radiation, conduction, convection, evaporation

    c. sensible, insensible, heat loss, heat gain

    d. thermogenesis, dynamic action, pyrexia, thermalphasic

OBJ. 10   _____ 29. The primary mechanisms for increasing heat loss in the body
                         include:

                            a. sensible and insensible
                            b. acclimatization and pyrexia
                            c. physiological mechanisms and behavioral modifications
                            d. vasomotor and respiratory

[L1]   Completion:

Using the terms below, complete the following statements.

glucagon	glycolysis	triglycerides
minerals	hypervitaminosis	cathepsins
liver	malnutrition	avitaminosis
pyrexia	thermoregulation	nutrition
insulin	beta oxidation	glycogen
lipemia	hypodermis	anabolism
TCA cycle	calorie	oxidative phosphorylation

OBJ. 1   30. The synthesis of new organic components which involves the formation
                of new chemical bonds is _____.

OBJ. 2   31. The major pathway of carbohydrate metabolism is _____.

OBJ. 3   32. The process that produces over 90 percent of the ATP used by our cells
                is _____.

OBJ. 4   33. The most abundant lipids in the body are the _____.

OBJ. 4   34. Fatty acid molecules are broken down into two-carbon fragments by
                means of a sequence of reactions known as _____.

OBJ. 4   35. The presence of chylomicrons, which give the plasma a milky appear-
                ance, is called _____.

OBJ. 5   36. Muscle proteins are hydrolyzed by special enzymes called _____.

OBJ. 6   37. During RNA catabolism the pyrimidines, cytosine and uracil, are
                converted to acetyl-CoA and metabolized via the _____.

OBJ. 7   38. In the liver and in skeletal muscle, glucose molecules are stored as
                _____.

OBJ. 7   39. The focal point for metabolic regulation and control in the body is the
                _____.

OBJ. 7   40. The greatest concentration of adipose tissue in the body is found in
                the _____.

OBJ. 7   41. During the absorptive state, increased glucose uptake and utilization
                is affected by the hormone _____.

OBJ. 7   42. During the postabsorptive state, decreased levels of stored glycogen
                are affected by the hormone _____.

OBJ. 8   43. The absorption of nutrients from food is called _____.

OBJ. 8   44. An unhealthy state resulting from inadequate intake of one or more
                nutrients that becomes life-threatening as the deficiencies accumulate
                is called _____.

OBJ. 8   45. Inorganic ions released through the dissociation of electrolytes are
                _____.

OBJ. 8   46. The deficiency disease resulting from a vitamin deficiency is referred to as _____.

OBJ. 8   47. The dietary intake of vitamins that exceeds the ability to store, utilize, or excrete a particular vitamin is called _____.

OBJ. 9   48. The amount of energy required to raise the temperature of 1 gram of water one degree centigrade is the _____.

OBJ. 10   49. The homeostatic process that keeps body temperatures within acceptable limits regardless of environmental conditions is called _____.

OBJ. 10   50. The nonpathological term for an elevated body temperature is _____.

[L1]   Matching:

Match the terms in column B with the terms in column A. Use letters for answers in the spaces provided.

PART I   **COLUMN A**   **COLUMN B**

OBJ. 1	____	51. catabolism	A. protein catabolism
OBJ. 1	____	52. anabolism	B. substrate-level phosphorylation
OBJ. 2	____	53. glycolysis	
OBJ. 3	____	54. GTP from TCA cycle	C. urea formation
OBJ. 4	____	55. lipolysis	D. lipid storage
OBJ. 4	____	56. lipogenesis	E. formation of new chemical bonds
OBJ. 4	____	57. adipose tissue	
OBJ. 5	____	58. ketone bodies	F. energy release
OBJ. 5	____	59. amino acid metabolism	G. lipid synthesis
OBJ. 5	____	60. carbohydrates	H. lipid metabolism
			I. anaerobic reaction sequence
			J. protein sparing

PART II   **COLUMN A**   **COLUMN B**

OBJ. 6	____	61. purines and pyrimidines	K. glycogen reserves
			L. B complex and C
OBJ. 7	____	62. gluconeogenesis	M. ADEK
OBJ. 7	____	63. glycogenesis	N. N compounds
OBJ. 7	____	64. skeletal muscle	O. increasing muscle tone
OBJ. 8	____	65. fat-soluble vitamins	P. glucose synthesis
OBJ. 8	____	66. water-soluble vitamins	Q. release of hormones; increasing metabolism
OBJ. 9	____	67. basal metabolic rate	
OBJ. 10	____	68. shivering thermogenesis	R. glycogen formation
			S. resting energy expenditure
OBJ. 10	____	69. non-shivering thermogenesis	

When you have successfully completed the exercises in L1 proceed to L2.

☐ LEVEL 2   CONCEPT SYNTHESIS

Concept Map I:

Using the following terms, fill in the circled, numbered, blank spaces to complete the concept map. Follow the numbers that comply with the organization of the map.

Vitamins	Metabolic regulators	Vegetables
Meat	Tissue growth and repair	9 cal/gram
Proteins	Carbohydrates	

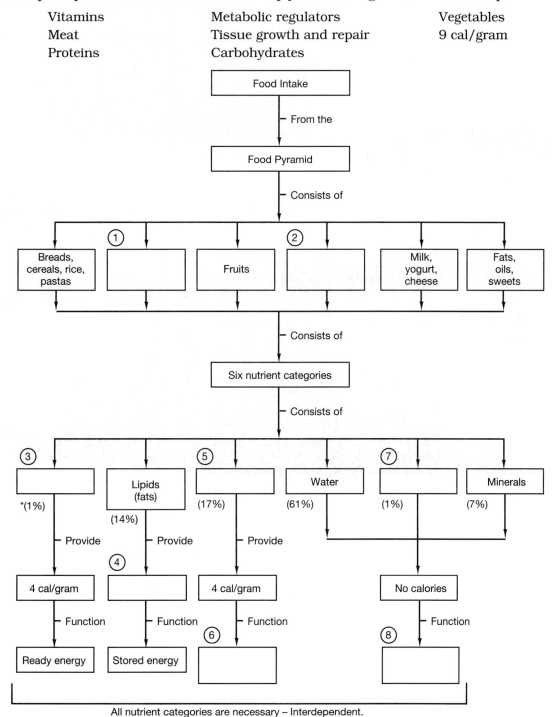

All nutrient categories are necessary – Interdependent.

*% of nutrients found in 25 year old male weighing 65 kg. (143 lbs.)

Concept Map II:

Using the following terms, fill in the circled, numbered, blank spaces to complete the concept map. Follow the numbers that comply with the organization of the map.

Gluconeogenesis   Lipogenesis   Lipolysis
Amino acids   Beta oxidation   Glycolysis

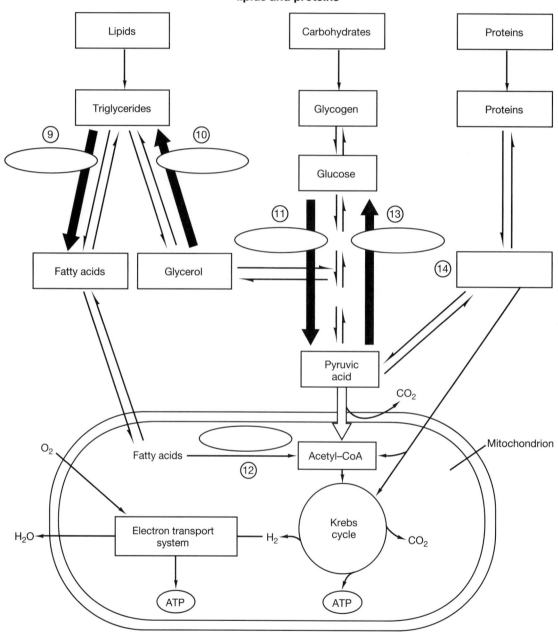

**Anabolism–Catabolism of carbohydrates
lipids and proteins**

Concept Map III:

Using the following terms, fill in the circled, numbered, blank spaces to complete the concept map. Follow the numbers that comply with the organization of the map.

"Good" cholesterol   Chylomicrons   Excess LDL
Atherosclerosis   Eliminated via feces   Liver
Cholesterol   "Bad" cholesterol

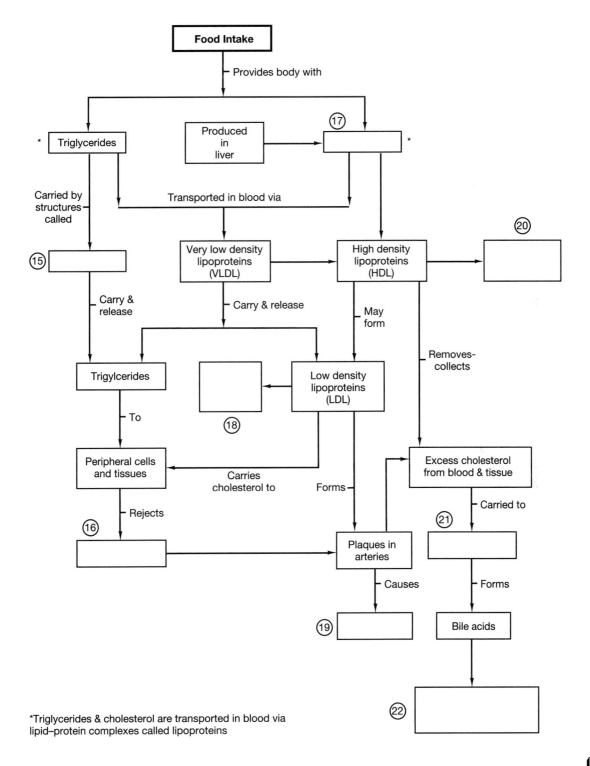

*Triglycerides & cholesterol are transported in blood via lipid–protein complexes called lipoproteins

Concept Map IV:

Using the following terms, fill in the circled, numbered, blank spaces to complete the concept map. Follow the numbers that comply with the organization of the map.

Glycogen → glucose   Fat cells   Liver   Glucagon
Epinephrine   Protein   Pancreas

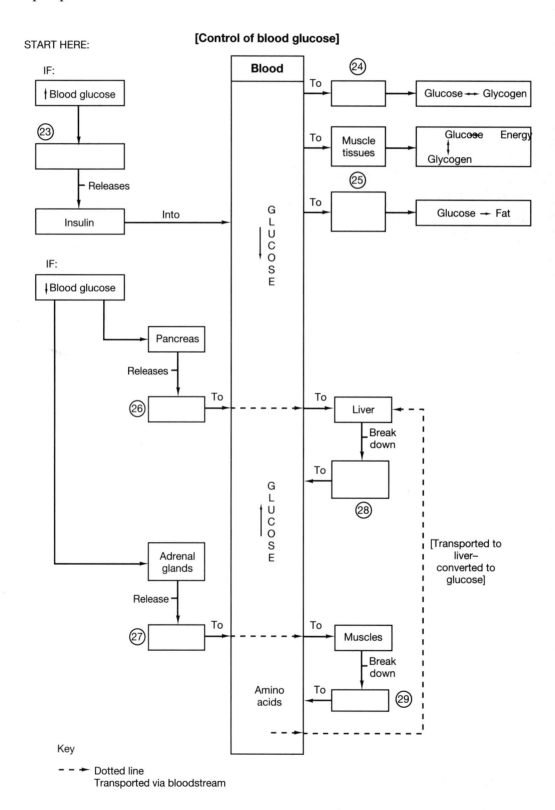

[L2]   Multiple Choice:

Place the letter corresponding to the correct answer in the space provided.

_____ 30. The major metabolic pathways that provide most of the ATP used by typical cells are:

      a. anabolism and catabolism

      b. gluconeogenesis and glycogenesis

      c. glycolysis and aerobic respiration

      d. lipolysis and lipogenesis

_____ 31. A cell with excess carbohydrates, lipids, and amino acids will break down carbohydrates to:

      a. provide tissue growth and repair

      b. obtain energy

      c. provide metabolic regulation

      d. a, b, and c are correct

_____ 32. Fatty acids and many amino acids cannot be converted to glucose because:

      a. their catabolic pathways produce acetyl-CoA

      b. they are not used for energy production

      c. other organic molecules cannot be converted to glucose

      d. glucose is necessary to start the metabolic processes

_____ 33. Lipogenesis can use almost any organic substrate because:

      a. lipid molecules contain carbon, hydrogen, and oxygen

      b. triglycerides are the most abundant lipids in the body

      c. lipids can be converted and channeled directly into the Kreb's cycle

      d. lipids, amino acids, and carbohydrates can be converted to acetyl-CoA

_____ 34. When more nitrogen is absorbed than excreted, an individual is said to be in a state of:

      a. protein sparing

      b. negative nitrogen balance

      c. positive nitrogen balance

      d. ketoacidosis

_____ 35. The catabolism of _lipids_ results in the release of:

      a. 4.18 C/g

      b. 4.32 C/g

      c. 9.46 C/g

      d. a, b, and c are incorrect

_____ 36. The catabolism of _carbohydrates_ and _proteins_ results in the release of:

      a. equal amounts of C/g

      b. 4.18 C/g and 4.32 C/g, respectively

      c. 9.46 C/g

      d. 4.32 C/g and 4.18 C/g, respectively

Level
=2=

_____ 37. The efficiency rate for the complete catabolism of glucose is about:

      a. 12 percent

      b. 32 percent

      c. 42 percent

      d. 90 percent

_____ 38. The three essential fatty acids that cannot be synthesized by the body but must be included in the diet are:

      a. cholesterol, glycerol, and butyric acid

      b. leucine, lysine, and phenylalanine acids

      c. arachidonic, linoleic, and linolenic acids

      d. LDL, IDL, and HDL

_____ 39. Lipids circulate through the bloodstream as:

      a. lipoproteins and free fatty acids

      b. saturated fats

      c. unsaturated fats

      d. polyunsaturated fat

_____ 40. The primary function of very low-density lipoproteins (VLDLs) is:

      a. transporting cholesterol from peripheral tissues back to the liver

      b. transporting cholesterol to peripheral tissues

      c. to carry absorbed lipids from the intestinal tract to circulation

      d. transport of triglycerides to peripheral tissues

_____ 41. LDLs are absorbed by cells through the process of:

      a. simple diffusion

      b. receptor-mediated endocytosis

      c. active transport

      d. facilitated diffusion

_____ 42. In the beta-oxidation of an 18-carbon fatty acid molecule, the cell gains:

      a. 36 ATP

      b. 38 ATP

      c. 78 ATP

      d. 144 ATP

_____ 43. Most of the lipids absorbed by the digestive tract are immediately transferred to the:

      a. liver

      b. red blood cells

      c. venous circulation via the left lymphatic duct

      d. hepatocytes for storage

_____ 44. Under normal circumstances, if you eat three meals a day, the body will spend approximately 12 hours in the:

      a. digestion of food

      b. absorptive state

      c. excretion of waste

      d. postabsorptive state

Level
2

_____ 45. The most important factors in good nutrition are to obtain nutrients:

    a. from the food pyramid

    b. that meet the recommended daily intakes

    c. that are low in fats and high in dietary fiber

    d. in sufficient quantity and quality

_____ 46. An important energy source during periods of starvation, when glucose supplies are limited, is:

    a. glycerol

    b. lipoproteins

    c. cholesterol

    d. free fatty acids

_____ 47. Of the following selections, the one that includes *only* essential amino acids is:

    a. aspartic acid, glutamic acid, tyrosine, glycine, alanine

    b. leucine, lysine, valine, tryptophan, arginine

    c. proline, serine, cysteine, glutamine, histidine

    d. valine, tyrosine, leucine, serine, alanine

_____ 48. Milk and eggs are complete proteins because they contain:

    a. more protein than fat

    b. the recommended intake for vitamin B_{12}

    c. all the essential amino acids in sufficient quantities

    d. all the essential fatty acids and amino acids

_____ 49. A nitrogen compound important in energy storage in muscle tissue is:

    a. porphyrins

    b. pyrimidines

    c. glycoprotein

    d. creatine

[L2]   Completion:

Using the terms below, complete the following statements.

convection	nutrient pool	lipoproteins
HDLs	transamination	chylomicrons
uric acid	sensible perspiration	deamination
LDLs	insensible perspiration	calorie
acclimatization	metabolic rate	BMR

50. All of the anabolic and catabolic pathways in the cell utilize a collection of organic substances known as the _____.

51. Complexes that contain large insoluble glycerides and cholesterol and a coating dominated by phospholipids and protein are called _____.

52. The largest lipoproteins in the body that are produced by the intestinal epithelial cells are the _____.

53. Lipoproteins that deliver cholesterol to peripheral tissues are _____.

54. Lipoproteins that transport excess cholesterol from peripheral tissues back to the liver are _____.

55. The amino group of an amino acid is attached to a keto acid by the process of _____.

Level
=2=

56. The removal of an amino group in a reaction that generates an ammonia molecule is called _____.

57. The purines (adenine and guanine) cannot be catabolized for energy but are deaminated and excreted as _____.

58. The amount of energy required to raise the temperature of 1 gram of water one degree centigrade defines a _____.

59. The value of the sum total of the anabolic and catabolic processes occurring in the body is the _____.

60. The minimum resting energy expenditure of an awake, alert person is the individual's _____.

61. The result of conductive heat loss to the air that overlies the surface of the body is called _____.

62. Water losses that occur via evaporation from alveolar surfaces and the surface of the skin are referred to as _____.

63. Loss of water from the sweat glands is responsible for _____.

64. Making physiological adjustment to a particular environment over time is referred to as _____.

[L2]   Short Essay:

Briefly answer the following questions in the spaces provided below.

65. What are the three (3) essential fatty acids (EFA) that cannot be synthesized by the body?

66. List the five (5) different classes of lipoproteins and give the relative proportions of lipid vs. protein.

67. What is the difference between *transamination* and *deamination*?

68. (a) What is the difference between an essential amino acid (EAA) and a nonessential amino acid?

(b) What two EAAs that are necessary for children can be synthesized in adults in amounts that are insufficient for growing children?

69. What four (4) factors make protein catabolism an impractical source of quick energy?

70. Why are nucleic acids insignificant contributors to the total energy reserves of the cell?

71. From a metabolic standpoint, what five (5) distinctive components are found in the human body?

72. What is the primary difference between the absorptive and postabsorptive states?

73. What four (4) food groups in sufficient quantity and quality provide the basis for a balanced diet?

74. What is the meaning of the term *nitrogen balance?*

75. Even though minerals do not contain calories, why are they important in good nutrition?

76. (a) List the fat-soluble vitamins.

    (b) List the water-soluble vitamins.

77. What four (4) basic processes are involved in heat exchange with the environment?

78. What two (2) primary general mechanisms are responsible for increasing heat loss in the body?

Level
=2=   **When you have successfully completed the exercises in L2 proceed to L3.**

☐ **LEVEL 3   CRITICAL THINKING/APPLICATION**

Using principles and concepts learned about metabolism and energetics, answer the following questions. Write your answers on a separate sheet of paper.

1. Greg S. received a blood test assessment that reported the following values: *total cholesterol*—200 mg/dl (normal); *HDL cholesterol*—21 mg/dl (below normal); *triglycerides*—125 mg/dl (normal). If the minimal normal value for HDLs is 35 mg/dl, what are the physiological implications of his test results?

2. Charlene S. expresses a desire to lose weight. She decides to go on a diet, selecting food from the four food groups; however, she limits her total caloric intake to approximately 300-400 calories daily. Within a few weeks Charlene begins to experience symptoms marked by a fruity, sweetish breath odor, excessive thirst, weakness, and at times nausea accompanied by vomiting. What are the physiological implications of this low-calorie diet?

3. Ron S. offers to cook supper for the family. He decides to bake potatoes to make home fries. He puts three tablespoons of fat in a skillet to fry the cooked potatoes. How many calories does he add to the potatoes by frying them? (1 tbsp = 13 grams of fat)

4. Steve S., age 24, eats a meal that contains 40 grams of protein, 50 grams of fat, and 69 grams of carbohydrates.

    (a) How many calories does Steve consume at this meal?

    (b) What are his percentages of carbohydrate, fat, and protein intake?

    (c) According to the Recommended Dietary Intake, what is wrong with this meal?

5. On a recent visit to her doctor, Renee C. complained that she felt sick. She had a temperature of 101°F and was feeling nauseous. The doctor prescribed an antibiotic, advised her to go home, drink plenty of fluids, and rest in bed for a few days. What is the physiological basis for drinking plenty of fluids when one is sick?

26

The Urinary System

■ Overview

As you have learned in previous chapters, the body's cells break down the food we eat, release energy from it, and produce chemical byproducts that collect in the bloodstream. The kidneys, playing a crucial homeostatic role, cleanse the blood of these waste products and excess water by forming urine, which is then transferred to the urinary bladder and eventually removed from the body via urination. The kidneys receive more blood from the heart than any other organ of the body. They handle 1½ quarts of blood every minute through a complex blood circulation housed in an organ 4 inches high, 2 inches wide, 1 inch thick, and weighing 5 to 6 ounces.

In addition to the kidneys, which produce the urine, the urinary system consists of the ureters, urinary bladder, and urethra, components that are responsible for transport, storage, and conduction of the urine to the exterior.

The activities for Chapter 26 focus on the structural and functional organization of the urinary system, the major regulatory mechanisms that control urine formation and modification, and urine transport, storage, and elimination.

☐ LEVEL 1 REVIEW OF CHAPTER OBJECTIVES

1. Identify the components of the urinary system and describe the vital functions performed by the system.

2. Describe the structural features of the kidney.

3. Describe the structure of the nephron and the processes involved in the formation of urine.

4. Identify the major blood vessels associated with the kidney and trace the path of blood flow through the kidney.

5. List and describe the factors that influence filtration pressure and the rate of filtrate formation.

6. Identify the types of transport mechanisms found along the nephron and discuss the reabsorptive or secretory functions of each segment of the nephron and collecting system.

7. Explain the role of countercurrent multiplication in the formation of a concentration gradient in the medulla.

8. Describe how antidiuretic hormone and aldosterone levels influence the flume and concentration of urine.

9. Describe the normal characteristics, composition, and solute concentrations of a representative urine sample.

10. Describe the structures and functions of the ureters, urinary bladder and urethra.

11. Discuss the voluntary and involuntary regulation of urination and details of the micturition reflex.

12. Describe the effects of aging on the urinary system.

[L1]   Multiple choice:

Place the letter corresponding to the correct answer in the space provided.

OBJ. 1   _____   1. Urine leaving the kidneys travels along the following sequential pathway to the exterior.

     a. ureters, urinary bladder, urethra
     b. urethra, urinary bladder, ureters
     c. urinary bladder, ureters, urethra
     d. urinary bladder, urethra, ureters

OBJ. 1   _____   2. Which organ or structure does *not* belong to the urinary system?

     a. urethra
     b. gall bladder
     c. kidneys
     d. ureters

OBJ. 1   _____   3. The openings of the urethra and the two ureters mark an area on the internal surface of the urinary bladder called the:

     a. internal urethral sphincter
     b. external urethral sphincter
     c. trigone
     d. renal sinus

OBJ. 1   _____   4. The initial factor which determines if urine production occurs is:

     a. secretion
     b. absorption
     c. sympathetic activation
     d. filtration

OBJ. 1   _____   5. Along with the urinary system, the other systems of the body that affect the composition of body fluids are:

     a. nervous, endocrine, and cardiovascular
     b. lymphatic, cardiovascular, and respiratory
     c. integumentary, respiratory, and digestive
     d. muscular, digestive, and lymphatic

OBJ. 2   _____   6. Seen in section, the kidney is divided into:

     a. renal columns and renal pelves
     b. an outer cortex and an inner medulla
     c. major and minor calyces
     d. a renal tubule and renal corpuscle

OBJ. 2   _____   7. The basic functional unit in the kidney is the:

       a. glomerulus
       b. loop of Henle
       c. Bowman's capsule
       d. nephron

OBJ. 2   _____   8. The three concentric layers of connective tissue that protect and anchor the kidneys are the:

       a. hilus, renal sinus, renal corpuscle
       b. cortex, medulla, papillae
       c. renal capsule, adipose capsule, renal fascia
       d. major calyces, minor calyces, renal pyramids

OBJ. 3   _____   9. In a nephron, the long tubular passageway through which the filtrate passes includes:

       a. collecting tubule, collecting duct, papillary duct
       b. renal corpuscle, renal tubule, renal pelvis
       c. proximal and distal convoluted tubules and loop of Henle
       d. loop of Henle, collecting and papillary duct

OBJ. 3   _____   10. The primary site of regulating water, sodium, and potassium ion loss in the nephron is the:

       a. distal convoluted tubule
       b. loop of Henle and collecting duct
       c. proximal convoluted tubule
       d. glomerulus

OBJ. 3   _____   11. The three processes involved in urine formation are:

       a. diffusion, osmosis, and filtration
       b. co-transport, countertransport, facilitated diffusion
       c. regulation, elimination, micturition
       d. filtration, reabsorption, secretion

OBJ. 3   _____   12. The primary site for secretion of substances into the filtrate is the:

       a. renal corpuscle
       b. loop of Henle
       c. distal convoluted tubule
       d. proximal convoluted tubule

OBJ. 4   _____   13. Dilation of the afferent arteriole and glomerular capillaries and constriction of the efferent arteriole causes:

       a. elevation of glomerular blood pressure to normal levels
       b. a decrease in glomerular blood pressure
       c. a decrease in the glomerular filtration rate
       d. an increase in the secretion of renin and erythropoietin

Level
-1-

OBJ. 4 ___ 14. Blood supply to the proximal and distal convoluted tubules of the nephron is provided by the:

    a. peritubular capillaries
    b. afferent arterioles
    c. segmental veins
    d. interlobular veins

OBJ. 5 ___ 15. The glomerular filtration rate is regulated by:

    a. autoregulation
    b. hormonal regulation
    c. autonomic regulation
    d. a, b, and c are correct

OBJ. 5 ___ 16. The pressure that represents the resistance to flow along the nephron and conducting system is the:

    a. blood colloid osmotic pressure (BCOP)
    b. glomerular hydrostatic pressure (GHP)
    c. capsular hydrostatic pressure (CHP)
    d. capsular colloid osmotic pressure (CCOP)

OBJ. 6 ___ 17. The mechanism important in the reabsorption of glucose and amino acids when their concentrations in the filtrate are relatively high is:

    a. active transport
    b. facilitated transport
    c. co-transport
    d. countertransport

OBJ. 6 ___ 18. Countertransport resembles co-transport in all respects except:

    a. calcium ions are exchanged for sodium ions
    b. chloride ions are exchanged for bicarbonate ions
    c. the two transported ions move in opposite directions
    d. a, b, and c are correct

OBJ. 6 ___ 19. The primary site of nutrient reabsorption in the nephron is the:

    a. proximal convoluted tubule
    b. distal convoluted tubule
    c. loop of Henle
    d. renal corpuscle

OBJ. 7 ___ 20. In countercurrent multiplication, the *countercurrent* refers to the fact that an exchange occurs between:

    a. sodium ions and chloride ions
    b. fluids moving in opposite directions
    c. potassium and chloride ions
    d. solute concentrations in the Loop of Henle

Level
-1-

OBJ. 7

_____ 21. The result of the countercurrent multiplication mechanism is:

    a. decreased solute concentration in descending limb of Loop of Henle

    b. decreased transport of sodium and chloride in ascending limb of Loop of Henle

    c. increased solute concentration in descending limb of Loop of Henle

    d. osmotic flow of water from peritubular fluid into descending limb of Loop of Henle

OBJ. 8

_____ 22. When antidiuretic hormone levels rise the distal convoluted tubule becomes:

    a. less permeable to water; reabsorption of water decreases

    b. more permeable to water; water reabsorption increases

    c. less permeable to water; reabsorption of water increases

    d. more permeable to water; water reabsorption decreases

OBJ. 8

_____ 23. The results of the effect of aldosterone along the DCT, the collecting tubule, and the collecting duct are:

    a. increased conservation of sodium ions and water

    b. increased sodium ion excretion

    c. decreased sodium ion reabsorption in the DCT

    d. increased sodium ion and water excretion

OBJ. 9

_____ 24. The _concentration_ of components in a given urine sample depends on the:

    a. identities and amount of materials eliminated in the urine

    b. degree to which evaporation of small compounds occurs

    c. individual's food intake

    d. osmotic movement of water across the walls of the tubules and collecting ducts

OBJ. 9

_____ 25. The _average_ pH for normal urine is about:

    a. 5.0

    b. 6.0

    c. 7.0

    d. 8.0

OBJ. 10

_____ 26. When urine leaves the kidney it travels to the urinary bladder via the:

    a. urethra

    b. ureters

    c. renal hilus

    d. renal calyces

Level
–1–

OBJ. 10   _____ 27. The expanded, funnel-shaped upper end of the ureter in the kidney is the:

      a. renal pelvis
      b. urethra
      c. renal hilus
      d. renal calyces

OBJ. 10   _____ 28. Contraction of the muscular bladder forces the urine out of the body through the:

      a. ureter
      b. urethra
      c. penis
      d. a, b, and c are correct

OBJ. 11   _____ 29. During the micturition reflex, increased afferent fiber activity in the pelvic nerves facilitates:

      a. parasympathetic motor neurons in the sacral spinal cord
      b. sympathetic sensory neurons in the sacral spinal cord
      c. the action of stretch receptors in the wall of the bladder
      d. urine ejection due to internal and external sphincter contractions

OBJ. 11   _____ 30. Urine reaches the urinary bladder by the:

      a. action of stretch receptors in the bladder wall
      b. fluid pressures in the renal pelvis
      c. peristaltic contractions of the ureters
      d. sustained contractions and relaxation of the urinary bladder

OBJ. 12   _____ 31. A reduction in the glomerular filtration rate (GFR) due to aging results from:

      a. decreased numbers of glomeruli
      b. cumulative damage to the filtration apparatus
      c. reductions in renal blood flow
      d. a, b, and c are correct

OBJ. 12   _____ 32. Reduced sensitivity to antidiuretic hormone (ADH) due to aging causes:

      a. decreasing reabsorption of water and sodium ions
      b. decreasing potassium ion loss in the urine
      c. decreasing numbers of glomeruli
      d. a, b, and c are correct

OBJ. 12   _____ 33. CNS problems that affect the cerebral cortex or hypothalamus may interfere with the ability to:

      a. avoid inflammation of the prostate gland
      b. control micturition
      c. control swelling and distortion of prostatic tissues compressing the urethra
      d. a, b, and c are correct

Level
-1-

[L1] Completion:

Using the terms below, complete the following statements.

cerebral cortex	urethra	micturition reflex
composition	rugae	concentration
incontinence	filtrate	parathormone
glomerulus	neck	nephrolithiasis
renal threshold	antidiuretic hormone	internal sphincter
kidneys	interlobar veins	countertransport
countercurrent multiplication		glomerular hydrostatic

OBJ. 1   34. The excretory functions of the urinary system are performed by the _____.

OBJ. 2   35. The renal corpuscle contains a capillary knot referred to as the _____.

OBJ. 3   36. The outflow across the walls of the glomerulus produces a protein-free solution known as the _____.

OBJ. 4   37. In a mirror image of arterial distribution, the interlobular veins deliver blood to arcuate veins that empty into _____.

OBJ. 5   38. The pressure that tends to drive water and solute molecules across the glomerular wall is the _____ pressure.

OBJ. 6   39. The hormone that stimulates carrier proteins to reduce the urinary loss of calcium ions is _____.

OBJ. 6   40. To keep intracellular calcium concentrations low, most cells in the body use sodium-calcium _____.

OBJ. 7   41. The mechanism that operates efficiently to reabsorb solutes and water before the tubular fluid reaches the DCT and collecting system is _____.

OBJ. 8   42. Passive reabsorption of water from urine in the collecting system is regulated by circulating levels of _____.

OBJ. 9   43. The plasma concentration at which a specific component or ion will begin appearing in the urine is called the _____.

OBJ. 9   44. The filtration, absorption, and secretion activities of the nephrons reflect the urine's _____.

OBJ. 9   45. The osmotic movement of water across the walls of the tubules and collecting ducts determines the urine's _____.

OBJ. 10   46. The structure that carries urine from the urinary bladder to the exterior of the body is the _____.

OBJ. 10   47. The muscular ring that provides involuntary control over the discharge of urine from the urinary bladder is the _____.

OBJ. 10   48. In a relaxed condition the epithelium of the urinary bladder forms a series of prominent folds called _____.

OBJ. 10   49. The area surrounding the urethral entrance of the urinary bladder is the _____.

OBJ. 11   50. The process of urination is coordinated by the _____.

Level
-1-

OBJ. 11   51. We become consciously aware of the fluid pressure in the urinary bladder because of sensations relayed to the _____.

OBJ. 12   52. A slow leakage of urine due to loss of sphincter muscle tone is referred to as _____.

OBJ. 12   53. The presence of kidney stones results in a condition referred to as _____.

[L1]   Matching:

Match the terms in column B with the terms in column A. Use letters for answers in the spaces provided.

PART I

		COLUMN A		COLUMN B
OBJ. 1	_____	54. juxtaglomerular apparatus	A.	blood to kidney
OBJ. 2	_____	55. renal capsule	B.	blood colloid osmotic pressure
OBJ. 3	_____	56. glomerular epithelium	C.	sodium-bicarbonate ion exchange
OBJ. 4	_____	57. renal artery	D.	ion pump — K^+ channels
OBJ. 4	_____	58. renal vein	E.	podocytes
OBJ. 5	_____	59. opposes filtration	F.	countercurrent multiplication
OBJ. 6	_____	60. countertransport	G.	fibrous tunic
OBJ. 7	_____	61. loop of Henle	H.	blood from kidney
OBJ. 8	_____	62. aldosterone	I.	renin and erythropoietin

PART II

		COLUMN A		COLUMN B
OBJ. 9	_____	63. glycosuria	J.	voluntary control
OBJ. 10	_____	64. urinary bladder	K.	urge to urinate
OBJ. 10	_____	65. middle umbilical ligament	L.	kidney failure
			M.	internal sphincter forced open
OBJ. 11	_____	66. internal sphincter		
OBJ. 11	_____	67. external sphincter	N.	no urine production
OBJ. 11	_____	68. 200 ml of urine in bladder	O.	urachus
			P.	glucose in the urine
OBJ. 11	_____	69. 500 ml of urine in bladder	Q.	involuntary control
OBJ. 12	_____	70. uremia	R.	detrusor muscle
OBJ. 12	_____	71. anuria		

[L1]   Drawing/Illustration Labeling:

Identify each numbered structure by labeling the following figures:

OBJ. 1   **FIGURE 26.1**   Components of the Urinary System

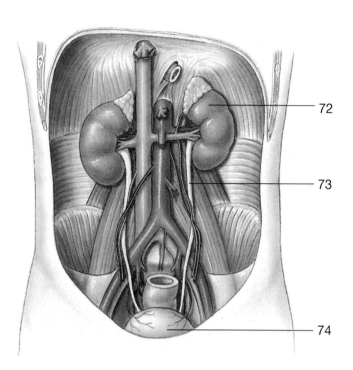

72 _____

73 _____

74 _____

OBJ. 2   **FIGURE 26.2**   Sectional Anatomy of the Kidney

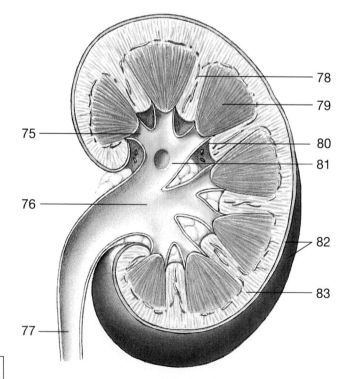

75 _____

76 _____

77 _____

78 _____

79 _____

80 _____

81 _____

82 _____

83 _____

Level
-1-

OBJ. 3 **FIGURE 26.3** Structure of a Typical Nepron Including Circulation

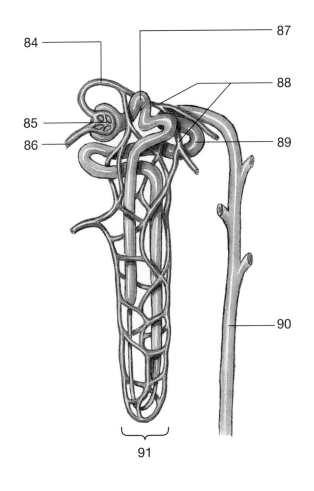

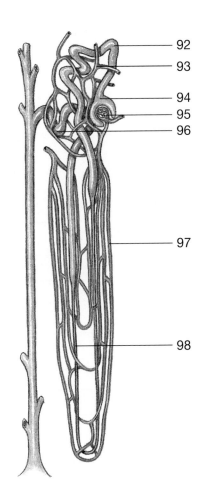

84 _____

85 _____

86 _____

87 _____

88 _____

89 _____

90 _____

91 _____

92 _____

93 _____

94 _____

95 _____

96 _____

97 _____

98 _____

When you have successfully completed the exercises in L1 proceed to L2.

Level
–1–

☐ LEVEL 2   CONCEPT SYNTHESIS

Concept Map I:

Using the following terms, fill in the circled, numbered, blank spaces to complete the concept map. Follow the numbers that comply with the organization of the map.

Renal sinus   Urinary bladder   Medulla
Ureters   Minor calyces   Nephrons
Glomerulus   Proximal convoluted   Collecting tubules
                      tubule

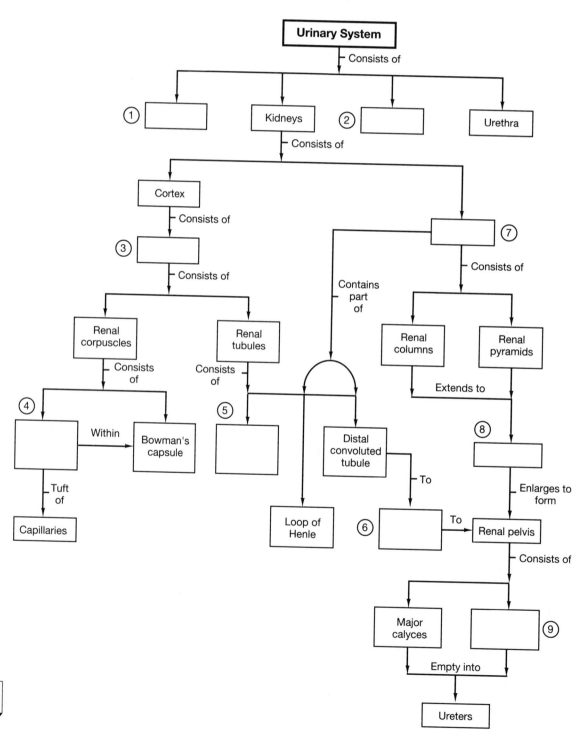

Concept Map II:

Using the following terms, fill in the circled, numbered, blank spaces to complete the concept map. Follow the numbers that comply with the organization of the map.

Efferent artery   Afferent artery   Interlobar vein
Interlobular vein   Renal artery   Arcuate artery

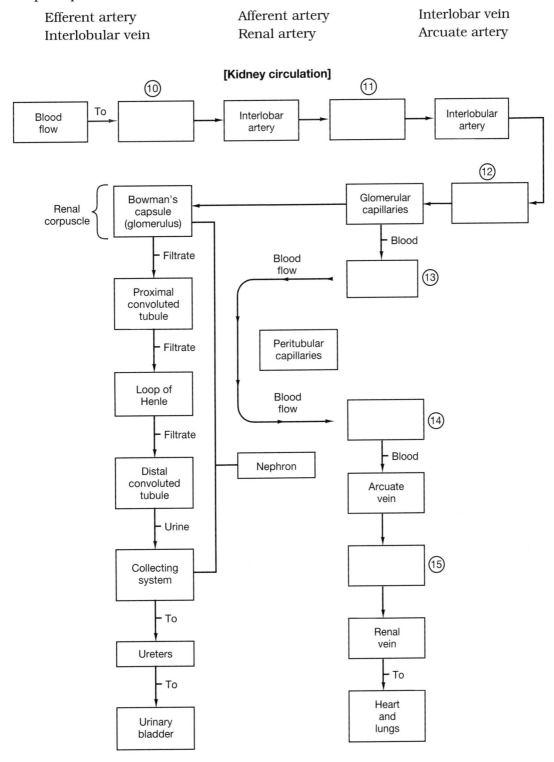

[Kidney circulation]

Concept Map III:

Using the following terms, fill in the circled, numbered, blank spaces to complete the concept map. Follow the numbers that comply with the organization of the map.

Adrenal cortex   ↓ Na⁺ excretion   ↓ H₂O excretion
liver   ↓ plasma volume   ↑ Na⁺ reabsorption
angiotensin I   Renin

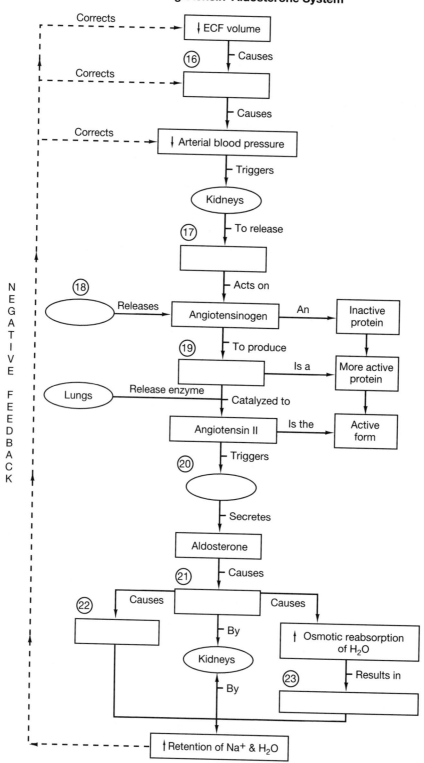

Renin–Angiotensin–Aldosterone System

Body Trek:

Using the terms below, fill in the blanks to complete the trek through the urinary system.

aldosterone	protein-free	ions
ascending limb	active transport	ADH
glomerulus	urine	urinary bladder
collecting	proximal	descending limb
urethra	filtrate	ureters
distal		

Robo's trek through the urinary system begins as the tiny robot is inserted into the urethra and with the aid of two mini-rockets is propelled through the urinary tract into the region of the renal corpuscle in the cortex of the kidney. The use of the mini-rockets eliminates the threat of the countercurrent flow of the filtrate and the urine and the inability of the robot to penetrate the physical barriers imposed in the capsular space.

Rocket shutdown occurs just as Robo comes into contact with a physical barrier, the filtration membrane. The robot's energizers are completely turned off because the current produced by the fluid coming through the membrane is sufficient to "carry" the robot as it treks through the tubular system.

As the fluid crosses the membrane it enters into the first tubular conduit, the (24) _____ convoluted tubule, identifiable because of its close proximity to the vascular pole of the (25) _____. Robo's chemosensors detect the presence of a (26) _____ filtrate in this area. Using the current in the tubule to move on, the robot's speed increases because of a sharp turn and descent into the (27) _____ of Henle, where water reabsorption and concentration of the (28) _____ are taking place. The trek turns upward after a "hairpin" turn into the (29) _____ of Henle where, in the thickened area, (30) _____ mechanisms are causing the reabsorption of (31) _____.

The somewhat "refined" fluid is delivered to the last segment of the nephron, the (32) _____ convoluted tubule, which is specially adapted for reabsorption of sodium ions due to the influence of the hormone (33) _____, and in its most distal position, the effect of (34) _____ , causing an osmotic flow of water that assists in concentrating the filtrate.

Final adjustments to the sodium ion concentration and the volume of the fluid are made in the (35) _____ tubules, which deliver the waste products in the form of (36) _____ to the renal pelvis. It will then be conducted into the (37) _____ on its way to the (38) _____ for storage until it leaves the body through the (39) _____ — Robo's way of escaping the possibility of toxicity due to the "polluted" environment of the urinary tract.

[L2] Multiple Choice:

Place the letter corresponding to the correct answer in the space provided.

_____ 40. The vital function(s) performed by the nephrons in the kidneys is (are):

        a. production of filtrate

        b. reabsorption of organic substrates

        c. reabsorption of water and ions

        d. a, b, and c are correct

_____ 41. The renal corpuscle consists of:

        a. renal columns and renal pyramids

        b. major and minor calyces

        c. Bowman's capsule and the glomerulus

        d. the cortex and the medulla

Level
=2=

_____ 42. The filtration process within the renal corpuscle involves passage across three physical barriers, which include the:

- a. podocytes, pedicels, slit pores
- b. capillary endothelium, basement membrane, glomerular epithelium
- c. capsular space, tubular pole, macula densa
- d. collecting tubules, collecting ducts, papillary ducts

_____ 43. The thin segments in the loop of Henle are:

- a. relatively impermeable to water; freely permeable to ions and other solutes
- b. freely permeable to water, ions, and other solutes
- c. relatively impermeable to water, ions, and other solutes
- d. freely permeable to water; relatively impermeable to ions and other solutes

_____ 44. The thick segments in the loop of Henle contain:

- a. ADH-regulated permeability
- b. an aldosterone-regulated pump
- c. transport mechanisms that pump materials out of the filtrate
- d. diffusion mechanisms for getting rid of excess water

_____ 45. The collecting system in the kidney is responsible for:

- a. active secretion and reabsorption of sodium ions
- b. absorption of nutrients, plasma proteins, and ions from the filtrate
- c. creation of the medullary concentration gradient
- d. making final adjustments to the sodium ion concentration and volume of urine

_____ 46. Blood reaches the vascular pole of each glomerulus through a(n):

- a. efferent arteriole
- b. afferent arteriole
- c. renal artery
- d. arcuate arteries

_____ 47. Sympathetic innervation into the kidney is responsible for:

- a. regulation of glomerular blood flow and pressure
- b. stimulation of renin release
- c. direct stimulation of water and sodium ion reabsorption
- d. a, b, and c are correct

_____ 48. When plasma glucose concentrations are higher than the renal threshold, glucose concentrations in the filtrate exceed the tubular maximum (T_m) and:

- a. glucose is transported across the membrane by countertransport
- b. the glucose is filtered out at the glomerulus
- c. glucose appears in the urine
- d. the individual has eaten excessive amounts of sweets

Level
=2=

_____ 49. The outward pressure forcing water and solute molecules across the glomerulus wall is the:

         a. capsular hydrostatic pressure

         b. glomerular hydrostatic pressure

         c. blood osmotic pressure

         d. filtration pressure

_____ 50. The opposing forces of the filtration pressure at the glomerulus are the:

         a. glomerular hydrostatic pressure and blood osmotic pressure

         b. capsular hydrostatic pressure and glomerular hydrostatic pressure

         c. blood pressure and glomerular filtration rate

         d. capsular hydrostatic pressure and blood osmotic pressure

_____ 51. The amount of filtrate produced in the kidneys each minute is the:

         a. glomerular filtration rate

         b. glomerular hydrostatic pressure

         c. capsular hydrostatic pressure

         d. osmotic gradient

_____ 52. Inadequate ADH secretion results in the inability to reclaim the water entering the filtrate, causing:

         a. glycosuria

         b. dehydration

         c. anuria

         d. dysuria

_____ 53. Under normal circumstances virtually all the glucose, amino acids, and other nutrients are reabsorbed before the filtrate leaves the:

         a. distal convoluted tubule

         b. glomerulus

         c. collecting ducts

         d. proximal convoluted tubule

_____ 54. Aldosterone stimulates ion pumps along the distal convoluted tubule (DCT), the collecting tubule, and the collecting duct, causing a(n):

         a. increase in the number of sodium ions lost in the urine

         b. decrease in the concentration of the filtrate

         c. reduction in the number of sodium ions lost in the urine

         d. countercurrent multiplication

_____ 55. The high osmotic concentrations found in the kidney medulla are primarily due to:

         a. presence of sodium ions, chloride ions, and urea

         b. presence of excessive amounts of water

         c. hydrogen and ammonium ions

         d. a, b, and c are correct

_____ 56. The hormones that affect the glomerular filtration rate (GFR) by regulating blood pressure and volume are:

   a. aldosterone, epinephrine, oxytocin
   b. insulin, glucagon, glucocorticoids
   c. renin, erythropoietin, ADH
   d. a, b, and c are correct

_____ 57. Angiotensin II is a potent hormone that:

   a. causes constriction of the efferent arteriole at the nephron
   b. triggers the release of ADH in the CNS
   c. stimulates secretion of aldosterone by the adrenal cortex and epinephrine by the adrenal medulla
   d. a, b, and c are correct

_____ 58. Sympathetic innervation of the afferent arterioles causes a(n):

   a. decrease in GFR and slowing of filtrate production
   b. increase in GFR and an increase in filtrate production
   c. decrease in GFR and an increase in filtrate production
   d. increase in GFR and a slowing of filtrate production

_____ 59. During periods of strenuous exercise sympathetic activation causes the blood flow to:

   a. decrease to skin and skeletal muscles; increase to kidneys
   b. cause an increase in GFR
   c. increase to skin and skeletal muscles; decrease to kidneys
   d. be shunted toward the kidneys

[L2] Completion:

Using the terms below, complete the following statements.

aldosterone	vasa recta	angiotensin II
filtration	diabetes insipidus	osmotic gradient
macula densa	renin	glomerular filtration rate
secretion	transport maximum	cortical
retroperitoneal	reabsorption	glomerular filtration

60. The kidneys lie between the muscles of the dorsal body wall and the peritoneal lining in a position referred to as _____.

61. The amount of filtrate produced in the kidneys each minute is the _____.

62. The process responsible for concentrating the filtrate as it travels through the collecting system toward the renal pelvis is the _____.

63. When all available carrier proteins are occupied at one time, the saturation concentration is called the _____.

64. The region in the distal convoluted tubule (DCT) where the tubule cells adjacent to the tubular pole are taller and contain clustered nuclei is called the _____.

65. Approximately 85 percent of the nephrons in a kidney are called _____ nephrons.

66. A capillary that accompanies the loop of Henle into the medulla is termed the _____.

67. When hydrostatic pressure forces water across a membrane the process is referred to as _____.

68. The removal of water and solute molecules from the filtrate is termed _____.

69. Transport of solutes across the tubular epithelium and into the filtrate is _____.

70. The vital first step essential to all kidney function is _____.

71. The ion pump and the potassium ion channels are controlled by the hormone _____.

72. When a person produces relatively large quantities of dilute urine the condition is termed _____.

73. The hormone that causes a brief but powerful vasoconstriction in peripheral capillary beds is _____.

74. Sympathetic activation stimulates the juxtaglomerular apparatus to release _____.

[L2]   Short Essay:

Briefly answer the following questions in the spaces provided below.

75. What are six (6) essential functions of the urinary system?

76. Trace the pathway of a drop of urine from the time it leaves the kidney until it is urinated from the body. (Use arrows to show direction.)

77. What three (3) concentric layers of connective tissue protect and anchor the kidneys?

78. Trace the pathway of a drop of filtrate from the time it goes through the filtration membrane in the glomerulus until it enters the collecting tubules as concentrated urine. (Use arrows to indicate direction of flow.)

79. What vital functions of the kidneys are performed by the nephrons?

Level =2=

80. (a) What three (3) physical barriers are involved with passage of materials during the filtration process?

    (b) Cite one structural characteristic of each barrier.

81. What two (2) hormones are secreted by the juxtaglomerular apparatus?

82. What known functions result from sympathetic innervation into the kidneys?

83. What are the three (3) processes involved in urine formation?

84. Write a formula to show the following relationship and define each component of the formula:

    The *filtration pressure* (P_f) at the glomerulus is the difference between the blood pressure and the opposing capsular and osmotic pressures.

85. Why does hydrogen ion secretion accelerate during starvation?

86. What three (3) basic concepts describe the mechanism of countercurrent multiplication?

87. What three (3) control mechanisms are involved with regulation of the glomerular filtration rate (GFR)?

88. What four (4) hormones affect urine production? Describe the role of each one.

When you have successfully completed the exercises in L2 proceed to L3.

☐ LEVEL 3   CRITICAL THINKING/APPLICATION

Using principles and concepts learned about the urinary system, answer the following questions. Write your answers on a separate sheet of paper.

1. I. O. Yew spends an evening with the boys at the local tavern. During the course of the night he makes numerous trips to the rest room to urinate. In addition to drinking an excessive amount of fluid, what effect does the consumption of alcohol have on the urinary system to cause urination?

2. Lori C. has had a series of laboratory tests, including a CBC, lipid profile series, and a urinalysis. The urinalysis revealed the presence of an abnormal amount of plasma proteins and white blood cells. (a) What is your diagnosis? (b) What effect does her condition have on her urine output?

3. On an anatomy and physiology examination your instructor asks the following question:

   What is the *principal function* associated with each of the following components of the nephron?

       (a)   renal corpuscle

       (b)   proximal convoluted tubule

       (c)   distal convoluted tubule

       (d)   loop of Henle and collecting system

4. (a) Given the following information, determine the effective filtration pressure (EFP) at the glomerulus:

glomerular blood hydrostatic pressure	G_{hp}	60 mm Hg
capsular hydrostatic pressure	C_{hp}	18 mm Hg
glomerular blood osmotic pressure	OP_b	32 mm Hg
capsular osmotic pressure	C_{op}	5 mm Hg

   (b) How does the effective filtration pressure affect the GFR?

   (c) What are the implications of an effective filtration pressure (EFP) of 15 mm Hg?

5. Cindy F. is a marathon runner who, from time to time, experiences acute kidney dysfunction during the event. Laboratory tests have shown elevated serum concentrations of K^+, lowered serum concentrations of Na^+, and a decrease in the GFR. At one time she experienced renal failure. Explain why these symptoms and conditions may occur while running a marathon.

27

Fluid, Electrolyte, and Acid-Base Balance

■ Overview

In many of the previous chapters we examined the various organ systems, focusing on the structural components and the functional activities necessary to *support* life. This chapter details the roles the organ systems play in *maintaining* life in the billions of individual cells in the body. Early in the study of anatomy and physiology we established that the cell is the basic unit of structure and function of all living things, no matter how complex the organism is as a whole. The last few chapters concentrated on the necessity of meeting the nutrient needs of cells and getting rid of waste products that might interfere with normal cell functions. To meet the nutrient needs of cells, substances must move in; and to get rid of wastes, materials must move out. The result is a constant movement of materials into and out of the cells.

Both the external environment (interstitial fluid) and the internal environment of the cell (intracellular fluid) comprise an "exchange system" operating within controlled and changing surroundings, producing a dynamic equilibrium by which homeostasis is maintained.

Chapter 27 considers the mechanics and dynamics of fluid balance, electrolyte balance, acid-base balance, and the interrelationships of functional patterns that operate in the body to support and maintain a constant state of equilibrium.

☐ LEVEL 1   REVIEW OF CHAPTER OBJECTIVES

1. Compare the composition of the intracellular and extracellular fluids.
2. Explain the basic concepts involved in fluid and electrolyte regulation.
3. Identify the hormones that play important roles in regulating fluid and electrolyte balance and describe their effects.
4. Discuss the mechanisms by which sodium, potassium, calcium, and chloride are regulated to maintain electrolyte balance in the body.
5. Explain the buffering systems that balance the pH of the intracellular and extracellular fluid.

6. Describe the compensatory mechanisms involved in the maintenance of acid-base balance.

7. Identify the most frequent threats to acid-base balance and explain how the body responds when the pH of body fluids varies outside normal limits.

[L1]   Multiple Choice:

Place the letter corresponding to the correct answer in the space provided.

OBJ. 1   _____   1. Nearly two-thirds of the total body water content is:

   a. extracellular fluid (ECF)
   b. intracellular fluid (ICF)
   c. tissue fluid
   d. interstitial fluid (IF)

OBJ. 1   _____   2. Extracellular fluids in the body consist of:

   a. interstitial fluid, blood plasma, lymph
   b. cerebrospinal fluid, synovial fluid, serous fluids
   c. aqueous humor, perilymph, endolymph
   d. a, b, and c are correct

OBJ. 1   _____   3. The principal ions in the extracellular fluid (ECF) are:

   a. sodium, chloride, and bicarbonate
   b. potassium, magnesium, and phosphate
   c. phosphate, sulfate, magnesium
   d. potassium, ammonium, and chloride

OBJ. 2   _____   4. If the ECF is *hypertonic* with respect to the ICF, water will move:

   a. from the ECF into the cell until osmotic equilibrium is restored
   b. from the cells into the ECF until osmotic equilibrium is restored
   c. in both directions until osmotic equilibrium is restored
   d. in response to the pressure of carrier molecules

OBJ. 2   _____   5. When water is lost but electrolytes are retained, the osmolarity of the ECF rises and osmosis then moves water:

   a. out of the ECF and into the ICF until isotonicity is reached
   b. back and forth between the ICF and the ECF
   c. out of the ICF and into the ECF until isotonicity is reached
   d. directly into the blood plasma until equilibrium is reached

OBJ. 2   _____   6. When pure water is consumed, the extracellular fluid becomes:

   a. hypotonic with respect to the ICF
   b. hypertonic with respect to the ICF
   c. isotonic with respect to the ICF
   d. the ICF and the ECF are in equilibrium

OBJ. 3   _____ 7. Physiological adjustments affecting fluid and electrolyte balance are mediated primarily by:

        a. antidiuretic hormone (ADH)
        b. aldosterone
        c. atrial natriuretic peptide (ANP)
        d. a, b, and c are correct

OBJ. 3   _____ 8. The two important effects of increased release of ADH are:

        a. increased rate of sodium absorption and decreased thirst
        b. reduction of urinary water losses and stimulation of the thirst center
        c. decrease in the plasma volume and elimination of the source of stimulation
        d. decrease in plasma osmolarity and alteration of composition of tissue fluid

OBJ. 3   _____ 9. Secretion of aldosterone occurs in response to:

        a. a fall in plasma volume at the juxtaglomerular apparatus
        b. a fall in blood pressure at the juxtaglomerular apparatus
        c. potassium ion concentrations
        d. a, b, and c are correct

OBJ. 3   _____ 10. Atrial natriuretic peptide hormone:

        a. reduces thirst
        b. blocks the release of ADH
        c. blocks the release of aldosterone
        d. a, b, and c are correct

OBJ. 3   _____ 11. The major contributors to the osmolarities of the ECF and the ICF are:

        a. ADH and aldosterone
        b. sodium and potassium
        c. renin and angiotensin
        d. chloride and bicarbonate

OBJ. 4   _____ 12. The concentration of potassium in the ECF is controlled by adjustments in the rate of active secretion:

        a. in the proximal convoluted tubule of the nephron
        b. in the loop of Henle
        c. along the distal convoluted tubule of the nephron
        d. along the collecting tubules

OBJ. 4   _____ 13. The activity that occurs in the body to maintain calcium homeostasis occurs primarily in the:

        a. bone
        b. digestive tract
        c. kidneys
        d. a, b, and c are correct

OBJ. 5 ____ 14. The hemoglobin buffer system helps prevent drastic alterations in pH when:

    a. the plasma Pco_2 is rising or falling

    b. there is an increase in hemoglobin production

    c. there is a decrease in RBC production

    d. the plasma of Pco_2 is constant

OBJ. 5 ____ 15. The primary role of the carbonic acid-bicarbonate buffer system is in preventing pH changes caused by:

    a. rising or falling Pco_2

    b. organic acid and fixed acids in the ECF

    c. increased production of sulfuric acid and phosphoric acid

    d. a, b, and c are correct

OBJ. 5 ____ 16. Pulmonary and renal mechanisms support the buffer systems by:

    a. secreting or generating hydrogen ions

    b. controlling the excretion of acids and bases

    c. generating additional buffers when necessary

    d. a, b, and c are correct

OBJ. 5 ____ 17. The lungs contribute to pH regulation by their effects on the:

    a. hemoglobin buffer system

    b. phosphate buffer system

    c. carbonic acid-bicarbonate buffer system

    d. protein buffer system

OBJ. 5 ____ 18. Increasing or decreasing the rate of respiration can have a profound effect on the buffering capacity of body fluids by:

    a. lowering or raising the Po_2

    b. lowering or raising the Pco_2

    c. increasing the production of lactic acid

    d. a, b, and c are correct

OBJ. 6 ____ 19. The renal response to acidosis is limited to:

    a. reabsorption of H^+ and secretion of HCO_3^-

    b. reabsorption of H^+ and HCO_3^-

    c. secretion of H^+ and generation or reabsorption of HCO_3^-

    d. secretion of H^+ and HCO_3^-

OBJ. 6 ____ 20. When carbon dioxide concentrations rise, additional hydrogen ions and bicarbonate ions are excreted and the:

    a. pH goes up

    b. pH goes down

    c. pH remains the same

    d. pH is not affected

OBJ. 7 _____ 21. Disorders that have the potential for disrupting pH balance in the body include:

    a. emphysema, renal failure

    b. neural damage, CNS disease

    c. heart failure, hypotension

    d. a, b, and c are correct

OBJ. 7 _____ 22. Respiratory alkalosis develops when respiratory activity:

    a. raises plasma Pco_2 to above-normal levels

    b. lowers plasma Pco_2 to below-normal levels

    c. decreases plasma Po_2 to below-normal levels

    d. when Pco_2 levels are not affected

OBJ. 7 _____ 23. The most frequent cause of metabolic acidosis is:

    a. production of a large number of fixed or organic acids

    b. a severe bicarbonate loss

    c. an impaired ability to excrete hydrogen ions at the kidneys

    d. generation of large quantities of ketone bodies

[L1]   Completion:

Using the terms below, complete the following statements.

respiratory compensation	fluid	osmoreceptors
hypotonic	alkalosis	acidosis
hypertonic	fluid shift	aldosterone
renal compensation	electrolyte	antidiuretic hormone
kidneys	calcium	buffers
hemoglobin		

OBJ. 1 24. When the amount of water gained each day is equal to the amount lost to the environment, a person is in _____ balance.

OBJ. 1 25. Water movement between the ECF and the ICF is termed a _____.

OBJ. 2 26. When there is neither a net gain nor a net loss of any ion in the body fluid, an _____ balance exists.

OBJ. 2 27. Osmotic concentrations of the plasma are monitored by special cells in the hypothalamus called _____.

OBJ. 2 28. If the ECF becomes more concentrated with respect to the ICF, the ECF is _____.

OBJ. 2 29. If the ECF becomes more dilute with respect to the ICF, the ECF is _____.

OBJ. 3 30. The rate of sodium absorption along the DCT (distal convoluted tubule) and collecting system of the kidneys is regulated by the secretion of _____.

OBJ. 3 31. The hormone that stimulates water conservation at the kidneys and the thirst center to promote the drinking of fluids is _____.

OBJ. 4 32. The most important site of sodium ion regulation is the _____.

OBJ. 4 33. Calcitonin from the C cells of the thyroid gland promotes a loss of _____.

Level -1-

OBJ. 5   34. Dissolved compounds that can provide or remove hydrogen ions and thereby stabilize the pH of a solution are _____.

OBJ. 5   35. The only intracellular buffer system that can have an immediate effect on the pH of the SCF is the _____ buffer system.

OBJ. 6   36. A change in the respiratory rate that helps stabilize pH is called _____.

OBJ. 6   37. A change in the rates of hydrogen ion and bicarbonate ion secretion or absorption in response to changes in plasma pH is called _____.

OBJ. 7   38. When the pH in the body falls below 7.35, the condition is called _____.

OBJ. 7   39. When the pH in the body increases above 7.45, the condition is called _____.

[L1] Matching:

Match the terms in column B with the terms in column A. Use letters for answers in the spaces provided.

PART I

		COLUMN A	COLUMN B
OBJ. 1	____ 40.	tissue fluid	A. dominant cation—ECF
OBJ. 1	____ 41.	fluid compartments	B. concentration of dissolved solutes
OBJ. 1	____ 42.	potassium	C. dominant cation—ICF
OBJ. 1	____ 43.	sodium	D. ICF and ECF
OBJ. 2	____ 44.	key components of ECF	E. posterior pituitary gland
OBJ. 2	____ 45.	osmolarity	F. interstitial fluid
OBJ. 3	____ 46.	cardiac muscle fiber	G. plasma, cerebrospinal fluid
OBJ. 3	____ 47.	antidiuretic hormone	H. atrial natriuretic peptide

PART II

		COLUMN A	COLUMN B
OBJ. 4	____ 48.	ADH secretion decrease	I. buffers pH of ICF
OBJ. 4	____ 49.	increased venous return	J. respiratory alkalosis
OBJ. 5	____ 50.	phosphate buffer system	K. buffers pH of ECF
OBJ. 5	____ 51.	carbonic acid-bicarbonate buffer	L. respiratory acidosis
			M. water loss at kidneys increases
OBJ. 6	____ 52.	decreased pCO_2	N. release of ANP
OBJ. 7	____ 53.	hypercapnia	O. pH increases
OBJ. 7	____ 54.	hypocapnia	

[L1]   Drawing/Illustration Labeling:

Using the terms and figures below, fill in each numbered box to complete the information regarding pH.

OBJ. 7   **FIGURE 27.1**   The pH Scale

alkalosis	pH 6.80	pH 7.80
acidosis	pH 7.35	pH 7.45

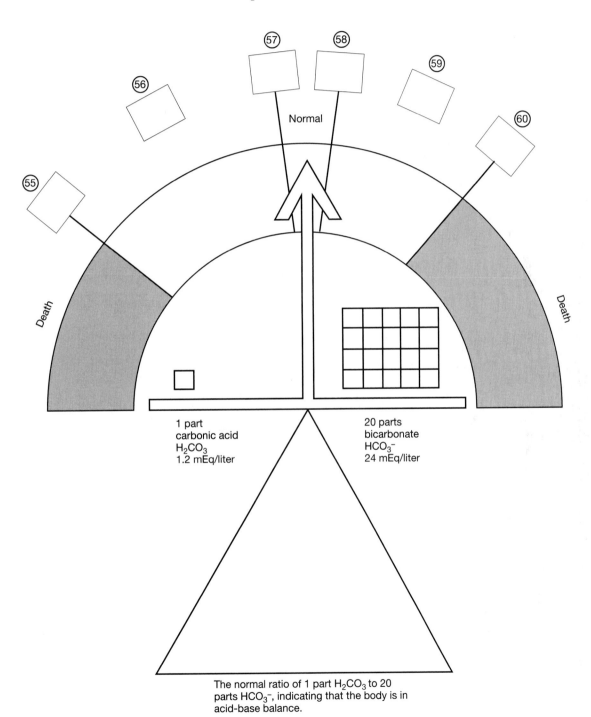

The normal ratio of 1 part H_2CO_3 to 20 parts HCO_3^-, indicating that the body is in acid-base balance.

OBJ. 7   *FIGURE 27.2*   Relationships among pH, Pco$_2$, and HCO$_3^-$

Using arrows and the letter N (normal), fill in the chart below to show the relationships among pH, Pco$_2$, and HCO$_3^-$. (Assume no compensation.)

Key: ↑ — increased or higher than
↓ — decreased or lower than
N — normal

	pH		pCO$_2$	HCO$_3^-$
Respiratory acidosis	7.35	㉑	㉕	㉙
Metabolic acidosis	7.35	㉒	㉖	㉚
Respiratory acidosis	7.45	㉓	㉗	㉛
Metabolic acidosis	7.45	㉔	㉘	㉜

When you have successfully completed the exercises in L1 proceed to L2.

☐ LEVEL 2   CONCEPT SYNTHESIS

Concept Map I:

Using the following terms, fill in the circled, numbered, blank spaces to complete the concept map. Follow the numbers that comply with the organization of the map.

↑ H_2O retention at kidneys   ↓ Volume of body H_2O
Aldosterone secretion by adrenal cortex

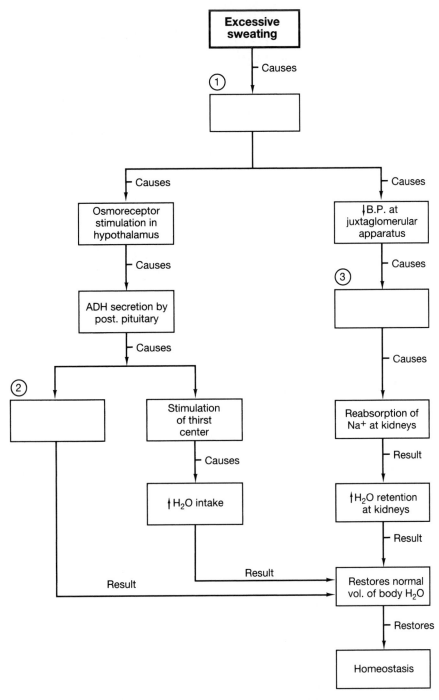

Homeostasis of total volume of body water

Concept Map II:

Using the following terms, fill in the circled, numbered, blank spaces to complete the concept map. Follow the numbers that comply with the organization of the map.

↓ ECF volume   ↓ pH
↑ ICF volume   ECF hypotonic to ICF

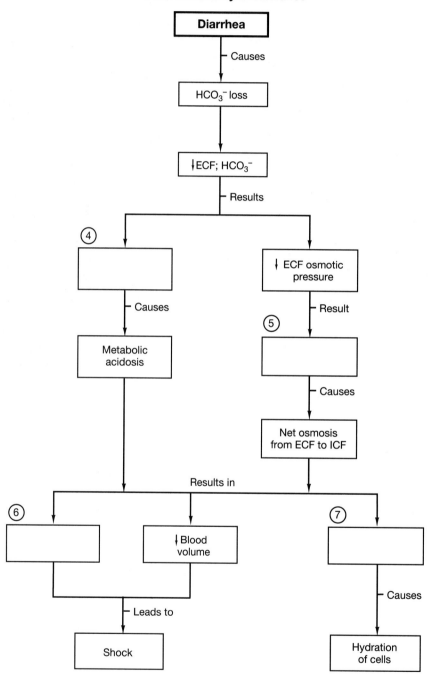

Fluid & electrolyte imbalance

Concept Map III:

Using the following terms, fill in the circled, numbered, blank spaces to complete the concept map. Follow the numbers that comply with the organization of the map.

↑ Blood pH   ↑ Depth of breathing   ↑ Blood CO$_2$
Hyperventilation   Normal blood pH

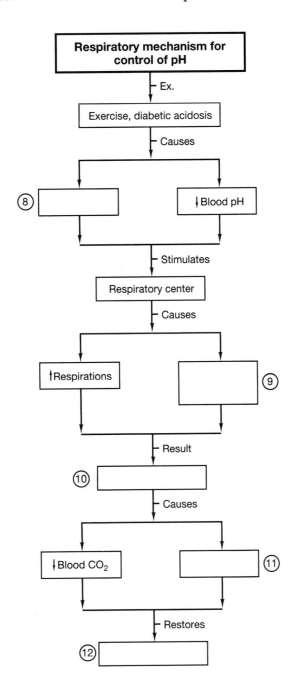

Concept Map IV:

Using the following terms, fill in the circled, numbered, blank spaces to complete the concept map. Follow the numbers that comply with the organization of the map.

HCO_3^- $\uparrow$ Blood pH $\downarrow$ Blood pH

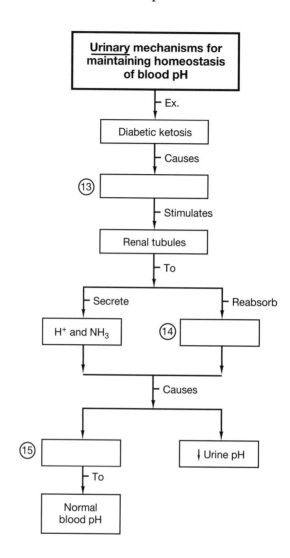

Concept Map V:

Using the following terms, fill in the circled, numbered, blank spaces to complete the concept map. Follow the numbers that comply with the organization of the map.

↑ Plasma volume   ↓ H_2O loss
↑ ANP release   ↓ ADH release
↑ H_2O loss   ↓ B.P. at kidneys
↑ Aldosterone release   ↓ Aldosterone release

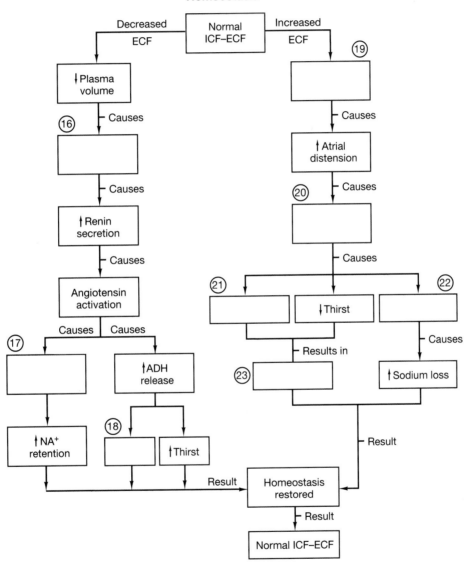

Fluid volume regulation–sodium ion concentrations Homeostasis

[L2]   Multiple Choice:

Place the letter corresponding to the correct answer in the space provided.

_____ 24. All of the homeostatic mechanisms that monitor and adjust the composition of body fluids respond to changes in the:

      a. intracellular fluid
      b. extracellular fluid
      c. regulatory hormones
      d. fluid balance

_____ 25. Important homeostatic adjustments occur in response to changes in:

      a. cell receptors that respond to ICF volumes
      b. hypothalamic osmoreceptors
      c. hormone levels
      d. plasma volume or osmolarity

_____ 26. All water transport across cell membranes and epithelia occur passively, in response to:

      a. active transport and co-transport
      b. countertransport and facilitated diffusion
      c. osmotic gradients and hydrostatic pressure
      d. co-transport and endocytosis

_____ 27. Whenever the rate of sodium intake or output changes, there is a corresponding gain or loss of water that tends to:

      a. keep the sodium concentration constant
      b. increase the sodium concentration
      c. decrease the sodium concentration
      d. alter the sodium concentration

_____ 28. Angiotensin II produces a coordinated elevation in the extracellular fluid volume by:

      a. stimulating thirst
      b. causing the release of ADH
      c. triggering the secretion of aldosterone
      d. a, b, and c are correct

_____ 29. The rate of tubular secretion of potassium ions changes in response to:

      a. alterations in the potassium ion concentration in the ECF
      b. changes in pH
      c. aldosterone levels
      d. a, b, and c are correct

_____ 30. The most important factor affecting the pH in body tissues is:

      a. the protein buffer system
      b. carbon dioxide concentration
      c. the bicarbonate reserve
      d. the presence of ammonium ions

_____ 31. The body content of water or electrolytes will rise if:

        a. losses exceed gains

        b. intake is less than outflow

        c. outflow exceeds intake

        d. intake exceeds outflow

_____ 32. When an individual loses body water:

        a. plasma volume decreases and electrolyte concentrations rise

        b. plasma volume increases and electrolyte concentrations decrease

        c. plasma volume increases and electrolyte concentrations increase

        d. plasma volume decreases and electrolyte concentrations decrease

_____ 33. The most common problems with electrolyte balance are caused by:

        a. shifts of the bicarbonate ion

        b. an imbalance between sodium gains and losses

        c. an imbalance between chloride gains and losses

        d. a, b, and c are correct

_____ 34. Sodium ions enter the ECF by crossing the digestive epithelium via:

        a. diffusion

        b. active transport

        c. diffusion and active transport

        d. facilitated diffusion

_____ 35. Deviations outside of the normal pH range due to changes in hydrogen ion concentrations:

        a. disrupt the stability of cell membranes

        b. alter protein structure

        c. change the activities of important enzymes

        d. a, b, and c are correct

_____ 36. When the P_{CO_2} increases and additional hydrogen ions and bicarbonate ions are released into the plasma, the:

        a. pH goes up; $\uparrow$ alkalinity

        b. pH goes down; $\uparrow$ acidity

        c. pH goes up; $\downarrow$ acidity

        d. pH is not affected

_____ 37. Important examples of organic acids found in the body are:

        a. sulfuric acid and phosphoric acid

        b. hydrochloric acid and carbonic acid

        c. lactic acid and ketone bodies

        d. a, b, and c are correct

Level
=2=

_____ 38. In a protein buffer system, if the pH increases, the carboxyl group (COOH) of the amino acid dissociates and releases:

      a. a hydroxyl ion

      b. a molecule of carbon monoxide

      c. a hydrogen ion

      d. a molecule of carbon dioxide

_____ 39. Normal pH values are limited to the range between:

      a. 7.35 to 7.45

      b. 6.8 to 7.0

      c. 6.0 to 8.0

      d. 6.35 to 8.35

_____ 40. Under normal circumstances, during respiratory acidosis the chemoreceptors monitoring the P_{CO_2} of the plasma and CSF will eliminate the problem by calling for:

      a. a decrease in the breathing rate

      b. a decrease in pulmonary ventilation rates

      c. an increase in pulmonary ventilation rates

      d. breathing into a small paper bag

_____ 41. When a normal pulmonary response does not reverse respiratory acidosis, the kidneys respond by:

      a. increasing the reabsorption of hydrogen ions

      b. increasing the rate of hydrogen ion secretion into the filtrate

      c. decreasing the rate of hydrogen ion secretion into the filtrate

      d. increased loss of bicarbonate ions

_____ 42. Chronic diarrhea causes a severe loss of bicarbonate ions resulting in:

      a. respiratory acidosis

      b. respiratory alkalosis

      c. metabolic acidosis

      d. metabolic alkalosis

_____ 43. Compensation for metabolic alkalosis involves:

      a. ↑ pulmonary ventilation; ↑ loss of bicarbonates in the urine

      b. ↑ pulmonary ventilation; ↓ loss of bicarbonates in the urine

      c. ↓ pulmonary ventilation; ↓ loss of bicarbonates in the urine

      d. ↓ pulmonary ventilation; ↑ loss of bicarbonates in the urine

[L2]   Completion:

Using the terms below, complete the following statements.

buffer system	organic acids	fixed acids
lactic acidosis	hypoventilation	angiotensin II
volatile acid	kidneys	respiratory acidosis
alkaline tide	ketoacidosis	hyperventilation

44. The most important sites of sodium ion regulation are the _____.

45. Renin release by kidney cells initiates a chain of events leading to the activation of _____.

46. An acid that can leave solution and enter the atmosphere is referred to as a _____.

47. Acids that remain in body fluids until excreted at the kidneys are called _____.

48. Acid participants in or byproducts of cellular metabolism are referred to as _____.

49. A combination of a weak acid and its dissociation products comprise a _____.

50. When the respiratory system is unable to eliminate normal amounts of CO_2 generated by peripheral tissues, the result is the development of _____.

51. The usual cause of respiratory acidosis is _____.

52. Physical or psychological stresses or conscious effort may produce an increased respiratory rate referred to as _____.

53. Severe exercise or prolonged oxygen starvation of cells may develop into _____.

54. Generation of large quantities of ketone bodies during the postabsorptive state results in a condition known as _____.

55. An influx of large numbers of bicarbonate ions into the ECF due to secretion of HCl by the gastric mucosa is known as the _____.

[L2]   Short Essay:

Briefly answer the following questions in the spaces provided below.

56. What three (3) different, interrelated types of homeostasis are involved in the maintenance of normal volume and composition in the ECF and the ICF?

57. What three (3) primary hormones mediate physiological adjustments that affect fluid and electrolyte balance?

58. What two (2) major effects does ADH have on maintaining homeostatic volumes of water in the body?

Level
=2=

59. What three (3) adjustments control the rate of tubular secretion of potassium ions along the DCT of the nephron?

60. What two (2) primary steps are involved in the regulation of sodium ion concentrations?

61. Write the chemical equation to show how CO_2 interacts with H_2O in solution to form molecules of carbonic acid. Continue the equation to show the dissociation of carbonic acid molecules to produce hydrogen ions and bicarbonate ions.

62. (a) What three (3) *chemical* buffer systems represent the first line of defense against pH shift?

    (b) What two (2) *physiological* buffers represent the second line of defense against pH shift?

63. How do pulmonary and renal mechanisms support the chemical buffer systems?

64. What is the difference between hypercapnia and hypocapnia?

65. What are the three (3) major causes of metabolic acidosis?

When you have successfully completed the exercises in L2 proceed to L3.

☐ LEVEL 3   CRITICAL THINKING/APPLICATION

Using principles and concepts learned in Chapter 27, answer the following questions. Write your answers on a separate sheet of paper. Some of the questions in this section will require the following information for your reference.

> *Normal* arterial blood gas values:
>
> pH: 7.35–7.45
> Pco_2:   35 to 45 mm Hg
> Po_2:   80 to 100 mm Hg
> HCO_3^-:   22 to 26 mEq/liter

1. A comatose teenager is taken to the nearby hospital emergency room by the local rescue squad. His friends reported that he had taken some drug with a large quantity of alcohol. His arterial blood gas (ABG) studies reveal: pH 7.17; Pco_2 73 mm Hg; HCO_3^- 26 mEq/liter.

   Identify the teenager's condition and explain what the clinical values reveal.

2. A 62-year-old woman has been vomiting and experiencing anorexia for several days. After being admitted to the hospital, her ABG studies are reported as follows: pH 7.65; Pco_2 52 mm Hg; HCO_3^- 55 mEq/liter.

   Identify the woman's condition and explain what the clinical values reveal.

3. After analyzing the ABG values below, identify the condition in each one of the following four cases.

   (a) pH 7.30; Pco_2 37 mm Hg; HCO_3^- 16 mEq/liter
   (b) pH 7.52; Pco_2 32 mm Hg; HCO_3^- 25 mEq/liter
   (c) pH 7.36; Pco_2 67 mm Hg; HCO_3^- 23 mEq/liter
   (d) pH 7.58; Pco_2 43 mm Hg; HCO_3^- 42 mEq/liter

4. Tom S. has just completed his first marathon. For a few days following the event, he experiences symptoms related to dehydration due to excessive fluid and salt losses and a decrease in his ECF volume. What physiological processes must take place to restore his body to normal ECF and ICF values? (Use arrows to indicate the progression of processes that take place to restore homeostasis.)

CHAPTER

28

The Reproductive System

■ Overview

The structures and functions of the reproductive system are notably different from any other organ system in the human body. The other systems of the body are functional at birth or shortly thereafter; however, the reproductive system does not become functional until it is acted on by hormones during puberty.

Most of the other body systems function to support and maintain the individual, but the reproductive system is specialized to ensure survival, not of the individual but of the species.

Even though major differences exist between the reproductive organs of the male and female, both are primarily concerned with propagation of the species and passing genetic material from one generation to another. In addition, the reproductive system produces hormones that allow for the development of secondary sex characteristics.

This chapter provides a series of exercises that will assist you in reviewing and reinforcing your understanding of the anatomy and physiology of the male and female reproductive systems, the effects of male and female hormones, and changes that occur during the aging process.

❏ LEVEL 1   REVIEW OF CHAPTER OBJECTIVES

1. Summarize the functions of the human reproductive system and its principal components.
2. Describe the components of the male reproductive system.
3. Detail the process of meiosis, spermatogenesis, and spermiogenesis.
4. Describe the roles the male reproductive tract and the accessory glands play in the functional maturation, nourishment, storage and transport of spermatozoa.
5. Discuss the normal composition of semen.
6. Describe the male external genitalia.
7. Describe the hormonal mechanisms that regulate male reproductive functions.
8. Describe the components of the female reproductive system.
9. Detail the processes of meiosis and oogenesis in the ovary.
10. Define the phases and event of the ovarian and uterine cycles.
11. Describe the structure, histology, and functions of the vagina.

12. Name and describe the parts of the female external genitalia and mammary glands.
13. Describe the anatomical physiological, and hormonal aspects of the female reproductive cycle.
14. Discuss the physiology of sexual intercourse as it affects the reproductive systems of males and females.
15. Describe the changes in the reproductive system that occur at puberty and with aging.

[L1]   Multiple Choice:

Place the letter corresponding to the correct answer in the space provided.

OBJ. 1   _____   1. The systems involved in an adequate sperm count, correct pH and nutrients, and erection and ejaculation are:

   a. reproductive and digestive
   b. endocrine and nervous
   c. cardiovascular and urinary
   d. a, b, and c are correct

OBJ. 1   _____   2. The reproductive organs that produce gametes and hormones are the:

   a. accessory glands
   b. gonads
   c. vagina and penis
   d. a, b, and c are correct

OBJ. 2   _____   3. In the male the important function(s) of the epididymis is (are):

   a. monitors and adjusts the composition of the tubular fluid
   b. acts as a recycling center for damaged spermatozoa
   c. the site of physical maturation of spermatozoa
   d. a, b, and c are correct

OBJ. 2   _____   4. Beginning inferior to the urinary bladder, sperm travel to the exterior through the urethral regions that include the:

   a. membranous urethra, prostatic urethra, penile urethra
   b. prostatic urethra, penile urethra, membranous urethra
   c. prostatic urethra, membranous urethra, penile urethra
   d. penile urethra, prostatic urethra, membranous urethra

OBJ. 3   _____   5. In the process of spermatogenesis, the developmental sequence includes:

   a. spermatids, spermatozoon, spermatogonia, spermatocytes
   b. spermatogonia, spermatocytes, spermatids, spermatozoon
   c. spermatocytes, spermatogonia, spermatids, spermatozoon
   d. spermatogonia, spermatids, spermatocytes, spermatozoon

OBJ. 3 _____ 6. An individual spermatozoan completes its development in the seminiferous tubules and its physical maturation in the epididymis in approximately:

    a. 2 weeks

    b. 3 weeks

    c. 5 weeks

    d. 8 weeks

OBJ. 4 _____ 7. In the male, sperm cells, before leaving the body, travel from the testes to the:

    a. ductus deferens → epididymis → urethra → ejaculatory duct

    b. ejaculatory duct → epididymis → ductus deferens → urethra

    c. epididymis → ductus deferens → ejaculatory duct → urethra

    d. epididymis → ejaculatory duct → ductus deferens → urethra

OBJ. 4 _____ 8. The accessory organs in the male that secrete into the ejaculatory ducts and the urethra are:

    a. epididymis, seminal vescicles, vas deferens

    b. prostate gland, inguinal canals, raphe

    c. adrenal glands, bulbourethral glands, seminal glands

    d. seminal vesicles, prostate gland, bulbourethral glands

OBJ. 5 _____ 9. Semen, the volume of fluid called the ejaculate, contains:

    a. spermatozoa, seminalplasmin, and enzymes

    b. alkaline and acid secretions and sperm

    c. mucus, sperm, and enzymes

    d. spermatozoa, seminal fluid, and enzymes

OBJ. 5 _____ 10. The correct average characteristics and composition of semen include:

    a. vol., 3.4 ml; specific gravity, 1.028; pH, 7.19; sperm count, 20-600 million/ml

    b. vol., 1.2 ml; specific gravity, 0.050; pH, 7.50; sperm count, 10-100 million/ml

    c. vol., 4.5 ml; specific gravity, 2.010; pH, 6.90; sperm count, 5-10 million/ml

    d. vol., 0.5 ml; specific gravity, 3.010; pH, 7.0; sperm count, 2-4 million/ml

OBJ. 6 _____ 11. The external genitalia of the male includes the:

    a. scrotum and penis

    b. urethra and bulbourethral glands

    c. raphe and dartos

    d. prepuce and glans

Level
-1-

OBJ. 6   _____ 12. The three masses of erectile tissue that comprise the body of the penis are:

    a. two cylindrical corpora cavernosa and a slender corpus spongiosum

    b. two slender corpora spongiosa and a cylindrical corpus cavernosum

    c. preputial glands, a corpus cavernosum, and a corpus spongiosum

    d. two corpora cavernosa and a preputial gland

OBJ. 7   _____ 13. The hormone synthesized in the hypothalamus that initiates release of pituitary hormones is:

    a. FSH (follicle-stimulating hormone)

    b. ICSH (interstitial cell-stimulating hormone)

    c. LH (lutenizing hormone)

    d. GnRH (gonadotropin-releasing hormone)

OBJ. 7   _____ 14. The hormone that promotes spermatogenesis along the seminiferous tubules is:

    a. ICSH

    b. FSH

    c. GnRH

    d. LH

OBJ. 8   _____ 15. The function of the uterus in the female is to:

    a. ciliate the sperm into the uterine tube for possible fertilization

    b. provide ovum transport via peristaltic contractions of the uterine wall

    c. provide mechanical protection and nutritional support to the developing embryo

    d. encounter spermatozoa during the first 12–24 hours of its passage

OBJ. 8   _____ 16. In the female, after leaving the ovaries, the ovum travels in the uterine tubes to the uterus via:

    a. ampulla → infundibulum → isthmus → intramural portion

    b. infundibulum → ampulla → isthmus → intramural portion

    c. isthmus → ampulla → infundibulum → intramural portion

    d. intramural portion → ampulla → isthmus → infundibulum

OBJ. 8   _____ 17. Starting at the superior end, the uterus in the female is divided into:

    a. body, isthmus, cervix

    b. isthmus, body, cervix

    c. body, cervix, isthmus

    d. cervix, body, isthmus

OBJ. 8   _____ 18. Ovum transport in the uterine tubes presumably involves a combination of:

      a. flagellar locomotion and ciliary movement
      b. active transport and ciliary movement
      c. ciliary movement and peristaltic contractions
      d. movement in uterine fluid and flagellar locomotion

OBJ. 9   _____ 19. The ovarian cycle begins as activated follicles develop into:

      a. ova
      b. primary follicles
      c. Graafian follicles
      d. primordial follicles

OBJ. 9   _____ 20. The process of oogenesis produces three non-functional polar bodies that eventually disintegrate and:

      a. a primordial follicle
      b. a granulosa cell
      c. one functional ovum
      d. a zona pellucida

OBJ. 10   _____ 21. The proper sequence that describes the ovarian cycle involves the formation of:

      a. primary follicles, secondary follicles, tertiary follicles, ovulation, and formation and destruction of the corpus luteum
      b. primary follicles, secondary follicles, tertiary follicles, corpus luteum, and ovulation
      c. corpus luteum; primary, secondary, and tertiary follicles; and ovulation
      d. primary and tertiary follicles, secondary follicles, ovulation, and formation and destruction of the corpus luteum

OBJ. 10   _____ 22. Under normal circumstances, in a 28-day cycle, ovulation occurs on _____, and the menses begins on _____.

      a. day 1; day 14
      b. day 28; day 14
      c. day 14; day 1
      d. day 6; day 14

OBJ. 11   _____ 23. Engorgement of erectile tissues of the clitoris and increased secretion of the greater vestibular glands during arousal is caused by:

      a. erotic sights and sounds
      b. a tall, dark, handsome man
      c. sympathetic activation
      d. parasympathetic activation

OBJ. 11   _____ 24. The reproductive function(s) of the vagina is (are):

      a. passageway for the elimination of menstrual fluids
      b. receives the penis during coitus
      c. forms the lower portion of the birth canal during childbirth
      d. a, b, and c are correct

Level
—1—

OBJ. 12 _____ 25. The outer limits of the vulva are established by the:

a. vestibule and the labia minora

b. monis pubis and labia majora

c. lesser and greater vestibular glands

d. prepuce and vestibule

OBJ. 12 _____ 26. Engorgement of the erectile tissues of the clitoris and increased secretion of the greater vestibular glands involve neural activity that includes:

a. somatic motor neurons

b. sympathetic activation

c. parasympathetic activation

d. a, b, and c are correct

OBJ. 13 _____ 27. Whether or not fertilization occurs the final destination of the ovum is the:

a. myometrium

b. endometrium

c. serosa

d. placenta

OBJ. 13 _____ 28. Hormones produced by the placenta include:

a. estrogen and progesterone

b. relaxin and human placental lactogen

c. human chorionic gonadotrophin (HCG)

d. a, b, and c are correct

OBJ. 13 _____ 29. The two hormones that help prepare the mammary glands for milk production are:

a. prolactin and human placental lactogen

b. colostrum and relaxin

c. human chorionic gonadotrophin and estrogen

d. estrogen and progesterone

OBJ. 14 _____ 30. Peristaltic contractions of the ampulla, pushing fluid and spermatozoa into the prostatic urethra, is called:

a. emission

b. ejaculation

c. detumescence

d. subsidence

OBJ. 15 _____ 31. Menopause is accompanied by a sharp and sustained rise in the production of _____ while circulating concentrations of _____ decline.

a. estrogen and progesterone; GnRH, FSH, LH

b. GnRH, FSH, LH; estrogen and progesterone

c. LH, estrogen; progesterone, GnRH, and FSH

d. FSH, LH, progesterone; estrogen and GnRH

Level
-1-

OBJ. 15   _____ 32. In the male, between the ages of 50 and 60 circulating _____ levels begin to decline, coupled with increases in circulating levels of _____.

       a. FSH and LH; testosterone

       b. FSH; testosterone and LH

       c. testosterone; FSH and LH

       d. a, b, and c are correct

[L1]   Completion:

Using the terms below, complete the following statements.

menopause	infundibulum	climacteric
ICSH	ovaries	fertilization
testes	placenta	implantation
digestive	ovulation	cardiovascular
oogenesis	seminiferous tubules	progesterone
ductus deferens	Bartholin's	Graafian
fructose	spermiogenesis	pudendum
areola		

OBJ. 1   33. The fusion of a sperm contributed by the father and an egg, or ovum, from the mother is called _____.

OBJ. 2   34. Sperm production occurs within the slender, tightly coiled _____.

OBJ. 2   35. The male gonads are the _____.

OBJ. 3   36. The physical transformation of a spermatid to a spermatozoon is called _____.

OBJ. 4   37. After passing along the tail of the epididymis the spermatozoa arrive at the _____.

OBJ. 4   38. The major systems involved in providing nutrients and the proper pH for semen are the urinary and _____ systems.

OBJ. 5   39. The primary energy source for the mobilization of sperm is _____.

OBJ. 6   40. The external genitalia in the male are collectively referred to as the _____.

OBJ. 7   41. A peptide hormone that causes the secretion of androgens by the interstitial cells of the testes is _____.

OBJ. 8   42. The female gonads are the _____.

OBJ. 9   43. By the tenth day of the ovarian cycle, the mature tertiary follicle is called the _____ follicle.

OBJ. 10   44. Egg release from the ovary into the uterine tube is called _____.

OBJ. 10   45. After ovulation the ovum passes from the ovary into the expanded funnel called the _____.

OBJ. 10   46. When the blastocyst contacts the endometrial wall, erodes the epithelium, and buries itself in the endometrium, the process is known as _____.

OBJ. 10   47. Ovum production, which occurs on a monthly basis as part of the ovarian cycle, is referred to as _____.

Level
–1–

OBJ. 11   48. The mucous glands that discharge secretions into the vestibule near the vaginal entrance are known as _____ glands.

OBJ. 12   49. The reddish-brown coloration surrounding the nipples of the breasts is referred to as the _____.

OBJ. 13   50. The principal hormone that prepares the uterus for pregnancy is _____.

OBJ. 13   51. Support for embryonic and fetal development occurs in a special organ called the _____.

OBJ. 14   52. The major systems involved in the process of erection are the nervous and _____.

OBJ. 15   53. In females, the time that ovulation and menstruation cease is referred to as _____.

OBJ. 15   54. Changes in the male reproductive system that occur gradually over a period of time are known as the male _____.

[L1]   Matching:

Match the terms in column B with the terms in column A. Use letters for answers in the spaces provided.

PART I   **COLUMN A**   **COLUMN B**

OBJ. 1   _____ 55. gametes   A. sperm production
OBJ. 1   _____ 56. gonadotropin-releasing   B. produce alkaline secretion
                    hormone   C. anterior pituitary
OBJ. 1   _____ 57. FSH, LH, ICSH   D. hormone—male secondary
OBJ. 2   _____ 58. sustentacular cells   sex characteristics
OBJ. 2   _____ 59. interstitial cells   E. perineal structures
OBJ. 3   _____ 60. seminiferous tubules   F. cells of Leydig
OBJ. 3   _____ 61. puberty in male   G. hypothalamus
OBJ. 4   _____ 62. prostate glands   H. sertoli cells
OBJ. 5   _____ 63. seminal plasmin   I. fallopian tubes
OBJ. 6   _____ 64. external genitalia   J. spermatogenesis
OBJ. 7   _____ 65. testosterone   K. reproductive cells
OBJ. 8   _____ 66. oviducts   L. antibiotic enzyme in
                                                 semen

PART II   **COLUMN A**   **COLUMN B**

OBJ. 9   _____ 67. ovaries   M. follicular degeneration
OBJ. 9   _____ 68. puberty in female   N. hormone—postovulatory
OBJ. 9   _____ 69. immature eggs   period
OBJ. 10 _____ 70. atresia   O. indicates pregnancy
OBJ. 11 _____ 71. vaginal folds   P. egg production
OBJ. 12 _____ 72. clitoris   Q. milk ejection
OBJ. 13 _____ 73. human chorionic   R. menarche
                    hormone   S. female equivalent of penis
OBJ. 13 _____ 74. oxytocin   T. aygote
OBJ. 13 _____ 75. progesterone   U. rugae
OBJ. 14 _____ 76. fertilized egg   V. cessation of ovulation
OBJ. 15 _____ 77. menopause   and menstruation

                                              W. oocytes

Level
—1—

[L1]   Drawing/Illustration Labeling:

Identify each numbered structure by labeling the following figures:

OBJ. 2   **_FIGURE 28.1_**   Male Reproductive Organs

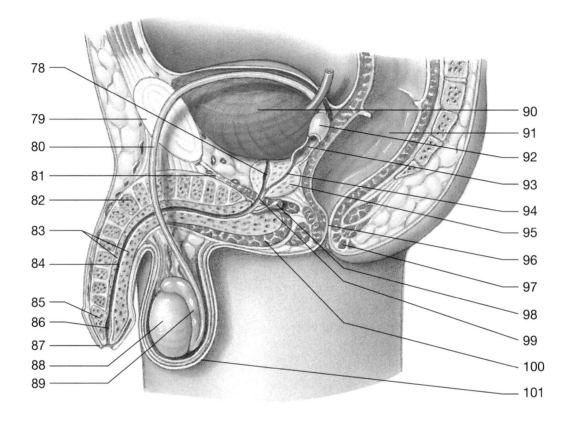

78 _____		90 _____
79 _____		91 _____
80 _____		92 _____
81 _____		93 _____
82 _____		94 _____
83 _____		95 _____
84 _____		96 _____
85 _____		97 _____
86 _____		98 _____
87 _____		99 _____
88 _____		100 _____
89 _____		101 _____

Level
-1-

OBJ. 2 **FIGURE 28.2** The Testes

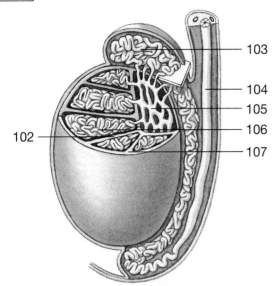

102 _____

103 _____

104 _____

105 _____

106 _____

107 _____

OBJ. 12 **FIGURE 28.3** Female External Genitalia

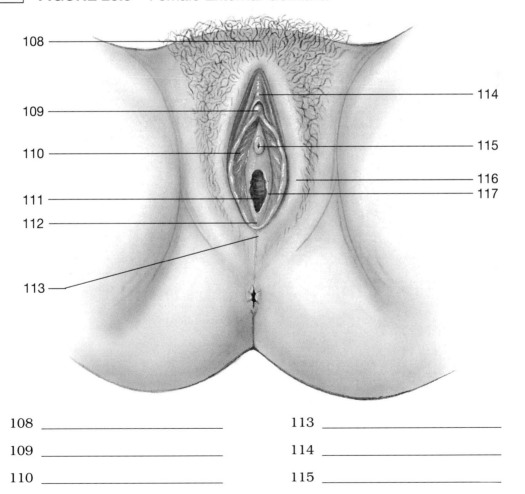

108 _____   113 _____

109 _____   114 _____

110 _____   115 _____

111 _____   116 _____

112 _____   117 _____

OBJ. 8   *FIGURE 28.4*   Female Reproductive Organs (sagittal section)

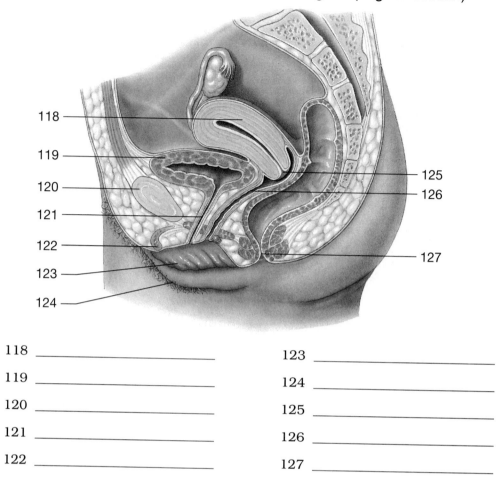

118 _____	123 _____
119 _____	124 _____
120 _____	125 _____
121 _____	126 _____
122 _____	127 _____

OBJ. 8   *FIGURE 28.5*   Female Reproductive Organs (frontal section)

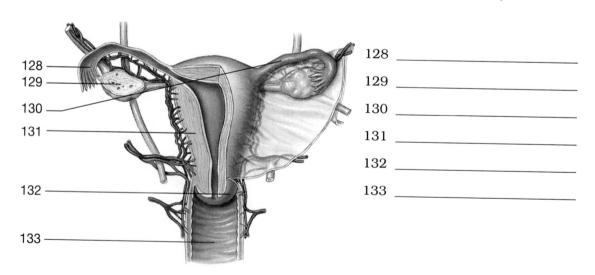

128 _____

129 _____

130 _____

131 _____

132 _____

133 _____

When you have successfully completed the exercises in L1 proceed to L2.

☐ LEVEL 2 CONCEPT SYNTHESIS

Concept Map I:

Using the following terms, fill in the circled, numbered, blank spaces to complete the concept map. Follow the numbers that comply with the organization of the map.

Urethra   Seminiferous tubules   Penis

Produce testosterone   FSH   Seminal vesicles

Ductus deferens   Bulbourethral glands

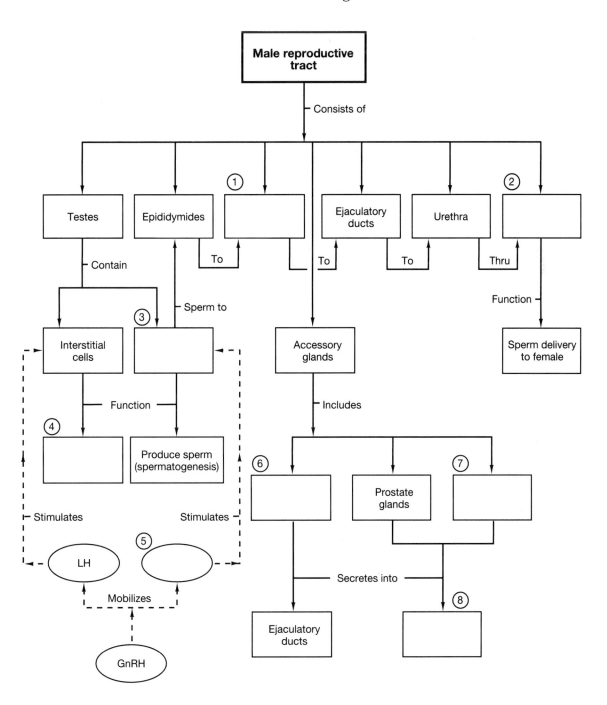

Concept Map II:

Using the following terms, fill in the circled, numbered, blank spaces to complete the concept map. Follow the numbers that comply with the organization of the map.

External urinary meatus   Shaft   Crus
Corpora spongiosum   Prepuce   Frenulum

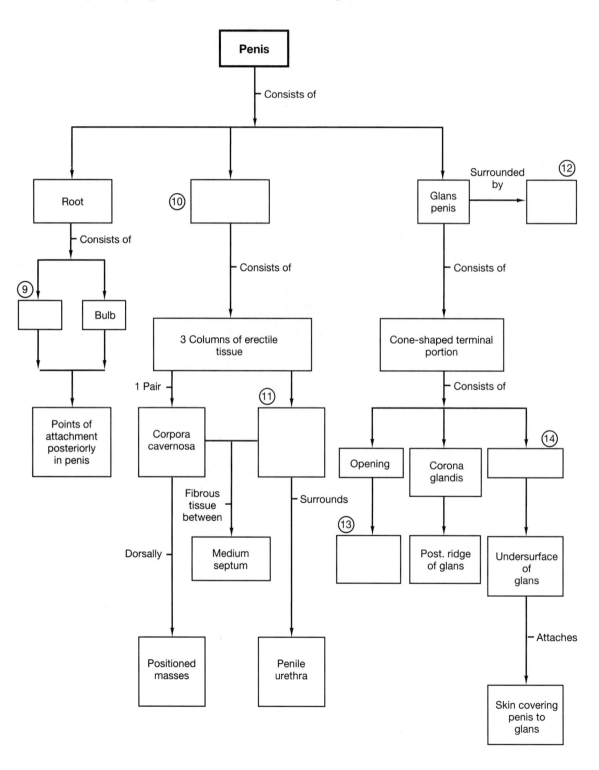

Concept Map III:

Using the following terms, fill in the circled, numbered, blank spaces to complete the concept map. Follow the numbers that comply with the organization of the map.

Inhibin   Testes   Male secondary sex characteristics
FSH   CNS   Anterior pituitary
Interstitial cells

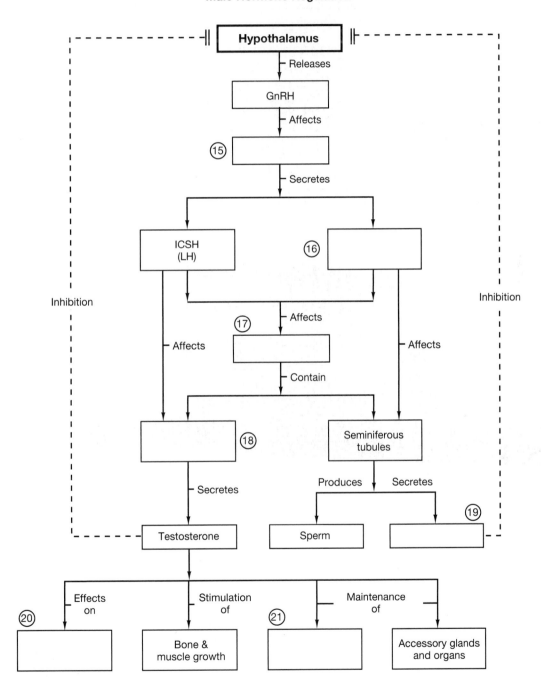

Male Hormone Regulation

Concept Map IV:

Using the following terms, fill in the circled, numbered, blank spaces to complete the concept map. Follow the numbers that comply with the organization of the map.

nutrients   supports fetal development   endometrium

vulva   granulosa and thecal cells   clitoris

vagina   uterine tubes   labis majora and minora

follicles

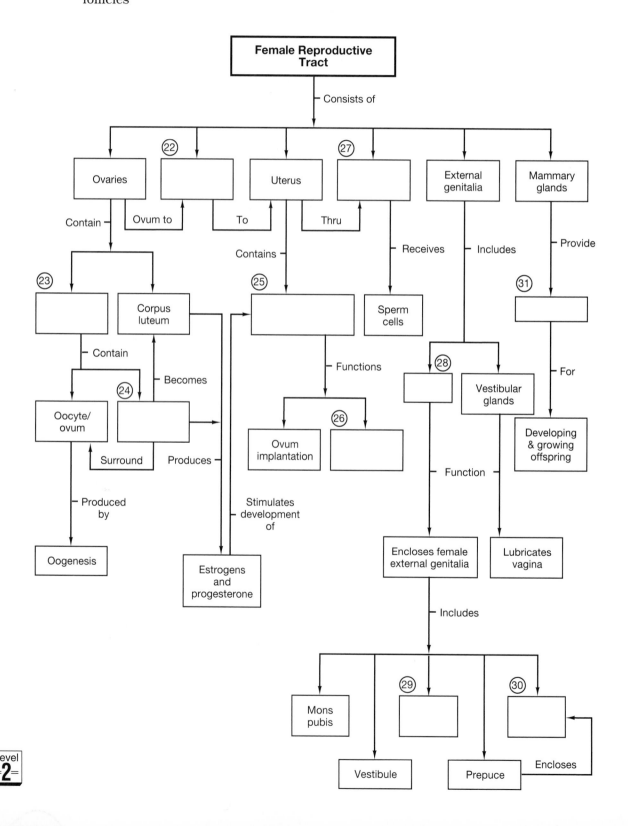

Concept Map V:

Using the following terms, fill in the circled, numbered, blank spaces to complete the concept map. Follow the numbers that comply with the organization of the map.

Bone and muscle growth Follicles GnRH
Accessory glands and organs Progesterone LH

Female Hormone Regulation

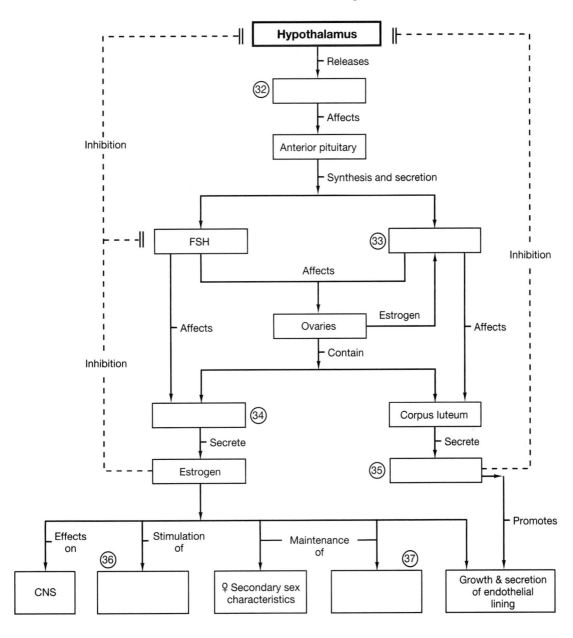

Concept Map VI:

Using the following terms, fill in the circled, numbered, blank spaces to complete the concept map. Follow the numbers that comply with the organization of the map.

Placenta   Ovary   Anterior pituitary
Relaxin   Progesterone   Mammary gland development

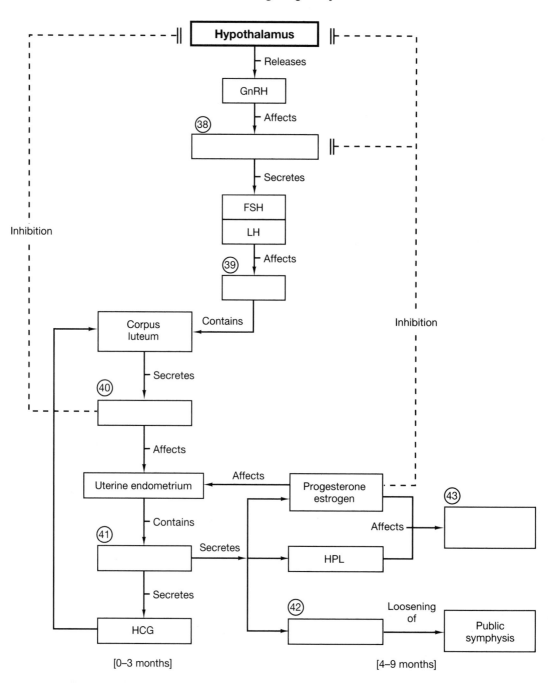

Hormones During Pregnancy

Body Trek:

Using the terms below, identify the numbered locations on the trek map through the reproductive system of the male.

penile urethra

ductus deferens

body of epididymis

external urethral meatus

seminiferous tubules and
    rete testis

ejaculatory duct

urethra

For the trek through the male reproductive system you will follow the path Robo takes as the micro-robot follows a population of sperm from the time they are formed in the testes until they leave the body of the male. You will assist the robot by identifying the numbered locations along the route and recording them in the spaces provided below. If you need assistance from Robo, refer to the Answer Key in the back of the Study Guide.

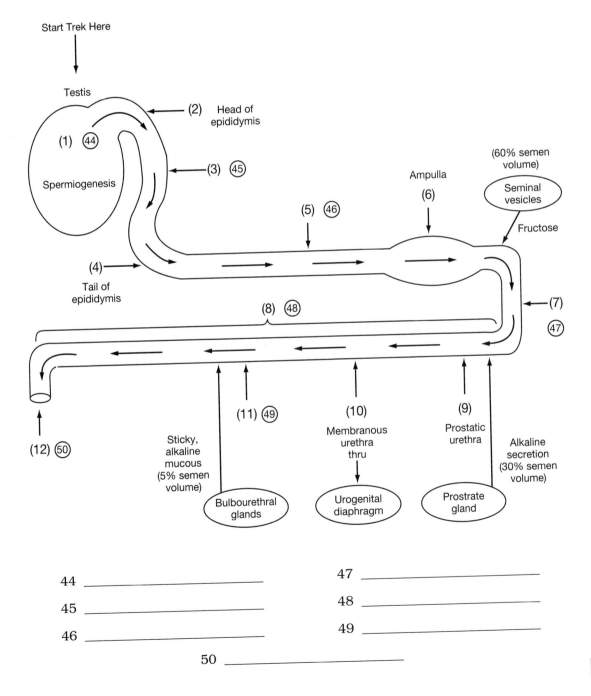

44 _____

45 _____

46 _____

47 _____

48 _____

49 _____

50 _____

[L2] Multiple Choice:

Place the letter corresponding to the correct answer in the space provided.

_____ 51. The part of the endometrium that undergoes cyclical changes in response to sexual hormone levels is the:

        a. functional zone

        b. basilar zone

        c. serosa

        d. muscular myometrium

_____ 52. In a 28-day cycle, estrogen levels peak at:

        a. day 1

        b. day 7

        c. day 14

        d. day 28

_____ 53. The rupture of the follicular wall and ovulation is caused by:

        a. a sudden surge in LH (luteinizing hormone) concentration

        b. a sudden surge in the secretion of estrogen

        c. an increase in the production of progesterone

        d. increased production and release of GnRH (gonadotrophin-releasing hormone)

_____ 54. The body of the spermatic cord is a structure that includes:

        a. epididymis, ductus deferens, blood vessels, and nerves

        b. vas deferens, prostate gland, blood vessels, and urethra

        c. ductus deferens, blood vessels, nerves, and lymphatics

        d. a, b, and c are correct

_____ 55. The function(s) of the sustentacular cells (Sertoli cells) in the male is (are):

        a. maintenance of the blood-testis barrier

        b. support of spermiogenesis

        c. secretion of inhibin and androgen-binding protein

        d. a, b, and c are correct

_____ 56. The tail of the sperm has the unique distinction of being the:

        a. only flagellum in the body that contains chromosomes

        b. only flagellum in the human body

        c. only flagellum in the body that contains mitochondria

        d. only flagellum in the body that contains centrioles

_____ 57. The function(s) of the prostate gland is (are):

        a. to produce acid and alkaline secretions

        b. to produce alkaline secretions and a compound with antibiotic properties

        c. to secrete a thick, sticky, alkaline mucus for lubrication

        d. to secrete seminal fluid with a distinctive ionic and nutrient composition

_____ 58. Powerful, rhythmic contractions in the ischiocavernosus and bulbocavernosus muscles of the pelvic floor result in:

     a. erection

     b. emission

     c. ejaculation

     d. a, b, and c are correct

_____ 59. The process of erection involves complex neural processes that include:

     a. increased sympathetic outflow over the pelvic nerves

     b. increased parasympathetic outflow over the pelvic nerves

     c. decreased parasympathetic outflow over the pelvic nerves

     d. somatic motor neurons in the upper sacral segments of the spinal cord

_____ 60. Impotence, a common male sexual dysfunction, is:

     a. the inability to produce sufficient sperm for fertilization

     b. the term used to describe male infertility

     c. the inability of the male to ejaculate

     d. the inability to achieve or maintain an erection

_____ 61. The ovaries, uterine tubes, and uterus are enclosed within an extensive mesentery known as the:

     a. rectouterine pouch

     b. suspensory ligament

     c. ovarian ligament

     d. broad ligament

_____ 62. If fertilization is to occur, the ovum must encounter spermatozoa during the first:

     a. 2 – 4 hours of its passage

     b. 6 – 8 hours of its passage

     c. 12 – 24 hours of its passage

     d. 30 – 36 hours of its passage

_____ 63. The three pairs of suspensory ligaments that stabilize the position of the uterus and limit its range of movement are:

     a. broad, ovarian, suspensory

     b. uterosacral, round, lateral

     c. anteflexure, retroflexure, tunica albuginea

     d. endometrium, myometrium, serosa

_____ 64. The histological composition of the uterine wall consists of the:

     a. body, isthmus, cervix

     b. mucosa, epithelial lining, lamina propria

     c. mucosa, functional zone, basilar layer

     d. endometrium, myometrium, serosa

Level
=2=

_____ 65. The hormone that acts to reduce the rate of GnRH and FSH production by the anterior pituitary is:

        a. inhibin

        b. LH

        c. ICSH

        d. testosterone

_____ 66. The three sequential stages of the menstrual cycle include:

        a. menses, luteal phase, postovulatory phase

        b. menses, follicular phase, preovulatory phase

        c. menses, proliferative phase, follicular phase

        d. menses, proliferative phase, secretory phase

_____ 67. At the time of ovulation, basal body temperature:

        a. declines sharply

        b. increases sharply

        c. stays the same

        d. is one-half a degree Fahrenheit lower than normal

_____ 68. The seminal vesicles:

        a. store semen

        b. secrete a fructose-rich mucoid substance

        c. conduct spermatozoa into the epididymis

        d. secrete a thin watery fluid

_____ 69. A recently marketed drug is said to "remind the pituitary" to produce gonadotropins, FSH, and LH. This drug might be useful as:

        a. a male contraceptive

        b. a female contraceptive

        c. both a male and female contraceptive

        d. a fertility drug

[L2] Completion:

Using the terms below, complete the following statements.

cervical os	ampulla	inguinal canals
detumescence	androgens	smegma
raphe	prepuce	fimbriae
hymen	tunica albuginea	rete testis
zona pellucida	ejaculation	corona radiata
clitoris	acrosomal sac	corpus luteum
cremaster	corpus albicans	menopause
menses	mesovarium	fornix

70. The interstitial cells are responsible for the production of male sex hormones, which are called _____.

71. The narrow canals linking the scrotal chambers with the peritoneal cavity are called the _____.

72. The scrotum is divided into two separate chambers, and the boundary between the two is marked by a raised thickening in the scrotal surface known as the _____.

73. The layer of skeletal muscle that contracts and tenses the scrotum, pulling the testes closer to the body, is the _____ muscle.

74. Semen is expelled from the body by a process called _____.

75. The maze of interconnected passageways within the mediastinum of the seminiferous tubules is known as the _____.

76. The tip of the sperm containing an enzyme that plays a role in fertilization is the _____.

77. The fold of skin that surrounds the tip of the penis is the _____.

78. Preputial glands in the skin of the neck and the inner surface of the prepuce secrete a waxy material known as _____.

79. Subsidence of erection mediated by the sympathetic nervous system is referred to as _____.

80. The last menstrual cycle of the female is known as _____.

81. Microvilli are present in the *space* between the developing oocyte and the innermost follicular cells called the _____.

82. Follicular cells surrounding the oocyte prior to ovulation are known as the _____.

83. Degenerated follicular cells proliferate to create an endocrine structure known as the _____.

84. A knot of pale scar tissue produced by fibroblasts invading a degenerated corpus luteum is called a _____.

85. The thickened fold of mesentery that supports and stabilizes the position of each ovary is the _____.

86. The thickened mesothelium that overlies a layer of dense connective tissue covering the exposed surfaces of each ovary is the _____.

87. The fingerlike projections on the infundibulum that extend into the pelvic cavity are called _____.

88. The uterine cavity opens into the vagina at the _____.

89. The shallow recess surrounding the cervical protrusion is known as the _____.

90. Prior to sexual activity, the thin epithelial fold that partially or completely blocks the entrance to the vagina is the _____.

91. The female equivalent of the penis derived from the same embryonic structure is the _____.

92. The period marked by the wholesale destruction of the functional zone of the endometrium is the _____.

93. Just before it reaches the prostate and seminal vesicles, the ductus deferens becomes enlarged, and the expanded portion is known as the _____.

[L2]   Short Essay:

Briefly answer the following questions in the spaces provided below.

94. What are the four (4) important functions of the sustentacular cells (Sertoli cells) in the testes?

95. What are the three (3) important functions of the epididymis?

Level
=2=

96. (a) What three (3) glands secrete their products into the male reproductive tract?

   (b) What are the four (4) primary functions of these glands?

97. What is the difference between seminal fluid and semen?

98. What is the difference between emission and ejaculation?

99. What are the five (5) primary functions of testosterone in the male?

100. What are the three (3) reproductive functions of the vagina?

101. What are the three (3) phases of female sexual function and what occurs in each phase?

102. What are the five (5) steps involved in the ovarian cycle?

103. What are the five (5) primary functions of the estrogens?

104. What are the three (3) stages of the menstrual cycle?

105. What hormones are secreted by the placenta?

106. What is colostrum and what are its major contributions to the infant?

When you have successfully completed the exercises in L2 proceed to L3.

☐ LEVEL 3   CRITICAL THINKING/APPLICATION

Using principles and concepts learned about the reproductive system, answer the following questions. Write your answers on a separate sheet of paper.

1. I. M. Hurt was struck in the abdomen with a baseball bat while playing in a young men's baseball league. As a result, his testes ascend into the abdominopelvic region quite frequently, causing sharp pains. He has been informed by his urologist that he is sterile due to his unfortunate accident.
   (a) Why does his condition cause sterility?
   (b) What primary factors are necessary for fertility in males?
2. A 19-year-old female has been accused by her brothers of looking and acting like a male. They make fun of her because she has a low-pitched voice, growth of hair on her face, and they tease her about her small breast size. Physiologically speaking, what is her problem related to?
3. A contraceptive pill "tricks the brain" into thinking you are pregnant. What does this mean?
4. Sexually transmitted diseases in males do not result in inflammation of the peritoneum (peritonitis) as they sometimes do in females. Why?

29

Development and Inheritance

■ Overview

It is difficult to imagine that today there is a single cell that 38 weeks from now, if fertilized, will develop into a complex organism with over 200 million cells organized into tissues, organs, and organ systems — so it is the miracle of life!

The previous chapter considered the male and female reproductive tracts that lead to the bridge of life that spans the generations through which gametes are transported and by which both gametes and developing offspring are housed and serviced.

A new life begins in the tubes of these systems, and it is here that new genetic combinations, similar to the parents, yet different, are made and nourished until they emerge from the female tract to take up life on their own — at first highly dependent on extrinsic support but growing independent with physical maturation.

All of these events occur as a result of the complex, unified process of development — an orderly sequence of progressive changes that begin at fertilization and have profound effects on the individual for a lifetime.

Chapter 29 highlights the major aspects of development and development processes, regulatory mechanisms, and how developmental patterns can be modified for the good or ill of the individual. Appropriately, the chapter concludes by addressing the topic of death and dying.

☐ LEVEL 1   REVIEW OF CHAPTER OBJECTIVES

1. Describe the process of fertilization.
2. Explain how developmental processes are regulated.
3. List the three prenatal periods and describe the major events associated with each period.
4. Explain how the germ layers participate in the formation of extraembryonic membranes.
5. Discuss the importance of the placenta as an endocrine organ.
6. Describe the interplay between the maternal organ systems and the developing fetus.
7. Discuss the structural and functional changes in the uterus during gestation.
8. Describe the events that occur during labor and delivery.
9. Identify the features and functions associated with the various life stages.
10. Relate basic principles of genetics to the inheritance of human traits.

[L1]   Multiple Choice:

Place the letter corresponding to the correct answer in the space provided.

OBJ. 1 _____ 1. The normal male genotype is _____ , and the normal female genotype is _____ .

    a. XX; XY

    b. X; Y

    c. XY; XX

    d. Y; X

OBJ. 1 _____ 2. Normal fertilization occurs in the:

    a. lower part of the uterine tube

    b. upper one-third of the uterine tube

    c. upper part of the uterus

    d. antrum of a tertiary follicle

OBJ. 1 _____ 3. Sterility in males may result from a sperm count of less than:

    a. 20 million sperm/ml

    b. 40 million sperm/ml

    c. 60 million sperm/ml

    d. 100 million sperm/ml

OBJ. 1 _____ 4. Fertilization is completed with the:

    a. formation of a gamete containing 23 chromosomes

    b. formation of the male and female pronuclei

    c. completion of the meiotic process

    d. formation of a zygote containing 46 chromosomes

OBJ. 2 _____ 5. Alterations in genetic activity during development occur as a result of:

    a. the maturation of the sperm and ovum

    b. the chromosome complement in the nucleus of the cell

    c. differences in the cytoplasmic composition of individual cells

    d. conception

OBJ. 2 _____ 6. As development proceeds, the differentiation of other embryonic cells is affected by small zygotic cells that:

    a. produce the initial stages of implantation

    b. release RNAs, polypeptides, and small proteins

    c. are involved with the placental process

    d. produce the three germ layers during gastrulation

OBJ. 3 _____ 7. The most dangerous period in prenatal or postnatal life is the:

    a. first trimester

    b. second trimester

    c. third trimester

    d. expulsion stage

Level
—1—

OBJ. 3 _____ 8. The four general processes that occur during the first trimester include:

   a. dilation, expulsion, placental, labor
   b. blastocyst, blastomere, morula, trophoblast
   c. cleavage, implantation, placentation, embryogenesis
   d. yolk sac, amnion, allantois, chorion

OBJ. 3 _____ 9. Organs and organ systems complete most of their development during the:

   a. first trimester
   b. second trimester
   c. third trimester
   d. time of placentation

OBJ. 4 _____ 10. Germ-layer formation results from the process of:

   a. embryogenesis
   b. organogenesis
   c. gastrulation
   d. parturition

OBJ. 4 _____ 11. The extraembryonic membranes that develop from the endoderm and mesoderm are:

   a. amnion and chorion
   b. yolk sac and allantois
   c. allantois and chorion
   d. yolk sac and amnion

OBJ. 4 _____ 12. The chorion develops from the:

   a. endoderm and mesoderm
   b. ectoderm and mesoderm
   c. trophoblast and endoderm
   d. mesoderm and trophoblast

OBJ. 5 _____ 13. Blood flows to and from the placenta via:

   a. paired umbilical veins and a single umbilical artery
   b. paired umbilical arteries and a single umbilical vein
   c. a single umbilical artery and a single umbilical vein
   d. two umbilical arteries and two umbilical veins

OBJ. 5 _____ 14. The hormone(s) produced by the placenta include(s):

   a. human chorionic gonadotrophin hormone
   b. estrogen, progesterone
   c. relaxin, human placental lactogen
   d. a, b, and c are correct

OBJ. 5 _____ 15. Throughout embryonic and fetal development metabolic wastes generated by the fetus are eliminated by transfer to the:

   a. maternal circulation
   b. amniotic fluid
   c. chorion
   d. allantois

Level
-1-

OBJ. 6 ____ 16. The umbilical cord or umbilical stalk contains:

      a. the amnion, allantois, and chorion

      b. paired umbilical arteries and the amnion

      c. a single umbilical vein and the chorion

      d. the allantois, blood vessels, and yolk stalk

OBJ. 7 ____ 17. The stretching of the myometrium during gestation is associated with a gradual increase in the:

      a. increased size of the developing fetus

      b. excessive secretion of progesterone

      c. rates of spontaneous smooth muscle contractions

      d. amount of fluid in the uterus

OBJ. 7 ____ 18. Prostaglandins in the endometrium:

      a. stimulate smooth muscle contractions

      b. cause the mammary glands to begin secretory activity

      c. cause an increase in the maternal blood volume

      d. initiate the process of organogenesis

OBJ. 8 ____ 19. The sequential stages of labor include:

      a. dilation, expulsion, placental

      b. fertilization, cleavage, implantation

      c. fertilization, implantation, placental

      d. expulsion, placental, birth

OBJ. 8 ____ 20. During gestation the primary major compensatory adjustment(s) is (are):

      a. increasing respiratory rate and tidal volume

      b. increasing maternal requirements for nutrients

      c. increasing glomerular filtration rate

      d. a, b, and c are correct

OBJ. 9 ____ 21. The sequential stages that identify the features and functions associated with the human experience are:

      a. neonatal, childhood, infancy, maturity

      b. neonatal, postnatal, childbirth, adolescence

      c. infancy, childhood, adolescence, maturity

      d. prenatal, neonatal, postnatal, infancy

OBJ. 9 ____ 22. The systems that were relatively nonfunctional during the fetus's prenatal period that must become functional at birth are the:

      a. circulatory, muscular, skeletal

      b. integumentary, reproductive, nervous

      c. endocrine, nervous, circulatory

      d. respiratory, digestive, excretory

OBJ. 10 ____ 23. The normal chromosome complement of a typical somatic, or body, cell is:

      a. 23

      b. N or haploid

      c. 46

      d. 92

Level
-1-

OBJ. 10   _____ 24. Gametes are different from ordinary somatic cells because:

    a. they contain only half the normal number of chromosomes

    b. they contain the full complement of chromosomes

    c. the chromosome number doubles in gametes

    d. gametes are diploid, or 2N

OBJ. 10   _____ 25. During gamete formation, meiosis splits the chromosome pairs, producing:

    a. diploid gametes

    b. haploid gametes

    c. gametes with a full chromosome complement

    d. duplicate gametes

OBJ. 10   _____ 26. The first meiotic division:

    a. results in the separation of the duplicate chromosomes

    b. yields four functional spermatids in the male

    c. produces one functional ovum in the female

    d. reduces the number of chromosomes from 46 to 23

OBJ. 10   _____ 27. Spermatogenesis produces:

    a. four functional spermatids for every primary spermatocyte undergoing meiosis

    b. functional spermatozoan with the diploid number of chromosomes

    c. secondary spermatocytes with the 2N number of chromosomes

    d. a, b, and c are correct

OBJ. 10   _____ 28. Oogenesis produces:

    a. an oogonium with the haploid number of chromosomes

    b. one functional ovum and three nonfunctional polar bodies

    c. a secondary oocyte with the diploid number of chromosomes

    d. a, b, and c are correct

OBJ. 10   _____ 29. If an allele is _dominant_ it will be expressed in the phenotype:

    a. if both alleles agree on the outcome of the phenotype

    b. by the use of lower-case abbreviations

    c. regardless of any conflicting instructions carried by the other allele

    d. by the use of capitalized abbreviations

OBJ. 10   _____ 30. If a female X chromosome of an allelic pair contains sex-linked character for color blindness, the individual would be:

    a. normal

    b. color blind

    c. color blind in one eye

    d. a, b, or c could occur

[L1]   Completion:

Using the terms below, complete the following statements.

childhood   first trimester   placenta
capacitation   thalidomide   autosomal
meiosis   gametogenesis   induction
dilation   polyspermy   heterozygous
chorion   infancy   human chorionic
homozygous   second trimester   gonadotropin
fertilization   true labor

OBJ. 1   31. One chromosome in each pair is contributed by the sperm and the other by the egg at _____.

OBJ. 1   32. Sperm cannot fertilize an egg until they have undergone an activation in the vagina called _____.

OBJ. 1   33. The process of fertilization by more than one sperm that produces a nonfunctional zygote is known as _____.

OBJ. 2   34. As development proceeds, the chemical interplay among developing cells is called _____.

OBJ. 2   35. During the 1960s the drug prescribed for women in early pregnancy that interfered with the induction process responsible for limb development was _____.

OBJ. 3   36. The time during prenatal development when the fetus begins to look distinctively human is referred to as the _____.

OBJ. 3   37. The period of time during which the rudiments of all the major organ systems appear is referred to as the _____.

OBJ. 4   38. The extraembryonic membrane formed from the mesoderm and trophoblast is the _____.

OBJ. 5   39. The placental hormone present in blood or urine samples that provides a reliable indication of pregnancy is _____.

OBJ. 6   40. The vital link between maternal and embryonic systems that support the fetus during development is the _____.

OBJ. 7   41. When the biochemical and mechanical factors reach the point of no return in the uterus, it indicates the beginning of _____.

OBJ. 8   42. Rupturing of the amnion or "having the water break" occurs during the late _____ stage.

OBJ. 9   43. The life stage characterized by events that occur prior to puberty is called _____.

OBJ. 9   44. The life stage that follows the neonatal period and continues to 2 years of age is referred to as _____.

OBJ. 10   45. The special form of cell division leading to the production of sperm or eggs is _____.

OBJ. 10   46. The formation of gametes is called _____.

OBJ. 10   47. Chromosomes with genes that affect only somatic characteristics are referred to as _____ chromosomes.

OBJ. 10   48. If both chromosomes of a homologous pair carry the same allele of a particular gene, the individual is _____ for that trait.

OBJ. 10   49. When an individual has two different alleles carrying different instructions, the individual is _____ for that trait.

[L1] Matching:

Match the terms in column B with the terms in column A. Use letters for answers in the spaces provided.

PART I

			COLUMN A		COLUMN B
OBJ. 1	_____	50.	fertilization	A.	cellular chemical interplay
OBJ. 1	_____	51.	amphimixis	B.	softens symphysis pubis
OBJ. 1	_____	52.	N	C.	prenatal development
OBJ. 2	_____	53.	induction	D.	milk production
OBJ. 3	_____	54.	gestation	E.	pronuclei fusion
OBJ. 4	_____	55.	chorion	F.	gamete
OBJ. 5	_____	56.	human placental lactogen	G.	stimulates smooth muscle contractions
OBJ. 6	_____	57.	maternal organ systems	H.	zygote formation
				I.	mesoderm and trophoblast
OBJ. 7	_____	58.	prostaglandin	J.	supports developing fetus
OBJ. 7	_____	59.	relaxin	K.	reductional division
OBJ. 8	_____	60.	parturition		

PART II

			COLUMN A		COLUMN B
OBJ. 8	_____	61.	true labor	L.	somatic cell
OBJ. 9	_____	62.	neonate	M.	visible characteristics
OBJ. 9	_____	63.	adolescence	N.	chromosomes and component genes
OBJ. 10	_____	64.	2N		
OBJ. 10	_____	65.	meiosis I	O.	begins at puberty
OBJ. 10	_____	66.	meiosis II	P.	forcible expulsion of the fetus
OBJ. 10	_____	67.	phenotype	Q.	newborn infant
OBJ. 10	_____	68.	genotype	R.	positive feedback
				S.	equational division

Level
-1-

[L1]   Drawing/Illustration Labeling:

Identify each numbered structure by labeling the following figures.

OBJ. 1   *FIGURE 29.1*   Spermatogenesis

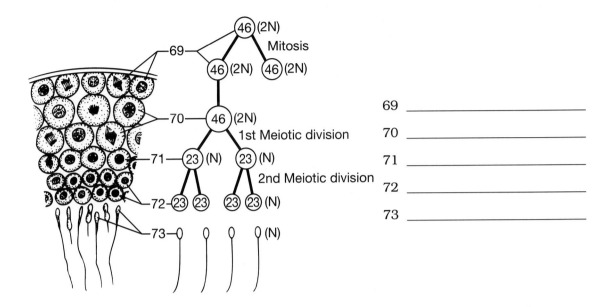

69 _____

70 _____

71 _____

72 _____

73 _____

OBJ. 2   *FIGURE 29.2*   Oogenesis

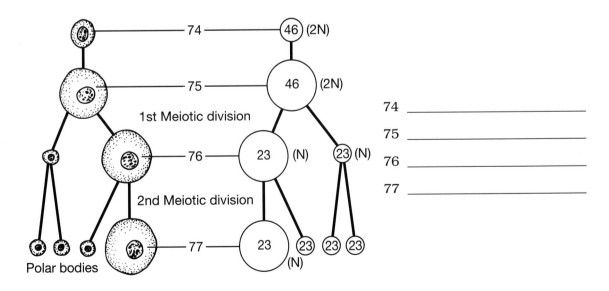

74 _____

75 _____

76 _____

77 _____

When you have successfully completed the exercises in L1 proceed to L2.   Level -1-

☐ LEVEL 2   CONCEPT SYNTHESIS

Concept Map I:

Using the following terms, fill in the circled, numbered, blank spaces to complete the concept map. Follow the numbers that comply with the organization of the map.

↑ Prostaglandin production   Relaxin   Estrogen
Positive feedback   Parturition

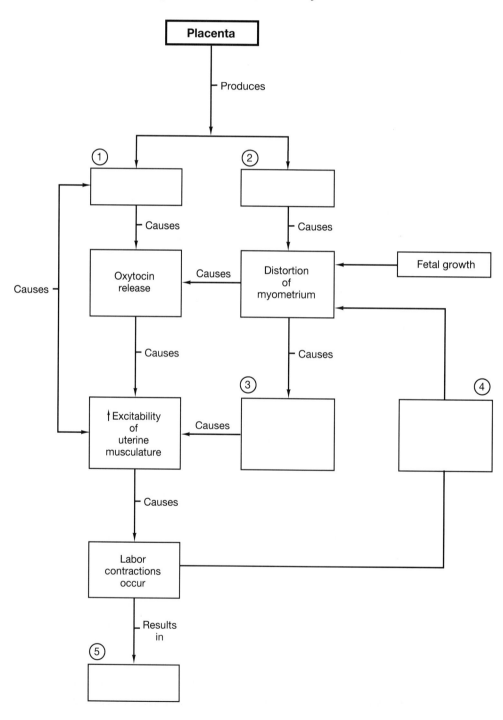

Interacting Factors – Labor & Delivery

Concept Map II:

Using the following terms, fill in the circled, numbered, blank spaces to complete the concept map. Follow the numbers that comply with the organization of the map.

Oxytocin   Prolactin   ↑ milk secretion
Posterior pituitary   Milk ejection   Anterior pituitary

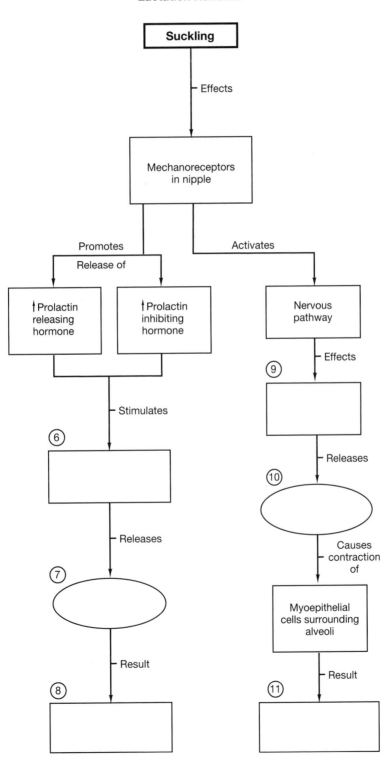

Lactation Reflexes

Body Trek:

Using the terms below, identify the numbered locations on the trek map through the female reproductive tract.

early blastocyst	morula	2-cell stage
fertilization	implantation	8-cell stage
secondary oocyte	zygote	

For the trek through the female reproductive system follow the path Robo takes as the tiny robot follows a developing ovum from the ovary into the uterine tube where it is fertilized and undergoes cleavage until implantation takes place in the uterine wall. Your task is to specify Robo's location by identifying structures or processes at the numbered locations on the trek map. Record your answers in the spaces below the map.

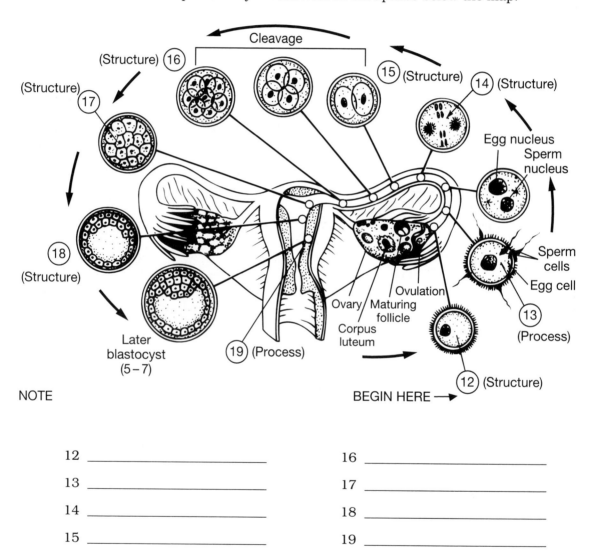

12 _____   16 _____

13 _____   17 _____

14 _____   18 _____

15 _____   19 _____

Level
=2=

[L2]   Multiple Choice:

Place the letter corresponding to the correct answer in the space provided.

_____ 20. The completion of metaphase II, anaphase II, and telophase II produces:

        a. four gametes, each containing 46 chromosomes

        b. four gametes, each containing 23 chromosomes

        c. one gamete containing 46 chromosomes

        d. one gamete containing 23 chromosomes

_____ 21. The primary function of the spermatozoa is to:

        a. nourish and program the ovum

        b. support the development of the ovum

        c. carry paternal chromosomes to the site of fertilization

        d. a, b, and c are correct

_____ 22. For a given trait, if the possibilities are indicated by _AA_, the individual is:

        a. homozygous recessive

        b. heterozygous dominant

        c. heterozygous recessive

        d. homozygous dominant

_____ 23. For a given trait, if the possibilities are indicated by _Aa_, the individual is:

        a. homozygous dominant

        b. homozygous recessive

        c. heterozygous

        d. homozygous

_____ 24. For a given trait, if the possibilities are indicated by _aa_, the individual is:

        a. homozygous recessive

        b. homozygous dominant

        c. homozygous

        d. heterozygous

_____ 25. If albinism is a recessive trait and an albino mother and a normal father with the genotype _AA_ have an offspring, the child will:

        a. be an albino

        b. have normal coloration

        c. have abnormal pigmentation

        d. have blue eyes due to lack of pigmentation

_____ 26. During oocyte activation the process that is important in preventing penetration by more than one sperm is the:

        a. process of amphimixis

        b. process of capacitation

        c. process of monospermy

        d. cortical reaction

_____ 27. The sequential developmental stages that occur during cleavage include:

        a. blastocyst → blastomeres → trophoblast → morula

        b. blastomeres → morula → blastocyst → trophoblast

        c. yolk sac → amnion → allantois → chorion

        d. amnion → allantois → chorion → yolk sac

Level
=2=

_____ 28. The zygote arrives in the uterine cavity as a:

      a. morula

      b. blastocyst

      c. trophoblast

      d. chorion

_____ 29. During implantation the inner cell mass of the blastocyst separates from the trophoblast, creating a fluid-filled chamber called the:

      a. blastodisk

      b. blastocoele

      c. allantois

      d. amniotic cavity

_____ 30. The formation of extraembryonic membranes occurs in the correct sequential steps, which include:

      a. blastomeres, morula, blastocyst, trophoblast

      b. amnion, allantois, chorion, yolk sac

      c. yolk sac, amnion, allantois, chorion

      d. blastocyst, trophoblast, amnion, chorion

_____ 31. In the event of "fraternal" or dizygotic twins:

      a. the blastomeres separate early during cleavage

      b. the inner cell mass splits prior to gastrulation

      c. two separate eggs are ovulated and fertilized

      d. a, b, and c are correct

_____ 32. The enzyme hyaluronidase is necessary to:

      a. induce labor

      b. permit fertilization

      c. promote gamete formation

      d. support the process of gastrulation

_____ 33. The important and complex development event(s) that occur during the first trimester is (are):

      a. cleavage

      b. implantation and placentation

      c. embryogenesis

      d. a, b, and c are correct

_____ 34. Exchange between the embryonic and maternal circulations occur by diffusion across the syncytial and cellular trophoblast layers via:

      a. the umbilicus

      b. the allantois

      c. the chorionic blood vessels

      d. the yolk sac

_____ 35. Karyotyping is the determination of:

      a. an individual's chromosome complement

      b. the life stages of an individual

      c. the stages of prenatal development

      d. an individual's stage of labor

Level =2=

_____ 36. Colostrum, produced and secreted by the mammary glands, contains proteins which help the infant:

     a. activate the digestive system to become fully functional

     b. get adequate amounts of fat to help control body temperature

     c. to initiate the milk let-down reflex

     d. ward off infections until its own immune system becomes fully functional

_____ 37. During fertilization, the process of _cortical reaction_ is important in:

     a. ensuring the fusion of the sperm nucleus with the egg nucleus

     b. preventing penetration by additional sperm

     c. producing a condition known as polyspermy

     d. helping the sperm penetrate the corona radiata

_____ 38. Embryogenesis is the process that establishes the foundation for:

     a. the formation of the blastocyst

     b. implantation to occur

     c. the formation of the placenta

     d. all the major organ systems

_____ 39. An ectopic pregnancy refers to:

     a. implantation occurring within the endometrium

     b. implantation occurring somewhere other than within the uterus

     c. the formation of a gestational neoplasm

     d. the formation of extraembryonic membranes in the uterus

[L2]   Completion:

Using the terms below, complete the following statements.

cleavage	oogenesis	synapsis
alleles	chromatids	spermiogenesis
tetrad	X-linked	differentiation
inheritance	chorion	hyaluronidase
activation	spermatogenesis	simple inheritance
corona radiata	genetics	polygenic inheritance

40. Transfer of genetically determined characteristics from generation to generation refers to _____.

41. The study of the mechanisms responsible for inheritance is called _____.

42. Attached duplicate, doubled chromosomes resulting from the first meiotic division are called _____.

43. When corresponding maternal and paternal chromosomes pair off, the event is known as _____.

44. Paired, duplicated chromosomes visible at the start of meiosis form a combination of four chromatids called a _____.

45. The process of sperm formation is termed _____.

46. The process of ovum production is called _____.

47. Spermatids are transformed into sperm through the process of _____.

48. The various forms of any one gene are called _____.

49. Characteristics carried by genes on the X chromosome that affect somatic structures are termed _____.

50. Phenotypic characters determined by interactions between a single pair of alleles are known as _____.

51. Interactions between alleles on several genes involve _____.

52. When the oocyte leaves the ovary it is surrounded by a layer of follicle cells, the _____.

53. The enzyme in the acrosomal cap of sperm that is used to break down the follicular cells surrounding the oocyte is _____.

54. The sequence of cell division that begins immediately after fertilization and ends at the first contact with the uterine wall is called _____.

55. The fetal contribution to the placenta is the _____.

56. The creation of different cell types during development is called _____.

57. Conditions inside the oocyte resulting from the sperm entering the ooplasm initiate _____.

[L2]   Short Essay:

Briefly answer the following questions in the spaces provided below.

58. What is the primary difference between simple inheritance and polygenic inheritance?

59. What does the term *capacitation* refer to?

60. What are the four (4) general processes that occur during the first trimester?

61. What three (3) germ layers result from the process of gastrulation?

Level
=2=

62. (a) What are the four (4) extraembryonic membranes that are formed from the three germ layers?

    (b) From which given layer(s) does each membrane originate?

63. What are the major compensatory adjustments necessary in the maternal systems to support the developing fetus?

64. What three (3) major factors oppose the inhibitory effect of progesterone on the uterine smooth muscle?

65. What primary factors interact to produce labor contractions in the uterine wall?

66. What are the three (3) stages of labor?

67. What are the identifiable life stages that comprise postnatal development of distinctive characteristics and abilities?

68. What is an Apgar rating and for what is it used?

69. What three (3) events interact to promote increased hormone production and sexual maturation at adolescence?

70. What four (4) processes are involved with aging that influence the genetic programming of individual cells?

When you have successfully completed the exercises in L2 proceed to L3.

☐ LEVEL 3   CRITICAL THINKING/APPLICATION

Using principles and concepts learned about development and inheritance, answer the following questions. Write your answers on a separate sheet of paper.

1. A common form of color blindness is associated with the presence of a dominant or recessive gene on the X chromosome. Normal color vision is determined by the presence of a dominant gene (C), and color blindness results from the presence of the recessive gene (c). Suppose a heterozygous normal female marries a normal male. Is it possible for any of their children to be color blind? Show the possibilities by using a Punnett square.

2. Albinism (aa) is inherited as a homozygous recessive trait. If a homozygous-recessive mother and a heterozygous father decide to have children, what are the possibilities of their offspring inheriting albinism? Use a Punnett square to show the possibilities.

3. Tongue rolling is inherited as a dominant trait. Even though a mother and father are tongue rollers (T), show how it would be possible to bear children who do not have the ability to roll the tongue. Use a Punnett square to show the possibilities.

4. Sharon S. has been married for two years and is pregnant with her first child. She has lived in the Harrisburg, Pennsylvania area all her life. During the nuclear power disaster at Three Mile Island she was exposed to radiation. Her gynecologist has advised her to have amniocentesis to determine whether her exposure to radiation has had any ill effects on the developing fetus. As a friend, would you encourage or discourage her from having the procedure? Why?

Level
3

Chapter 1

An Introduction to Anatomy and Physiology

[L1]   MULTIPLE CHOICE—pp. 2–5

(1.) B	(6.) C	(11.) A	(16.) A	(21.) B
(2.) C	(7.) C	(12.) D	(17.) D	(22.) B
(3.) A	(8.) D	(13.) C	(18.) D	(23.) C
(4.) D	(9.) C	(14.) D	(19.) C	(24.) D
(5.) B	(10.) D	(15.) C	(20.) D	(25.) A

[L1]   COMPLETION—pp. 5–6

(26.) responsiveness	(34.) organs	(42.) endocrine
(27.) digestion	(35.) Urinary	(43.) liver
(28.) excretion	(36.) digestive	(44.) medial
(29.) histologist	(37.) integumentary	(45.) distal
(30.) embryology	(38.) regulation	(46.) transverse
(31.) physiology	(39.) extrinsic	(47.) pericardial
(32.) tissues	(40.) autoregulation	(48.) mediastinum
(33.) molecules	(41.) positive feedback	(49.) peritoneal

[L1]   MATCHING—p. 6

(50.) F	(53.) A	(56.) C	(59.) L	(62.) O
(51.) D	(54.) B	(57.) J	(60.) I	(63.) M
(52.) G	(55.) E	(58.) N	(61.) H	(64.) K

[L1]   DRAWING/ILLUSTRATION—LABELING—p. 7

Figure 1.1   Planes of the Body—p. 7

(65.) frontal (coronal)	(66.) transverse	(67.) midsagittal

Figure 1.2   Human Body Orientation and Direction—p. 7

(68.) superior (cephalad)	(70.) inferior (caudal)	(72.) proximal
(69.) posterior (dorsal)	(71.) anterior (ventral)	(73.) distal

Figure 1.3   Regional Body References—p. 8

(74.) occipital

(75.) deltoid

(76.) scapular

(77.) lumbar

(78.) gluteal

(79.) popliteal

(80.) calf

(81.) axillary

(82.) brachial

(83.) abdominal

(84.) femoral

(85.) orbital

(86.) buccal

(87.) cervical

(88.) thorax

(89.) cubital

(90.) umbilical

(91.) pubic

(92.) palmar

(93.) patellar

Figure 1.4   Body Cavities, Sagittal View—p. 9

(94.) ventral cavity

(95.) thoracic cavity

(96.) diaphragm

(97.) abdominal cavity

(98.) pelvic cavity

(99.) cranial cavity

(100.) spinal cavity

(101.) dorsal cavity

Figure 1.5   Body Cavities, Anterior View—p. 9

(102.) left thoracic cavity

(103.) pericardial cavity

(104.) right thoracic cavity

(105.) diaphragm

(106.) abdominal cavity

[L2]   CONCEPT MAPS—pp. 10–13

I   ANATOMY—p. 10

(1.) macroscopic anatomy

(2.) regional anatomy

(3.) structure of organ systems

(4.) surgical anatomy

(5.) embryology

(6.) cytology

(7.) tissues

II   PHYSIOLOGY—p. 11

(8.) functions of anatomical structures

(9.) functions of living cells

(10.) histophysiology

(11.) specific organ systems

(12.) pathological physiology

(13.) body function response to changes in atmospheric pressure

(14.) exercise physiology

(15.) body function response to athletics

III   BODY CAVITIES—p. 12

(16.) cranial cavity

(17.) spinal cord

(18.) two pleural cavities

(19.) heart

(20.) abdominopelvic cavity

(21.) pelvic cavity

IV   CLINICAL TECHNOLOGY—p. 13

(22.) radiologist

(23.) high-energy radiation

(24.) angiogram

(25.) CT scans

(26.) radio waves

(27.) echogram

[L2]   BODY TREK—p. 14

(28.) subatomic particles

(29.) atoms

(30.) molecules

(31.) protoplasm

(32.) organelles

(33.) cells

(34.) tissues

(35.) organs

(36.) systems

(37.) organisms

[L2]   MULTIPLE CHOICE—pp. 14–16

(38.) D

(39.) A

(40.) C

(41.) C

(42.) A

(43.) B

(44.) B

(45.) C

(46.) D

(47.) B

(48.) D

(49.) A

(50.) B

(51.) C

(52.) D

[L2]   COMPLETION—p. 17

(53.) adaptability

(54.) cardiovascular

(55.) lymphatic

(56.) extrinsic regulation

(57.) nervous

(58.) knee

(59.) appendicitis

(60.) elbow

(61.) stethoscope

(62.) sternum

[L2]   SHORT ESSAY—pp. 17–18

63. Any one of the following might be listed: responsiveness, adaptability, growth, reproduction, movement, absorption, respiration, excretion, digestion, circulation

64. Subatomic particles ↔ atoms ↔ molecules ↔ protoplasm ↔ organelle ↔ cell(s) ↔ tissue(s) ↔ organ(s) ↔ system(s) ↔ organism

65. In the anatomical position, the body is erect, feet are parallel and flat on the floor, eyes are directed forward, and the arms are at the sides of the body with the palm of the hands turned forward.

66. In negative feedback a variation outside of normal limits triggers an automatic response that corrects the situation. In positive feedback the initial stimulus produces a response that exaggerates the stimulus.

67. ventral, dorsal, cranial (cephalic), caudal

68. A sagittal section separates right and left portions. A transverse, or horizontal section separates superior and inferior portions of the body.

69. (a) They protect delicate organs from accidental shocks, and cushion them from the thumps and bumps that occur during walking, jumping, and running.

    (b) They permit significant changes in the size and shape of visceral organs.

70.

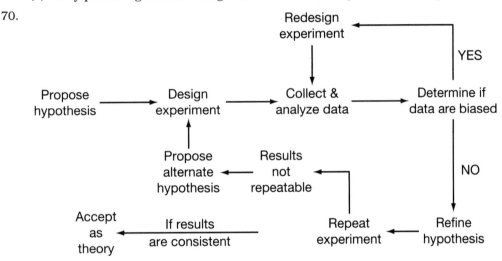

[L3]   CRITICAL THINKING/APPLICATION—p. 19

1. Since there is a lung in each compartment, if one lung is diseased or infected the other lung may remain functional. Also, if one lung is traumatized due to injury, the other one may be spared and function sufficiently to save the life of the injured person.

2. Stretching of the uterus by the developing embryo stimulates the start of contractions. Contractions push the baby toward the opening of the uterus, causing additional stretching which initiates more contractions. The cycle continues until the baby is delivered and the stretching stimulation is eliminated.

3. pericardial, mediastinal, abdominal, pelvic

4. Extrinsic regulation. The nervous and endocrine systems can control or adjust the activities of many different systems simultaneously. This usually causes more extensive and potentially more effective adjustments in system activities.

5. No adverse effects have been attributed to the sound waves, and fetal development can be monitored without a significant risk of birth defects. Ultrasound machines are relatively inexpensive and portable.

6. Barium is very *radiodense*, and the contours of the gastric and intestinal lining can be seen against the white of the barium solution.

7. At no. 1 on the graph there will be *increased blood flow to the skin* and *increased sweating* if the body temperature rises above 99° F. The *body surface cools* and the *temperature declines*. If the temperature falls below 98° F there is *a decrease in blood flow to the skin* and *shivering* occurs. These activities help to *conserve body heat* and the body temperature rises.

Chapter 2:

The Chemical Level
of Organization

[L1]   MULTIPLE CHOICE—pp. 21–24

(1.) D	(6.) C	(11.) C	(16.) C	(21.) C
(2.) B	(7.) C	(12.) A	(17.) B	(22.) B
(3.) C	(8.) D	(13.) D	(18.) C	(23.) D
(4.) B	(9.) D	(14.) D	(19.) B	(24.) B
(5.) A	(10.) A	(15.) B	(20.) A	(25.) C

[L1]   COMPLETION—pp. 24–25

(26.) protons	(33.) exergonic, endergonic	(40.) buffers
(27.) mass number	(34.) catalysts	(41.) salt
(28.) covalent bonds	(35.) organic	(42.) carbonic acid
(29.) ionic bond	(36.) inorganic	(43.) glucose
(30.) H_2	(37.) solvent, solute	(44.) dehydration synthesis
(31.) 2H	(38.) electrolytes	(45.) isomers
(32.) decomposition	(39.) acidic	

[L1]   MATCHING—p. 25

(46.) C	(49.) B	(52.) D	(55.) O	(58.) L
(47.) E	(50.) A	(53.) H	(56.) I	(59.) K
(48.) F	(51.) E	(54.) N	(57.) M	(60.) J

[L1]   DRAWING/ILLUSTRATION—LABELING—pp. 26–27

Figure 2.1—p. 26

(61.) nucleus	(63.) electron
(62.) orbital (shell)	(64.) protons and neutrons

Figure 2.2—p. 26

(65.) nonpolar covalent bond	(66.) polar covalent bond	(67.) ionic bond

Figure 2.3—p. 27

(68.) monosaccharide	(71.) saturated fatty acid	(73.) amino acid
(69.) disaccharide	(72.) polyunsaturated fatty acid	(74.) cholesterol
(70.) polysaccharide		(75.) DNA

[L2]   CONCEPT MAPS—pp. 28–31

I   CARBOHYDRATES—p. 28

(1.) monosaccharides	(4.) sucrose	(6.) glycogen
(2.) glucose	(5.) complex carbohydrates	(7.) bulk, fiber
(3.) disaccharide		

II   LIPIDS—p. 29

(8.) saturated	(11.) glycerol + 3 fatty acids	(14.) carbohydrates + diglyceride
(9.) glyceride	(12.) local hormones	
(10.) diglyceride (Di)	(13.) phospholipid	(15.) steroids

III PROTEINS—p. 30

(16.) amino acids
(17.) amino group
(18.) – COOH
(19.) variable group

(20.) structural proteins
(21.) elastin
(22.) enzymes
(23.) primary

(24.) alpha helix
(25.) globular proteins
(26.) quaternary

IV NUCLEIC ACIDS—p. 31

(27.) deoxyribose nucleic acid
(28.) deoxyribose
(29.) purines

(30.) thymine
(31.) ribonucleic acid
(32.) N bases

(33.) ribose
(34.) pyrimidines
(35.) adenine

[L2] BODY TREK—pp. 32–33

(36.) electrons
(37.) orbitals
(38.) nucleus
(39.) protons
(40.) neutrons
(41.) deuterium
(42.) isotope
(43.) oxygen
(44.) six
(45.) eight
(46.) double covalent bond
(47.) molecule

(48.) oxygen gas
(49.) polar covalent
(50.) negatively
(51.) water
(52.) 66
(53.) zero
(54.) 100
(55.) heat capacity
(56.) lowering
(57.) solvent
(58.) hydrophilic
(59.) hydrophobic

(60.) glucose
(61.) monosaccharide
(62.) glycogen
(63.) dehydration synthesis
(64.) hydrolysis
(65.) DNA
(66.) RNA
(67.) polypeptide
(68.) protein
(69.) phosphorylation
(70.) ATP

[L2] MULTIPLE CHOICE—pp. 33–36

(71.) A
(72.) B
(73.) D
(74.) D
(75.) A

(76.) A
(77.) C
(78.) A
(79.) D
(80.) C

(81.) B
(82.) D
(83.) C
(84.) B
(85.) D

(86.) D
(87.) B
(88.) D
(89.) A
(90.) C

(91.) D
(92.) C
(93.) A
(94.) D
(95.) B

[L2] COMPLETION—p. 37

(96.) alpha particles
(97.) mole
(98.) molecule
(99.) ions
(100.) ionic bond

(101.) molecular weight
(102.) inorganic compounds
(103.) hydrophobic
(104.) alkaline
(105.) nucleic acids

(106.) isomers
(107.) hydrolysis
(108.) dehydration synthesis
(109.) saturated
(110.) peptide bond

[L2] SHORT ESSAY—pp. 37–39

111.

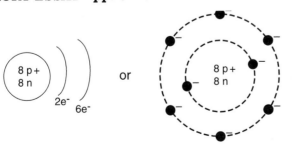

112. It is much easier to keep track of the relative numbers of atoms or molecules in chemical samples and processes.

113. Their outer energy levels contain the maximum number of electrons and they will not react with one another nor combine with atoms of other elements.

114. The oxygen atom has a much stronger attraction for the shared electrons than do the hydrogen atoms, so the electrons spend most of this time in the vicinity of the oxygen nucleus. Because the oxygen atom has two extra electrons part of the time, it develops a slight negative charge. The hydrogens develop a slight positive charge because their electrons are away part of the time.

115. $6 \times 12 = \quad 72$
$12 \times 1 = \quad 12$
$6 \times 16 = \quad \underline{96}$
$MW = \quad 180$

116. (1) molecular structure, (2) freezing point 0° C; boiling point 100° C, (3) capacity to absorb and distribute heat, (4) heat absorbed during evaporation, (5) solvent properties, (6) 66 percent total body weight

117. *carbohydrates*, ex. glucose; *lipids*, ex. steroids; *proteins*, ex. enzymes; *nucleic acid*, ex. DNA

118. In a saturated fatty acid each carbon atom in the hydrocarbon tail has four single covalent bonds. If some of the carbon-to-carbon bonds are double covalent bonds, the fatty acid is unsaturated.

119. (1) adenine nucleotide
(2) thymine nucleotide
(3) cytosine nucleotide
(4) quanine nucleotide

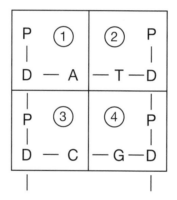

120. adenine, ribose, 3 phosphates

[L3]   CRITICAL THINKING/APPLICATION—p. 39

1. AB + CD → AD + CB or
AB + CD → AC + BD

2. Baking soda is sodium bicarbonate ($NaHCO_3$). In solution sodium bicarbonate reversibly dissociates into a sodium ion (Na^+) and a bicarbonate ion (HCO_3^-) [$NaHCO \leftrightarrow NA^+ + HCO_3^-$] The bicarbonate ion will remove an excess hydrogen ion (H^+) from the solution and form a weak acid, carbonic acid [$H^+ + HCO_3^- \leftrightarrow H_2CO_3$].

3. $C_6H_{12}O_6 + C_6H_{12}O_6 \leftrightarrow C_{12}H_{22}O_{11} + H_2O$

4. Carbohydrates are an immediate source of energy and can be metabolized quickly. Those not immediately metabolized may be stored as glycogen or excesses may be stored as fat.

5. Cannabinol, the active ingredient in marijuana, is a lipid-soluble molecule; therefore, it slowly diffuses out of the body's lipids after administration has ceased.

6. A, D, E and K are fat-soluble vitamins and are stored in the fats of our bodies. Excessive levels of these vitamins may cause toxicity to such vital organs as the liver and the brain.

7. The RDA for fat intake per day is 30 percent of the total caloric intake. Fats are necessary for good health. They are involved in many body functions and serve as energy reserves, provide insulation, cushion delicate organs, and are essential structural components of all cells.

8. Proteins perform a variety of essential functions in the human body. In addition to providing tissue growth and repair, some are involved as transport proteins, buffers, enzymes, antibodies, hormones, and other physiological processes.

Chapter 3:

The Cellular Level of Organization: Cell Structure

[L1] MULTIPLE CHOICE—pp. 41–45

(1.) C	(7.) A	(13.) B	(19.) B	(25.) A
(2.) D	(8.) B	(14.) C	(20.) C	(26.) B
(3.) C	(9.) D	(15.) A	(21.) B	(27.) B
(4.) C	(10.) A	(16.) D	(22.) B	(28.) B
(5.) A	(11.) D	(17.) C	(23.) C	(29.) C
(6.) B	(12.) D	(18.) A	(24.) B	(30.) C

[L1] COMPLETION—p. 45

(31.) phospholipid bilayer	(36.) ribosomes	(41.) translation
(32.) cytoskeleton	(37.) fixed ribosomes	(42.) interphase
(33.) endocytosis	(38.) nuclear pores	(43.) anaphase
(34.) transmembrane	(39.) nucleus	(44.) differentiation
(35.) resting	(40.) gene	

[L1] MATCHING—p. 46

(45.) H	(49.) J	(53.) C	(57.) T	(61.) L
(46.) G	(50.) D	(54.) F	(58.) K	(62.) M
(47.) I	(51.) B	(55.) N	(59.) S	(63.) O
(48.) A	(52.) E	(56.) Q	(60.) R	(64.) P

[L1] DRAWING/ILLUSTRATION—LABELING—pp. 47–49

Figure 3.1—p. 47

(65.) cilia	(70.) rough endoplasmic reticulum	(75.) lusosome
(66.) secretory vesicles		(76.) cell membrane
(67.) centriole	(71.) nuclear envelope	(77.) smooth endoplasmic reticulum
(68.) Golgi apparatus	(72.) nuclear pores	
(69.) mitochondrion	(73.) microvilli	(78.) free ribosomes
	(74.) cytosol	(79.) nucleolus
		(80.) chromatin

Figure 3.2—p. 48

(81.) glycolipids	(83.) cholesterol	(85.) integral protein with pore
(82.) integral proteins	(84.) peripheral protein	

Figure 3.3—p. 48

(86.) interphase	(88.) late prophase	(90.) anaphase
(87.) early prophase	(89.) metaphase	(91.) telophase

[L2] CONCEPT MAPS—pp. 49–51

I TYPICAL ANIMAL CELL—p. 49

(1.) lipid bilayer	(4.) membranes	(7.) centrioles
(2.) proteins	(5.) nucleolus	(8.) ribosomes
(3.) organelles	(6.) lysosomes	(9.) fluid component

II   MEMBRANE PERMEABILITY—p. 50

(10.) lipid solubility
(11.) osmosis
(12.) carrier proteins

(13.) filtration
(14.) energy
(15.) ion pumps

(16.) pinocytosis
(17.) ligands
(18.) cell eating

III   CELL CYCLE—p. 51

(19.) somatic cells
(20.) G_1 phase
(21.) G_2 phase

(22.) DNA replicaton
(23.) mitosis
(24.) metaphase

(25.) telophase
(26.) cytokinesis

[L2]   BODY TREK—pp. 52–53

(27.) extracellular fluid
(28.) cilia
(29.) cell membrane
(30.) integral proteins
(31.) channels
(32.) phospholipid
(33.) peripheral proteins
(34.) intracellular fluid
(35.) ions
(36.) proteins
(37.) cytoskeleton
(38.) organelles

(39.) cytosol
(40.) nonmembranous
(41.) protein synthesis
(42.) cell division
(43.) mitochondrion
(44.) cristae
(45.) matrix
(46.) respiratory enzymes
(47.) ATP
(48.) nucleus
(49.) nuclear envelope
(50.) nuclear pores

(51.) nucleoli
(52.) ribosomes
(53.) nucleoplasm
(54.) chromosomes
(55.) endoplasmic reticulum
(56.) rough endoplasmic reticulum
(57.) Golgi apparatus
(58.) saccules
(59.) lysosomes

[L2]   MULTIPLE CHOICE—pp. 53–57

(60.) A
(61.) B
(62.) C
(63.) A
(64.) B
(65.) D

(66.) C
(67.) A
(68.) C
(69.) D
(70.) B

(71.) A
(72.) D
(73.) A
(74.) B
(75.) C

(76.) D
(77.) B
(78.) A
(79.) A
(80.) D

(81.) C
(82.) B
(83.) D
(84.) A
(85.) C

[L2]   COMPLETION—p. 57

(86.) hydrophobic
(87.) channels
(88.) microvilli
(89.) cilia
(90.) rough ER

(91.) exocytosis
(92.) endocytosis
(93.) peroxisomes
(94.) permeability
(95.) diffusion

(96.) isotonic
(97.) hormones
(98.) phagocytosis
(99.) mitosis
(100.) cytokinesis

[L2]   SHORT ESSAY—p. 58

101.   (a) Circulatory—RBC; (b) Muscular—muscle cell; (c) Reproductive—sperm cell; (d) Skeletal—bone cell; (e) Nervous—neuron (Other systems and cells could be listed for this answer.)

102.   (a) Physical isolation; (b) Regulation of exchange with the environment; (c) Sensitivity; (d) Structural support

103.   Cytosol—high concentration of K^+; Extra Cellular Fluid (E.C.F.)—high concentration of Na^+
Cytosol—higher concentration of dissolved proteins
Cytosol—smaller quantities of carbohydrates and lipids

104.   Cytoskeleton, centrioles, ribosomes, mitochondria, nucleus, nucleolus, ER, Golgi apparatus, lysosomes

105.   Centrioles—move DNA during cell division
Cilia—move fluids or solids across cell surfaces
Flagella—move cell through fluids

106.   (a) Lipid solubility; (b) Channel size; (c) Electrical interactions

[L3]   CRITICAL THINKING/APPLICATION—p. 59

1. Kartagener's syndrome is an inherited disorder identified by an inability to synthesize normal microtubules. Cilia and flagella containing these abnormal microtubules are unable to move. Individuals who have this syndrome may experience chronic respiratory infections because the immobile cilia cannot keep dust, bacteria, etc. from the exchange surfaces of the lungs. Males may be sterile because the flagella in their sperm do not beat, and non-mobile sperm cannot reach and fertilize an egg.

2. Mitochondria contain RNA, DNA, and enzymes needed to synthesize proteins. The mitochondrial DNA and associated enzymes are unlike those found in the nucleus of the cell. The synthetic capability enables mitochondrial DNA to control their own maintenance, growth and reproduction. The muscle cells of bodybuilders have high rates of energy consumption and over time their mitochondria respond to increased energy demand by reproducing mitochondria.

3. The stimulus is magnified in that any stimulus that opens the gated ion channel will produce a sudden rush of ions into or out of the cell. Even if a relatively weak stimulus can affect a channel protein, it can have a significant impact on the cell. The stimulus opens the floodgates through which ion movement occurs.

4. In salt water drowning, the body fluid becomes hypertonic to cells. The cells dehydrate and shrink, which causes the victim to go into shock, thus allowing time for resuscitation to occur. In freshwater drowning, the body fluid is hypotonic to cells causing the cells to swell and burst. The intracellular K^+ are released into circulation and when contacting the heart in excessive amounts cause cardiac arrest, thus allowing little time for the possibility of resuscitation.

5. $MgSO_4$—Epsom salts—increase the solute concentration in the lumen of the large intestine making the intestine hypertonic to surrounding tissues. The osmosis of water occurs from the surrounding tissues into the intestinal lumen. The fluid helps to soften the stool and the watery environment prepares the intestine for eventual evacuation of the stool from the bowel.

6. Gatorade is a combination fruit drink that is slightly hypotonic to body fluids after strenuous exercise. The greater solute concentration inside the cells causes increased osmosis of water into the cells to replenish the fluid that has been lost during exercise—hydration.

7. Because of the increase of solute concentration in the body fluid, it becomes hypertonic to the RBCs. The red blood cells dehydrate and shrink—crenation. The crenated RBCs lose their oxygen-carrying capacity and body tissues are deprived of the oxygen necessary for cellular metabolism.

8. The vegetables contain a greater solute concentration than does the watery environment surrounding them. The vegetables are hypertonic to the surrounding fluid; therefore, the osmosis of water into the vegetables causes crispness, or turgidity.

Chapter 4:

The Tissue Level of Organization

[L1]   MULTIPLE CHOICE—pp. 61–66

(1.) D	(10.) C	(19.) C	(27.) D	(35.) B
(2.) C	(11.) D	(20.) A	(28.) C	(36.) D
(3.) D	(12.) D	(21.) C	(29.) D	(37.) B
(4.) B	(13.) D	(22.) A	(30.) B	(38.) C
(5.) C	(14.) B	(23.) B	(31.) A	(39.) A
(6.) A	(15.) C	(24.) A	(32.) D	(40.) C
(7.) D	(16.) A	(25.) C	(33.) D	(41.) D
(8.) A	(17.) D	(26.) D	(34.) A	(42.) D
(9.) B	(18.) B			

[L1]   COMPLETION—pp. 66–67

(43.) Neural	(49.) Collagen	(54.) Skeletal
(44.) Epithelial	(50.) Reticular	(55.) Neuroglia
(45.) Mesothelium	(51.) Areolar	(56.) Necrosis
(46.) Endothelium	(52.) Lamina propria	(57.) Abscess
(47.) Exocytosis	(53.) Aponeurosis	(58.) Genetic
(48.) Connective		

[L1]   MATCHING—p. 67

(59.) E	(63.) A	(66.) D	(69.) P	(72.) N
(60.) H	(64.) C	(67.) M	(70.) I	(73.) O
(61.) G	(65.) F	(68.) K	(71.) J	(74.) L
(62.) B				

[L1]   DRAWING/ILLUSTRATION—LABELING

Figure 4.1   Types of Epithelial Tissue—p. 68

(75.) simple squamous epithelium	(78.) simple cuboidal epithelium	(80.) simple ciliated columnar epithelium
(76.) stratified squamous epithelium	(79.) transitional epithelium	(81.) pseudostratified columnar epithelium
(77.) simple columnar epithelium		

Figure 4.2   Types of Connective Tissue (proper)—p. 69

(82.) loose connective tissue	(86.) dense fibrous connective tissue	(89.) elastic cartilage
(83.) reticular connective tissue	(87.) hyaline cartilage	(90.) osseous (bone) tissue
(84.) adipose connective tissue	(88.) fibrocartilage	(91.) vascular (fluid) connective tissue
(85.) elastic connective tissue		

Figure 4.3   Types of Muscle Tissue—p. 70

(92.) smooth muscle tissue	(93.) cardiac muscle tissue	(94.) skeletal muscle tissue

Figure 4.4   Identify the Type of Tissue—p. 70

(95.) neural (nervous) tissue

[L2]   CONCEPT MAPS—pp. 71–77

I   HUMAN BODY TISSUES—p. 71

(1.) epithelial

(2.) connective

(3.) muscle

(4.) neural

II   EPITHELIAL TYPES—p. 72

(5.) stratified squamous

(6.) lining

(7.) transitional

(8.) stratified cuboidal

(9.) male urethra mucosa

(10.) pseudostratified columnar

(11.) exocrine glands

(12.) endocrine glands

III   CONNECTIVE TISSUES—p. 73

(13.) loose connective tissue

(14.) adipose

(15.) regular

(16.) tendons

(17.) ligaments

(18.) fluid connective tissue

(19.) blood

(20.) hyaline

(21.) chondrocytes in lacunae

(22.) bone

IV   CONNECTIVE TISSUE PROPER—p. 74

(23.) fixed

(24.) adipose cells

(25.) phagocytosis

(26.) mast cells

(27.) collagen

(28.) branching interwoven framework

(29.) ground substance

V   MUSCLE TISSUE—p. 75

(30.) cardiac

(31.) involuntary

(32.) nonstriated

(33.) viscera

(34.) multinucleated

(35.) voluntary

VI   NEURAL TISSUE—p. 76

(36.) soma

(37.) dendrites

(38.) neuroglia

VII   MEMBRANES—p. 77

(39.) mucous

(40.) goblet cells

(41.) pericardium

(42.) fluid formed on membrane surface

(43.) skin

(44.) thick, waterproof, dry

(45.) synovial

(46.) no basement membrane

(47.) phagocytosis

[L2]   BODY TREK—pp. 78–79

(48.) stratified squamous

(49.) trachea mucosa—(ciliated)

(50.) simple squamous

(51.) simple cuboidal

(52.) transitional

(53.) layers of column-like cells

(54.) adipose

(55.) tendons, ligaments

(56.) irregular dense fibrous

(57.) hyaline cartilage

(58.) chondrocytes in lacunae

(59.) external ear, epiglottis

(60.) bone or osseous

(61.) cardiovascular system

(62.) skeletal

(63.) heart

(64.) nonstriated, uninucleated cells

(65.) neurons—axons, dendrites, neuroglia—support cells

[L2]   MULTIPLE CHOICE—pp. 80–83

(66.) B	(71.) B	(76.) A	(81.) C	(86.) A
(67.) C	(72.) C	(77.) D	(82.) D	(87.) B
(68.) D	(73.) D	(78.) B	(83.) B	(88.) C
(69.) A	(74.) B	(79.) C	(84.) C	(89.) D
(70.) C	(75.) C	(80.) A	(85.) D	(90.) B

[L2]   COMPLETION—pp. 83–84

(91.) lamina propria

(92.) transudate

(93.) adhesions

(94.) subserous fascia

(95.) cancer

(96.) remission

(97.) caution

(98.) chemotherapy

(99.) anaplasia

(100.) dysplasia

(101.) goblet cells

(102.) alveolar

(103.) stoma

(104.) fibrosis

(105.) dense regular connective tissue

[L2]   SHORT ESSAY—pp. 84–85

106. Epithelial, connective, muscle, and neural tissue

107. Provides physical protection, controls permeability, provides sensations, and provides specialized secretion.

108. Microvilli are abundant on epithelial surfaces where absorption and secretion take place. A cell with microvilli has at least twenty times the surface area of a cell without them. A typical ciliated cell contains about 250 cilia that beat in coordinated fashion. Materials are moved over the epithelial surface by the synchronized beating of cilia.

109. Merocrine secretion—the product is released through exocytosis.

Apocrine secretion—loss of cytoplasm as well as the secretory produce.

Holocrine secretion—product is released, cell is destroyed

Merocrine and apocrine secretions leave the cell intact and are able to continue secreting; holocrine does not.

110. Serous—watery solution containing enzymes.

Mucous—viscous mucous.

Mixed glands—serous and mucous secretions.

111. Specialized cells, extracellular protein fibers, and ground substance.

112. Connective tissue proper, fluid connective tissues, supporting connective tissues.

113. Collagen, reticular, elastic.

114. Mucous membranes, serous membranes, cutaneous membranes, and synovial membranes.

115. Skeletal, cardiac, smooth.

116. Neurons—transmission of nerve impulses from one region of the body to another.

Neuroglia—support framework for neural tissue.

[L3]   CRITICAL THINKING/APPLICATION—p. 86

1. Tissues play an important part in diagnosis, which is necessary prior to treatment. As a part of the Allied Health Professional team, the histologist synthesizes anatomical and histological observations to determine the nature and severity of the disease or illness.

2. During cornification the outer layer of the epidermis dies and the cells are moved toward the surface. Cornification is an adaptation process of the stratum corneum where the skin is subjected to friction or pressure.

The stratum corneum consists of dead cells surrounded by a hard protein envelope and filled with the protein keratin. The keratin is responsible for the structural strength of the stratum corneum, permitting it to withstand bacterial invasion, deterioration, and abrasion.

3. Integumentary system—sebaceous, sweat, mammary.

Digestive system—salivary, pancreatic.

4. The leukocytes (WBC) in the blood and the reticular tissue of lymphoid organs are involved with the process of phagocytosis.

5. *Staph aureus* secretes the enzyme hyaluronidase which breaks down hyaluronic acid and other proteoglycans. These bacteria are dangerous because they can spread rapidly by liquefying the ground substance of connective tissues and dissolving the intercellular cement that holds epithelial cells together.

6. Because chondrocytes produce a chemical that discourages the formation of blood vessels, cartilage is avascular. All nutrient and waste product exchange must occur by diffusion through the matrix. This may cause the healing process to be slow, if it occurs at all.

7. Collagen is a protein substance that forms the framework of connective tissue. The connective tissue is the building material for bones, cartilage, teeth, tendons, ligaments, and blood vessels. Without vitamin C, collagen cannot be synthesized; consequently, many body systems are affected.

8. Histamine—dilates blood vessels, thus increasing the blood flow to the area and providing nutrients and oxygen necessary for the regenerative process.

   Heparin—prevents blood clotting that might oppose the increased flow of blood to the injured area.

9. Dysplasia—smoke paralyzes the cilia of the pseudostratified, ciliated, columnar epithelium.

   Metaplasia—epithelial cells produced by stem cell division no longer differentiates into ciliated columnar cells. A lack of cilia eliminates the moisturizing and cleaning function of the epithelium, therefore the smoke has a greater effect on more delicate portions of the respiratory tract.

   Anaplasia—tissue cells change size and shape, often becoming unusually large or abnormally small. Tissue organization breaks down, cell divisions are frequent, and cancerous tumors form in which the cells have abnormal chromosomes.

10. **C**hange in bowel or bladder habits

    A sore that does not heal

    Unusual bleeding or discharge

    Thickening or lump in breast or elsewhere

    Indigestion or difficulty in swallowing

    Obvious change in wart or mole

    Nagging cough or hoarseness.

Chapter 5:

The Integumentary System

[L1]   MULTIPLE CHOICE—pp. 88–91

(1.) C	(6.) C	(11.) D	(16.) D	(21.) D
(2.) D	(7.) D	(12.) B	(17.) A	(22.) D
(3.) A	(8.) B	(13.) D	(18.) C	(23.) C
(4.) D	(9.) B	(14.) C	(19.) A	(24.) D
(5.) A	(10.) D	(15.) C	(20.) D	

[L1]   COMPLETION—pp. 91–92

(25.) stratum lucidum	(32.) sebaceous glands	(39.) eponychium
(26.) stratum corneum	(33.) follicle	(40.) cyanosis
(27.) langerhans cells	(34.) iron	(41.) blisters
(28.) MSH	(35.) sebum	(42.) contraction
(29.) melanin	(36.) eccrine	(43.) melanocyte
(30.) connective	(37.) apocrine	(44.) glandular
(31.) vellus	(38.) decrease	

[L1]   MATCHING—p. 92

(45.) H	(47.) A	(49.) C	(51.) D	(53.) B
(46.) J	(48.) G	(50.) I	(52.) E	(54.) F

[L1]   DRAWING/ILLUSTRATION—LABELING—pp. 93–94

Figure 5.1   Organization of the Integument—p. 93

(55.) dermis	(59.) stratum corneum	(63.) stratum germinativum
(56.) papillary layer	(60.) stratum lucidum	(64.) epidermis
(57.) reticular layer	(61.) stratum granulosum	
(58.) hypodermis	(62.) stratum spinosum	

Figure 5.2   Hair Follicle, Sebaceous Gland, Arrector Pili Muscle—p. 94

(65.) epidermis	(69.) sebaceous gland	(73.) external root sheath
(66.) bulb	(70.) arrector pili muscle	(74.) matrix
(67.) hair shaft	(71.) root	(75.) dermal papilla
(68.) sebaceous gland duct	(72.) internal root sheath	(76.) hair follicle

Figure 5.3   Nail Structure (surface and sectional view)—p. 94

(77.) free edge	(81.) lunula	(86.) nail body
(78.) hyponychium (underneath)	(82.) eponychium	(87.) hyponychium
	(83.) nail root	(88.) phalanx (bone of fingertip)
(79.) nail bed (underneath)	(84.) eponychium	
(80.) lateral nail groove	(85.) lunula	

[L2]   CONCEPT MAPS—pp. 95–97

I   INTEGUMENTARY SYSTEM—p. 95

(1.) dermis	(3.) sensory reception	(5.) lubrication
(2.) Vitamin D synthesis	(4.) exocrine glands	(6.) produce secretions

II   INTEGUMENT—p. 96

(7.) skin
(8.) epidermis
(9.) granulosum

(10.) papillary layer
(11.) nerves
(12.) collagen

(13.) hypodermis
(14.) connective
(15.) fat

III   ACCESSORY STRUCTURES—p. 97

(16.) terminal
(17.) "peach fuzz"
(18.) arms and legs
(19.) glands

(20.) sebaceous
(21.) odors
(22.) merocrine or "eccrine"
(23.) cerumen (Ear wax)

(24.) cuticle
(25.) lunula
(26.) thickened stratum corneum

[L2]   BODY TREK—p. 98

(27.) accessory
(28.) hyponychium
(29.) lunula
(30.) cuticle
(31.) keratin
(32.) stratum lucidum
(33.) eleidin

(34.) keratohyalin
(35.) stratum granulosum
(36.) desmosomes
(37.) stratum spinosum
(38.) stem cells
(39.) mitosis
(40.) epidermal ridges

(41.) papillary
(42.) dermal papillae
(43.) sebaceous
(44.) reticular layer
(45.) collagen
(46.) elastin
(47.) hypodermis

[L2]   MULTIPLE CHOICE—pp. 98–100

(48.) C
(49.) D
(50.) C

(51.) A
(52.) B

(53.) C
(54.) D

(55.) B
(56.) D

(57.) D
(58.) A

[L2]   COMPLETION—p. 100

(59.) hives
(60.) hirsutism
(61.) contusions
(62.) merkel cells

(63.) papillary region
(64.) hemangiomas
(65.) alopedia areata

(66.) seborrheic dermatitis
(67.) keloid
(68.) langerhans cells

[L2]   SHORT ESSAY—pp. 101–102

69. Palms of hand; soles of feet. These areas are covered by 30 or more layers of cornified cells. The epidermis in these locations may be six times thicker than the epidermis covering the general body surface.

70. Bacteria break down the organic secretions of the apocrine glands.

71. Stratum corneum, stratum granulosum, stratum spinosium, stratum germinativum, papillary layer, reticular layer.

72. Long term damage can result from chronic exposure, and an individual attempting to acquire a deep tan places severe stress on the skin. Alterations in underlying connective tissue lead to premature wrinkling and skin cancer can result from chromosomal damage or breakage.

73. Hairs are dead, keratinized structures, and no amount of oiling or shampooing with added ingredients will influence either the exposed hair or the follicles buried in the dermis.

74. "Whiteheads" contain accumulated, stagnant secretions. "Blackheads" contain more solid material that has been invaded by bacteria.

75. Bedsores are a result of circulatory restrictions. They can be prevented by frequent changes in body position that vary the pressure applied to specific blood vessels.

76. Special smooth muscles in the dermis, the arrector pili muscles, pull on the hair follicles and elevate the hair, producing "goose bumps."

[L3] CRITICAL THINKING/APPLICATION—p. 102

1. The ocean is a hypertonic solution, thus causing water to leave the body by crossing the epidermis from the underlying tissues.

2. Blood with abundant O_2 is bright red, and blood vessels in the dermis normally give the skin a reddish tint. During the frightening experience, the circulatory supply to the skin is temporarily reduced because of the constriction of the blood vessels, thus the skin becomes relatively pale.

3. Retin-A increases blood flow to the dermis and stimulates dermal repairs resulting in a decrease in wrinkle formation and existing wrinkles become smaller.

4. Basal cells of the matrix divide; overlying layers undergo keratinization. Layers closest to center of papilla form the medulla (soft core), while mitoses farther away produce the cortex. Medulla contains soft keratin, the cortex contains hard keratin. Matrix cells form the cuticle, a layer of hard keratin that coats the hair.

5. (a) Protects the scalp from ultraviolet light.

    (b) Cushions a blow to the head.

    (c) Prevents entry of foreign particles (eyelashes, nostrils, ear canals).

    (d) Sensory nerves surrounding the base of each hair provide an early warning system that can help prevent injury.

6. 365 days/yr. @ 3 yrs. = 1095 days

    1095 days @ .33 mm/day = 361.35 mm/3 yrs.

    361.35 mm = 36.14 cm

    36.14 cm ÷ 2.54 cm/in. = 14.2 in. long

7. Sensible perspiration produced by merocrine (eccrine) glands plus a mixture of electrolytes, metabolites, and waste products is a clear secretion that is more than 99% water. When all of the merocrine sweat glands are working at maximum, the rate of perspiration may exceed a gallon per hour, and dangerous fluid and electrolyte losses occur.

8. When the skin is subjected to mechanical stresses, stem cells in the germinativum divide more rapidly, and the depth of the epithelium increases.

9. To reach the underlying connective tissues, a bacterium must:

    (a) Survive the bactericidal components of sebum.

    (b) Avoid being flushed from the surface by sweat gland secretions.

    (c) Penetrate the stratum corneum.

    (d) Squeeze between the junctional complexes of deeper layers.

    (e) Escape the Langerhans cells.

    (f) Cross the basement membrane.

Chapter 6:

Osseous Tissue and Skeletal Structure

[L1] MULTIPLE CHOICE—pp. 104–107

(1.) D	(6.) C	(11.) B	(15.) A	(19.) A
(2.) B	(7.) B	(12.) C	(16.) D	(20.) B
(3.) B	(8.) D	(13.) A	(17.) C	(21.) C
(4.) C	(9.) A	(14.) B	(18.) A	(22.) A
(5.) A	(10.) D			

[L1] COMPLETION—pp. 107–108

(23.) yellow Marrow	(29.) intramembranous	(35.) minerals
(24.) support	(30.) endochondral	(36.) calcitriol
(25.) osteoblasts	(31.) ossification	(37.) comminuted
(26.) osteocytes	(32.) calcium	(38.) compound
(27.) osteon	(33.) remodeling	(39.) irregular
(28.) epiphysis	(34.) osteoclasts	(40.) wormian

[L1] MATCHING—p. 108

(41.) G	(44.) F	(47.) D	(50.) N	(53.) J
(42.) H	(45.) C	(48.) E	(51.) I	(54.) L
(43.) A	(46.) B	(49.) O	(52.) K	(55.) M

[L1] DRAWING/ILLUSTRATION—LABELING—pp. 109–110

Figure 6.1 Structural Organization of Bone—p. 109

(56.) small vein	(60.) trabeculae	(64.) Haversian canal
(57.) capillary	(61.) spongy bone	(65.) periosteum
(58.) lamellae	(62.) compact bone	(66.) canaliculi
(59.) osteon	(63.) lacunae	

Figure 6.2 Structure of a Long Bone-L.S., X.S., 3-D Views—p. 109

(67.) proximal epiphysis	(72.) medullary cavity	(77.) periosteum
(68.) diaphysis	(73.) epiphyseal line	(78.) spongy bone
(69.) distal epiphysis	(74.) endosteum	(79.) compact bone
(70.) epiphyseal line	(75.) compact bone	(80.) periosteum
(71.) spongy bone	(76.) yellow marrow	

Figure 6.3 Types of Fractures—p. 110

(81.) spiral fracture	(83.) simple fracture	(85.) comminuted fracture
(82.) greenstick fracture	(84.) compound fracture	(86.) depressed fracture

[L2] CONCEPT MAPS—pp. 111–112

I BONE FORMATION—p. 111

(1.) intramembranous ossification	(2.) collagen	(4.) hyaline cartilage
	(3.) osteocytes	(5.) periosteum

II   REGULATION OF CALCIUM ION CONCENTRATION—p. 112

(6.) calcitonin

(7.) ↓ intestinal absorption of Ca

(8.) ↓ calcium level

(9.) parathyroid

(10.) releases stored Ca from bone

(11.) homeostasis

[L2]   BODY TREK—p. 113

(12.) compound

(13.) yellow marrow

(14.) endosteum

(15.) cancellous or spongy

(16.) trabeculae

(17.) lamella

(18.) osteocytes

(19.) canaliculi

(20.) red blood cells

(21.) red marrow

(22.) Volkmann's canal

(23.) periosteum

(24.) compact bone

(25.) haversian canal

(26.) blood vessels

(27.) osteon

(28.) lacunae

(29.) osteoclasts

[L2]   MULTIPLE CHOICE—pp. 113–116

(30.) B

(31.) C

(32.) D

(33.) A

(34.) D

(35.) A

(36.) B

(37.) D

(38.) C

(39.) C

(40.) D

(41.) A

(42.) D

(43.) C

(44.) B

(45.) A

(46.) D

(47.) B

(48.) C

(49.) A

(50.) B

[L2]   COMPLETION—p. 116

(51.) canaliculi

(52.) rickets

(53.) osteoblasts

(54.) intramembranous

(55.) endochondral

(56.) epiphyseal plates

(57.) osteomalacia

(58.) osteopenia

(59.) depressed

(60.) diaphysis

[L2]   SHORT ESSAY—pp. 117–118

61.   *Support*; storage of calcium salts and lipids; blood cell *production*; *protection* of delicate tissues and organs; and *leverage*

62.   *Compact Bone*:   basic functional unit is the osteon (Haversian system)

lamella (bony matrix) fills in spaces between the osteons

Spongy Bone: no osteons

lamella from bony plates (trabeculae)

63.   *Periosteum*:   covers the *outer surface* of a bone

isolates bone from surrounding tissues

consists of fibrous outer layer and cellular inner layer

provides a route for circulatory and nervous supply

Endosteum: cellular layer *inside* bone

lines the narrow cavity

covers the trabeculae of spongy bone

lines inner surfaces of the central canals

64.   *Calcification*:   refers to the deposition of calcium salts within a tissue

Ossification: refers specifically to the formation of bone

65.   *Intramembranous* ossification begins when osteoblasts differentiate within a connective tissue. *Endochondral* ossification begins with the formation of a cartilaginous model.

66.   Excessive or inadequate hormone production has a pronounced effect on activity at epiphyseal plates. *Gigantism* results from an overproduction of growth hormone before puberty. *Pituitary dwarfism* results from inadequate growth hormone production which leads to reduced epiphyseal activity and abnormally short bones.

67. (1) Long bones – humerus, femur, tibia, fibula...

    (2) Short bones – carpals, tarsals

    (3) Flat bones – sternum, ribs, scapula

    (4) Irregular bones – spinal vertebrae

    (5) Sesamoid – patella (kneecap)

    (6) Sutural (Wormian) bones – between flat bones of skull (sutures).

68. The bones of the skeleton are attached to the muscular system, extensively connected to the cardiovascular and lymphatic systems, and largely under the physiological control of the endocrine system. The digestive and excretory systems provide the calcium and phosphate minerals needed for bone growth. The skeleton represents a reserve of calcium, phosphate, and other minerals that can compensate for changes in the dietary supply of these ions.

[L3] CRITICAL THINKING/APPLICATION—p. 118

1. In premature or precocious puberty, production of sex hormones escalates early, usually by age 8 or earlier. This results in an abbreviated growth spurt, followed by premature closure of the epiphyseal plates. Estrogen and testosterone both cause early uniting of the epiphyses of long bones; thus, the growth of long bones is completed prematurely.

2. Reduction in bone mass and excessive bone fragility are symptoms of osteoporosis, which can develop as a secondary effect of many cancers. Cancers of the bone marrow, breast, or other tissues release a chemical known as osteoclast-activating factor. This compound increases both the number and activity of osteoclasts and produces a severe osteoporosis.

3. The bones of the skeleton become thinner and relatively weaker as a normal part of the aging process. Inadequate ossification, called osteopenia, causes a reduction in bone mass between the ages of 30 and 40. Once this reduction begins, women lose about 8 percent of their skeletal mass every decade, and men lose about 3 percent per decade. The greatest areas of loss are in the epiphyses, vertebrae, and the jaws. The result is fragile limbs, reduction in height, and loss of teeth.

4. Foods or supplements containing vitamins A, C, and D are necessary to maintain the integrity of bone. Vitamin D plays an important role in calcium metabolism by stimulating the absorption and transport of calcium and phosphate ions. Vitamins A and C are essential for normal bone growth and remodeling. Any exercise that is weight-bearing or that exerts a pressure on the bones is necessary to retain the minerals in the bones, especially calcium salts. As the mineral content of a bone decreases, the bone softens and skeletal support decreases.

5. Your new tennis racket is 2 to 3 times stiffer than the wooden racket. As a result, less energy is absorbed by the racket and more is imparted to the ball. It is lighter, and easier to swing faster, placing greater stress on the arm and shoulder. In time the increased shock means development of tendinitis, or "tennis elbow," which affects muscles and tendons around the elbow. The healing process for tendons is very slow owing to the rate at which blood, nutrients, and chemicals involved in tissue repair enter the tissue of the tendons.

6. Continual excessive force causes chronic knee pain, a condition referred to as *chondromalacia patella*. The condition is exemplified by irritation and inflammation of cartilage under the kneecap. Studies have shown that the compressive forces on the knee while using the stairmaster are equal to 5 times the body weight on the way up and 7 times the body weight on the way down.

7. Each bone in the body has a distinctive shape and characteristic external and internal features. The bone markings (surface features), depressions, grooves, and tunnels can be used to determine the size, weight, sex and general appearance of an individual.

Chapter 7:

The Axial Skeleton

[L1]   MULTIPLE CHOICE—pp. 120–124

(1.) B	(8.) A	(15.) D	(21.) D	(27.) B
(2.) D	(9.) D	(16.) B	(22.) C	(28.) A
(3.) C	(10.) C	(17.) C	(23.) B	(29.) D
(4.) A	(11.) C	(18.) C	(24.) C	(30.) B
(5.) B	(12.) A	(19.) B	(25.) C	(31.) D
(6.) D	(13.) D	(20.) A	(26.) A	(32.) B
(7.) C	(14.) B			

[L1]   COMPLETION—pp. 124–125

(33.) axial	(38.) paranasal	(43.) centrum
(34.) muscles	(39.) mucus	(44.) cervical
(35.) cranium	(40.) fontanels	(45.) costal
(36.) foramen magnum	(41.) microcephaly	(46.) capitulum
(37.) inferior concha	(42.) compensation	(47.) floating

[L1]   MATCHING—p. 125

(48.) H	(51.) F	(54.) C	(57.) O	(60.) L
(49.) G	(52.) D	(55.) B	(58.) I	(61.) M
(50.) A	(53.) E	(56.) N	(59.) K	(62.) J

[L1]   DRAWING/ILLUSTRATION—LABELING—pp. 126–129

Figure 7.1   Bones of the Axial Skeleton—p. 126

(63.) cranium	(66.) sternum (manubrium)	(68.) vertebral column
(64.) face (maxilla)	(67.) ribs	(69.) sacrum
(65.) hyoid		

Figure 7.2   Anterior View of the Skull—p. 127

(70.) coronal suture	(75.) zygomatic bone	(80.) sphenoid bone
(71.) parietal bone	(76.) infraorbital foramen	(81.) ethnoid bone
(72.) frontal bone	(77.) alveolar margins	(82.) vomer
(73.) nasal bone	(78.) mental foramen	(83.) maxilla
(74.) lacrimal bone	(79.) temporal bone	(84.) mandible

Figure 7.3   Lateral View of the Skull—p. 127

(85.) coronal suture	(91.) mandibular condyle	(97.) external auditory canal
(86.) frontal bone	(92.) mandible	(98.) occipital bone
(87.) nasal bone	(93.) parietal bone	(99.) mastoid process
(88.) zygomatic arch	(94.) squamous suture	(100.) styloid process
(89.) zygomatic bone	(95.) temporal bone	
(90.) maxilla	(96.) lambdoidal suture	

Figure 7.4   Inferior View of the Skull—p. 127

(101.) sphenoid bone

(102.) vomer

(103.) styloid process

(104.) foramen magnum

(105.) occipital bone

(106.) palatine process (maxillae)

(107.) maxilla

(108.) zygomatic bone

(109.) carotid foramen

(110.) jugular foramen

(111.) occipital condyle

(112.) inferior nuchal line

(113.) superior nuchal line

Figure 7.5   Paranasal Sinuses—p. 128

(114.) frontal sinus

(115.) ethmoid sinus

(116.) sphenoid sinus

(117.) maxillary sinus

Figure 7.6   Fetal Skull – Lateral view—p. 128

(118.) sphenoidal fontanel

(119.) anterior fontanel

(120.) posterior fontanel

(121.) mastoid fontanel

Figure 7.7   Fetal Skull – Lateral view—p. 128

(122.) anterior fontanel

(123.) posterior fontanel

Figure 7.8   The Vertebral Column—p. 129

(124.) cervical vertebrae

(125.) thoracic vertebrae

(126.) lumbar vertebrae

(127.) sacrum

(128.) coccyx

(129.) intevertebral discs

Figure 7.9   The Ribs—p. 129

(130.) sternum

(131.) manubrium

(132.) body of sternum

(133.) xiphoid process

(134.) vertebrochondraribs

(135.) floating ribs

(136.) vertebrosternal ribs (true ribs)

(137.) false ribs

(138.) costal cartilage

[L2]   CONCEPT MAP—p. 130

I   SKELETON—AXIAL DIVISION—p. 130

(1.) skull

(2.) mandible

(3.) lacrimal

(4.) occipital

(5.) temporal

(6.) sutures

(7.) coronal

(8.) hyoid

(9.) vertebral column

(10.) thoracic

(11.) sacral

(12.) floating ribs

(13.) sternum

(14.) xiphoid process

[L2]   BODY TREK—p. 131

(15.) axial

(16.) skull

(17.) cranium

(18.) ribs

(19.) sternum

(20.) vertebrae

(21.) zygomatic

(22.) nasal

(23.) lacrimal

(24.) mandible

(25.) occipital

(26.) sphenoid

(27.) parietal

(28.) sutures

(29.) sagittal

(30.) hyoid

(31.) cervical

(32.) thoracic

(33.) lumbar

(34.) sacrum

(35.) true

(36.) false

(37.) floating

(38.) manubrium

(39.) xiphoid process

[L2]   MULTIPLE CHOICE—pp. 132–133

(40.) C

(41.) A

(42.) A

(43.) D

(44.) C

(45.) B

(46.) C

(47.) A

(48.) D

(49.) B

(50.) B

[L2]   COMPLETION—pp. 133–134

(51.) pharyngotympanic

(52.) metopic

(53.) tears

(54.) auditory ossicles

(55.) alveolar processes

(56.) mental foramina

(57.) compensation

(58.) kyphosis

(59.) lordosis

(60.) scoliosis

[L2]   SHORT ESSAY—pp. 134–135

61.   (1) create a framework that supports and protects organ systems in the dorsal and ventral body cavities.

(2) provide a surface area for attachment of muscles that adjust the positions of the head, neck, and trunk.

(3) performs respiratory movements.

(4) stabilize or position elements of the appendicular system.

62.   A. *Paired bones*: parietal, temporal

B. *Unpaired bones*: occipital, frontal, sphenoid, ethmoid

63.   A. *Paired bones*: zygomatic, maxilla, nasal, palatine, lacrimal, inferior nasal concha

B. *Unpaired bones*: mandible, vomer

64.   The auditory ossicles consist of three (3) tiny bones on each side of the skull that are enclosed by the temporal bone. The hyoid bone lies below the skull suspended by the stylohyoid ligaments.

65.   Craniostenosis is premature closure of one or more fontanels, which results in unusual distortions of the skull.

66.   The thoracic and sacral curves are called primary curves because they begin to appear late in fetal development. They accommodate the thoracic and abdominopelvic viscera.

The lumbar and cervical curves are called secondary curves because they do not appear until several months after birth. They help position the body weight over the legs.

67.   *Kyphosis*: normal thoracic curvature becomes exaggerated, producing "roundback" appearance.

Lordosis: exaggerated lumbar curvature produces "swayback" appearance.

Scoliosis: abnormal lateral curvature.

68.   The *true* ribs reach the anterior body wall and are connected to the sternum by separate cartilaginous extensions. The *false* ribs do not attach directly to the sternum.

[L3]   CRITICAL THINKING/APPLICATION—p. 135

1.   Vision, hearing, balance, olfaction (smell), and gustation (taste).

2.   TMJ is temporomandibular joint syndrome. It occurs in the joint where the *mandible* articulates with the *temporal* bone. The mandibular condyle fits into the mandibular fossa of the temporal bone. The condition generally involves pain around the joint and its associated muscles, a noticeable clicking within the joint, and usually a pronounced malocclusion of the lower jaw.

3.   The crooked nose may be a result of a deviated septum, a condition in which the nasal septum has a bend in it, most often at the junction between the bony and cartilaginous portions of the septum. Septal deviation often blocks drainage of one or more sinuses, producing chronic bouts of infection and inflammation.

4.   During whiplash the movement of the head resembles the cracking of a whip. The head is relatively massive, and it sits on top of the cervical vertebrae. Small muscles articulate with the bones and can produce significant effects by tipping the balance one way or another. If the body suddenly changes position, the balancing muscles are not strong enough to stabilize the head. As a result, a partial or complete dislocation of the cervical vertebrae can occur, with injury to muscles and ligaments and potential injury to the spinal cord.

5.   A plausible explanation would be that the child has scoliosis, a condition in which the hips are abnormally tilted sideways, making one of the lower limbs shorter than the other.

6.   When the gelatinous interior (nucleus pulposus) of the disc leaks through the fibrous outer portion (annulus fibrosis) of the disc, the affected disc balloons out from between the bony parts of the vertebrae. If the bulging or herniated area is large enough, it may press on a nerve, causing severe or incapacitating pain. Usually, the sciatic nerve is affected. Sciatica is generally located in the lumbar region and can radiate over the buttock, rear thigh, and calf, and can extend into the foot.

Chapter 8:

Appendicular Division

[L1] MULTIPLE CHOICE—pp. 137–139

(1.) D	(5.) B	(9.) C	(13.) D	(17.) D
(2.) C	(6.) B	(10.) A	(14.) C	(18.) D
(3.) A	(7.) C	(11.) C	(15.) D	
(4.) A	(8.) B	(12.) D	(16.) C	

[L1] COMPLETION—pp. 139–140

(19.) upper and lower extremities	(24.) wrist	(29.) knee
	(25.) coxae	(30.) childbearing
(20.) clavicle	(26.) pubic symphysis	(31.) teeth
(21.) pectoral girdle	(27.) malleolus	(32.) pelvis
(22.) glenoid fossa	(28.) acetabulum	(33.) age
(23.) styloid		

[L1] MATCHING—p. 140

(34.) F	(37.) A	(40.) D	(42.) K	(44.) H
(35.) J	(38.) C	(41.) E	(43.) B	(45.) I
(36.) G	(39.) L			

[L1] DRAWING/ILLUSTRATION—LABELING—pp. 141–145

Figure 8.1 Appendicular Skeleton—p. 141

(46.) scapula	(52.) phalanges	(58.) ischium
(47.) humerus	(53.) femur	(59.) patella
(48.) ulna	(54.) fibula	(60.) tibia
(49.) radius	(55.) clavicle	(61.) tarsals
(50.) carpals	(56.) ilium	(62.) metatarsals
(51.) metacarpals	(57.) pubis	(63.) phalanges

Figure 8.2 The Scapula—p. 142

(64.) coracoid process	(68.) acromion	(71.) spine
(65.) glenoid fossa	(69.) coracoid process	(72.) body
(66.) axillary border	(70.) superior border	(73.) medial border
(67.) inferior angle		

Figure 8.3 The Humerus—p. 142

(74.) greater tubercle	(78.) capitulum	(82.) greater tubercle
(75.) lesser tubercle	(79.) head	(83.) radial groove
(76.) deltoid tuberosity	(80.) coronoid fossa	(84.) olecranon fossa
(77.) lateral epicondyle	(81.) medial epicondyle	(85.) lateral epicondyle

Figure 8.4 The Radius and Ulna—p. 143

(86.) olecranon	(90.) radius	(93.) coronoid process
(87.) ulna	(91.) styloid process of radius	(94.) ulna
(88.) ulnar head	(92.) trochlear notch	(95.) styloid process of ulna
(89.) head of radius		

Figure 8.5   BONES OF THE WRIST AND HAND—p. 143

(96.) phalanges
(97.) carpals
(98.) distal
(99.) middle
(100.) proximal
(101.) hamate

(102.) pisiform
(103.) triangular
(104.) lunate
(105.) ulna
(106.) metacarpals
(107.) trapezoid

(108.) trapezium
(109.) scaphoid
(110.) capitate
(111.) radius

Figure 8.6   The Pelvis (anterior view)—p. 144

(112.) acetabulum
(113.) pubic symphysis
(114.) obturator foramen

(115.) iliac crest
(116.) sacroiliac joint
(117.) ilium

(118.) pubis
(119.) ischium

Figure 8.7   The Pelvis (lateral view)—p. 144

(120.) greater sciatic notch
(121.) ischial spine
(122.) lesser sciatic notch
(123.) ischial tuberosity

(124.) anterior superior iliac spine
(125.) anterior inferior iliac spine

(126.) inferior iliac notch
(127.) ischial ramus

Figure 8.8   The Femur—p. 144

(128.) greater trochanter
(129.) neck
(130.) lateral epicondyle
(131.) lateral condyle
(132.) head

(133.) lesser trochanter
(134.) medial epicondyle
(135.) medial condyle
(136.) trochanteric crest
(137.) gluteal tuberosity

(138.) linea aspera
(139.) lateral condyle
(140.) lateral epicondyle

Figure 8.9   The Tibia and Fibula—p. 145

(141.) lateral condyle
(142.) head of the fibula
(143.) lateral malleolus

(144.) medial condyle
(145.) tibial tuberosity
(146.) anterior crest

(147.) medial malleolus

Figure 8.10   Bones of the Ankle and Foot—p. 145

(148.) medial cuneiform
(149.) intermediate cuneiform
(150.) navicular
(151.) talus
(152.) distal

(153.) middle
(154.) proximal
(155.) lateral cuneiform
(156.) cuboid
(157.) calcaneus

(158.) phalanges
(159.) metatarsals
(160.) tarsals

[L2]   CONCEPT MAP—p. 146

I   SKELETON—APPENDICULAR DIVISION—p. 146

(1.) pectoral girdle
(2.) humerus
(3.) radius

(4.) metacarpals
(5.) ischium
(6.) femur

(7.) fibula
(8.) tarsals
(9.) phalanges

[L2]   BODY TREK—p. 147

(10.) clavicle
(11.) pectoral
(12.) ball and socket
(13.) scapula
(14.) humerus
(15.) radius
(16.) ulna
(17.) brachium
(18.) hinge

(19.) carpals
(20.) metacarpals
(21.) finger bones
(22.) ilium
(23.) ischium
(24.) pubis
(25.) acetabulum
(26.) pelvic
(27.) femur

(28.) shoulder
(29.) fibula
(30.) tibia
(31.) patella
(32.) knee
(33.) elbow
(34.) tarsals
(35.) metatarsals
(36.) phalanges

[L2]   MULTIPLE CHOICE—pp. 148–150

(37.) C	(40.) D	(43.) C	(46.) D	(49.) A
(38.) B	(41.) D	(44.) A	(47.) D	(50.) C
(39.) A	(42.) B	(45.) B	(48.) D	(51.) B

[L2]   COMPLETION—p. 150

(52.) clavicle	(57.) bursitis	(62.) fibula
(53.) scapula	(58.) acetabulum	(63.) calcaneus
(54.) ulna	(59.) ilium	(64.) pollex
(55.) metacarpals	(60.) pelvis	(65.) hallux
(56.) bursae	(61.) femur	(66.) arthroscopy

[L2]   SHORT ESSAY—pp. 150–151

67. Provides control over the immediate environment; changes your position in space, and makes you an active, mobile person.

68. Bones of the arms and legs and the supporting elements that connect the limbs to the trunk.

69. Pectoral girdle: scapula, clavicle

    Pelvic girdle: ilium, ischium, pubis

70. Both have long shafts and styloid processes for support of the wrist. The radius has a radial tuberosity and the ulna has a prominent olecranon process for muscle attachments. Both the radius and the ulna articulate proximally with the humerus and distally with the carpal bones.

71. The knee joint and elbow joint are both hinge joints.

72. The tibia is the massive, weight-bearing bone of the lower leg.

    The fibula is the long narrow bone lateral to the tibia that is more important for muscle attachment than for support. The tibia articulates proximally with the femur and distally with the talus.

73. The shoulder and hip joints are both ball-and-socket joints.

74. The articulations of the carpals and tarsals are gliding diarthroses, which permit a limited gliding motion.

75. Muscle contractions can occur only when the extracellular concentration of calcium remains within relatively narrow limits. Most of the body's calcium is tied up in the skeleton.

[L3]   CRITICAL THINKING/APPLICATION—p. 152

1. The bones and ligaments that form the walls of the carpal tunnel do not stretch; therefore, any trauma or activity that applies pressure against the nerves or blood vessels passing through the tunnel will cause a fluid buildup (edema) or connective tissue deposition within the carpal tunnel, resulting in the described symptoms.

2. Doctors explained that the cause of death was fat embolism syndrome. Fatty droplets from the yellow marrow of the fractured bone got into the bloodstream and eventually passed through the heart to the lungs. The droplets triggered immune mechanisms in the lungs, filling the lungs with fluid and blocking the lungs' ability to take in oxygen. Hemorrhaging occurred and physicians were unable to save him.

3. The female pelvis has a wide oval pelvic inlet and widely spaced ischial spines that are ideal for delivery. The pelvic outlet is broader and more shallow in the female. The male pelvis has a heart-shaped pelvic inlet and the ischial spines are closer together. The pelvic outlet is deeper and narrower in the male.

4. Gout is a metabolic disorder in which there is an increase in uric acid in the body with precipitation of monosodium urate crystals in the kidneys and joint capsules. The presence of uric acid crystals in the joints can lead to an inflammatory response in the joints. Usually the great toe and other foot and leg joints are affected, and kidney damage from crystal formation occurs in more advanced cases.

5. A bunion is a common pressure-related bursitis. Bunions form over the base of the great toe as a result of the friction and distortion of the joint caused by tight shoes, especially those with pointed toes. There is chronic inflammation of the region, and as the wall of the bursa thickens, fluid builds up in the surrounding tissues. The result is a firm, tender nodule.

6. All three are conditions of bursitis, which indicate the occupations associated with them. "Housemaid's knee," which accompanies prolonged kneeling, affects the bursa, which lies between the patella and the skin. "Weaver's bottom" is produced by pressure on the posterior and inferior tip of the pelvic girdle and can result from prolonged sitting on hard surfaces. "Student's elbow" is a form of bursitis resulting from constant and excessive pressure on the elbows. A student propping his head above a desk while studying can trigger this condition.

Chapter 9:

Articulations

[L1]   MULTIPLE CHOICE—pp. 154–157

(1.) D	(6.) D	(11.) D	(16.) D	(21.) B
(2.) A	(7.) B	(12.) B	(17.) C	(22.) C
(3.) C	(8.) C	(13.) D	(18.) A	(23.) B
(4.) B	(9.) A	(14.) D	(19.) C	
(5.) D	(10.) C	(15.) B	(20.) D	

[L1]   COMPLETION—pp. 157–158

(24.) suture	(29.) flexion	(34.) anulus fibrosus
(25.) synostosis	(30.) supination	(35.) scapulohumeral
(26.) symphysis	(31.) synovial	(36.) hip
(27.) accessory ligaments	(32.) ellipsoidal	(37.) knee
(28.) bursae	(33.) gliding	(38.) elbow

[L1]   MATCHING—p. 158

(39.) H	(42.) A	(45.) E	(48.) L	(51.) O
(40.) F	(43.) B	(46.) D	(49.) J	(52.) I
(41.) G	(44.) C	(47.) M	(50.) N	(53.) K

[L1]   DRAWING/ILLUSTRATION — LABELING—pp. 159–162

Figure 9.1   Structure of a Synovial Joint—p. 159

(54.) marrow cavity	(57.) synovial membrane	(60.) articular capsule
(55.) spongy bone	(58.) articular cartilage	(61.) compact bone
(56.) periosteum	(59.) joint cavity	

Figure 9.2   Structure of the Shoulder Joint—p. 159

(62.) clavicle	(67.) subscapularis muscle	(71.) articular capsule
(63.) coracoacromial ligament	(68.) acromioclavicular ligament	(72.) glenoid fossa
(64.) tendon of biceps muscle		(73.) glenoid labrum
(65.) subcoracoid bursa	(69.) acromion	(74.) glenohumeral ligament
(66.) subscapular bursa	(70.) subacromial bursa	

Figure 9.3   Movements of the Skeleton—p. 160

(75.) elevation	(83.) opposition	(91.) extension
(76.) depression	(84.) pronation	(92.) inversion
(77.) flexion	(85.) abduction	(93.) eversion
(78.) extension	(86.) adduction	(94.) retraction
(79.) hyperextension	(87.) adduction	(95.) protraction
(80.) rotation	(88.) abduction	(96.) dorsiflexion
(81.) abduction	(89.) circumduction	(97.) plantar flexion
(82.) adduction	(90.) flexion	

Figure 9.4   Joint Movement and Types of Diarthrotic Joints—p. 161

(98.) saddle joint

(99.) D

(100.) pivot joint

(101.) B

(102.) ellipsoidal joint

(103.) F

(104.) hinge joint

(105.) A

(106.) ball and socket joint

(107.) C

(108.) gliding joint

(109.) E

Figure 9.5   Sectional View of Knee Joint—p. 162

(110.) bursa

(111.) extracapsular ligament

(112.) fat pad

(113.) joint capsule

(114.) meniscus

(115.) joint cavity

(116.) intracapsular ligament

Figure 9.6   Anterior View of a Flexed Knee—p. 162

(117.) femoral lateral condyle

(118.) anterior cruciate ligament

(119.) lateral meniscus

(120.) fibular collateral ligament

(121.) anterior ligament – head of fibula

(122.) fibula

(123.) patellar surface of femur

(124.) posterior cruciate ligament

(125.) medial condyle of femur

(126.) medial meniscus

(127.) transverse ligament

(128.) tibial tuberosity

(129.) tibial collateral ligament

(130.) tibia

[L2]   CONCEPT MAP—p. 163

I   ARTICULATIONS/JOINTS—p. 163

(1.) no movement

(2.) sutures

(3.) cartilaginous

(4.) amphiarthrosis

(5.) fibrous

(6.) symphysis

(7.) synovial

(8.) monoaxial

(9.) wrist

[L2]   BODY TREK—p. 164

(10.) capsule

(11.) synovial

(12.) synovial fluid

(13.) condyles

(14.) femur

(15.) tibia

(16.) menisci

(17.) cushions

(18.) seven

(19.) stability

(20.) patellar

(21.) popliteal

(22.) fibula

(23.) cruciate

(24.) femoral

(25.) tibial

(26.) tibial collateral

(27.) fibular collateral

[L2]   MULTIPLE CHOICE—pp. 164–167

(28.) A

(29.) C

(30.) B

(31.) D

(32.) C

(33.) A

(34.) C

(35.) C

(36.) C

(37.) A

(38.) D

(39.) B

(40.) C

(41.) D

(42.) D

(43.) A

(44.) C

[L2]   COMPLETION—p. 167

(45.) amykylosis

(46.) arthritis

(47.) gomphosis

(48.) synchondrosis

(49.) syndesmosis

(50.) menisci

(51.) fat pad

(52.) articular cartilage

(53.) hyperextension

(54.) rheumatism

(55.) tendons

(56.) luxation

[L2]   SHORT ESSAY—pp. 167–169

57.   (1) synarthrosis – immovable joint

(2) amphiarthrosis – slightly movable joint

(3) diarthrosis – freely movable joint

58.   A *synostosis* is a totally rigid immovable joint in which two separate bones actually fuse together forming a synarthrosis

A *symphysis* is an amphiarthrotic joint in which bones are separated by a wedge or pad of fibrocartilage.

59. A tendon attaches muscle to bone. Ligaments attach bones to bones.

60. (a) Bursa are small, synovial-filled pockets in connective tissue that form where a tendon or ligament rubs against other tissues. Their function is to reduce friction and act as a shock absorber.

    (b) Menisci are articular discs that: (1) subdivide a synovial cavity (2) channel the flow of synovial fluid, and (3) allow variations in the shape of the articular surfaces.

61. (1) Provides lubrication

    (2) Acts as a shock absorber

    (3) Nourishes the chondrocytes

62. (a) monoaxial – hinge joint (elbow, knee)

    (b) biaxial – condyloid (radiocarpal), saddle (base of thumb)

    (c) triaxial – ball and socket joints (shoulder, hip)

63. (1) Gliding joints – intercarpals

    (2) Hinge joints – elbow, knee

    (3) Pivot joints – between atlas and axis

    (4) Ellipsoidal joints – between metacarpals and phalanges

    (5) Saddle joints – base of thumb

    (6) Ball-and-socket joints – shoulder, hip

64. The knee joint and elbow joint are both hinge joints.

65. The shoulder joint and hip joint are both ball-and-socket joints.

66. The articulations of the carpals and tarsals are gliding diarthroses, which permit a limited gliding motion.

67. Flexion, extension, abduction, adduction

68. The scapulohumeral joint has a relatively oversized loose articular capsule that extends from the scapular neck to the humerus. The loose oversized capsule permits an extensive range of motion, but does not afford the stability found in the ball and socket joints.

69. Intervertebral discs are not found between the first and second cervical vertebrae (atlas and axis), the sacrum and the coccyx. C, the atlas, sits on top of C, the axis; the den of the axis provides for rotation of the first cervical vertebrae which supports the head. An intervertebral head would prohibit rotation. The sacrum and the coccyx are fused bones.

70. (1) inversion   (2) opposition   (3) plantar flexion

    (4) protraction   (5) depression   (6) elevation

71. Both the elbow and knee are extremely stable joints. Stability for the elbow joint is provided by the:

    (a) interlocking of the bony surfaces of the humerus and ulna

    (b) the thickness of the articular capsule

    (c) strong ligaments reinforcing the capsule

    Stability for the hip joint is provided by the:

    (a) almost complete bony socket

    (b) strong articular capsule

    (c) supporting ligaments

    (d) muscular padding

72. The most vulnerable structures subject to possible damage if a locked knee is struck from the side are torn menisci and serious damage to the supporting ligaments.

[L3] **CRITICAL THINKING/APPLICATION—p. 170**

1. In the hip joint, the femur articulates with a deep, complete bony socket, the acetabulum. The articular capsule is unusually strong; there are numerous supporting ligaments, and there is an abundance of muscular padding. The joint is extremely stable and mobility is sacrificed for stability.

   In the shoulder joint, the head of the humerus articulates with a shallow glenoid fossa and there is an oversized loose articular capsule that contributes to the weakness of the joint. The joint contains bursae that serve to reduce friction, and ligaments and muscles contribute to minimal strength and stability of the area. In this joint, stability is sacrificed for mobility.

2. Diarthroses, or synovial joints such as the knee joint, contain small pockets of synovial fluid which are called bursae. The "water on the knee" is the synovial fluid that has been released from the bursae due to ligament damage in or around the joint capsule. The synovial fluid leaks out of the bursae and fills the cavities in and around the region of the knee.

3. Continual excessive force causes chronic knee pain, a condition referred to as chrondromalacia patella. The condition is exemplified by irritation and inflammation of cartilage under the kneecap. Studies have shown that the compressive forces on the knee while using the stairmaster are equal to 5 times the body weight on the way up and 7 times the body weight on the way down.

4. His pain is probably caused by bursitis, an inflammation of the bursae. Bursitis can result from repetitive motion, infection, trauma, chemical irritation, and pressure over the joint. Given the location of the pain, his case probably results from the repetitive motion of practicing pitches.

5. Steve probably tore the medial meniscus. This is the most common knee injury, and is caused by the lateral surface of the lower leg being forced medially. The torn cartilage is painful, usually restricts joint mobility, and may lead to chronic problems. It is possible to examine the interior of a joint without major surgery, by using an arthroscope. An arthroscope uses fiber optics—thin threads of glass or plastic that conduct light—to investigate inside a joint, and if necessary perform surgical modification at that time (arthroscopic surgery). A totally non-invasive method of examination is MRI (magnetic resonance imaging).

6. The first patient probably sustained a trochanteric fracture, which usually heals well if the hip joint can stabilized (not always an easy matter because the powerful muscles surrounding the hip joint can prevent proper alignment of the bone fragments). Steel frames, pins, and/or screws may be used to preserve alignment and encourage healing. The second patient probably suffered a fracture of the femoral neck; these fractures have a higher complication rate because the blood supply to that area is more delicate. This means that the procedures which succeed in stabilizing trochanteric fractures are often unsuccessful in stabilizing femoral neck fractures. If pinning fails, the entire joint can be surgically replaced in a procedure which removes the damaged portion of the femur. An artificial femoral head and neck are attached by a spike that extends into the marrow cavity of the shaft. Special cement anchors it and provides a new articular surface to the acetabulum.

7. Bunions are the most common form of pressure-related bursitis; they form over the base of the big toe as a result of friction and distortion of the joint. A frequent cause of this friction and distortion is tight-fitting shoes with pointed toes. This style is certainly more common in women's shoes than in men's, so one would expect that women probably have more trouble with bunions.

Chapter 10:

Muscle Tissue

[L1] MULTIPLE CHOICE—pp. 172–176

(1.) D	(7.) A	(13.) D	(19.) D	(25.) B
(2.) B	(8.) B	(14.) C	(20.) D	(26.) B
(3.) D	(9.) A	(15.) D	(21.) B	(27.) C
(4.) C	(10.) D	(16.) C	(22.) B	(28.) A
(5.) D	(11.) D	(17.) A	(23.) D	(29.) D
(6.) C	(12.) D	(18.) B	(24.) B	

[L1] COMPLETION—pp. 176–177

(30.) contraction	(38.) action potential	(46.) oxygen debt
(31.) epimysium	(39.) recruitment	(47.) endurance
(32.) fascicles	(40.) troponin	(48.) atrophy
(33.) tendon	(41.) treppe	(49.) flaccid
(34.) sarcolemma	(42.) twitch	(50.) hypertrophy
(35.) T tubules	(43.) ATP	(51.) pacemaker
(36.) sarcomeres	(44.) white muscles	(52.) plasticity
(37.) cross-bridges	(45.) red muscles	

[L1] MATCHING—p. 177

(53.) J	(56.) A	(59.) B	(62.) F	(64.) D
(54.) E	(57.) C	(60.) M	(63.) H	(65.) I
(55.) L	(58.) K	(61.) G		

[L1] DRAWING/ILLUSTRATION—LABELING—pp. 178–180

Figure 10.1   Organization of Skeletal Muscles—p. 178

(66.) sarcolemma	(72.) muscle fibers	(77.) epimysium
(67.) sarcoplasm	(73.) endomysium	(78.) muscle fasciculus
(68.) myofibril	(74.) perimysium	(79.) body of muscle
(69.) nucleus	(75.) endomysium	(80.) tendon
(70.) muscle fiber	(76.) blood vessels	(81.) bone
(71.) muscle fasciculus		

Figure 10.2   The Histological Organization of Skeletal Muscles—p. 179

(82.) myofibril	(85.) nucleus	(88.) sarcomere
(83.) mitochondrion	(86.) sarcoplasmic reticulum	(89.) thin filament (actin)
(84.) muscle fiber	(87.) T tubules	(90.) thick filament (myosin)

Figure 10.3   Types of Muscle Tissue—p. 179

(91.) smooth	(92.) cardiac	(93.) skeletal

Figure 10.4   Structure of a Sarcomere—p. 180

(94.) sarcomere	(97.) Z line	(100.) myosin (thick filaments)
(95.) Z line	(98.) I band	(101.) actin (thin filaments)
(96.) H zone	(99.) A band	

Figure 10.5   Neuromuscular Junction—p. 180

(102.) nerve impulse

(103.) neurilemma

(104.) sarcolemma

(105.) neuron (axon)

(106.) vesicle containing acetylcholine

(107.) synaptic cleft

[L2]   CONCEPT MAPS—pp. 180–183

I   MUSCLE TISSUE—p. 181

(1.) heart

(2.) striated

(3.) smooth

(4.) involuntary

(5.) non-striated

(6.) bones

(7.) multi-nucleated

II   MUSCLE—p. 182

(8.) muscle bundles (fascicles)

(9.) myofibrils

(10.) sarcomeres

(11.) actin

(12.) Z lines

(13.) thick filaments

(14.) H zone

III   MUSCLE CONTRACTION—p. 183

(15.) release of Ca^{++} from sacs of sarcoplasmic reticulum

(16.) cross-bridging (heads of myosin attach to turned-on thin filaments)

(17.) shortening, i.e., contraction of myofibrils and muscle fibers they comprise

(18.) energy + ADP + phosphate

[L2]   BODY TREK—p. 184

(19.) epimysium

(20.) perimysium

(21.) fascicles

(22.) endomysium

(23.) satellite

(24.) sarcolemma

(25.) T tubules

(26.) nuclei

(27.) myofibrils

(28.) myofilaments

(29.) actin

(30.) myosin

(31.) sarcomeres

(32.) Z line

(33.) I band

(34.) A band

(35.) thick

(36.) sliding filament

(37.) contraction

[L2]   MULTIPLE CHOICE—pp. 184–187

(38.) A

(39.) B

(40.) B

(41.) C

(42.) D

(43.) C

(44.) B

(45.) B

(46.) D

(47.) D

(48.) C

(49.) A

(50.) B

(51.) C

(52.) C

(53.) A

(54.) A

(55.) A

(56.) C

(57.) C

[L2]   COMPLETION—pp. 187–188

(58.) satellite

(59.) myoblasts

(60.) muscle tone

(61.) rigor

(62.) absolute refractory period

(63.) fatigue

(64.) A bands

(65.) motor unit

(66.) tetanus

(67.) isotonic

[L2]   SHORT ESSAY—pp. 188–189

68.   (a) produce skeletal movement

(b) maintain posture and body position

(c) support soft tissues

(d) guard entrances and exits

(e) maintain body temperature

69.   (a) an outer *epimysium*, (b) a central *perimysium*, and (c) an inner *endomysium*.

70.

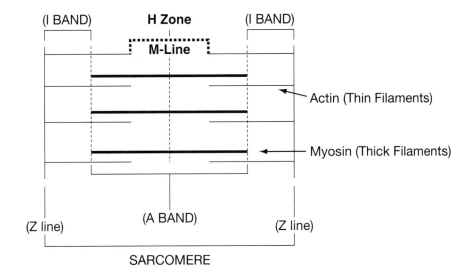

SARCOMERE

71. (a) active site exposure
     (b) cross-bridge attachment
     (c) pivoting
     (d) cross-bridging detachment
     (e) myosin activation

72. (a) release of acetylcholine
     (b) depolarization of the motor end plate
     (c) generation of an action potential
     (d) conduction of an action potential
     (e) release of calcium ions

73. A—twitch; B—incomplete tetanus; C—complete tetanus

74. *Isometric* contraction—tension rises to a maximum but the length of the muscle remains constant.

     Isotonic contraction—tension in the muscle builds until it exceeds the amount of resistance and the muscle shortens. As the muscle shortens, the tension in the muscle remains constant, at a value that just exceeds the applied resistance.

75. When fatigue occurs and the oxygen supply to muscles is depleted, aerobic respiration ceases owing to the decreased oxygen supply. Anaerobic glycolysis supplies the needed energy for a short period of time. The amount of oxygen needed to restore normal pre-exertion conditions is the oxygen debt.

76. Fast fiber muscles produce powerful contractions, which use ATP in massive amounts. Prolonged activity is primarily supported by anaerobic glycolysis, and fast fibers fatigue rapidly. Slow fibers are specialized to enable them to continue contracting for extended periods. The specializations include an extensive network of capillaries so supplies of O_2 are available, and the presence of myoglobin, which binds O_2 molecules and which results in the buildup of O_2 reserves. These factors improve mitochondrial performance.

[L3] CRITICAL THINKING/APPLICATION—p. 190

1. Alcohol is a drug that interferes with the release of acetylcholine at the neuromuscular junction. When acetylcholine is not released, action potentials in the muscle cell membrane are prohibited and the release of calcium ions necessary to begin the contractile process does not occur.

2. Training to improve *aerobic* endurance usually involves sustained low levels of muscular activity. Jogging, distance swimming, biking, and other cardiovascular activities that do not require peak tension production are appropriate for developing aerobic endurance. Training to develop *anaerobic* endurance involves frequent, brief, intensive workouts. Activities might include running sprints; fast, short-distance swimming; pole vaulting; or other exercises requiring peak tension production in a short period of time.

3. *Rigor mortis* occurs several hours after death. It is caused by the lack of ATP in muscles to sustain the contractile process or to allow the cross-bridge to release. The muscles remain stiff until degeneration occurs. *Physiological contracture* occurs in the living individual as a result of a lack of ATP in muscle fibers. The lack of available energy does not allow cross-bridging to occur and prohibits the release of previously formed cross-bridges, resulting in physiological contracture.

4. This diet probably lacks dietary calcium and vitamin D. The resulting condition is hypo-calcemia, a lower-than-normal concentration of calcium ions in blood or extracellular fluid. Because of the decreased number of calcium ions, sodium ion channels open and the sodium ions diffuse into the cell, causing depolarization of the cell membranes to threshold and initiate action potentials. Spontaneous reactions of nerves and muscles result in nervousness and muscle spasms.

5. Both attack motor neurons in the nervous system. The Botulinus toxin prevents the release of acetylcholine by the motor neuron, producing a severe and potentially fatal paralysis. The poliovirus attacks and kills motor neurons in the spinal cord and brain. The infection causes degeneration of the motor neurons and atrophy of the muscles they innervate.

Chapter 11:

The Muscular System: Organization

[L1]   MULTIPLE CHOICE—pp. 192–195

(1.) C	(6.) D	(11.) A	(16.) C	(21.) D
(2.) B	(7.) B	(12.) C	(17.) A	(22.) D
(3.) D	(8.) B	(13.) B	(18.) A	(23.) C
(4.) C	(9.) C	(14.) C	(19.) D	(24.) C
(5.) B	(10.) D	(15.) A	(20.) B	(25.) A

[L1]   COMPLETION—p. 196

(26.) biceps brachii	(30.) origin	(34.) diaphragm
(27.) sphincters	(31.) synergist	(35.) deltoid
(28.) third-class	(32.) popliteus	(36.) rectus femoris
(29.) second-class	(33.) cervicis	

[L1]   MATCHING—p. 196

(37.) D	(40.) A	(43.) B	(46.) O	(49.) J
(38.) G	(41.) H	(44.) C	(47.) N	(50.) L
(39.) F	(42.) E	(45.) K	(48.) I	(51.) M

[L1]   DRAWING/ILLUSTRATION—LABELING—pp. 197–202

Figure 11.1   Major Superficial Skeletal Muscles (anterior view)—p. 197

(52.) sternocleidomastoid	(57.) rectus femoris	(62.) rectus abdominis
(53.) deltoid	(58.) vastus lateralis	(63.) gracilis
(54.) biceps brachii	(59.) vastus medialis	(64.) sartorius
(55.) external obliques	(60.) tibialis anterior	(65.) gastrocnemius
(56.) brachioradialis	(61.) pectoralis major	(66.) soleus

Figure 11.2   Major Superficial Skeletal Muscles (posterior view)—p. 198

(67.) sternocleidomastoid	(72.) biceps femoris	(77.) gluteus medius
(68.) teres major	(73.) tibialis posterior	(78.) gluteus maximus
(69.) latissimus dorsi	(74.) trapezius	(79.) gastrocnemius
(70.) external oblique	(75.) deltoid	(80.) Achilles tendon
(71.) semitendinosus	(76.) triceps brachii	

Figure 11.3   Facial Muscles (lateral view)—p. 199

(81.) occipitalis	(85.) orbicularis oculi	(89.) orbicularis oris
(82.) masseter	(86.) levator labii	(90.) risorius
(83.) temporalis	(87.) zygomaticus major	(91.) platysma
(84.) frontalis	(88.) buccinator	(92.) sternocleidomastoid

Figure 11.4   Muscles of the Neck (anterior view)—p. 199

(93.) mylohyoid	(96.) sternohyoid	(98.) digastric
(94.) stylohyoid	(97.) sternothyroid	(99.) sternocleidomastoid
(95.) thyrohyoid		

Figure 11.5   Muscles of the Forearm (anterior view)—p. 200

(100.) biceps brachii

(101.) pronator teres

(102.) brachioradialis

(103.) radialis longus

(104.) flexor carpi radialis

(105.) palmaris longus

(106.) flexor carpi ulnaris

(107.) pronator quadratus

Figure 11.6   Muscles of the Hand—p. 200

(108.) abductor pollicis brevis

(109.) flexor pollicis brevis

(110.) tendon flexor pollicis longus

(111.) adductor pollicis

(112.) tendon flexor digitorium profundus

(113.) flexor retinaculum

(114.) palmaris brevis

(115.) lumbricales

Figure 11.7   Muscles of the Thigh (anterior view)—p. 201

(116.) sartorius

(117.) rectus femoris

(118.) vastus lateralis

(119.) quadriceps femoris tendon

(120.) gracilis

(121.) vastus medialis

Figure 11.8   Muscles of the Thigh (posterior view)—p. 201

(122.) adductor magnus

(123.) semitendinosus

(124.) biceps femoris

(125.) semimembranosus

Figure 11.9   Muscles of the Lower Leg (lateral view)—p. 202

(126.) gastrocnemius

(127.) soleus

(128.) extensor digitorium longus

(129.) tibialis anterior

(130.) peroneus longus

(131.) peroneus brevis

Figure 11.10   Muscles of the Foot (plantar surface)—p. 202

(132.) lumbricales

(133.) abductor digiti minimi

(134.) abductor hallicus

(135.) flexor digitorium brevis

[L2]   CONCEPT MAPS—pp. 203–204

I   AXIAL MUSCULATURE—p. 203

(1.) buccinator

(2.) oculomotor muscles

(3.) masseter

(4.) hypoglossus

(5.) stylohyoid

(6.) splenius

(7.) capitis

(8.) thoracis

(9.) scalenes

(10.) external-internal intercostals

(11.) external obliques

(12.) sternohyoid

(13.) diaphragm

(14.) rectus abdominis

(15.) bulbocavernosus

(16.) urethral sphincter

(17.) external anal sphincter

II   APPENDICULAR MUSCULATURE—p. 204

(18.) trapezius

(19.) deltoid

(20.) biceps brachii

(21.) triceps

(22.) flexor carpi radialis

(23.) extensor carpi ulnaris

(24.) extensor digitorum

(25.) gluteus maximus

(26.) psoas

(27.) obturators

(28.) biceps femoris

(29.) rectus femoris

(30.) gastrocnemius

(31.) flexor digitorum longus

[L2]   BODY TREK: Origins – Insertion Report—p. 205

(32.) skin of eyebrow, bridge of nose

(33.) medial margin of orbit

(34.) lips

(35.) manubrium and clavicle

(36.) clavicle, scapula

(37.) ribs (8–12), thoracic-lumbar vertebrae

(38.) deltoid tuberosity of humerus

(39.) scapula

(40.) ulna-olecranon process

(41.) humerus-lateral epicondyle

(42.) humerus-greater tubercle

(43.) ribs

(44.) symphyis pubis

(45.) lower eight ribs

(46.) iliotibial tract and femur

(47.) ischium, femur, pubis, ilium

(48.) tibia

(49.) femoral condyles

(50.) first metatarsal

[L2]   MULTIPLE CHOICE—pp. 206–208

(51.) B	(54.) C	(57.) C	(60.) C	(63.) B
(52.) A	(55.) D	(58.) D	(61.) A	(64.) C
(53.) D	(56.) A	(59.) B	(62.) D	(65.) A

[L2]   COMPLETION—p. 208

(66.) biomechanics

(67.) cervicis

(68.) innervation

(69.) sartorius

(70.) risorius

(71.) perineum

(72.) hamstrings

(73.) quadriceps

(74.) strain

(75.) charley horse

[L2]   SHORT ESSAY—pp. 208–209

76. (a) parallel muscle—biceps brachii

(b) convergent muscle—pectoralis major

(c) pennate muscle—deltoid

(d) circular muscle—pyloric sphincter

77. The origin remains stationary while the insertion moves.

78. prime mover or agonist, synergist, antagonists

79. (a) muscles of the head and neck

(b) muscles of the spine

(c) oblique and rectus muscles

(d) muscles of the pelvic floor

80. (a) muscles of the shoulders and arms

(b) muscles of the pelvic girdle and legs

81. first-class lever, see-saw, muscles that extend the neck

second-class lever, wheelbarrow, person standing on one's toes

third-class lever, shovel, flexing the forearm

82. Biceps femoris, semimembranosus, semitendinosus, gracilis, sartorius

83. Rectus femoris, vastus intermedius, vastus lateralis, vastus medialis

[L3]   CRITICAL THINKING/APPLICATIONpp. 209

1. A lever is a rigid structure that moves on a fixed point, called the *fulcrum*. In the human body, each bone is a lever and each joint is a fulcrum. Levers can change (a) the direction of an applied force, (b) the distance and speed of movement produced by a force, and (c) the strength of a force.

2. The rotator cuff muscles are located in the shoulder region and are responsible for acting on the arm. They include the infraspinatus, subscapularis, supraspinatus, and teres minor.

3. The "hamstrings" are found in the posterior region of the upper leg, or thigh. The muscles include the biceps femoris, semimembranosus, and semitendinosus.

4. When a crushing injury, severe contusion, or a strain occurs, the blood vessels within one or more compartments may be damaged. The compartments become swollen with blood and fluid leaked from damaged vessels. Because the connective tissue partitions are very strong, the accumulated fluid cannot escape, and pressures rise within the affected compartments. Eventually compartment pressure becomes so high that they compress the regional blood vessels and eliminate the circulatory supply to the muscles and nerves of the compartment.

5. (a) Injecting into tissues rather than circulation allows a large amount of a drug to be introduced at one time because it will enter the circulation gradually. The uptake is usually faster and produces less irritation than when drugs are administered through intradermal or subcutaneous routes; also, multiple injections are possible.

(b) Bulky muscles containing few large blood vessels or nerves are ideal targets. The muscles that may serve as targets are the posterior, lateral, and superior portion of the gluteus maximus; the deltoid of the arm; and the vastus lateralis of the thigh. The vastus lateralis of the thigh is the preferred site in infants and young children whose gluteal and deltoid muscles are relatively small.

Chapter 12:

Neural Tissue

[L1] MULTIPLE CHOICE—pp. 211–215

(1.) A	(7.) C	(13.) D	(19.) A	(25.) D
(2.) C	(8.) C	(14.) B	(20.) B	(26.) C
(3.) D	(9.) D	(15.) C	(21.) B	(27.) A
(4.) D	(10.) B	(16.) D	(22.) D	(28.) A
(5.) A	(11.) D	(17.) C	(23.) B	(29.) C
(6.) B	(12.) C	(18.) C	(24.) A	(30.) C

[L1] COMPLETION—pp. 215–216

(31.) autonomic nervous system

(32.) microglia

(33.) collaterals

(34.) afferent

(35.) electrochemical gradient

(36.) threshold

(37.) saltatory

(38.) cholinergic

(39.) adrenergic

(40.) neuromodulators

(41.) temporal summation

(42.) spatial summation

(43.) postsynaptic potentials

(44.) proprioceptors

(45.) divergence

[L1] MATCHING—p. 216

(46.) H	(50.) D	(54.) B	(57.) O	(60.) K
(47.) F	(51.) C	(55.) Q	(58.) J	(61.) P
(48.) G	(52.) I	(56.) L	(59.) M	(62.) N
(49.) A	(53.) E			

[L1] DRAWING/ILLUSTRATION—LABELING—pp. 217–218

Figure 12.1 Structure & Classification of Neurons—p. 217

(63.) dendrites

(64.) neurilemma

(65.) soma (cell body)

(66.) axon hillock

(67.) soma

(68.) nucleus

(69.) nucleus of Schwann cell

(70.) axon

(71.) Schwann cell

(72.) nodes of Ranvier

(73.) myelin

(74.) telodendria

(75.) motor neuron

(76.) sensory neuron

Figure 12.2 Neuron Classification (based on structure)—p. 218

(77.) multipolar neuron

(78.) unipolar neuron

(79.) anaxonic

(80.) bipolar neuron

Figure 12.3 Organization of Neuronal Pools—p. 218

(81.) convergence

(82.) parallel processing

(83.) serial processing

(84.) reverberation

(85.) divergence

[L2] CONCEPT MAPS—pp. 219–220

I NEURAL TISSUE—p. 219

(1.) Schwann cells

(2.) glial cells

(3.) transmit nerve impulses

(4.) surround peripheral ganglia

(5.) central nervous system

II STRUCTURAL ORGANIZATION OF NERVOUS SYSTEM—p. 220

(6.) brain

(7.) motor system

(8.) somatic nervous system

(9.) sympathetic nervous system

(10.) smooth muscle

(11.) peripheral nervous system

(12.) afferent division

[L2]   MULTIPLE CHOICE—pp. 221–223

(13.) C	(17.) A	(21.) A	(25.) D	(28.) C
(14.) D	(18.) C	(22.) B	(26.) B	(29.) D
(15.) A	(19.) D	(23.) A	(27.) B	(30.) A
(16.) B	(20.) B	(24.) C		

[L2]   COMPLETION—p. 224

(31.) perikaryon	(36.) hyperpolarization	(41.) association
(32.) ganglia	(37.) nerve impulse	(42.) convergence
(33.) current	(38.) stroke	(43.) parallel
(34.) voltage	(39.) preganglionic fibers	(44.) nuclei
(35.) gated	(40.) postganglionic fibers	(45.) tracts

[L2]   SHORT ESSAY—pp. 224–225

46. (a) providing sensation of the internal and external environments

(b) integrating sensory information

(c) coordinating voluntary and involuntary activities

(d) regulating or controlling peripheral structures and systems

47. CNS consists of the brain and the spinal cord

PNS consists of the somatic nervous system and the autonomic nervous system

48. astrocytes, oligodendrocytes, microglia, ependymal cells

49. *Neurons* are responsible for information transfer and processing in the nervous system.
Neuroglia are specialized cells that provide support throughout the nervous system.

50. (a) activation of sodium channels and membrane depolarization

(b) sodium channel inactivation

(c) potassium channel activation

(d) return to normal permeability

51. A node of Ranvier represents an area along the axon where there is an absence of myelin. Because ions can cross the membrane only at the nodes, only a node can respond to a depolarizing stimulus. Action potentials appear to "leap" or "jump" from node to node, a process called *saltatory conduction.* The process conducts nerve impulses along an axon five to seven times faster than continuous conduction.

52. In an adrenergic synapse at the postsynaptic membrane surface, norepinephrine activates an enzyme, adenyl cyclase, that catalyzes the conversion of ATP to cyclic AMP (cAMP) on the inner surface of the membrane. cAMP then activates cytoplasmic enzymes that open ion channels and produce depolarization. The cAMP is called a *second messenger.*

53. An EPSP is a depolarization produced by the arrival of a neurotransmitter at the postsynaptic membrane. A typical EPSP produces a depolarization of around 0.5 mV, much less than the 15–20 mV depolarization needed to bring the axon hillock to threshold.

An IPSP is a transient hyperpolarization of the postsynaptic membrane. A hyperpolarized membrane is inhibited because a larger-than-usual depolarization stimulus must be provided to bring the membrane potential to threshold.

54. (a) sensory neurons—carry nerve impulses to the CNS

(b) motor neurons—carry nerve impulses from CNS to PNS

(c) association neurons—situated between sensory and motor neurons within the brain and spinal cord

55. *Divergence* is the spread of information from one neuron to several neurons, or from one neuronal pool to multiple pools.

Convergence—several neurons synapse on the same postsynaptic neuron.

[L3]   CRITICAL THINKING/APPLICATION—p. 226

1. The combination of coffee and cigarette has a strong stimulatory effect, making the person appear to be "nervous" or "jumpy" or feeling like he or she is "on edge." The caffeine in the coffee lowers the threshold at the axon hillock, making the neurons more sensitive to depolarizing stimuli. Nicotine stimulates ACh (acetylcholine) receptors by binding to the ACh receptor sites.

2. In myelinated fibers, saltatory conduction transmits nerve impulses at rates over 300 mph, allowing the impulses to reach the neuromuscular junctions fast enough to initiate muscle contraction and promote normal movements. In unmyelinated fibers, continuous conduction transmits impulses at rates of approximately 2 mph. The impulses do not reach the peripheral neuromuscular junctions fast enough to initiate muscle contractions, which promote normal movements. Eventually the muscles atrophy because of a lack of adequate activity involving contraction.

3. Microglia do not develop in neural tissue; they are phagocytic WBC that have migrated across capillary walls in the neural tissue of the CNS. They engulf cellular debris, waste products, and pathogens. In times of infection or injury their numbers increase dramatically, as other phagocytic cells are attracted to the damaged area.

4. Acetylcholine is released at synapses in the central and peripheral nervous systems, and in most cases ACh produces a depolarization in the postsynaptic membrane. The ACh released at neuromuscular junctions in the heart produces a transient hyperpolarization of the membrane, moving the transmembrane potential farther from threshold.

5. A subthreshold membrane potential is referred to as an excitatory postsynaptic potential (EPSP), and the membrane is said to be *facilitated*. EPSPs may combine to reach threshold and initiate an action potential in two ways: (a) by spatial summation, during which several presynaptic neurons simultaneously release neurotransmitter to a single postsynaptic neuron; and (b) by temporal summation, during which the EPSPs result from the rapid and successive discharges of neurotransmitter from the same presynaptic knob.

Chapter 13:

The Spinal Cord and Spinal Nerves

[L1]   MULTIPLE CHOICE—pp. 228–231

(1.) D	(6.) C	(11.) B	(16.) D	(21.) A
(2.) A	(7.) A	(12.) C	(17.) C	(22.) B
(3.) A	(8.) A	(13.) B	(18.) A	
(4.) C	(9.) C	(14.) D	(19.) A	
(5.) A	(10.) C	(15.) B	(20.) C	

[L1]   COMPLETION—p. 231

(23.) conus medullaris

(24.) filum terminale

(25.) nuclei

(26.) columns

(27.) epineurium

(28.) dorsal ramus

(29.) dermatome

(30.) neural reflexes

(31.) receptor

(32.) innate reflexes

(33.) acquired reflexes

(34.) cranial reflexes

(35.) somatic reflexes

(36.) crossed extensor reflex

(37.) flexor reflex

(38.) reinforcement

(39.) Babinski sign

[L1]   MATCHING—p. 232

(40.) G	(43.) E	(46.) D	(49.) I	(52.) N
(41.) F	(44.) B	(47.) K	(50.) H	(53.) O
(42.) A	(45.) C	(48.) J	(51.) L	(54.) M

[L1]   DRAWING/ILLUSTRATION—LABELING—p. 233–234

Figure 13.1   Organizaion of the Spinal Cord—p. 233

(55.) white matter

(56.) central canal

(57.) gray matter (anterior horn)

(58.) spinal nerve

(59.) arachnoid

(60.) dura mater

(61.) posterior gray horn

(62.) pia mater

(63.) gray commissure

(64.) subarachnoid space

Figure 13.2   Structure of the Reflex Arc—p. 234

(65.) white matter

(66.) gray matter

(67.) synapse

(68.) interneuron

(69.) sensory (afferent) neuron

(70.) motor (efferent) neuron

(71.) effector (muscle)

(72.) receptor (skin)

[L2]   CONCEPT MAPS—pp. 235–236

I   SPINAL CORD—p. 235

(1.) gray matter

(2.) glial cells

(3.) posterior gray horns

(4.) sensory information

(5.) viscera

(6.) somatic motor nuclei

(7.) peripheral effectors

(8.) anterior white columns

(9.) ascending tract

(10.) brain

(11.) motor commands

II   SPINAL CORD—MENINGES—p. 236

(12.) dura mater

(13.) stability, support

(14.) lymphatic fluid

(15.) pia mater

(16.) subarachnoid space

(17.) shock absorber

(18.) diffusion medium

(19.) blood vessels

[L2]   BODY TREK—p. 237

(20.) exteroceptors

(21.) proprioceptors

(22.) interoceptors

(23.) peripheral sensory ganglion

(24.) sensory

(25.) afferent fibers (sensory)

(26.) interneurons

(27.) somatic motor neurons

(28.) motor

(29.) efferent fibers (motor)

(30.) skeletal muscles

(31.) visceral motor neurons

(32.) preganglionic efferent fibers

(33.) postganglionic efferent fibers

(34.) peripheral motor ganglion

(35.) glands

(36.) smooth muscle

(37.) fat cells

Note: Numbers 20 and 21 could be in reverse order. Numbers 35, 36, and 37 can be in any order.

[L2]   MULTIPLE CHOICE—pp. 238–241

(38.) C	(43.) B	(48.) B	(52.) A	(56.) B
(39.) B	(44.) A	(49.) D	(53.) C	(57.) C
(40.) A	(45.) C	(50.) B	(54.) A	(58.) A
(41.) D	(46.) A	(51.) D	(55.) D	(59.) C
(42.) D	(47.) C			

[L2]   COMPLETION—p. 241

(60.) perineurium

(61.) descending tracts

(62.) dura mater

(63.) spinal meninges

(64.) pia mater

(65.) cauda equina

(66.) sensory nuclei

(67.) motor nuclei

(68.) myelography

(69.) gray commissures

(70.) gray ramus

(71.) dorsal ramus

(72.) nerve plexus

(73.) brachial plexus

(74.) reflexes

(75.) gamma efferents

[L2]   SHORT ESSAY—pp. 242–243

76.   (a) dura mater, (b) arachnoid, (c) pia mater

77.   Mixed nerves contain both afferent (sensory) and efferent (motor) fibers.

78.   The white matter contains large numbers of myelinated and unmyelinated axons. The gray matter is dominated by bodies of neurons and glial cells.

79.   A dermatome is a specific region of the body surface monitored by a pair of spinal nerves. Damage to a spinal nerve or dorsal root ganglion will produce characteristic loss of sensation in the skin.

80.   (a) Arrival of a stimulus and activation of a receptor

(b) Activation of a sensory neuron

(c) Information processing

(d) Activation of a motor neuron

(e) Response of a peripheral effector

81.   (a) their development

(b) the site where information processing occurs

(c) the nature of the resulting motor response

(d) the complexity of the neural circuit involved

82.   (a) The interneurons can control several different muscle groups.

(b) The interneurons may produce either excitatory or inhibitory postsynaptic potentials at CNS motor nuclei; thus the response can involve the stimulation of some muscles and the inhibition of others.

83.   When one set of motor neurons is stimulated, those controlling antagonistic muscles are inhibited. The term reciprocal refers to the fact that the system works both ways. When the flexors contract, the extensors relax; when the extensors contract, the flexors relax.

84. In a contralateral reflex arc, the motor response occurs on the side opposite the stimulus. In an ipsilateral reflex arc, the sensory stimulus and the motor response occur on the same side of the body.

85. The Babinski reflex in an adult is a result of an injury in the CNS. Usually the higher centers or the descending tracts are damaged.

[L3]   CRITICAL THINKING/APPLICATION—p. 243

1. Because the spinal cord extends to the L2 level of the vertebral column and the meninges extend to the end of the vertebral column, a needle can be inserted through the meninges inferior to the medullary cone into the subarachnoid space with minimal risk to the cauda equina.

2. Meningitis is an inflammation of the meninges. It is caused by viral or bacterial infection and produces symptoms that include headache and fever. Severe cases result in paralysis, coma, and sometimes death.

3. Multiple sclerosis (MS) results from the degeneration of the myelin sheaths that surround the axons in the brain and the spinal cord. This demyelination decreases the rate of action potential conduction along the neurons, which ultimately inhibits normal muscle contraction and leads to eventual muscle degeneration and atrophy.

4. The positive patellar reflex will confirm that the spinal nerves and spinal segments $L_2 - L_4$ are not damaged.

5. (a) Eliminates sensation and motor control of the arms and legs. Usually results in extensive paralysis—quadriplegia.

   (b) Motor paralysis and major respiratory muscles such as the diaphragm—patient needs mechanical assistance to breathe.

   (c) The loss of motor control of the legs—paraplegia.

   (d) Damage to the elements of the cauda equina causes problems with peripheral nerve function.

Chapter 14:

The Brain
and the Cranial Nerves

[L1]   MULTIPLE CHOICE—pp. 245–248

(1.) D	(6.) A	(11.) A	(16.) B	(21.) D
(2.) B	(7.) D	(12.) B	(17.) A	(22.) A
(3.) A	(8.) D	(13.) C	(18.) D	(23.) B
(4.) D	(9.) C	(14.) B	(19.) C	(24.) C
(5.) A	(10.) B	(15.) C	(20.) B	(25.) D

[L1]   COMPLETION—pp. 248–249

(26.) epithalamus	(31.) tentorium cerebelli	(36.) amygdaloid body
(27.) neural cortex	(32.) choroid plexus	(37.) corpora quadrigemina
(28.) third ventricle	(33.) shunt	(38.) hypoglossal
(29.) aqueduct of Sylvius	(34.) postcentral gyrus	(39.) spinal accessory
(30.) falx cerebri	(35.) pyramidal cells	(40.) vestibuloocular

[L1]   MATCHING—p. 249

(41.) H	(44.) F	(47.) A	(50.) J	(53.) K
(42.) C	(45.) G	(48.) D	(51.) L	(54.) I
(43.) E	(46.) B	(49.) O	(52.) N	(55.) M

[L1]   DRAWING/ILLUSTRATION—LABELING—p. 250

Figure 14.1   Lateral View of the Human Brain—p. 250

(56.) central sulcus	(60.) cerebellum	(64.) frontal lobe
(57.) parietal lobe	(61.) medulla oblongata	(65.) lateral fissure
(58.) parieto-occipital fissure	(62.) postcentral gyrus	(66.) temporal lobe
(59.) occipital lobe	(63.) precentral gyrus	(67.) pons

Figure 14.2   Sagittal View of the Human Brain—p. 250

(68.) choroid plexus	(74.) fourth ventricle	(80.) optic chiasma
(69.) cerebral hemispheres	(75.) cerebellum	(81.) pituitary gland
(70.) corpus callosum	(76.) thalamus	(82.) mammilary body
(71.) pineal body	(77.) fornix	(83.) pons
(72.) cerebral peduncle	(78.) third ventricle	(84.) medulla oblongata
(73.) cerebral aqueduct	(79.) corpora quadrigemina	

[Figure 14.3   Origins of Cranial Nerves–Inferior View of the Brain–p. 250

(85.) olfactory bulb	(89.) N VI glossopharyngeal	(93.) N VI abducens
(86.) N III oculomotor	(90.) N X vagus	(94.) N VIII vestibulocochlear
(87.) N V trigeminal	(91.) N II optic	(95.) N XII hypoglossal
(88.) N VII facial	(92.) N IV trochlear	(96.) N XI accessory

L2]   CONCEPT MAPS—pp. 251–257

I   MAJOR REGIONS OF THE BRAIN—p. 251

(1.) diencephalon   (3.) corpora quadrigemina   (5.) pons
(2.) hypothalamus   (4.) 2 cerebellar hemispheres   (6.) medulla oblongata

II   REGION I: CEREBRUM—p. 252

(7.) temporal   (10.) cerebral cortex   (12.) commissural
(8.) somesthetic cortex   (11.) projections   (13.) nuclei
(9.) occipital

III   REGION II: DIENCEPHALON—p. 253

(14.) thalamus   (17.) geniculates   (20.) pineal gland
(15.) anterior   (18.) preoptic area   (21.) cerebrospinal fluid
(16.) ventral   (19.) autonomic centers

IV   REGION III: MESENCEPHALON—p. 254

(22.) cerebral peduncles   (24.) corpora quadrigemina   (26.) red nucleus
(23.) gray matter   (25.) inferior colliculi   (27.) substantia nigra

V   REGION IV: CEREBELLUM—p. 255

(28.) vermis   (30.) gray matter   (32.) cerebellar nuclei
(29.) arbor vitae   (31.) Purkinje cells

VI   REGION V: PONS—p. 256

(33.) respiratory centers   (35.) white matter   (37.) inferior
(34.) apneustic   (36.) superior   (38.) transverse fibers

VII   REGION VI: MEDULLA OBLONGATA—p. 257

(39.) white matter   (42.) olivary   (45.) distribution of blood flow
(40.) brain and spinal cord (43.) reflex centers   (46.) respiratory
(41.) cuneatus   (44.) cardiac   rhythmicity center

[L2]   MULTIPLE CHOICE—pp. 258–261

(47.) D   (52.) C   (57.) B   (62.) A   (67.) C
(48.) A   (53.) A   (58.) A   (63.) C   (68.) D
(49.) D   (54.) D   (59.) D   (64.) D   (69.) B
(50.) D   (55.) C   (60.) C   (65.) B   (70.) B
(51.) B   (56.) B   (61.) B   (66.) A   (71.) C

[L2]   COMPLETION—pp. 261–262

(72.) thalamus   (77.) hippocampus   (82.) spinal cord
(73.) pituitary gland   (78.) fornix   (83.) sulci
(74.) arcuate   (79.) drives   (84.) fissures
(75.) commissural   (80.) third ventricle   (85.) extrapyramidal system
(76.) corpus striatum   (81.) aqueduct of Sylvius   (86.) hypothalamus

[L2]   REVIEW OF CRANIAL NERVES—p. 262

(87.) olfactory   (91.) glossopharyngeal   (95.) M
(88.) oculomotor   (92.) spinal accessory   (96.) S
(89.) trigeminal   (93.) S   (97.) B
(90.) facial   (94.) M   (98.) M

[L2]   SHORT ESSAY—pp. 263–264

99.   The brain's versatility results from (a) the tremendous number of neurons and neuronal pools

in the brain, and (b) the complexity of the interconnections between the neurons and neuronal pools.

100. (a) cerebrum; (b) diencephalon; (c) mesencephalon; (d) cerebellum; (e) pons; (f) medulla oblongata

101. "Higher centers" refers to nuclei, centers, and cortical areas of the cerebrum, cerebellum, diencephalon, and mesencephalon.

102. The limbic system includes nuclei and tracts along the border between the cerebrum and diencephalon. It is involved in the processing of memories, creation of emotional states, drives, and associated behaviors.

103. Most endocrine organs are under direct or indirect hypothalamic control. Releasing hormones and inhibiting hormones secreted by nuclei in the tuberal area of the hypothalamus promote or inhibit secretion of hormones by the anterior pituitary gland. The hypothalamus also secretes the hormones ADH (antidiuretic hormone) and oxytocin.

104. (a) The cerebellum oversees the postural muscles of the body, making rapid adjustments to maintain balance and equilibrium. (b) The cerebellum programs and times voluntary and involuntary movements.

105. (a) Sensory and motor nuclei for four of the cranial nerves; (b) nuclei concerned with the involuntary control of respiration; (c) tracts that link the cerebellum with the brain stem, cerebrum, and spinal cord; (d) ascending and descending tracts.

106. It isolates neural tissue in the CNS from the general circulation.

107. (a) It provides cushioning, (b) it provides support, and (c) it transports nutrients, chemical messengers, and waste products.

108. N I—olfactory, N II—optic, N VIII—vestibulocochlear.

109. Cranial reflexes provide a quick and easy method for checking the condition of cranial nerves and specific nuclei and tracts in the brain.

[L3]   CRITICAL THINKING/APPLICATION—p. 264

1. Even though cerebrospinal fluid exits in the brain are blocked, CSF continues to be produced. The fluid continues to build inside the brain, causing pressure that compresses the nerve tissue and causes the ventricles to dilate. The compression of the nervous tissue causes irreversible brain damage.

2. Smelling salts (i.e., ammonia) stimulate trigeminal nerve endings in the nose, sending impulses to the reticular formation in the brain stem and the cerebral cortex. The axons in the reticular formation cause arousal and maintenance of consciousness. The reticular formation and its connections form the reticular activating system that is involved with the sleep-wake cycle.

3. Activity in the thirst center produces the conscious urge to take a drink. Hypothalamic neurons in this center detect changes in the osmotic concentration of the blood. When the concentration rises, the thirst center is stimulated. The thirst center is also stimulated by ADH (antidiuretic hormone). Stimulation of the thirst center in the hypothalamus triggers a behavior response (drinking) that complements the physiological response to ADH (water conservation by the kidneys).

4. The neurotransmitter dopamine is manufactured by neurons in the substantia nigra and carried to synapses in the cerebral nuclei where it has an inhibitory effect. If the dopamine-producing neurons are damaged, inhibition is lost and the excitatory neurons become increasingly active. This increased activity produces the motor symptoms of spasticity and/or tremor associated with Parkinson's disease.

5. The objects serve as stimuli initiating tactile sensations that travel via sensory neurons to the spinal cord. From the ascending tracts in the spinal cord the impulses are transmitted to the somesthetic association cortex of the left cerebral hemisphere where the objects are "recognized." The impulses are then transmitted to the speech comprehension area (Wernicke's area) where the objects are given names. From there the impulses travel to Broca's area for formulation of the spoken words, and finally the impulses arrive at the premotor and motor cortex for programming and for producing the movements to form and say the words that identify the objects.

Chapter 15:

Integrative Functions

[L1]   MULTIPLE CHOICE—pp. 266–269

(1.) C	(6.) C	(11.) D	(16.) A	(21.) D
(2.) B	(7.) D	(12.) B	(17.) D	(22.) B
(3.) A	(8.) A	(13.) C	(18.) B	(23.) A
(4.) D	(9.) B	(14.) C	(19.) B	(24.) C
(5.) B	(10.) C	(15.) A	(20.) D	(25.) D

[L1]   COMPLETION—pp. 269–270

(26.) sensation	(32.) cerebellum	(38.) REM
(27.) first-order	(33.) general senses	(39.) Parkinson's disease
(28.) pyramidal system	(34.) special senses	(40.) depression
(29.) pyramids	(35.) declarative	(41.) Alzheimer's disease
(30.) cerebral nuclei	(36.) reflexive	(42.) arteriosclerosis
(31.) acetylcholine	(37.) reticular activating system	

[L1]   MATCHING—p. 270

(43.) E	(47.) G	(51.) A	(54.) N	(57.) M
(44.) H	(48.) C	(52.) O	(55.) L	(58.) Q
(45.) B	(49.) I	(53.) J	(56.) K	(59.) P
(46.) F	(50.) D			

[L1]   DRAWING/ILLUSTRATION—LABELING—p. 271

Figure 15.1   Ascending and Descending Tracts of the Spinal Cord—p. 271

(Descending)
(60.) lateral corticospinal
(61.) rubrospinal
(62.) vestibulospinal
(63.) reticulospinal

(64.) tectospinal
(65.) anterior corticospinal
(Ascending)
(66.) fasciculus gracilis
(67.) fasciculus cuneatus

(68.) posterior spinocerebellar
(69.) lateral spinothalamic
(70.) anterior spinocerebellar
(71.) anterior spinothalamic

[L2]   CONCEPT MAPS—pp. 272–274

I   SPINAL CORD—SENSORY TRACTS—p. 272

(1.) posterior column
(2.) medulla (nucleus gracilis)
(3.) fasciculus cuneatus
(4.) posterior tract

(5.) cerebellum
(6.) proprioceptors
(7.) spinothalamic

(8.) thalamus (ventral nuclei)
(9.) anterior tract
(10.) interneurons

II   SPINAL CORD—MOTOR TRACTS—p. 273

(11.) pyramidal tracts
(12.) primary motor cortex
(13.) lateral corticospinal tracts
(14.) voluntary motor control of skeletal muscles

(15.) rubrospinal tract
(16.) posture, muscle tone
(17.) reticular formation (brain stem)

(18.) vestibulospinal tract
(19.) balance, muscle tone
(20.) tectum (midbrain)

III   EXTRA PYRAMIDAL SYSTEM—p. 274

(21.) cerebrum
(22.) equilibrium sensations
(23.) superior colliculi

(24.) auditory information
(25.) red nucleus
(26.) thalamus

(27.) sensory information
(28.) reticular formation

[L2] BODY TREK—p. 275

(29.) receptor

(30.) dorsal root ganglion

(31.) posterior horn

(32.) primary

(33.) secondary

(34.) commissural

(35.) medulla

(36.) pons

(37.) midbrain

(38.) diencephalon

(39.) thalamus

(40.) synapse

(41.) tertiary

(42.) somesthetic

(43.) cerebral cortex

(44.) pyramidal

[L2] MULTIPLE CHOICE—pp. 275–279

(45.) D	(50.) A	(55.) B	(60.) C	(65.) C
(46.) A	(51.) B	(56.) C	(61.) A	(66.) B
(47.) C	(52.) D	(57.) A	(62.) B	(67.) D
(48.) C	(53.) C	(58.) D	(63.) B	(68.) A
(49.) B	(54.) A	(59.) A	(64.) D	(69.) A

[L2] COMPLETION—p. 279

(70.) descending

(71.) ascending

(72.) GABA

(73.) anencephaly

(74.) cerebral nuclei

(75.) electroencephalogram

(76.) global aphasia

(77.) aphasia

(78.) dyslexia

(79.) medial lemniscus

(80.) arousal

(81.) dopamine

(82.) cerebellum

(83.) thalamus

(84.) homunculus

[L2] SHORT ESSAY—pp. 279–281

85. Posterior column, spinothalamic, and spinocerebellar pathways.

86. Corticobulbar tracts, lateral corticospinal tract, and anteriorcorticospinal tract.

87. Rubrospinal tract, reticulospinal tract, vestibulospinal tract, and tectospinal tract.

88. The pyramidal system provides a rapid and direct mechanism for voluntary somatic motor control of skeletal muscles. The extrapyramidal system provides less precise control of motor functions, especially ones associated with overall body coordination and cerebellar function.

89. (a) Overseeing the postural muscles of the body, and (b) adjusting voluntary and involuntary motor patterns.

90. (a) Cortical neurons continue to increase in number until at least age 1. (b) The brain grows in size and complexity until at least age 4. (c) Myelination of CNS neurons continues at least until puberty.

91. (a) Alpha waves are characteristic of normal resting adults. (b) Beta waves typically accompany intense concentration. (c) Theta waves are seen in children and in frustrated adults. (d) Delta waves occur during sleep and in certain pathological states.

92. (a) They are performed by the cerebral cortex. (b) They involve complex interactions between areas of the cortex and between the cerebral cortex and other areas of the brain. (c) They involve both conscious and unconscious information processing. (d) The functions are subject to modification and adjustment over time.

93. The general interpretive area is called Wernicke's area and the speech center is called Broca's area.

94. The *left* hemisphere contains the general interpretive and speech centers and is responsible for language-based skills. It is also important in performing analytical tasks, such as mathematical calculations and logical decision making. The *right* hemisphere is concerned with spatial relationships and analyses of sensory information involving touch, smell, taste, and feel.

95. (a) Increased neurotransmitter release; (b) Facilitation at synapses; (c) Formation of additional synaptic connections.

96. (a) A reduction in brain size and weight; (b) A decrease in blood flow to the brain; (c) Changes in synaptic organization of the brain; (d) Intracellular and extracellular changes in CNS neurons.

[L3]   CRITICAL THINKING/APPLICATION—p. 281

1. The organized distribution of sensory information on the sensory cortex enables us to determine what specific portion of the body has been affected by a stimulus. In the case of the arm pains during a heart attack, apparently the sensations of the attack arrived at the wrong part of the sensory cortex, causing the individual to reach an improper conclusion about the source of the stimulus.

2. The Babinski reflex and the abdominal reflex are two spinal reflexes dependent on lower motor neuron function. It would be assumed that the accident caused damage to the upper motor neurons of the descending motor tract, which facilitated or inhibited the activity in the lower motor neurons. The severance of the upper motor neurons facilitated the reappearance of the Babinski reflex and the disappearance of the abdominal reflex.

3. (a) One group of axons synapses with thalamic neurons, which then send their axons to the primary motor cortex. A feedback loop is created that changes the sensitivity of the pyramidal cells and alters the pattern of instructions carried by the corticospinal tracts.

    (b) A second group of axons innervates the red nucleus and alters the activity in the rubrospinal tracts.

    (c) The third group travels through the thalamus to reach centers in the reticular formation where they adjust the output of the reticulospinal tracts.

4. Injuries to the motor cortex eliminate the ability to exert fine control over motor units, but gross movements may still be produced by the cerebral nuclei using the reticulospinal or rubrospinal tracts.

5. The suspicious cause of death is anencephaly, a rare condition in which the brain fails to develop at levels above the midbrain or lower diencephalon. The brain stem, which controls complex involuntary motor patterns, is still intact; therefore, all the normal behavior patterns expected of a newborn can still occur.

6. Delta waves are seen in the brains of infants whose cortical development is incomplete, during deep sleep at all ages, and when a tumor, vascular blockage, or inflammation has damaged portions of the brain.

7. Initial symptoms might include moodiness, irritability, depression, and a general lack of energy. There may be difficulty in making decisions, and with the gradual deterioration of mental organization the individual loses memories, verbal and reading skills, and emotional control. She may experience difficulty in performing even the simplest motor tasks.

Chapter 16:

Autonomic Nervous System

[L1] MULTIPLE CHOICE—pp. 283–286

(1.) C	(6.) D	(11.) C	(16.) B	(21.) C
(2.) D	(7.) C	(12.) B	(17.) C	(22.) B
(3.) B	(8.) B	(13.) C	(18.) D	(23.) C
(4.) D	(9.) A	(14.) D	(19.) A	(24.) B
(5.) D	(10.) D	(15.) D	(20.) D	

[L1] COMPLETION—p. 287

(25.) synapses
(26.) involuntary
(27.) "fight or flight"
(28.) "rest and repose"
(29.) acetylcholine

(30.) epinephrine
(31.) excitatory
(32.) norepinephrine
(33.) opposing

(34.) autonomic tone
(35.) visceral reflexes
(36.) limbic
(37.) sympathiomimetic

[L1] MATCHING—p. 288

(38.) C	(41.) D	(44.) M	(47.) E	(49.) H
(39.) G	(42.) A	(45.) F	(48.) B	(50.) L
(40.) J	(43.) K	(46.) I		

[L1] DRAWING/ILLUSTRATION—LABELING—p. 289

**Figure 16.1   Preganglionic and Postganglionic
Cell Bodies of Sympathetic Neurons—p. 289**

(51.) superior cervical ganglion
(52.) T_1
(53.) L_1
(54.) sympathetic chain ganglion
(55.) spinal cord
(56.) autonomic nerves

(57.) splanchnic nerve
(58.) celiac ganglion
(59.) superior mesenteric ganglion
(60.) inferior mesenteric ganglion
(61.) preganglionic axon
(62.) postganglionic neuron
(63.) sympathetic

**Figure 16.2   Preganglionic and Postganglionic
Cell Bodies of Parasympathetic Neurons—p. 290**

(64.) postganglionic axon
(65.) ciliary ganglion
(66.) preganglionic axon
(67.) pterygopalatine ganglion
(68.) submandibular ganglion

(69.) N III
(70.) midbrain
(71.) N VII
(72.) otic ganglion
(73.) N IX

(74.) medulla
(75.) N X
(76.) S_2
(77.) S_4

[L2] CONCEPT MAPS—pp. 291–294

I   AUTONOMIC NERVOUS SYSTEM (ORGANIZATION)—p. 291

(1.) sympathetic
(2.) motor neurons

(3.) first-order neurons
(4.) smooth muscle

(5.) ganglia outside CNS
(6.) postganglionic

II   AUTONOMIC NERVOUS SYSTEM—SYMPATHETIC DIVISION—p. 292

(7.) thoracolumbar

(8.) spinal segments T_1–L_2

(9.) second-order neurons (postganglionic)

(10.) sympathetic chain of ganglia (paired)

(11.) visceral effectors

(12.) adrenal medulla (paired)

(13.) general circulation

III   AUTONOMIC NERVOUS SYSTEM—PARASYMPATHETIC DIVISION—p. 293

(14.) craniosacral

(15.) brain stem

(16.) ciliary ganglion

(17.) N VII

(18.) nasal, tear, salivary glands

(19.) otic ganglia

(20.) N X

(21.) segments S_2–S_4

(22.) intramural ganglia

(23.) lower-abdominopelvic cavity

IV   LEVELS OF AUTONOMIC CONTROL—p. 294

(24.) thalamus

(25.) pons

(26.) spinal cord T_1–L_2

(27.) respiratory

(28.) sympathetic visceral reflexes

(29.) parasympathetic visceral reflexes

[L2]   BODY TREK—p. 295

(30.) gray matter

(31.) axon

(32.) spinal nerve

(33.) sympathetic chain ganglion

(34.) white ramus

(35.) postganglionic

(36.) gray ramus

(37.) target organ

(38.) splanchnic

(39.) collateral

(40.) synapse

(41.) abdominopelvic

(42.) epinephrine

(43.) circulation

[L2]   MULTIPLE CHOICE—pp. 295–297

(44.) A

(45.) D

(46.) B

(47.) C

(48.) A

(49.) D

(50.) A

(51.) C

(52.) B

(53.) A

(54.) C

(55.) D

(56.) D

(57.) B

(58.) D

[L2]   COMPLETION—p. 298

(59.) postganglionic

(60.) norepinephrine

(61.) adrenal medulla

(62.) collateral

(63.) gray ramus

(64.) white ramus

(65.) splanchnic

(66.) hypothalamus

(67.) acetylcholine

(68.) blocking agents

[L2]   SHORT ESSAY—p. 298–300

69. The sympathetic division stimulates tissue metabolism, increases alertness, and generally prepares the body to deal with emergencies.

70. The parasympathetic division conserves energy and promotes sedentary activities, such as digestion.

71. (a) All preganglionic fibers are cholinergic; they release acetylcholine (Ach) at their synaptic terminals. The effects are always excitatory.

    (b) Postganglionic fibers are also cholinergic, but the effects may be excitatory or inhibitory, depending on the nature of the receptor.

    (c) Most postganglionic sympathetic terminals are adrenergic; they release norepinephrine (NE). The effects are usually excitatory.

72. (a) Preganglionic (first-order) neurons located between segments T_1 and L_2 of the spinal cord.

    (b) Ganglionic (second-order) neurons located in ganglia near the vertebral column (sympathetic chain ganglia, collateral ganglia).

    (c) Specialized second-order neurons in the interior of the adrenal gland.

73. (a) The reduction of blood flow and energy use by visceral organs such as the digestive tract that are not important to short-term survival.

    (b) The release of stored energy reserves.

74. (a) Preganglionic (first-order) neurons in the brain stem and in sacral segments of the spinal cord.

    (b) Ganglionic (second-order) neurons in peripheral ganglia located within or adjacent to the target organs.

75. III, VII, IX, X

76. Stimulation of the parasympathetic system leads to a general increase in the nutrient content of the blood. Cells throughout the body respond to this increase by absorbing nutrients and using them to support growth and other anabolic activities.

77. alpha-1, alpha-2; beta-1, beta-2

78. nicotinic, muscarinic

79. When dual innervation exists, the two divisions of the ANS *often* but not always have opposing effects. Sympathetic-parasympathetic opposition can be seen along the digestive tract, at the heart, in the lungs, and elsewhere throughout the body.

80. If a nerve maintains a background level of activity, it may either increase or decrease its activity, thus providing a range of control options.

81. (a) defecation and urination reflexes

    (b) pupillary and vasomotor reflexes

[L3]   CRITICAL THINKING/APPLICATION—p. 300

1. Sympathetic postganglionic fibers that enter the thoracic cavity in autonomic nerves cause the heart rate to accelerate (thus increasing the force of cardiac contractions) and dilate the respiratory passageways. Because of these functions the heart works harder, moving blood faster. The muscles receive more blood, and their utilization of stored and absorbed nutrients accelerates. Lipids, a potential energy source, are released. The lungs deliver more oxygen and prepare to eliminate the carbon dioxide produced by contracting muscles. Sweat glands become active and the eyes look for approaching dangers.

2. Alpha-2 receptors are found on the presynaptic surfaces at autonomic neuroeffector junctions. A synaptic terminal releases norepinephrine until concentrations within the synapse rise enough to stimulate the presynaptic alpha-2 receptors. This stimulation has an inhibitory effect, and neurotransmitter release stops.

3. Even though most sympathetic postganglionic fibers are adrenergic, releasing norepinephrine, a few are cholinergic, releasing acetylcholine. The distribution of the cholinergic fibers via the sympathetic division provides a method of regulating sweat gland secretion and selectively controlling blood flow to skeletal muscles while reducing the flow to other tissues in a body wall.

4. When beta-1 receptors are stimulated, they produce an increase in heart rate and force of contraction. Beta-blockers such as Propranolol and Metiprolol cause a decrease in heart rate and force of contraction, thus reducing the strain on the heart and simultaneously lowering peripheral blood pressure.

5. Nicotine stimulates the postganglionic neurons of both the sympathetic and parasympathetic division. In response to the nicotine contained in a cigarette, the heart rate may increase or decrease, and its rhythm tends to become less regular as a result of the simultaneous actions on the sympathetic division, which increases the heart rate, and the parasympathetic division, which decreases the heart rate. Blood pressure tends to increase because the sympathetic neurons that constrict blood vessels are more numerous than either the sympathetic neurons or the parasympathetic neurons that dilate blood vessels.

Chapter 17:

Sensory Function

[L1] MULTIPLE CHOICE—pp. 302–306

(1.) C	(7.) D	(13.) C	(19.) A	(25.) B
(2.) A	(8.) C	(14.) B	(20.) D	(26.) C
(3.) B	(9.) A	(15.) C	(21.) B	(27.) B
(4.) B	(10.) C	(16.) D	(22.) B	(28.) D
(5.) C	(11.) D	(17.) C	(23.) D	(29.) C
(6.) B	(12.) A	(18.) C	(24.) C	(30.) B

[L1] COMPLETION—pp. 306–307

(31.) somatosensory	(37.) taste buds	(42.) occipital
(32.) cerebral cortex	(38.) sclera	(43.) endolymph
(33.) sensitivity	(39.) pupil	(44.) round window
(34.) thermoreceptors	(40.) rods	(45.) saccule, utricle
(35.) mechanoreceptors	(41.) cones	(46.) midbrain
(36.) olfactory		

[L1] MATCHING—p. 307

(47.) D	(51.) F	(54.) E	(57.) K	(60.) O
(48.) A	(52.) H	(55.) N	(58.) J	(61.) M
(49.) B	(53.) C	(56.) I	(59.) P	(62.) L
(50.) G				

[L1] DRAWING/ILLUSTRATION—LABELING—pp. 308–310

Figure 17.1   Sensory Receptors of the Skin—p. 308

(63.) epidermis	(66.) free nerve endings	(69.) Pacinian corpuscle
(64.) dermis	(67.) Meissner's corpuscles	(70.) Ruffini corpuscle
(65.) Merkel's discs	(68.) root hair plexus	(71.) sensory nerves

Figure 17.2   Sectional Anatomy of the Eye—p. 308

(72.) vitreous chamber	(78.) sclera	(84.) lens
(73.) choroid	(79.) fornix	(85.) cornea
(74.) fovea	(80.) palpebral conjunction	(86.) limbus
(75.) optic nerve	(81.) ocular conjunction	(87.) suspensory ligaments
(76.) optic disc	(82.) ciliary body	(88.) ora serrata
(77.) retina	(83.) iris	

Figure 17.3   External, Middle, and Inner Ears—p. 309

(89.) pinna	(93.) tympanic membrane	(97.) facial nerve (VIII)
(90.) external auditory canal	(94.) auditory ossicles	(98.) inner ear
(91.) external auditory meatus	(95.) middle ear	(99.) pharyngotympanic tube
(92.) cartilage	(96.) Petrous portion of temporal bone	

Figure 17.4   Body Labyrinth of Inner Ear—p. 309

(100.) semicircular canals	(103.) tympanic duct	(105.) cochlea
(101.) vestibular duct	(104.) maculea	(106.) organ of Corti
(102.) cochlear duct		

Figure 17.5   Gross Anatomy of the Cochlea—p. 309

(107.) scala vestibuli

(108.) body cochlear wall

(109.) scala media

(110.) tectorial membrane

(111.) organ of Corti

(112.) basilar membrane

(113.) scala tympani

(114.) spiral ganglion

(115.) cochlear nerve

Figure 17.6   Organ of Corti and Tectorial Membrane in 3-D—p. 310

(116.) tectorial membrane

(117.) outer hair cell

(118.) basilar membrane

(119.) inner hair cell

(120.) nerve fibers

Figure 17.7   Gustatory Pathways—p. 310

(121.) bitter

(122.) sour

(123.) salt

(124.) sweet

(125.) cranial nerve VII

(126.) cranial nerve IX

(127.) cranial nerve X

[L2]   CONCEPT MAPS—pp. 311–313

I   GENERAL SENSES—p. 311

(1.) thermoreceptors

(2.) pain

(3.) tactile

(4.) pressure

(5.) dendritic processes

(6.) Merkel's discs

(7.) Pacinian corpuscles

(8.) baroreceptors

(9.) aortic sinus

(10.) proprioception

(11.) muscle spindles

II   SPECIAL SENSES—p. 312

(12.) olfaction

(13.) smell

(14.) tongue

(15.) taste buds

(16.) ears

(17.) balance and hearing

(18.) audition

(19.) retina

(20.) rods and cones

III   SOUND PERCEPTION—p. 313

(21.) external acoustic meatus

(22.) middle ear

(23.) incus

(24.) oval window

(25.) endolymph in cochlear duct

(26.) round window

(27.) hair cells of organ of corti

(28.) vestibulocochlear nerve VIII

[L2]   BODY TREK—p. 314

(29.) pinna

(30.) external auditory meatus

(31.) ceruminous

(32.) tympanic membrane

(33.) external ear

(34.) middle ear

(35.) ossicles

(36.) malleus

(37.) incus

(38.) stapes

(39.) pharyngotympanic

(40.) nasopharynx

(41.) oval window

(42.) endolymph

(43.) vestibular duct

(44.) scala vestibuli

(45.) scala tympani

(46.) basilar membrane

(47.) round window

(48.) organ of Corti

(49.) tectorial

(50.) stereocilia

(51.) cochlear

[L2]   MULTIPLE CHOICE—pp. 314–318

(52.) B	(57.) D	(62.) D	(67.) D	(72.) C
(53.) A	(58.) B	(63.) B	(68.) B	(73.) B
(54.) C	(59.) A	(64.) C	(69.) A	(74.) A
(55.) B	(60.) B	(65.) A	(70.) D	(75.) C
(56.) A	(61.) C	(66.) B	(71.) A	(76.) D

[L2]   COMPLETION—pp. 318–319

(77.) sensory receptor	(84.) adaptation	(91.) pupil
(78.) sensation	(85.) central adaptation	(92.) aqueous humor
(79.) receptive field	(86.) referred pain	(93.) retina
(80.) transduction	(87.) primary sensory cortex	(94.) cataract
(81.) receptor potential	(88.) ampulla	(95.) accommodation
(82.) afferent fiber	(89.) nystagmus	(96.) myopia
(83.) labeled line	(90.) Hertz	(97.) hyperopia

[L2]   SHORT ESSAY—pp. 319–321

98. Sensations of temperature, pain, touch, pressure, vibration, and proprioception.

99. Smell (olfaction), taste (gustation), balance (equilibrium), hearing (audition), vision (sight).

100. (a) The stimulus alters the permeability of the receptor membrane.
   (b) The receptor potential produces a generator potential.
   (c) The action potential travels to the CNS over an afferent fiber.

101. (a) Nociceptors—variety of stimuli usually associated with tissue damage.
   (b) Thermoreceptors—changes in temperature.
   (c) Mechanoreceptors—stimulated or inhibited by physical distortion, contact, or pressure on their cell membranes.
   (d) Chemoreceptors—respond to presence of specific molecules.

102. Baroreceptors monitor changes in pressure. Proprioceptors monitor the position of joints, the tension in tendons and ligaments, and the state of muscular contraction.

103. Sensations leaving the olfactory bulb travel along the olfactory tract (N I) to reach the olfactory cortex, the hypothalamus, and portions of the limbic system.

104. Sweet, salty, sour, bitter.

105. Receptors in the saccule and utricle provide sensations of gravity and linear acceleration.

106. (a) Provides mechanical support and physical protection.
   (b) Serves as an attachment site for the extrinsic eye muscles.
   (c) Assists in the focusing process.

107. (a) Provides a route for blood vessels and lymphatics to the eye.
   (b) Secretes and reabsorbs aqueous humor.
   (c) Controls the shape of the lens; important in focusing process.

108. The pigment layer: (a) absorbs light after it passes through the retina; (b) biochemically interacts with the photoreceptor layer of the retina.

   The retina contains: (a) photoreceptors that respond to light; (b) supporting cells and neurons that perform preliminary processing and integration of visual information; (c) blood vessels supplying tissues lining the vitreous chamber.

109. During accommodation the lens becomes rounder to focus the image of a nearby object on the retina.

110. As aging proceeds, the lens becomes less elastic, takes on a yellowish hue, and eventually begins to lose its transparency. Visual clarity begins to fade, and when the lens turns completely opaque, the person becomes functionally blind despite the fact that the retinal receptors are alive and well.

[L3] CRITICAL THINKING/APPLICATION—p. 321

1. The mechanical stimuli of smell are eventually converted into neural events that reach the olfactory cortex, the hypothalamus, and portions of the limbic system. The extensive limbic and hypothalamic connections help to explain the profound emotional and behavioral responses that can be produced by certain scents. The perfume industry understands the practical implications of these connections.

2. The total number of olfactory receptors declines with age, and elderly individuals have difficulty detecting odors unless excessive quantities are used. Younger individuals can detect the odor in lower concentrations because the olfactory receptor population is constantly producing new receptor cells by division and differentiation of basal cells in the epithelium.

3. The middle ear contains the auditory ossicles including the *malleus*, or "hammer," the *incus*, or "anvil," and the *stapes*, or "stirrups."

4. Contrary to popular belief, carrots do not contain vitamin A. Carrots contain carotene, an orange pigment, which is converted to retinol (vitamin A) in the body. Retinal, the pigment in the rhodopsin molecule, is synthesized from vitamin A. A deficiency of vitamin A can cause night blindness, keratinization of the cornea which can progress to xerosis (drying), and then to thickening and permanent blindness—xerophthalmia.

5. When light falls on the eye, it passes through the cornea and strikes the retina, bleaching many molecules of the pigment rhodopsin that lie within them. Vitamin A, a part of the rhodopsin molecule, is broken off when bleaching occurs. After an intense exposure to light, a photoreceptor cannot respond to further stimulation until its rhodopsin molecules have been regenerated.

Chapter 18:

The Endocrine System

(1.) A	(6.) D	(11.) C	(16.) B	(21.) D
(2.) C	(7.) B	(12.) C	(17.) C	(22.) D
(3.) D	(8.) A	(13.) A	(18.) D	(23.) B
(4.) B	(9.) B	(14.) C	(19.) D	(24.) D
(5.) B	(10.) D	(15.) B	(20.) D	(25.) D

(26.) neurotransmitters
(27.) adrenal gland
(28.) catecholamines
(29.) testosterone
(30.) G protein
(31.) reflex

(32.) hypothalamus
(33.) parathyroids
(34.) thymus
(35.) diabetes mellitus
(36.) cytoplasm

(37.) gigantism
(38.) exhaustive
(39.) general adaptation syndrome
(40.) aggressive

(41.) D	(45.) B	(49.) C	(53.) O	(57.) L
(42.) G	(46.) A	(50.) J	(54.) K	(58.) M
(43.) I	(47.) E	(51.) N	(55.) Q	(59.) R
(44.) H	(48.) F	(52.) S	(56.) T	(60.) P

Figure 18.1   The Endocrine System—p. 328

(61.) parathyroid gland
(62.) adrenal gland
(63.) hypothalamus

(64.) pituitary gland
(65.) thyroid
(66.) thymus

(67.) pancreas
(68.) gonads (female—ovary; male—testes)

Figure 18.2   Classification of Hormones—p. 328

(69.) amino acid derivatives

(70.) steroid hormone

(71.) peptide hormone

I   ENDOCRINE GLANDS—p. 329

(1.) hormones
(2.) epinephrine
(3.) peptide hormones
(4.) testosterone

(5.) pituitary
(6.) parathyroids
(7.) heart

(8.) male/female gonads
(9.) pineal
(10.) bloodstream

II   PITUITARY GLAND—p. 330

(11.) sella turcica
(12.) adenohypophysis
(13.) melanocytes
(14.) pars distalis

(15.) thyroid
(16.) adrenal cortex
(17.) gonadotropic hormones
(18.) mammary glands

(19.) posterior pituitary
(20.) pars nervosa
(21.) oxytocin
(22.) ADH-Vasopressin

III   ENDOCRINE SYSTEM FUNCTIONS—p. 331

(23.) cellular communication
(24.) homeostasis

(25.) target cells
(26.) contraction

(27.) ion channel opening
(28.) hormones

[L2]   BODY TREK—p. 332

(29.) pineal

(30.) melatonin

(31.) infundibulum

(32.) pituitary

(33.) sella turcica

(34.) thyroid

(35.) parathyroid

(36.) thymus

(37.) heart

(38.) pancreas

(39.) insulin

(40.) adrenal

(41.) kidneys

(42.) erythropoietin

(43.) RBC's

(44.) ovaries

(45.) estrogen

(46.) progesterone

(47.) testes

(48.) testosterone

[L2]   MULTIPLE CHOICE—p. 332–336

(49.) B

(50.) A

(51.) D

(52.) B

(53.) C

(54.) D

(55.) B

(56.) A

(57.) C

(58.) A

(59.) B

(60.) D

(61.) B

(62.) D

(63.) A

(64.) D

(65.) B

(66.) C

(67.) A

(68.) C

(69.) D

(70.) A

(71.) B

(72.) C

(73.) A

[L2]   COMPLETION—p. 336

(74.) hypothalamus

(75.) target cells

(76.) nucleus

(77.) parathormone

(78.) tyrosine

(79.) prehormone

(80.) cyclic-AMP

(81.) calmodulin

(82.) somatomedins

(83.) fenestrated

(84.) portal

(85.) aldosterone

(86.) reticularis

(87.) epinephrine

(88.) glucocorticoids

[L2]   SHORT ESSAY—p. 336–337

89.   (a) Amino acid derivatives; (b) peptide hormones; (c) steroids.

90.   (a) Activation of adenyl cyclase; (b) release or entry of calcium ions; (c) activation of phosphodiesterase; (d) activation of other intracellular enzymes.

91.   The hypothalamus: (a) Contains autonomic centers that exert direct neural control over the endocrine cells of the adrenal medulla. Sympathetic activation causes the adrenal medulla to release hormones into the bloodstream; (b) acts as an endocrine organ to release hormones into the circulation at the posterior pituitary; (c) secretes regulatory hormones that control activities of endocrine cells in the pituitary glands.

92.   (a) Control by releasing hormones; (b) control by inhibiting hormones; (c) regulation by releasing and inhibiting hormones.

93.   Thyroid hormones elevate oxygen consumption and rate of energy consumption in peripheral tissues, causing an increase in the metabolic rate. As a result, more heat is generated, replacing the heat lost to the environment.

94.   Erythropoietin stimulates the production of red blood cells by the bone marrow. The increase in the number of RBCs elevates the blood volume, causing an increase in blood pressure.

95.   The secretion of melatonin by the pineal gland is lowest during daylight hours and highest in the darkness of night. The cyclic nature of this activity parallels daily changes in physiological processes that follow a regular pattern.

96.   (a) The two hormones may have opposing, or *antagonistic*, effects; (b) the two hormones may have an additive, or *synergistic*, effect; (c) one can have a *permissive* effect on another. In such cases the first hormone is needed for the second to produce its effect; (d) the hormones may have *integrative* effects, i.e., the hormones may produce different but complementary results in specific tissues and organs.

[L3]   CRITICAL THINKING/APPLICATION—p. 338

1.

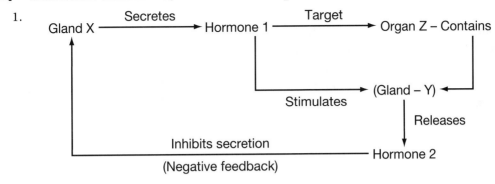

2.   (a)   *Patient A*:   Underproduction of ADH; diabetes insipidus.
         Patient B: Underproduction of thyroxine (T_4); myxedema.
         Patient C: Overproduction of glucocorticoids; Cushing's disease.

     (b)   *Patient A*:   Pituitary disorder.
         Patient B: Thyroid disorder.
         Patient C: Adrenal disorder.

3.   Structurally, the hypothalamus is a part of the diencephalon; however, this portion of the brain secretes ADH and oxytocin that target peripheral effectors; the hypothalamus controls secretory output of the adrenal medulla, and it releases hormones that control the pituitary gland.

4.   The anti-inflammatory activities of these steroids result from their effects on white blood cells and other components of the immune system. The glucocorticoids (steroid hormones) slow the migration of phagocytic cells into an injury site, and phagocytic cells already in the area become less active. In addition, most cells exposed to these steroids are less likely to release the chemicals that promote inflammation, thereby slowing the wound healing process. Because the region of the open wound's defenses are weakened, the area becomes an easy target for infecting organisms.

5.   Iodine must be available for thyroxine to be synthesized. Thyroxine deficiencies result in decreased rates of metabolism, decreased body temperature, poor response to physiological stress, and an increase in the size of the thyroid gland (goiter). People with goiter usually suffer sluggishness and weight gain.

6.   Anxiety, anticipation, and excitement may stimulate the sympathetic division of the ANS or release of epinephrine from the adrenal medulla. The peripheral effect of epinephrine results from interactions with alpha and beta receptors on cell membranes. The result is increased cellular energy utilization and mobilization of energy reserves. Catecholamine secretion triggers a mobilization of glycogen reserves in skeletal muscles and accelerates the breakdown of glucose to provide ATP. This combination increases muscular power and endurance.

7.   The androgens or "anabolic steroids" affect many tissues in a variety of ways. Extreme aggressive behavior and complications such as liver dysfunction, prostate enlargement, infertility, and testicular atrophy are common. The normal regulation of androgen production involves a feedback mechanism, and the administration of androgens may affect the normal production of testosterone in the testes and permanently suppress the manufacture of GnRH by the hypothalamus. In females these hormones alter muscular proportions and secondary sex characteristics.

Chapter 19:

Blood

[L1]   MULTIPLE CHOICE—p. 340–344

(1.) B	(8.) C	(15.) B	(22.) C	(28.) C
(2.) A	(9.) B	(16.) D	(23.) A	(29.) D
(3.) D	(10.) B	(17.) A	(24.) A	(30.) B
(4.) C	(11.) C	(18.) A	(25.) C	(31.) D
(5.) A	(12.) D	(19.) D	(26.) D	(32.) C
(6.) C	(13.) A	(20.) D	(27.) A	(33.) B
(7.) B	(14.) C	(21.) C		

[L1]   COMPLETION—pp. 345–346

(34.) plasma	(42.) transferrin	(49.) fixed macrophages
(35.) formed elements	(43.) vitamin B_{12}	(50.) diapedesis
(36.) venepuncture	(44.) erythroblasts	(51.) lymphopoiesis
(37.) viscosity	(45.) hematocrit	(52.) platelets
(38.) serum	(46.) agglutinogens	(53.) monocyte
(39.) hemopoiesis	(47.) agglutinins	(54.) coumadin
(40.) rouleaux	(48.) lymphocytes	(55.) vascular
(41.) hemoglobin		

[L1]   MATCHING—p. 346

(56.) J	(61.) G	(66.) D	(70.) S	(74.) T
(57.) C	(62.) A	(67.) R	(71.) O	(75.) V
(58.) K	(63.) E	(68.) U	(72.) Q	(76.) M
(59.) H	(64.) B	(69.) N	(73.) L	(77.) P
(60.) I	(65.) F			

[L1]   DRAWING/ILLUSTRATION—LABELING—pp. 347–348

Figure 19.1   Agranular Leukocytes—p. 347

(78.) small lymphocyte	(79.) monocyte	(80.) large lymphocyte

Figure 19.2   Granular Leukocytes—p. 347

(81.) basophil	(82.) neutrophil	(83.) eosinophil

Figure 19.3—p. 348

(84.) erythrocytes—RBCs

[L2]   CONCEPT MAPS—pp. 349–352

I   WHOLE BLOOD—p. 349

(1.) plasma	(4.) gamma	(6.) neutrophils
(2.) solutes	(5.) leukocytes	(7.) monocytes
(3.) albumins		

II   HEMOSTASIS—p. 350

(8.) vascular spasm	(11.) forms blood clot	(14.) plasminogen
(9.) platelet phase	(12.) clot retraction	(15.) plasmin
(10.) forms plug	(13.) clot dissolves	

III   ERYTHROCYTE DISORDERS—p. 351

(16.) polycythemia
(17.) abnormally large number of RBCs

(18.) polycythemia vera
(19.) hemorrhagic anemia
(20.) iron deficiency

(21.) sickle cell anemia
(22.) thalassemia

IV   CLOT FORMATION—p. 352

(23.) intrinsic pathway
(24.) tissue damage
(25.) platelets

(26.) inactive factor
(27.) active factor IX
(28.) factor VII & Ca^{++}

(29.) thrombin
(30.) fibrinogen

[L2] BODY TREK—p. 353

(31.) endothelium
(32.) vascular spasm
(33.) thrombocytes
(34.) platelet plug
(35.) vascular

(36.) platelet
(37.) coagulation
(38.) fibrinogen
(39.) fibrin
(40.) RBCs

(41.) clot
(42.) clot retraction
(43.) fibrinolysis
(44.) plasminogen
(45.) plasmin

[L2] MULTIPLE CHOICE—pp. 353–357

(46.) A	(51.) A	(56.) B	(61.) A	(66.) D
(47.) B	(52.) C	(57.) D	(62.) C	(67.) A
(48.) C	(53.) A	(58.) B	(63.) C	(68.) A
(49.) A	(54.) D	(59.) C	(64.) D	(69.) C
(50.) C	(55.) B	(60.) B	(65.) B	(70.) D

[L2] COMPLETION—p. 357

(71.) fractionated
(72.) hypovolemic
(73.) fibrin
(74.) metalloproteins
(75.) lipoproteins

(76.) hematocrit
(77.) hemoglobinuria
(78.) hemolysis
(79.) bilirubin

(80.) leukopenia
(81.) leukocytosis
(82.) thrombocytopenia
(83.) differential

[L2] SHORT ESSAY—pp. 357–359

84. (a) Blood *transports* dissolved gases.

(b) Blood *distributes* nutrients.

(c) Blood *transports* metabolic wastes.

(d) Blood *delivers* enzymes and hormones.

(e) Blood *regulates* the pH and electrolyte composition of interstitial fluid.

(f) Blood *restricts* fluid losses through damaged vessels.

(g) Blood *defends* the body against toxins and pathogens.

(h) Blood helps *regulate* body temperature by absorbing and redistributing heat.

85. (a) water; (b) electrolytes; (c) nutrients; (d) organic wastes; (e) proteins

86. (a) red blood cells; (b) white blood cells; (c) platelets

87. (a) albumins; (b) globulins; (c) fibrinogen

88. Mature red blood cells do not have mitochondria. They also lack ribosomes and a nucleus.

89. Granular leukocytes: neutrophils, eosinophils, basophils
Agranular WBC: monocytes, lymphocytes

90. (a) Transport of chemicals important to the clotting process

(b) Formation of a plug in the walls of damaged blood vessels

(c) Active contraction after clot formation has occurred

91. (a) Vascular phase: spasm in damaged smooth muscle.

    (b) Platelet phase: platelet aggregation and adhesion

    (c) Coagulation phase: activation of clotting system and clot formation

    (d) Clot retraction: contraction of blood clot

    (e) Clot destruction: enzymatic destruction of clot

92. Embolus: a drifting blood clot.

    Thrombosis: a blood clot that sticks to the wall of an intact blood vessel.

93. Erythropoietin:

    (a) Stimulates increased rates of mitotic divisions in erythroblasts and in stem cells that produce erythroblasts; (b) speeds up the maturation of RBCs by accelerating the rate of hemoglobin synthesis.

[L3] CRITICAL THINKING/APPLICATION—pp. 359–360

1. (a) 154 lb ÷ 2.2 kg/lb = 70 kg

     1 kg blood = approximately 1 liter

     therefore, 70 kg × 0.07 = 4.9 kg or 4.9 l

     4.9 l × 1 l = 4.9 l

   (b) Hematocrit: 45%

     therefore, total cell vol = 0.45 × 4.9 l = 2.2

   (c) plasma volume = 4.9 l – 2.2 l = 2.7 l

   (d) plasma volume in % = 2.7 l ÷ 4.9 l = 55%

   (e) % formed elements = 2.2 l ÷ 4.9 l = 45%

2. (a) increasing hematocrit

   (b) increasing hematocrit

   (c) decreasing hematocrit

   (d) decreasing hematocrit

3. The condition is called *polycythemia vera* or erythemia, which occurs when an excess of RBCs is produced as a result of tumorous abnormalities of the tissues that produce blood cells. The condition is usually linked to radiation exposure.

4. The patient with type B blood has type B antigens and type A antibodies. The donor's blood (type A) has type A antigens and type B antibodies. When the patient receives the type A blood, the B antibodies in the donor's blood will bind to the B antigen of the patient's blood and a transfusion reaction will occur. Also, the patient's type A antibodies can bind to the type A antigens of the donor's blood.

5. The RhoGAM is given to prevent the mother from producing anti-Rh-positive antibodies. The RhoGAM contains anti-Rh positive antibodies that remove fetal Rh positive antigens from the mother's circulation before the mother's immune system recognizes their presence and begins to produce Rh-positive antibodies.

6. The clinical test values would indicate that the condition is pernicious anemia. Normal red blood cell maturation ceases because of an inadequate supply of vitamin B_{12}. RBC production declines, and the red blood cells are abnormally large and may develop a variety of bizarre shapes.

7. (a) Vitamin K deficiencies are rare because vitamin K is produced by bacterial synthesis in the digestive tract; (b) dietary sources include liver, green leafy vegetables, cabbage-type vegetables, and milk; (c) adequate amounts of vitamin K must be present for the liver to be able to synthesize four of the clotting factors, including prothrombin. A deficiency of vitamin K leads to the breakdown of the common pathway, inactivating the clotting system.

Chapter 20:

The Cardiovascular System: The Heart

[L1]   MULTIPLE CHOICE—pp. 362–367

(1.) D	(8.) C	(15.) D	(22.) D	(29.) A
(2.) C	(9.) A	(16.) C	(23.) B	(30.) C
(3.) A	(10.) A	(17.) C	(24.) B	(31.) A
(4.) B	(11.) B	(18.) D	(25.) C	(32.) D
(5.) A	(12.) D	(19.) A	(26.) C	(33.) C
(6.) A	(13.) B	(20.) C	(27.) D	(34.) C
(7.) D	(14.) C	(21.) B	(28.) B	(35.) B

[L1]   COMPLETION—pp. 367–368

(36.) atria	(43.) pulmonary	(50.) auscultation
(37.) fibrous skeleton	(44.) myocardium	(51.) Bainbridge reflex
(38.) intercalated discs	(45.) coronary sinus	(52.) chemoreceptors
(39.) pericardium	(46.) repolarization	(53.) stroke volume
(40.) carbon dioxide	(47.) nodal cells	(54.) autonomic nervous system
(41.) left ventricle	(48.) automaticity	
(42.) pulmonary veins	(49.) electrocardiogram	(55.) hyperpolarization

[L1]   MATCHING—p. 368

(56.) E	(61.) J	(65.) D	(69.) P	(73.) T
(57.) H	(62.) A	(66.) O	(70.) R	(74.) U
(58.) I	(63.) C	(67.) L	(71.) K	(75.) N
(59.) F	(64.) G	(68.) S	(72.) M	(76.) Q
(60.) B				

[L1]   DRAWING/ILLUSTRATION—LABELING—p. 369

Figure 20.1   Anatomy of the Heart (ventral view)—p. 369

(77.) pulmonary semilunar valve	(85.) right ventricle	(94.) bicuspid valve
(78.) superior vena cava	(86.) papillary muscle	(95.) aortic semilunar valve
(79.) right pulmonary arteries	(87.) inferior vena cava	(96.) endocardium
(80.) right atrium	(88.) trabeculae carneae	(97.) interventricular system
(81.) right pulmonary veins	(89.) aortic arch	(98.) left ventricle
(82.) coronary sinus	(90.) pulmonary trunk	(99.) myocardium
(83.) tricuspid valve	(91.) left atrium	(100.) epicardium (visceral pericardium)
(84.) chordae tendinae	(92.) left pulmonary arteries	
	(93.) left pulmonary veins	(101.) dorsal aorta

[L2]   CONCEPT MAPS—p. 370–372

I   THE HEART—p. 370

(1.) two atria	(5.) oxygenated blood	(8.) deoxygenated blood
(2.) blood from atria	(6.) two semilunar	(9.) epicardium
(3.) endocardium	(7.) aortic	(10.) pacemaker cells
(4.) tricuspid		

II   FACTORS AFFECTING CARDIAC OUTPUT—p. 371

(11.) increasing atrial pressure

(12.) decreasing intrathoracic pressure

(13.) increasing end-diastolic ventricular volume

(14.) decreasing parasympathetic stimulation to heart

(15.) increasing stroke volume

(16.) increasing cardiac output

III   HEART RATE CONTROLS—p. 372

(17.) increasing blood pressure, decreasing blood pressure

(18.) carotid and aortic bodies

(19.) decreasing CO_2

(20.) increasing CO_2

(21.) increasing acetylcholine

(22.) increasing epinephrine, norepinephrine

(23.) decreasing heart rate

(24.) increasing heart rate

[L2]   BODY TREK—p. 373

(25.) superior vena cava

(26.) right atrium

(27.) tricuspid valve

(28.) right ventricle

(29.) pulmonary semi-lunar valve

(30.) pulmonary arteries

(31.) pulmonary veins

(32.) left atrium

(33.) bicuspid valve

(34.) left ventricle

(35.) aortic semi-lunar valve

(36.) L. common carotid artery

(37.) aorta

(38.) systemic arteries

(39.) systemic veins

(40.) inferior vena cava

[L2]   MULTIPLE CHOICE—pp. 374–377

(41.) A	(45.) B	(49.) A	(53.) D	(57.) D
(42.) B	(46.) C	(50.) C	(54.) C	(58.) A
(43.) B	(47.) D	(51.) D	(55.) B	(59.) C
(44.) A	(48.) B	(52.) B	(56.) C	(60.) B

[L2]   COMPLETION—p. 377

(61.) anastomoses

(62.) systemic

(63.) pectinate muscles

(64.) trabeculae carneae

(65.) endocardium

(66.) myocardium

(67.) bradycardia

(68.) tachycardia

(69.) nodal

(70.) ventricular fibrillation

(71.) cardiac reserve

(72.) diuretics

(73.) auricle

(74.) angina pectoris

(75.) infarct

[L2]   SHORT ESSAY—pp. 378–379

76.   CO = SV × HR

CO = 75 ml × 80 beats/min = 6000 ml/min

6000 ml = 6.0 l/min

77.   % increase = $\dfrac{5\ l/min.}{10\ l/min.}$ = .50 = 50%

78.   decreasing CO, decreasing SV, decreasing length of diastole, decreasing ventricular filling

79.   SV = EDV – ESV

SV = 140 ml – 60 ml = 80 ml

80.   CO = SV × HR

Therefore, $\dfrac{CO}{HR} = \dfrac{SV \times \cancel{HR}}{\cancel{HR}}$

Therefore, SV = $\dfrac{CO}{HR}$

SV = $\dfrac{5\ l/\cancel{min.}}{100\ B/\cancel{min.}}$ = 0.05 l/beat

81. The visceral pericardium, or epicardium, covers the outer surface of the heart. The parietal pericardium lines the inner surface of the pericardial sac that surrounds the heart.

82. The chordae tendinae and papillary muscles are located in the right ventricle, the trabeculae carneae are found on the interior walls of both ventricles, and the pectinate muscles are found on the interior of both atria and a part of the right atrial wall.

    The chordae tendinae are tendinous fibers that brace each cusp of the tricuspid valve and are connected to the papillary muscles. The trabeculae carneae of the ventricles contain a series of deep grooves and folds.

    The pectinate muscles are prominent muscular ridges that run along the surfaces of the atria and across the anterior atrial wall.

83. (a) epicardium; (b) myocardium; (c) endocardium

84. (a) Stabilizes positions of muscle fibers and valves in heart.
    (b) Provides support for cardiac muscle fibers and blood vessels and nerves in the myocardium.
    (c) Helps distribute the forces of contraction.
    (d) Adds strength and prevents overexpansion of the heart.
    (e) Helps to maintain the shape of the heart.
    (f) Provides elasticity that helps the heart return to original shape after each contraction.
    (g) Physically isolates the muscle fibers of the atria from those in the ventricles.

85. Cardiac muscle fibers are connected by gap junctions at intercalated discs, which allow ions and small molecules to move from one cell to another. This creates a direct electrical connection between the two muscle fibers, and an action potential can travel across an intercalated disk, moving quickly from one cardiac muscle fiber to another. Because the cardiac muscle fibers are mechanically, chemically, and electrically connected to one another, the entire tissue resembles a single, enormous muscle fiber. For this reason cardiac muscle has been called a functional syncytium.

86. SA node → AV node → bundle of His → bundle branches → Purkinje cells → contractile cells of ventricular myocardium

87. bradycardia: heart rate slower than normal
    tachycardia: faster than normal heart rate

88. (a) Ion concentrations in extracellular fluid:
       (*Note:* EC = extracellular)
       decreasing EC K^+ → decreasing HR
       increasing EC Ca^{++} → increasing excitability and prolonged contractions
    (b) Changes in body temperature:
       decreasing temp → decreasing HR and decreasing strength of contractions
       increasing temp → increasing HR and increasing strength of contraction
    (c) Autonomic activity:
       parasympathetic stimulation → releases ACh → decreasing heart rate
       sympathetic stimulation → releases norepinephrine → increasing heart rate

[L3]   CRITICAL THINKING/APPLICATION—p. 380

1. *Inhalation:*
   increasing size of thoracic cavity; air into lungs
   blood → vena cava and right atrium (RA) from systemic veins
   Exhalation:
   decreasing size of thoracic cavity; air out of lungs
   venous blood → RA
   On inhalation:
   increasing venous return (VR)
   On exhalation:
   decreasing venous return
   (Changes in venous return have a direct effect on cardiac output (CO). On inhalation the increasing VR stretches the atrial walls, causing increasing SV [Starling's law], increasing CO, and increasing heart rate (HR) (atrial reflex).

2. In cardiac muscle tissue there are no antagonistic muscle groups to extend the cardiac muscle fibers after each contraction. The necessary stretching force is provided by the blood pouring into the heart, aided by the elasticity of the fibrous skeleton. As a result, the amount of blood entering the heart is equal to the amount ejected during the next contraction.

3. If a coronary artery is obstructed by a clot, the myocardial cells that the artery supplies do not receive sufficient amounts of blood (ischemia). Chest pains (angina pectoris) are usually a symptom of ischemia. When the heart tissue is deprived of O_2 due to ischemia, the affected portion of the heart dies. This is called myocardial infarction (heart attack).

4. Cardiac muscle cells have an abundance of mitochondria, which provide enough adenosine triphosphate (ATP) energy to sustain normal myocardial energy requirements. Most of the ATP is provided from the metabolism of fatty acids when the heart is resting. During periods of strenuous exercise muscle cells use lactic acid as an additional source of energy.

5. Your friend's heart condition would probably be premature atrial contractions or premature ventricular contractions. On an ECG this condition causes the P wave to be superimposed on the Q-T complex because of shortened intervals between one contraction and the succeeding contraction.

   Another possibility could be premature ventricular contraction. In this condition there is a prolonged QRS complex, an inverted T wave, exaggerated voltage because of a single ventricle depolarizing, and an increased probability of fibrillation.

6. The diagnosis is most likely a stenosed atrioventricular valve. Blood flows through a stenosed valve in a turbulent fashion, causing a rushing sound preceding the valve closure.

   A stenosed semilunar valve results in a rushing sound immediately before the second heart sound.

7. Anastomoses are the interconnections between arteries in coronary circulation. Because of the interconnected arteries, the blood supply to the cardiac muscle remains relatively constant, regardless of pressure fluctuations within the left and right coronary arteries.

8. Both of the atria are responsible for pumping blood into the ventricles. The right ventricle normally does not need to push very hard to propel blood through the pulmonary circuit because of the close proximity of the lungs to the heart, and the pulmonary arteries and veins are relatively short and wide. The wall of the right ventricle is relatively thin. When it contracts it squeezes the blood against the mass of the left ventricle, moving blood efficiently with minimal effort, developing relatively low pressures. The left ventricle has an extremely thick muscular wall, generating powerful contractions and causing the ejection of blood into the ascending aorta and producing a bulge into the right ventricular cavity. The left ventricle must exert six to seven times as much force as the right ventricle to push blood around the systemic circuit.

Chapter 21:

Blood Vessels and Circulation

[L1]   MULTIPLE CHOICE—pp. 382–386

(1.) A	(7.) B	(13.) D	(19.) A	(25.) C
(2.) D	(8.) D	(14.) B	(20.) B	(26.) D
(3.) B	(9.) C	(15.) B	(21.) A	(27.) B
(4.) C	(10.) A	(16.) A	(22.) D	(28.) C
(5.) A	(11.) D	(17.) D	(23.) C	(29.) B
(6.) D	(12.) C	(18.) C	(24.) C	(30.) D

[L1]   COMPLETION—pp. 386–387

(31.) arterioles
(32.) venules
(33.) fenestrated
(34.) precapillary sphincter
(35.) pulse pressure
(36.) sphygmomanometer
(37.) circulatory pressure
(38.) total peripheral resistance
(39.) vasomotion
(40.) hydrostatic pressure
(41.) osmotic pressure
(42.) autoregulation
(43.) hepatic portal system
(44.) vasoconstriction
(45.) thoracoabdominal pump
(46.) shock
(47.) central ischemic
(48.) aortic arch
(49.) venous thrombus
(50.) arteriosclerosis

[L1]   MATCHING—p. 387

(51.) I	(55.) C	(59.) H	(63.) T	(67.) O
(52.) G	(56.) J	(60.) A	(64.) K	(68.) M
(53.) E	(57.) D	(61.) S	(65.) L	(69.) N
(54.) B	(58.) F	(62.) Q	(66.) R	(70.) P

[L1]   DRAWING/ILLUSTRATION—LABELING—pp. 388–390

Figure 21.1   The Arterial System—p. 388

(71.) brachiocephalic
(72.) aortic arch
(73.) ascending aorta
(74.) abdominal aorta
(75.) common iliac
(76.) internal iliac
(77.) external iliac
(78.) deep femoral
(79.) common carotid
(80.) descending aorta
(81.) subclavian
(82.) axillary
(83.) brachial
(84.) thoracic aorta
(85.) celiac
(86.) renal
(87.) superior mesenteric
(88.) gonadal
(89.) radial
(90.) inferior mesenteric
(91.) ulnar
(92.) femoral
(93.) popliteal
(94.) anterior tibial
(95.) posterior tibial
(96.) peroneal
(97.) dorsalis pedis
(98.) plantar anastomoses

Figure 21.2   The Venous System—p. 389

(99.) subclavian
(100.) axillary
(101.) brachial
(102.) cephalic
(103.) basilic
(104.) inferior vena cava
(105.) median cubital
(106.) accessory cephalic
(107.) cephalic
(108.) median antebrachial
(109.) basilic
(110.) palmar venous network
(111.) internal jugular
(112.) external jugular
(113.) brachiocephalic
(114.) superior vena cava
(115.) intercostals
(116.) left suprarenal
(117.) renal
(118.) gonadal
(119.) lumbar
(120.) common iliac
(121.) internal iliac
(122.) external iliac
(123.) deep femoral
(124.) femoral
(125.) great saphenous
(126.) popliteal
(127.) posterior tibial
(128.) small saphenous
(129.) anterior tibial
(130.) peroneal
(131.) plantar venous network

Figure 21.3   Major Arteries of the Head and Neck—p. 390

(132.) superficial temporal artery

(133.) circle of Willis

(134.) posterior cerebral arteries

(135.) basilar artery

(136.) internal carotid artery

(137.) vertebral artery

(138.) thyrocervical trunk

(139.) subclavian artery

(140.) internal thoracic artery

(141.) anterior cerebral arteries

(142.) maxillary artery

(143.) facial artery

(144.) external carotid artery

(145.) carotid sinus

(146.) common carotid artery

(147.) brachiocephalic artery

Figure 21.4   Major Veins Draining the Head and Neck—p. 390

(148.) superior sagitall sinus

(149.) straight sinus

(150.) transverse sinus

(151.) vertebral vein

(152.) external jugular vein

(153.) temporal vein

(154.) maxillary vein

(155.) facial vein

(156.) internal jugular vein

(157.) brachiocephalic vein

(158.) internal thoracic vein

[L2]   CONCEPT MAPS—pp. 391–396

I   THE CARDIOVASCULAR SYSTEM—p. 391

(1.) pulmonary veins

(2.) arteries and arterioles

(3.) veins and venules

(4.) pulmonary arteries

(5.) systemic circuit

II   ENDOCRINE SYSTEM AND CARDIOVASCULAR REGULATION—p. 392

(6.) epinephrine, norepinephrine

(7.) adrenal cortex

(8.) increasing blood pressure

(9.) ADH (vasopressin)

(10.) increasing plasma volume

(11.) kidneys

(12.) increasing fluid

(13.) erythropoietin

(14.) atrial natriuretic factor

III   ARTERIOLAR CONSTRICTION—p. 393

(15.) hyperemia

(16.) reactive

(17.) ADH (vasopressin)

(18.) epinephrine

(19.) vasomotor center

(20.) vasoconstriction

(21.) vasodilation

IV   ANF EFFECTS ON BLOOD VOLUME AND BLOOD PRESSURE—p. 394

(22.) increasing H_2O loss by kidneys

(23.) decreasing H_2O intake

(24.) decreasing blood pressure

(25.) increasing blood flow (l/min.)

V   MAJOR BRANCHES OF THE AORTA—p. 395

(26.) ascending aorta

(27.) brachiocephalic artery

(28.) L. subclavian artery

(29.) thoracic artery

(30.) celiac trunk

(31.) superior mesenteric artery

(32.) R. gonadal artery

(33.) L. common iliac artery

VI   MAJOR VEINS DRAINING INTO THE SUPERIOR AND INFERIOR VENAE CAVAE—p. 396

(34.) superior vena cava

(35.) azygous vein

(36.) L. hepatic veins

(37.) R. suprarenal vein

(38.) L. Renal vein

(39.) L. common iliac vein

[L2]   BODY TREK—p. 397

(40.) aortic valve

(41.) aortic arch

(42.) descending aorta

(43.) brachiocephalic

(44.) L. common carotid

(45.) L. subclavian

(46.) thoracic aorta

(47.) mediastinum

(48.) intercostal

(49.) superior phrenic

(50.) abdominal aorta

(51.) inferior phrenic

(52.) celiac

(53.) suprarenal

(54.) renal

(55.) superior mesenteric

(56.) gonadal

(57.) inferior mesenteric

(58.) lumbar

(59.) common iliacs

[L2]   MULTIPLE CHOICE—pp. 398–401

(60.) B	(65.) B	(70.) A	(75.) D	(80.) D
(61.) D	(66.) C	(71.) C	(76.) B	(81.) C
(62.) D	(67.) A	(72.) D	(77.) D	(82.) B
(63.) C	(68.) C	(73.) A	(78.) B	(83.) A
(64.) D	(69.) D	(74.) C	(79.) A	(84.) B

[L2]   COMPLETION—p. 401

(85.) circle of Willis

(86.) elastic rebound

(87.) recall of fluids

(88.) edema

(89.) precapillary sphincters

(90.) reactive hyperemia

(91.) endothelium

(92.) mesoderm

(93.) veins

(94.) venous return

(95.) aorta

(96.) brachial

(97.) radial

(98.) great saphenous

(99.) lumen

[L2]   SHORT ESSAY—pp. 402–404

100. tunica interna, tunica media, tunica externa

101. heart → arteries → arterioles → capillaries (gas exchange area) → venules → veins → heart

102. (a) Sinusoids are specialized fenestrated capillaries.

   (b) They are found in the liver, bone marrow, and the adrenal glands.

   (c) They form fattened, irregular passageways, so blood flows through the tissues slowly maximizing time for absorption and secretion and molecular exchange.

103. In the pulmonary circuit, oxygen stores are replenished, carbon dioxide is excreted, and the "reoxygenated" blood is returned to the heart for distribution in the systemic circuit.

   The systemic circuit supplies the capillary beds in all parts of the body with oxygenated blood, and returns deoxygenated blood to the heart of the pulmonary circuit for removal of carbon dioxide.

104. (a) vascular resistance, viscosity, turbulence

   (b) Only vascular resistance can be adjusted by the nervous and endocrine systems.

105. $F = \dfrac{BP}{PR}$

   Flow is directly proportional to the blood pressure and inversely proportional to peripheral resistance; i.e., increasing pressure, increasing flow; decreasing pressure, decreasing flow; increasing PR, decreasing flow; decreasing PR, increasing flow.

106. $\dfrac{120 \text{ mm Hg}}{80 \text{ mm Hg}}$ is a "normal" blood pressure reading.

   The top number, 120 mm Hg, is the *systolic* pressure, i.e., the peak blood pressure measured during ventricular systole.

   The bottom number, 80 mm Hg, is the *diastolic* pressure, i.e., the minimum blood pressure at the end of ventricular diastole.

107. MAP = 1/3 pulse pressure (p.p.) + diastolic pressure

   Therefore, p.p. = 110 mm Hg – 80 mm Hg = 30 mm Hg

   MAP = 1/3 (.30) = 10 + 80 mm Hg

   MAP = 90 mm Hg

108. (a) Distributes nutrients, hormones, and dissolved gases throughout tissues.

   (b) Transports insoluble lipids and tissue proteins that cannot enter circulation by crossing capillary linings.

   (c) Speeds removal of hormones and carries bacterial toxins and other chemical stimuli to cells of the immune system.

109. Cardiac output, blood volume, peripheral resistance.

110. Aortic baroreceptors, carotid sinus baroreceptors, atrial baroreceptors.

111. Epinephrine and norepinephrine, ADH, angiotensin II, erythropoietin, and atrial natriuretic peptide.

112. Decreasing hematocrit, venous thrombosus, pulmonary embolism, increasing pulmonary blood pressure, pooling of blood in the veins, edema.

113. Arteries lose their elasticity, the amount of smooth muscle they contain decreases, and they become stiff and relatively inflexible.

[L3]   CRITICAL THINKING/APPLICATION—p. 404

1. L. ventricle → aortic arch → L. subclavian artery → axillary artery → brachial artery → radial and ulnar arteries → palm and wrist arterial anastomoses → digital arteries → digital veins → palm and wrist venous anastomoses → cephalic and basilic veins → radius and ulnar veins → brachial vein → axillary vein → L. subclavian vein → brachiocephalic vein → superior vena cava → R. atrium

2. (a) superficial temporal artery

   (b) common carotid artery

   (c) facial artery

   (d) axillary artery

   (e) brachial artery

   (f) femoral artery

   (g) popliteal artery

   (h) dorsalis pedis artery

3. When a person rises rapidly from a lying position, a drop in blood pressure in the neck and thoracic regions occurs because of the pull of gravity on the blood. Owing to the sudden decrease in blood pressure, the blood flow to the brain is reduced enough to cause dizziness or loss of consciousness.

4. By applying pressure on the carotid artery at frequent intervals during exercise, the pressure to the region of the carotid sinus may be sufficient to stimulate the baroreceptors. The increased action potentials from the baroreceptors initiate reflexes in parasympathetic impulses to the heart, causing a decrease in the heart rate.

5. During exercise the following changes and benefits occur:

   (a) increased blood flow to tissues, supplying O_2 and nutrients, removing wastes

   (b) increased blood flow to skin—thermoregulatory—gets rid of excess heat in the body

   (c) increased venous return due to skeletal muscle movement and the thoracoabdominal pump

   (d) decreased oxygen tension resulting from increased muscular activity

   (e) increased blood pressure

   (f) increased skeletal blood vessel dilation

   (g) increased sympathetic stimulation to the heart—increased cardiac output

   (h) increased sympathetic innervation—this causes vasoconstriction in blood vessels of skin and viscera, shunting blood to skeletal muscles

   (i) increased vasodilation of capillaries because of presence of CO_2, K^+, and lactic acid

Chapter 22:

The Lymphatic System and Immunity

[L1]   MULTIPLE CHOICE—pp. 406–410

(1.) D	(7.) C	(13.) A	(19.) A	(25.) D
(2.) C	(8.) C	(14.) A	(20.) C	(26.) D
(3.) D	(9.) B	(15.) D	(21.) C	(27.) D
(4.) B	(10.) D	(16.) D	(22.) A	(28.) B
(5.) A	(11.) A	(17.) A	(23.) C	(29.) D
(6.) A	(12.) C	(18.) B	(24.) B	(30.) D

[L1]   COMPLETION—pp. 410–411

(31.) lacteals

(32.) cytotoxic T cells

(33.) antibodies

(34.) lymph capillaries

(35.) phagocytes

(36.) diapedesis

(37.) passive

(38.) cell-mediated

(39.) innate

(40.) active

(41.) suppressor T

(42.) helper T

(43.) plasma cells

(44.) neutralization

(45.) precipitation

(46.) haptens

(47.) immunological competence

(48.) Lymphokines

(49.) monokines

(50.) IgG

(51.) immunodeficiency disease

(52.) vaccinated

[L1]   MATCHING—p. 412

(53.) I	(58.) F	(63.) B	(68.) T	(73.) X
(54.) G	(59.) C	(64.) D	(69.) R	(74.) N
(55.) H	(60.) E	(65.) S	(70.) U	(75.) O
(56.) K	(61.) L	(66.) W	(71.) M	(76.) Q
(57.) A	(62.) J	(67.) V	(72.) P	

[L1]   DRAWING/ILLUSTRATION—LABELING—pp. 413–415

Figure 22.1   The Lymphatic System—p. 413

(77.) cervical lymph nodes

(78.) R. lymphatic ducts

(79.) Thymus

(80.) thoracic duct

(81.) L. lymphatic duct

(82.) cisterna chyli

(83.) spleen

Figure 22.2   The Lymphatic Ducts—p. 414

(84.) cervical nodes

(85.) R. lymphatic duct

(86.) superior vena cava

(87.) axillary nodes

(88.) cisterna chyli

(89.) para-aortic nodes

(90.) inguinal nodes

(91.) L. subclavian vein

(92.) thoracic duct

Figure 22.3   Nonspecific Defenses—p. 415

(93.) physical barriers

(94.) phagocytes

(95.) immunological surveillance

(96.) complement system

(97.) inflammatory response

(98.) fever

(99.) interferon

[L2] **CONCEPT MAPS—pp. 416–417**

I   IMMUNE SYSTEM—p. 416

(1.) nonspecific immunity

(2.) phagocytic cells

(3.) inflammation

(4.) specific immunity

(5.) innate

(6.) acquired

(7.) active

(8.) active immunization

(9.) transfer of antibodies via placenta

(10.) passive immunization

II   INFLAMMATION RESPONSE—p. 417

(11.) tissue damage

(12.) increasing vascular permeability

(13.) phagocytosis of bacteria

(14.) tissue repaired

(15.) bacteria not destroyed

[L2] **BODY TREK—p. 418**

(16.) viruses

(17.) macrophages

(18.) natural killer cells

(19.) helper T cells

(20.) B cells

(21.) antibodies

(22.) killer T cells

(23.) suppressor T cells

(24.) memory T and B cells

[L2] **MULTIPLE CHOICE—pp. 419–421**

(25.) B	(29.) A	(33.) D	(37.) A	(41.) B
(26.) D	(30.) C	(34.) D	(38.) C	(42.) C
(27.) C	(31.) A	(35.) C	(39.) D	(43.) B
(28.) B	(32.) D	(36.) B	(40.) C	(44.) D

[L2] **COMPLETION—pp. 421–422**

(45.) T cells

(46.) microglia

(47.) Kupffer cells

(48.) Langerhans cells

(49.) antigen

(50.) cytokines

(51.) properdin

(52.) mast

(53.) pyrogens

(54.) opsonins

(55.) NK cells

(56.) interferon

(57.) IgG

(58.) IgM

(59.) helper T

[L2] **SHORT ESSAY—pp. 422–424**

60. (a) lymphatic vessels

     (b) lymph

     (c) lymphatic organs

61. (a) production, maintenance, and distribution of lymphocytes

     (b) maintenance of normal blood volume

     (c) elimination of local variations in the composition of the interstitial fluid

62. (a) T cells (thymus)

     (b) B cells (bone-marrow)

     (c) NK cells—natural killer (bone marrow)

63. (a) Cytotoxic T cells—cell-mediated immunity

     (b) Helper T cells—release lymphokines that coordinate specific and nonspecific defenses

     (c) Suppressor T cells—depress responses of other T cells and B cells

64. Stimulated B cells differentiate into plasma cells that are responsible for production and secretion of antibodies. B cells are said to be responsible for humoral immunity.

65. (a) lymph nodes

     (b) thymus

     (c) spleen

66. (a) physical barriers

(b) phagocytes

(c) immunological surveillance

(d) complement system

(e) inflammatory response

(f) fever

67. NK (natural killer) cells are sensitive to the presence of abnormal cell membranes and respond immediately. When the NK cell makes contact with an abnormal cell it releases secretory vesicles that contain proteins called perforins. The perforins create a network of pores in the target cell membrane, releasing free passage of intracellular materials necessary for homeostasis, thus causing the cell to disintegrate.

T cells and B cells provide defenses against specific threats, and their activation requires a relatively complex and time-consuming sequence of events.

68. (a) destruction of target cell membranes

(b) stimulation of inflammation

(c) attraction of phagocytes

(d) enhancement of phagocytosis

69. (a) specificity

(b) versatility

(c) memory

(d) tolerance

70. Active immunity appears following exposure to an antigen, as a consequence of the immune response.

Passive immunity is produced by transfer of antibodies from another individual.

71. (a) direct attack by activated T cells (cellular immunity)

(b) attack by circulating antibodies released by plasma cells derived from activated B cells (humoral immunity)

72. (a) neutralization

(b) agglutination and precipitation

(c) activation of complement

(d) attraction of phagocytes

(e) opsonization

(f) stimulation of inflammation

(g) prevention of bacterial and viral adhesion

73. (a) IgG   (b) IgE   (c) IgD   (d) IgM   (e) IgA

74. (a) interleukins

(b) interferons

(c) tumor necrosis factor

(d) chemicals that regulate phagocytic activity

75. Autoimmune disorders develop when the immune response mistakenly targets normal body cells and tissues. When the immune recognition system malfunctions, activated B cells begin to manufacture antibodies against other cells and tissues.

[L3] CRITICAL THINKING/APPLICATION—p. 424

1. Tissue transplants normally contain protein molecules called human lymphocyte antigen (HLA) genes that are foreign to the recipient. These antigens trigger the recipient's immune responses, activating the cellular and humoral mediated responses that may act to destroy the donated tissue.

2.

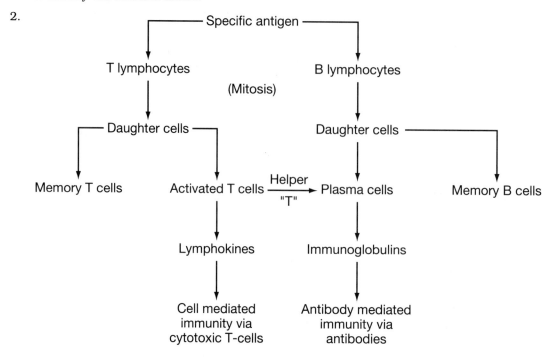

3. Neutrophils are leukocytes that comprise 50 to 70 percent of the circulating WBC population. Neutrophils are highly mobile and are usually the first white blood cells to arrive at an injury site. They are very active phagocytes specializing in attacking and digesting bacteria. Because their "weaponry" includes defenses and they are phagocytic specialists, they can segregate the microbe in a tiny fat-covered sac that merges with a pocket of defensins. The defensins pierce the membranes of the microbe, killing it by causing the intracellular components to leak out of the intruding microbe. Defensins provide additional weaponry for the body's immunological surveillance system similar to the NK cells and the action of perforins.

4. They are classified as infectious agents because they can enter cells and replicate themselves. Viruses contain a core of nucleic acid (DNA or RNA) surrounded by a protein coat. When a virus infects a cell, its nucleic acid enters the nucleus of the host and takes over the cell's metabolic machinery. The viral DNA replicates, forms new viruses, and causes the host's cell membrane to rupture or lyse, causing a disruption in normal cell function.

5. Viruses reproduce inside living cells, beyond the reach of lymphocytes. Infected cells incorporate viral antigens into their cell membranes. NK cells recognize these infected cells as abnormal. By destroying virally infected cells, NK cells slow or prevent the spread of viral infection.

6. High body temperatures may inhibit some viruses and bacteria, or may speed up their reproductive rates so that the disease runs its course more rapidly. The body's metabolic processes are accelerated, which may help mobilize tissue defenses and speed the repair process.

7. During times of stress the body responds by increasing the activity of the adrenal glands, which may cause a decrease in the inflammatory response, reduce the activities and numbers of phagocytes in peripheral tissues, and inhibit interleukin secretion. "Stress reduction" may cause a decrease in the adrenal gland secretions and maintain homeostatic control of the immune response.

Chapter 23:

The Respiratory System

[L1] **MULTIPLE CHOICE—pp. 426–431**

(1.) D	(8.) B	(15.) A	(22.) B	(29.) C
(2.) B	(9.) A	(16.) A	(23.) C	(30.) C
(3.) C	(10.) B	(17.) C	(24.) B	(31.) B
(4.) D	(11.) A	(18.) B	(25.) D	(32.) A
(5.) C	(12.) D	(19.) D	(26.) C	(33.) B
(6.) B	(13.) C	(20.) B	(27.) D	(34.) B
(7.) C	(14.) A	(21.) C	(28.) D	

[L1] **COMPLETION—pp. 431–432**

(35.) expiration	(42.) bronchioles	(49.) Bohr effect
(36.) inspiration	(43.) respiratory bronchioles	(50.) carbamino-hemoglobin
(37.) surfactant	(44.) external respiration	(51.) apnea
(38.) mucus escalator	(45.) internal respiration	(52.) mechanoreceptor
(39.) external nares	(46.) ductus arteriosus	(53.) chemoreceptor
(40.) glottis	(47.) vital capacity	(54.) apneustic
(41.) larynx	(48.) alveolar ventilation	(55.) pneumotaxic

[L1] **MATCHING—p. 432**

(56.) E	(60.) C	(64.) J	(68.) F	(72.) M
(57.) D	(61.) H	(65.) K	(69.) B	(73.) P
(58.) G	(62.) B	(66.) T	(70.) L	(74.) O
(59.) A	(63.) I	(67.) R	(71.) C	(75.) Q

[L1] **DRAWING/ILLUSTRATION—LABELING—pp. 433–435**

Figure 23.1 **Upper Respiratory Tract—p. 433**

(76.) internal nares	(82.) oropharynx	(88.) nasal concha
(77.) nasopharynx	(83.) epiglottis	(89.) nasal vestibule
(78.) eustachian tube opening	(84.) glottis	(90.) oral cavity
(79.) hard palate	(85.) vocal fold	(91.) tongue
(80.) soft palate	(86.) laryngopharynx	(92.) thyroid cartilage
(81.) palatine tonsil	(87.) trachea	(93.) cricoid cartilage

Figure 23.2 **Thorax and Lungs—p. 434**

(94.) larynx	(100.) alveolus	(105.) trachea
(95.) parietal pleura	(101.) diaphragm	(106.) tracheal cartilage
(96.) pleural cavity	(102.) epiglottis	(107.) apex of lung
(97.) visceral pleura	(103.) thyroid cartilage	(108.) mediastinum
(98.) alveolar sac	(104.) cricoid cartilage	(109.) L. primary bronchus
(99.) alveolar duct		

Figure 23.3 **Lower Respiratory Tract—p. 435**

(110.) epiglottis	(114.) terminal bronchioles	(118.) carina
(111.) "Adam's apple"	(115.) alveoli	(119.) L. primary bronchus
(112.) thyroid cartilage	(116.) larynx	(120.) secondary bronchi
(113.) cricoid cartilage	(117.) trachea	(121.) tertiary bronchi

[L2]   CONCEPT MAPS—pp. 433–437

I   EXTERNAL–INTERNAL RESPIRATION—p. 436

(1.) pharynx

(2.) trachea

(3.) secondary bronchi

(4.) respiratory bronchioles

(5.) alveoli

(6.) pulmonary veins

(7.) left ventricle

(8.) systemic capillaries

(9.) right atrium

(10.) pulmonary arteries

II   RESPIRATORY SYSTEM—p. 437

(11.) upper respiratory tract

(12.) pharynx and larynx

(13.) paranasal sinuses

(14.) speech production

(15.) lungs

(16.) ciliated mucous membrane

(17.) alveoli

(18.) blood

(19.) O_2 from alveoli into blood

[L2]   BODY TREK—p. 438

(20.) external nares

(21.) nasal cavity

(22.) vestibule

(23.) nasal conchae

(24.) hard palate

(25.) soft palate

(26.) nasopharynx

(27.) pharynx

(28.) epiglottis

(29.) larynx

(30.) vocal folds

(31.) phonation

(32.) trachea

(33.) cilia

(34.) oral cavity

(35.) primary bronchus

(36.) smooth muscle

(37.) bronchioles

(38.) alveolus

(39.) oxygen

(40.) carbon dioxide

(41.) pulmonary capillaries

(42.) mucus escalator

[L2]   MULTIPLE CHOICE—pp. 439–442

(43.) C	(48.) A	(53.) D	(58.) C	(63.) B
(44.) B	(49.) D	(54.) C	(59.) C	(64.) C
(45.) D	(50.) A	(55.) B	(60.) D	(65.) A
(46.) B	(51.) B	(56.) B	(61.) B	(66.) C
(47.) D	(52.) C	(57.) D	(62.) C	(67.) A

[L2]   COMPLETION—pp. 442–443

(68.) lamina propria

(69.) vestibule

(70.) pharynx

(71.) laryngopharynx

(72.) epiglottis

(73.) thyroid cartilage

(74.) phonation

(75.) hilus

(76.) cribiform plate

(77.) hypoxia

(78.) anoxia

(79.) atelectasis

(80.) pneumothorax

(81.) intrapleural

(82.) hypercapnia

(83.) carina

(84.) parenchyma

(85.) anatomic dead space

(86.) alveolar ventilation

(87.) Henry's law

(88.) Dalton's law

[L2]   SHORT ESSAY—pp. 443–444

89. (a) Provides an area for gas exchange between air and blood.

(b) To move air to and from exchange surfaces.

(c) Protects respiratory surfaces from abnormal changes or variations.

(d) Defends the respiratory system and other tissues from pathogenic invasion.

(e) Permits communication via production of sound.

(f) Participates in the regulation of blood volume, pressure, and body fluid pH.

90. The air is warmed to within 1 degree of body temperature before it enters the pharynx. This results from the warmth of rapidly flowing blood in an extensive vasculature of the nasal mucosa. For humidifying the air, the nasal mucosa is supplied with small mucus glands that secrete a mucoid fluid inside the nose. The warm and humid air prevents drying of the pharynx, trachea, and lungs, and facilitates the flow of air through the respiratory passageways without affecting environmental changes.

91. Surfactant cells are scattered among the simple squamous epithelial cells of the respiratory membranes. They produce an oily secretion containing a mixture of phospholipids that coat the alveolar epithelium and keep the alveoli from collapsing like bursted bubbles.

92. Hairs, cilia, and mucus traps in the nasal cavity; mucus escalator in the larynx, trachea, and bronchi. These areas have a mucus lining and contain cilia that beat toward the pharynx; phagocytes by alveolar macrophages.

93. External respiration includes the diffusion of gases between the alveoli and the circulating blood.

    Internal respiration is the exchange of dissolved gases between the blood and the interstitial fluids in peripheral tissues.

    Cellular respiration is the absorption and utilization of oxygen by living cells via biochemical pathways that generate carbon dioxide.

94. Increasing volume; decreasing pressure and decreasing volume; increasing pressure

95. A *pneumothorax* is an injury to the chest wall that penetrates the parietal pleura or damages the alveoli and the visceral pleura allowing air into the pleural cavity.

    An *atelectasis* is a collapsed lung.

    A pneumothorax breaks the fluid bond between the pleurae and allows the elastic fibers to contract and the result is a collapsed lung.

96. The vital capacity is the maximum amount of air that can be moved into and out of the respiratory system in a single respiratory cycle.

    Vital capacity = inspiratory reserve + expiratory reserve + tidal volume.

97. Alveolar ventilation, or V_e, is the amount of air reaching the alveoli each minute. It is calculated by subtracting the anatomic dead space, V_d, from the tidal volume (V_t) using the formula:

    $$V_e = f_x (V_t - V_d)$$

98. (a) CO_2 may be dissolved in the plasma (7 percent).
    (b) CO_2 may be bound to the hemoglobin of RBC (23 percent).
    (c) CO_2 may be converted to a molecule of carbonic acid (70 percent).

99. (a) mechanoreceptor reflexes
    (b) chemoreceptor reflexes
    (c) protective reflexes

100. A rise in arterial Pco_2 immediately elevates cerebrospinal fluid CO_2 levels and stimulates the chemoreceptor neurons of the medulla. These receptors stimulate the respiratory center causing an increase in the rate and depth of respiration, or hyperventilation.

[L3]   CRITICAL THINKING/APPLICATION—p. 445

1. Running usually requires breathing rapidly through the mouth, which eliminates much of the preliminary filtration, heating, and humidifying of the inspired air. When these conditions are eliminated by breathing through the mouth, the delicate respiratory surfaces are subject to chilling, drying out, and possible damage.

2. The lungs receive the entire cardiac output. Blood pressure in the pulmonary circuit is low; therefore, the pulmonary vessels can easily become blocked by blood clots or fatty plaques. If the blockage remains in place for several hours, the alveoli will permanently collapse and pulmonary resistance increases, placing an extra strain on the R. ventricle. With a decrease in O_2 supply and an increase in pulmonary vascular resistance, the R. ventricle cannot maintain the cardiac output. This is referred to as congestive heart failure.

3. The esophagus lies immediately posterior to the cartilage-free posterior wall of the trachea. Distention of the esophagus due to the passage of a bolus of food exerts pressure on surrounding structures. If the tracheal cartilages are complete rings, the tracheal region of the neck and upper thorax cannot accommodate the necessary compression to allow the food bolus to pass freely through the esophagus.

4. In emphysema, respiratory bronchi and alveoli are functionally eliminated. The alveoli expand and their walls become infiltrated with fibrous tissue. The capillaries deteriorate and gas exchange ceases. During the later stages of development bronchitis may develop, especially in the bronchioles. Emphysema and bronchitis produce a particularly dangerous combination known as chronic obstructive pulmonary disease (COPD). These conditions have been linked to the inhalation of air containing fine particulate matter or toxic vapors, such as those found in cigarette smoke.

5. Coughing and sneezing are both protective reflexes of the respiratory system. Mr. Bunting's injuries may involve the vagus nerve, which carries impulses to the medulla, initiating the cough reflex; or the trigeminal nerves, which carry impulses to the medulla, initiating the sneezing reflex. The reflex centers in the medulla could be damaged.

6. When respiratory acidosis is due to reduced elimination of CO_2 from the body (i.e., increasing arterial P_{CO_2}, decreasing blood pH), the respiratory center is stimulated and *hyperventilation* results. Hyperventilation attempts to *reduce blood CO_2* levels and to cause an *increase in blood pH*. When respiratory acidosis is due to depression of the respiratory center, *hypoventilation* results, causing a *buildup of CO_2* in the blood and a *decrease in blood pH*.

Chapter 24:

The Digestive System

[L1] MULTIPLE CHOICE—pp. 447–451

(1.) C	(8.) A	(15.) D	(22.) D	(29.) A
(2.) C	(9.) B	(16.) B	(23.) D	(30.) B
(3.) B	(10.) B	(17.) C	(24.) D	(31.) A
(4.) D	(11.) D	(18.) A	(25.) C	(32.) B
(5.) A	(12.) B	(19.) A	(26.) B	(33.) C
(6.) D	(13.) C	(20.) B	(27.) D	(34.) A
(7.) B	(14.) C	(21.) C	(28.) D	(35.) B

[L1] COMPLETION—pp. 452–453

(36.) esophagus
(37.) digestion
(38.) alveolar connective tissue
(39.) lamina propria
(40.) dense bodies
(41.) plasticity
(42.) peristalsis

(43.) gastrin
(44.) mucins
(45.) cuspids
(46.) carbohydrates
(47.) chyme
(48.) duodenum
(49.) cholecystokinin

(50.) Kupffer cells
(51.) acini
(52.) pyloric sphincter
(53.) haustrae
(54.) lipase
(55.) pepsin
(56.) vitamin B_{12}

[L1] MATCHING—p. 453

(57.) I	(62.) A	(67.) F	(72.) M	(77.) P
(58.) G	(63.) L	(68.) B	(73.) W	(78.) T
(59.) J	(64.) C	(69.) U	(74.) R	(79.) V
(60.) D	(65.) H	(70.) S	(75.) O	(80.) X
(61.) K	(66.) E	(71.) Q	(76.) N	

[L1] DRAWING/ILLUSTRATION—LABELING—pp. 454–455

Figure 24.1 The Digestive Tract—p. 454

(81.) sublingual gland
(82.) submandibular gland
(83.) liver
(84.) hepatic duct
(85.) gall bladder
(86.) cystic duct
(87.) common bile duct
(88.) pancreatic duct
(89.) duodenum
(90.) transverse colon

(91.) ascending colon
(92.) ileocecal valve
(93.) cecum
(94.) appendix
(95.) ileum
(96.) parotid gland
(97.) oropharynx
(98.) esophagus
(99.) lower esophageal sphincter

(100.) fundus of stomach
(101.) body of stomach
(102.) pyloric sphincter
(103.) pylorus
(104.) pancreas
(105.) jejunum
(106.) descending colon
(107.) sigmoid colon
(108.) rectum
(109.) anus

Figure 24.2 Histology of G.I. Tract Wall—p. 455

(110.) mesenteric artery and vein
(111.) mesentery
(112.) visceral peritoneum (serosa)

(113.) plica
(114.) mucosa
(115.) submucosa
(116.) lumen

(117.) circular muscle layer
(118.) muscularis externa
(119.) mucosal gland
(120.) visceral peritoneum (serosa)

Figure 24.3   Structure of a Typical Tooth—p. 455

(121.) crown

(122.) neck

(123.) root

(124.) enamel

(125.) dentin

(126.) pulp cavity

(127.) cementum

(128.) root canal

(129.) bone of alveolus

[L2]   CONCEPT MAPS—pp. 456–458

I   DIGESTIVE SYSTEM—p. 456

(1.) amylase

(2.) pancreas

(3.) bile

(4.) hormones

(5.) digestive tract movements

(6.) stomach

(7.) large intestine

(8.) hydrochloric acid

(9.) intestinal mucosa

II   CHEMICAL EVENTS IN DIGESTION—p. 457

(10.) esophagus

(11.) small intestine

(12.) polypeptides

(13.) amino acids

(14.) complex sugars and starches

(15.) disaccharides, trisaccharides

(16.) simple sugars

(17.) monoglycerides, fatty acids in micelles

(18.) triglycerides

(19.) lacteal

III   REGULATION OF GASTRIC EMPTYING—p. 458

(20.) decreasing sympathetic

(21.) gastrin

(22.) increasing stomach motility

(23.) increasing gastric emptying

(24.) increasing sympathetic

(25.) enterogastric reflex

(26.) decreasing stomach motility

(27.) decreasing gastric emptying

[L2]   BODY TREK—p. 459

(28.) oral cavity

(29.) lubrication

(30.) mechanical processing

(31.) gingivae (gums)

(32.) teeth

(33.) mastication

(34.) palates

(35.) tongue

(36.) saliva

(37.) salivary amylase

(38.) carbohydrates

(39.) pharynx

(40.) fauces

(41.) uvula

(42.) swallowing

(43.) esophagus

(44.) peristalsis

(45.) gastroesophygeal sphincter

(46.) stomach

(47.) gastric juices

(48.) chyme

(49.) rugae

(50.) mucus

(51.) hydrochloric

(52.) pepsin

(53.) protein

(54.) pylorus

(55.) duodenum

(56.) lipase

(57.) pancreas

(58.) large intestine

(59.) feces

[L2]   MULTIPLE CHOICE—pp. 460–462

(60.) A

(61.) D

(62.) B

(63.) C

(64.) C

(65.) A

(66.) D

(67.) D

(68.) D

(69.) B

(70.) C

(71.) A

(72.) B

(73.) C

(74.) B

(75.) D

(76.) A

(77.) C

(78.) B

(79.) D

[L2]   COMPLETION—p. 463

(80.) plexus of Meissner

(81.) visceral peritoneum

(82.) mesenteries

(83.) adventitia

(84.) segmentation

(85.) ankyloglossia

(86.) incisors

(87.) bicuspids

(88.) alkaline tide

(89.) duodenum

(90.) ileum

(91.) lacteals

(92.) Brunner's glands

(93.) Peyer's patches

(94.) enterocrinin

[L2] SHORT ESSAY—pp. 463–465

95. (a) ingestion
    (b) mechanical processing
    (c) digestion
    (d) secretion
    (e) absorption
    (f) compaction
    (g) excretion (defecation)

96. (a) epithelium
    (b) lamina propria
    (c) muscularis mucosa
    (d) submucosa
    (e) muscularis externa
    (f) adventitia

97. During a peristaltic movement, the circular muscles first contract behind the digestive contents. Longitudinal muscles contract next, shortening adjacent segments. A wave of contraction in the circular muscles then forces the materials in the desired direction. During segmentation, a section of intestine resembles a chain of sausages owing to the contractions of circular muscles in the muscularis externa. A given pattern exists for a brief moment before those circular muscles relax and others contract, subdividing the "sausages."

98. (a) parotid glands
    (b) sublingual glands
    (c) submandibular glands

99. (a) incisors—clipping or cutting
    (b) cuspids (canines)—tearing or slashing
    (c) bicuspids (premolars)—crushing, mashing, and grinding
    (d) molars—crushing, mashing, and grinding

100. (a) upper esophygeal sphincter
    (b) lower esophygeal sphincter
    (c) pyloric sphincter
    (d) ileocecal sphincter
    (e) internal anal sphincter
    (f) external anal sphincter

101. Parietal cells and chief cells are types of secretory cells found in the wall of the stomach. Parietal cells secrete intrinsic factors and hydrochloric acid. Chief cells secrete an inactive proenzyme, pepsinogen.

102. (a) cephalic, gastric, intestinal
    (b) CNS regulation; release of gastrin into circulation; enterogastric reflexes, secretion of cholecystokinin (CCK) and secretin

103. secretin, cholecystokinin, and glucose-dependent insulinotropic peptide (GIP)

104. (a) resorption of water and compaction of feces
    (b) the absorption of important vitamins liberated by bacterial action
    (c) the storing of fecal material prior to defecation

105. (a) metabolic regulation
    (b) hematological regulation
    (c) bile production

106. (a) Endocrine function—pancreatic islets secrete insulin and glucagon into the bloodstream.
    (b) Exocrine function—secrete a mixture of water, ions, and digestive enzymes into the small intestine.

[L3]   CRITICAL THINKING/APPLICATION—p. 465

1. Although actin and myosin filaments are present in smooth muscle, oriented along the axis of the cell, there are no sarcomeres. As a result, smooth muscle fibers do not have striations. The thin filaments are attached to dense bodies that are firmly attached to the inner surface of the sarcolemma. Dense bodies anchor the thin filaments so that when sliding occurs between thin and thick filaments, the cell shortens. When a contraction occurs, the muscle fiber twists like a corkscrew. In smooth muscle cells, neither troponin nor tropomyosin is involved. Calcium ions interact with calmodulin, a binding protein. The calmodulin activates myosin light chain kinase, which breaks down ATP and initiates the contraction. Smooth muscle can contract over a range of lengths four times greater than that of skeletal muscle.

2. Swelling in front of and below the ear is an obvious symptom of the mumps, a contagious viral disease. It is a painful enlargement of the parotid salivary glands and is far more severe in adulthood and can occasionally cause infertility in males. Today there is a safe mumps vaccine, MMR (measles, mumps, and rubella), which can be administered after the age of 15 months.

3. Heartburn, or acid indigestion, is not related to the heart, but is caused by a backflow, or reflux, of stomach acids into the esophagus, a condition called gastroesophygeal reflux. Caffeine and alcohol stimulate the secretion of HCl in the stomach. When the HCl is refluxed into the esophagus, it causes the burning sensation.

4. I. M. Nervous probably experienced a great deal of stress and anxiety while studying anatomy and physiology. If a person is excessively stressed, an increase in sympathetic nervous activity may inhibit duodenal gland secretion, increasing susceptibility to a duodenal ulcer. Decreased duodenal gland secretion reduces the duodenal wall's coating of mucus, which protects it against gastric enzymes and acid.

5. This young adolescent is lactose intolerant. This is a common disorder caused by a lack of lactase, an enzyme secreted in the walls of the small intestine that is needed to break down lactose, the sugar in cow's milk. The intestinal mucosa often stops producing lactase by adolescence, causing lactose intolerance. The condition is treated by avoiding cow's milk and other products that contain lactose. Babies may be given formula based on soy or other milk substitutes. There are also milk products in which the lactose is predigested. Yogurt and certain cheeses usually can be tolerated because the lactose already has been broken down.

Chapter 25:

Metabolism and Energetics

[L1]   MULTIPLE CHOICE—pp. 467–471

(1.) C	(7.) B	(13.) B	(19.) D	(25.) D
(2.) A	(8.) A	(14.) B	(20.) C	(26.) A
(3.) B	(9.) C	(15.) C	(21.) A	(27.) A
(4.) C	(10.) D	(16.) D	(22.) B	(28.) B
(5.) D	(11.) B	(17.) A	(23.) C	(29.) C
(6.) A	(12.) D	(18.) D	(24.) B	

[L1]   COMPLETION—pp. 471–472

(30.) anabolism
(31.) glycolysis
(32.) oxidative phosphorylation
(33.) triglycerides
(34.) beta oxidation
(35.) lipemia
(36.) cathepsins
(37.) TCA cycle
(38.) glycogen
(39.) liver
(40.) hypodermis
(41.) insulin
(42.) glucagon
(43.) nutrition
(44.) malnutrition
(45.) minerals
(46.) avitaminosis
(47.) hypervitaminosis
(48.) calorie
(49.) thermoregulation
(50.) pyrexia

[L1]   MATCHING—p. 472

(51.) F	(55.) H	(59.) C	(63.) R	(67.) S
(52.) E	(56.) G	(60.) J	(64.) K	(68.) O
(53.) I	(57.) D	(61.) N	(65.) M	(69.) Q
(54.) B	(58.) A	(62.) P	(66.) L	

[L2]   CONCEPT MAPS—pp. 473–476

I   FOOD INTAKE—p. 473

(1.) vegetables
(2.) meat
(3.) carbohydrates
(4.) 9 cal/gram
(5.) proteins
(6.) tissue growth and repair
(7.) vitamins
(8.) metabolic regulators

II   ANABOLISM-CATABOLISM OF CARBOHYDRATES, LIPIDS, AND PROTEINS—p. 474

(9.) lipolysis
(10.) lipogenesis
(11.) glycolysis
(12.) beta oxidation
(13.) gluconeogenesis
(14.) amino acids

III   FATE OF TRIGLYCERIDES/CHOLESTEROL—p. 475

(15.) chylmicrons
(16.) excess LDL
(17.) cholesterol
(18.) "bad" cholesterol
(19.) atherosclerosis
(20.) "good" cholesterol
(21.) liver
(22.) eliminated via feces

IV   CONTROL OF BLOOD GLUCOSE—p. 476

(23.) pancreas
(24.) liver
(25.) fat cells
(26.) glucagon
(27.) epinephrine
(28.) glycogen → glucose
(29.) protein

[L2]   MULTIPLE CHOICE—pp. 477–479

(30.) C	(34.) C	(38.) C	(42.) D	(46.) D
(31.) B	(35.) C	(39.) A	(43.) C	(47.) B
(32.) A	(36.) B	(40.) D	(44.) B	(48.) C
(33.) D	(37.) C	(41.) B	(45.) D	(49.) D

[L2]   COMPLETION—pp. 479–480

(50.) nutrient pool

(51.) lipoproteins

(52.) chylomicrons

(53.) LDLs

(54.) HDLs

(55.) transamination

(56.) deamination

(57.) uric acid

(58.) calorie

(59.) metabolic rate

(60.) BMR

(61.) convection

(62.) insensible perspiration

(63.) sensible perspiration

(64.) acclimatization

[L2]   SHORT ESSAY—pp. 480–482

65.   (a) linoleic acid, (b) arachidonic acid, (c) linolenic acid

66.   (a) Chylomicrons—95 percent triglycerides.

(b) Very low-density lipoproteins (VDL)—triglycerides + small amounts of phospholipids and cholesterol.

(c) Intermediate-density lipoproteins (IDL)—small amounts of triglycerides, more phospholipids and cholesterol.

(d) Low-density lipoproteins (LDL)—cholesterol, lesser amounts of phospholipids, very few triglycerides

(e) High-density lipoproteins (HDL)—equal amounts of lipid and protein.

67.   *Transamination*—attaches the amino group of an amino acid to a keto acid.

Deamination—the removal of an amino group in a reaction that generates an ammonia molecule.

68.   (a) Essential amino acids are necessary in the diet because they cannot be synthesized by the body. Nonessential amino acids can be synthesized by the body on demand.

(b) arginine and histidine

69.   (a) Proteins are difficult to break apart.

(b) Their energy yield is less than that of lipids.

(c) The byproduct, ammonia, is a toxin that can damage cells.

(d) Proteins form the most important structural and functional components of any cell. Extensive protein catabolism threatens homeostasis at the cellular and system levels.

70.   When nucleic acids are broken down, only the sugar and pyrimidines provide energy. Purines cannot be catabolized at all. Instead, they are deaminated and excreted as uric acid, a nitrogenous waste.

71.   (a) liver, (b) adipose tissue, (c) skeletal muscle, (d) neural tissue, (e) other peripheral tissues

72.   During the absorptive state the intestinal mucosa is busily absorbing the nutrients from the food you've eaten. Attention in the postabsorptive state is focused on the mobilization of energy reserves and the maintenance of normal blood glucose levels.

73.   (a) milk group

(b) meat group

(c) vegetable and fruit group

(d) bread and cereal group

74.   The amount of nitrogen absorbed from the diet balances the amount lost in the urine and feces. N-compound synthesis and breakdown are equivalent.

75.   Minerals (i.e., inorganic ions) are important because:

(a) they determine the osmolarities of fluids

(b) they play major roles in important physiological processes

(c) they are essential co-factors in a variety of enzymatic reactions

76.   (a) fat-soluble—ADEK

(b) water-soluble—B complex and C (ascorbic acid)

77.   (a) radiation (b) conduction (c) convection (d) evaporation

78.   (a) physiological mechanisms (b) behavioral modifications

[L3]   CRITICAL THINKING/APPLICATION—p. 483

1. Because of a normal total cholesterol and triglyceride level, Greg's low HDL level implies that excess cholesterol delivered to the tissues cannot be easily returned to the liver for excretion. The amount of cholesterol in peripheral tissues, and especially in arterial walls, is likely to increase.

2. Diets too low in calories, especially carbohydrates, will bring about physiological responses that are similar to fasting. During brief periods of fasting or low calorie intake, the increased production of ketone bodies resulting from lipid catabolism results in ketosis, a high concentration of ketone bodies in body fluids. When keto-acids dissociate in solution, they release a hydrogen ion. The appearance of ketone bodies in the circulation presents a threat to the plasma pH. During prolonged starvation or low-calorie dieting, a dangerous drop in pH occurs. This acidification of the blood is called ketoacidosis. Charlene's symptoms represent warning signals to what could develop into more disruptive normal tissue activities, ultimately causing coma, cardiac arrhythmias, and death, if left unchecked.

3. fats = 9 cal/g; 1 tbsp = 13 grams of fat

    3 tbsp $\times$ 13 grams of fat = 39 grams of fat

    39 g $\times$ 9 cal/g = 351 calories

4. (a)   40 grams protein: 40 g $\times$ 4 cal/g   =   160 cal

    50 grams fat: 50 g $\times$ 9 cal/g   =   450 cal

    60 grams carbohydrate: 69 g $\times$ 4 cal/g   =   276 cal

    Total   886 cal

    (b)   160 cal $\div$ 886 cal = 18% protein

    450 cal $\div$ 886 cal = 50% fat

    276 cal $\div$ 886 cal = 31% carbohydrate

    (c)   If Steve is an active 24-year-old male, the total number of calories is insufficient. At 175 lb, Steve's RDI is between 1800–2000 calories. The fat intake is too high (RDI should be between 25–30%). The carbohydrate intake is too low (RDI, 55–60%).

5. A temperature rise accompanying a fever increases the body's metabolic energy requirements and accelerates water losses stemming from evaporation and perspiration. For each degree the temperature rises above normal, the daily water loss increases by 200 mL. Drinking fluids helps to replace the water loss.

Chapter 26:

The Urinary System

[L1]   MULTIPLE CHOICE—pp. 485–489

(1.) A	(8.) C	(15.) D	(22.) B	(28.) B
(2.) B	(9.) C	(16.) C	(23.) A	(29.) A
(3.) C	(10.) B	(17.) B	(24.) D	(30.) C
(4.) D	(11.) D	(18.) D	(25.) B	(31.) D
(5.) C	(12.) C	(19.) A	(26.) B	(32.) A
(6.) B	(13.) A	(20.) B	(27.) C	(33.) B
(7.) D	(14.) A	(21.) C		

[L1]   COMPLETION—pp. 490–491

(34.) kidneys
(35.) glomerulus
(36.) filtrate
(37.) interlobar veins
(38.) glomerular hydrostatic
(39.) parathormone
(40.) countertransport

(41.) countercurrent multiplication
(42.) antidiuretic hormone
(43.) renal threshold
(44.) composition
(45.) concentration
(46.) urethra

(47.) internal sphincter
(48.) rugae
(49.) neck
(50.) micturition reflex
(51.) cerebral cortex
(52.) incontinence
(53.) nephrolithiasis

[L1]   MATCHING—p. 491

(54.) I	(58.) H	(62.) D	(66.) Q	(69.) M
(55.) G	(59.) B	(63.) P	(67.) J	(70.) L
(56.) E	(60.) C	(64.) R	(68.) K	(71.) N
(57.) A	(61.) F	(65.) O		

[L1]   DRAWING/ILLUSTRATION—LABELING—pp. 492–493

Figure 26.1   Components of the Urinary System—p. 492

(72.) kidney

(73.) ureter

(74.) urinary bladder

Figure 26.2   Sectional Anatomy of the Kidney—p. 492

(75.) minor calyx
(76.) renal pelvis
(77.) ureter

(78.) renal column
(79.) renal pyramid
(80.) renal vein

(81.) major calyx
(82.) renal capsule
(83.) cortex

Figure 26.3   Structure of a Typical Nephron Including Circulation—p. 493

(84.) efferent arteriole
(85.) glomerulus
(86.) afferent arteriole
(87.) proximal convoluted tubule (PCT)
(88.) peritubular capillaries

(89.) distal convoluted tubule (DCT)
(90.) collecting duct
(91.) loop of Henle
(92.) proximal convoluted tubule (PCT)
(93.) peritubular capillaries

(94.) Bowman's capsule
(95.) glomerulus
(96.) distal convoluted tubule (DCT)
(97.) vasa recta
(98.) loop of Henle

[L2]   CONCEPT MAPS—pp. 494–496

I   URINARY SYSTEM—p. 494

(1.) ureters
(2.) urinary bladder
(3.) nephrons
(4.) glomerulus

(5.) proximal convoluted tubule
(6.) collecting tubules

(7.) medulla
(8.) renal sinus
(9.) minor calyces

II   KIDNEY CIRCULATION—p. 495

(10.) renal artery   (12.) afferent artery   (14.) interlobular vein
(11.) arcuate artery   (13.) efferent artery   (15.) interlobar vein

III   RENIN-ANGIOTENSIN-ALDOSTERONE SECRETION—p. 496

(16.) $\downarrow$ plasma volume   (19.) angiotensin I   (22.) $\downarrow$ Na$^+$ excretion
(17.) renin   (20.) adrenal cortex   (23.) $\downarrow$ H$_2$O excretion
(18.) liver   (21.) $\uparrow$ Na$^+$ reabsorption

[L2]   BODY TREK—p. 497

(24.) proximal   (30.) active transport   (35.) collecting
(25.) glomerulus   (31.) ions   (36.) urine
(26.) protein-free   (32.) distal   (37.) ureters
(27.) descending limb   (33.) aldosterone   (38.) urinary bladder
(28.) filtrate   (34.) ADH   (39.) urethra
(29.) ascending limb

[L2]   MULTIPLE CHOICE—pp. 497–500

(40.) D   (44.) C   (48.) C   (52.) B   (56.) C
(41.) C   (45.) D   (49.) B   (53.) D   (57.) D
(42.) B   (46.) B   (50.) D   (54.) C   (58.) A
(43.) D   (47.) D   (51.) A   (55.) A   (59.) C

[L2]   COMPLETION—pp. 500–501

(60.) retroperitoneal   (65.) cortical   (70.) glomerular filtration
(61.) glomerular filtration rate (66.) vasa recta   (71.) aldosterone
(62.) osmotic gradient   (67.) filtration   (72.) diabetes insipidus
(63.) transport maximum   (68.) reabsorption   (73.) angiotensin II
(64.) macula densa   (69.) secretion   (74.) renin

[L2]   SHORT ESSAY—pp. 501–503

75. (a) Regulates plasma concentrations of ions.
     (b) Regulates blood volume and blood pressure.
     (c) Contributes to stabilization of blood pH.
     (d) Conserves valuable nutrients.
     (e) Eliminates organic wastes.
     (f) Assists liver in detoxification and deamination

76. kidney $\rightarrow$ ureters $\rightarrow$ urinary bladder $\rightarrow$ urethra

77. (a) renal capsule (fibrous tunic)
     (b) adipose capsule
     (c) renal fascia

78. glomerulus $\rightarrow$ proximal convoluted tubule $\rightarrow$ descending limb of loop of Henle $\rightarrow$ ascending limb of loop of Henle $\rightarrow$ distal convoluted tubule

79. (a) production of filtrate
     (b) reabsorption of organic substrates
     (c) reabsorption of water and ions

80.   (a)   (b)
     capillary endothelium   fenestrated capillaries
     basement membrane   dense and thick (lamina densa)
     glomerular epithelium   pedocytes with pedicels separated by slit pores

81. renin and erythropoietin

82. (a) Produces a powerful vasoconstriction of the afferent arteriole, thereby decreasing the GFR and slowing the production of filtrate.

    (b) Stimulation of renin release.

    (c) Direct stimulation of water and sodium ion reabsorption.

83. filtration, reabsorption, secretion

84. $P_f = G_{hp} - (C_{hp} + OP_b)$

$$\frac{\text{filtration}}{\text{pressure}} = \frac{\text{glomerular}}{\substack{\text{blood} \\ \text{(hydrostatic)} \\ \text{pressure}}} - \left\{ \substack{\text{capsular} \\ \text{hydrostatic} \\ \text{pressure}} + \substack{\text{blood} \\ \text{osmotic} \\ \text{pressure}} \right\}$$

85. Muscle fibers breaking down glycogen reserves release lactic acid.

    The number of circulating ketoacids increases.

    Adipose tissues are releasing fatty acids into the circulation.

86. (a) Sodium and chloride are pumped out of the filtrate in the ascending limb and into the peritubular fluid.

    (b) The pumping elevates the osmotic concentration in the peritubular fluid around the descending limb.

    (c) The result is an osmotic flow of water out of the filtrate held in the descending limb and into the peritubular fluid.

87. (a) autoregulation

    (b) hormonal regulation

    (c) autonomic regulation

88. (a) ADH—decreased urine volume

    (b) renin—causes angiotensin II production; stimulates aldosterone production

    (c) aldosterone—increased sodium ion reabsorption; decreased urine concentration and volume

    (d) Atrial Natriuretic Peptide (ANP)—inhibits ADH production; results in increased urine volume

[L3] CRITICAL THINKING/APPLICATION—p. 504

1. The alcohol acts as a diuretic. It inhibits ADH secretion from the posterior pituitary causing the distal convoluted tubule and the collecting duct to be relatively impermeable to water. Inhibiting the osmosis of water from the tubule along with the increased fluid intake results in an increase in urine production, and increased urination becomes necessary.

2. (a) If plasma proteins and numerous WBC are appearing in the urine there is obviously increased permeability of the filtration membrane. This condition usually results from inflammation of the filtration membrane within the renal corpuscle. If the condition is temporary it is probably an acute glomerular nephritis usually associated with a bacterial infection such as streptococcal sore throat. If the condition is long term, resulting in a nonfunctional kidney, it is referred to as chronic glomerular nephritis.

    (b) The plasma proteins in the filtrate increase the osmolarity of the filtrate, causing the urine volume to be greater than normal.

3. (a) Filtration

    (b) Primary site of nutrient reabsorption

    (c) Primary site for secretion of substances into the filtrate

    (d) Loop of Henle and collecting system interact to regulate the amount of water and the number of sodium and potassium ions lost in the urine.

4. (a)

$$\text{Effective filtration pressure (EFP)} = \left\{ \begin{array}{ccc} \text{Glomerular} & & \text{Capsular} \\ \text{hydrostatic} & + & \text{osmotic} \\ \text{pressure} & & \text{pressure} \end{array} \right\} - \left\{ \begin{array}{ccc} \text{Glomerular} & & \text{Capsular} \\ \text{osmotic} & + & \text{hydrostatic} \\ \text{pressure} & & \text{pressure} \end{array} \right\}$$

$$\text{EFP} = (G_{hp} + C_{op}) - (OP_b + C_{hp})$$

$$\text{EFP} = \left\{ \begin{array}{c} 60 + 5 \\ \text{(mm Hg)} \end{array} \right\} - \left\{ \begin{array}{c} 32 + 18 \\ \text{(mm Hg)} \end{array} \right\} = 15 \text{ mm Hg}$$

(b) An EFP of 10 mm Hg is normal. A change in the EFP produces a similar change in the GFR.

(c) A capsular osmotic pressure of 5 mm Hg develops in the capsular filtrate due to increased permeability of the glomerular endothelium, allowing blood proteins to filter out into the capsule. An EFP of 15 mm Hg indicates some type of kidney disease.

5. Strenuous exercise causes sympathetic activation to produce powerful vasoconstriction of the afferent arteriole, which delivers blood to the renal capsule. This causes a decrease in the GFR and alters the GFR by changing the required pattern of blood circulation. Dilation of peripheral blood vessels during exercise shunts blood away from the kidney and the GFR declines. A decreased GFR slows the production of filtrate. As the blood flow increases to the skin and skeletal muscles, kidney perfusion gradually declines and potentially dangerous conditions develop as the circulating concentration of metabolic wastes increases and peripheral water losses mount.

Chapter 27:

Fluid, Electrolyte, and Acid-Base Balance

[L1]   MULTIPLE CHOICE—pp. 506–509

(1.) B	(6.) A	(11.) B	(16.) D	(20.) A
(2.) D	(7.) D	(12.) C	(17.) C	(21.) D
(3.) A	(8.) B	(13.) D	(18.) B	(22.) B
(4.) B	(9.) D	(14.) A	(19.) C	(23.) C
(5.) C	(10.) D	(15.) B		

[L1]   COMPLETION—pp. 509–510

(24.) fluid
(25.) fluid shift
(26.) electrolyte
(27.) osmoreceptors
(28.) hypertonic
(29.) hypotonic

(30.) aldosterone
(31.) antidiuretic hormone
(32.) kidneys
(33.) calcium
(34.) buffers
(35.) hemoglobin

(36.) respiratory compensation
(37.) renal compensation
(38.) acidosis
(39.) alkalosis

[L1]   MATCHING—p. 510

(40.) F	(43.) A	(46.) H	(49.) N	(52.) O
(41.) D	(44.) G	(47.) E	(50.) I	(53.) L
(42.) C	(45.) B	(48.) M	(51.) K	(54.) J

[L1]   DRAWING/ILLUSTRATION—LABELING—pp. 511–512

Figure 27.1   The pH Scale—p. 511

(55.) pH 6.80
(56.) acidosis

(57.) pH 7.35
(58.) pH 7.45

(59.) alkalosis
(60.) pH 7.80

Figure 27.2   Relationships among pH, Pco$_2$, and HCO$_3^-$—p. 512

pH	Pco$_2$	HCO$_3^-$
(61.) ↓	(65.) ↑	(69.) N
(62.) ↓	(66.) N	(70.) ↓
(63.) ↑	(67.) ↓	(71.) N
(64.) ↑	(68.) N	(72.) ↑

[L2]   CONCEPT MAPS—pp. 513–517

I   HOMEOSTASIS—TOTAL VOLUME OF BODY WATER—p. 513

(1.) ↓ volume of body H$_2$O   (3.) aldosterone secretion by adrenal cortex
(2.) ↑ H$_2$O retention at kidneys

II   FLUID AND ELECTROLYTE IMBALANCE—p. 514

(4.) ↓ pH   (6.) ↓ ECF volume
(5.) ECF hypotonic to ICF   (7.) ↑ ICF volume

III   RESPIRATORY MECHANISMS FOR CONTROL OF PH—p. 515

(8.) ↑ blood CO$_2$ (10.) hyperventilation   (12.) normal blood pH
(9.) ↑ depth of breathing   (11.) ↑ blood pH

IV URINARY MECHANISMS FOR MAINTAINING HOMEOSTASIS OF BLOOD PH—p. 516

(13.) $\downarrow$ blood pH (14.) HCO_3^- (15.) $\uparrow$ blood pH

V HOMEOSTASIS – FLUID VOLUME REGULATION – SODIUM ION CONCENTRATIONS—p. 517

(16.) $\downarrow$ B.P. at kidneys (19.) $\uparrow$ plasma volume (22.) $\downarrow$ Aldosterone release

(17.) $\uparrow$ Aldosterone release (20.) $\uparrow$ ANP release (23.) $\uparrow$ H_2O loss

(18.) $\downarrow$ H_2O loss (21.) $\downarrow$ ADH release

[L2] MULTIPLE CHOICE—pp. 518–520

(24.) B	(28.) D	(32.) A	(36.) B	(40.) C
(25.) D	(29.) D	(33.) B	(37.) C	(41.) B
(26.) C	(30.) B	(34.) C	(38.) C	(42.) C
(27.) A	(31.) D	(35.) D	(39.) A	(43.) D

[L2] COMPLETION—p. 521

(44.) kidneys (48.) organic acids (52.) hyperventilation

(45.) angiotensin II (49.) buffer system (53.) lactic acidosis

(46.) volatile acid (50.) respiratory acidosis (54.) ketoacidosis

(47.) fixed acids (51.) hypoventilation (55.) alkaline tide

[L2] SHORT ESSAY—pp. 521–523

56. (a) fluid balance
     (b) electrolyte balance
     (c) acid-base balance

57. (a) antidiuretic hormone (ADH)
     (b) aldosterone
     (c) atrial natriuretic peptide (ANP)

58. (a) It stimulates water conservation at the kidney, reducing urinary water losses.
     (b) It stimulates the thirst center to promote the drinking of fluids. The combination of decreased water loss and increased water intake gradually restores normal plasma osmolarity.

59. (a) alterations in the potassium ion concentration in the ECF
     (b) changes in pH
     (c) aldosterone levels

60. (a) an initial fluid shift into or out of the ICF
     (b) "fine tuning" via changes in circulating levels of ADH

61. $CO_2 + H_2O \leftrightarrow H_2CO_3 \leftrightarrow H^+ + HCO_3^-$

62. (a) protein buffer system, phosphate buffer system, and carbonic acid-bicarbonate buffer system
     (b) respiratory mechanisms, renal mechanisms

63. (a) secrete or absorb hydrogen ions
     (b) control excretion of acids and bases
     (c) generate additional buffers when necessary

64. hypercapnia—$\uparrow$ plasma P_{CO_2}; $\downarrow$ plasma pH—respiratory acidosis
     hypocapnia—$\downarrow$ plasma P_{CO_2}; $\uparrow$ plasma pH—respiratory alkalosis

65. (a) impaired ability to excrete H^+ at the kidneys
     (b) production of a large number of fixed and/or organic acids
     (c) severe bicarbonate loss

[L3]   CRITICAL THINKING/APPLICATION—p. 523

1. The comatose teenager's ABG studies reveal a severe respiratory acidosis. A pH of 7.17 (low) and a Pco_2 of 73 mm Hg (high) cause his respiratory centers to be depressed, resulting in hypoventilation, CO_2 retention, and consequent acidosis. His normal HCO_3^- value indicates that his kidneys haven't had time to retain significant amounts of HCO_3^- to compensate for the respiratory condition.

2. The 62-year-old woman's ABG studies reveal that she has metabolic alkalosis. A pH of 7.65 (high) and an HCO_3^- of 55 mEq/liter (high) are abnormal values. Her Pco_2 of 52 mm Hg indicates that her lungs are attempting to compensate for the alkalosis by retaining CO_2 in an effort to balance the HCO_3^- value. Her vomiting caused a large acid loss from her body via HCl, which means a loss of H^+, the acid ion. Predictable effects include slow respirations, an overexcitable CNS, leading to irritability and, if untreated, possible tetany and convulsions.

3. (a) metabolic acidosis

   (b) respiratory alkalosis

   (c) respiratory acidosis

   (d) metabolic alkalosis

4.

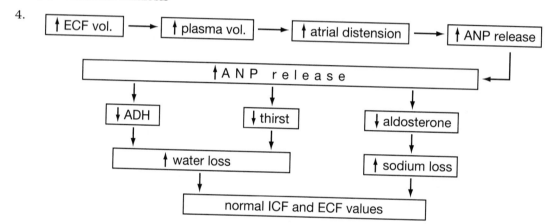

Chapter 28:

The Reproductive System

[L1]   MULTIPLE CHOICE—pp. 525–530

(1.) D	(8.) D	(15.) C	(21.) A	(27.) B
(2.) B	(9.) D	(16.) B	(22.) C	(28.) D
(3.) D	(10.) A	(17.) A	(23.) D	(29.) A
(4.) C	(11.) A	(18.) C	(24.) D	(30.) A
(5.) B	(12.) A	(19.) B	(25.) B	(31.) B
(6.) C	(13.) D	(20.) C	(26.) C	(32.) C
(7.) C	(14.) B			

[L1]   COMPLETION—pp. 530–531

(33.) fertilization	(41.) ICSH	(48.) Bartholin's
(34.) seminiferous tubules	(42.) ovaries	(49.) areola
(35.) testes	(43.) Graafian	(50.) progesterone
(36.) spermiogenesis	(44.) ovulation	(51.) placenta
(37.) ductus deferens	(45.) infundibulum	(52.) cardiovascular
(38.) digestive	(46.) implantation	(53.) menopause
(39.) fructose	(47.) oogenesis	(54.) climacteric
(40.) pudendum		

[L1]   MATCHING—p. 531

(55.) K	(60.) A	(65.) D	(70.) M	(74.) Q
(56.) G	(61.) J	(66.) I	(71.) U	(75.) N
(57.) C	(62.) B	(67.) P	(72.) S	(76.) T
(58.) H	(63.) L	(68.) R	(73.) O	(77.) V
(59.) F	(64.) E	(69.) W		

[L1]   DRAWING/ILLUSTRATION—LABELING—pp. 532–534

Figure 28.1   Male Reproductive Organs—p. 532

(78.) prostatic urethra	(86.) navicular fossa	(94.) ejaculatory duct
(79.) pubic symphysis	(87.) prepuce	(95.) prostate gland
(80.) ductus deferens	(88.) testis	(96.) anus
(81.) urogenital diaphragm	(89.) epididymis	(97.) anal sphincters
(82.) corpora cavernosa	(90.) urinary bladder	(98.) bulbourethral gland
(83.) corpus spongiosum	(91.) rectum	(99.) membranous urethra
(84.) penile urethra	(92.) seminal vesicle	(100.) bulb of penis
(85.) glans penis	(93.) seminal vesicle	(101.) scrotum

Figure 28.2   The Testis—p. 533

(102.) septum	(104.) ductus deferens	(106.) seminiferous tubule
(103.) epididymis (head)	(105.) rete testis	(107.) tunica albuginea

Figure 28.3   Female External Genitalia—p. 533

(108.) mons pubis	(112.) vestibule	(115.) urethral orifice
(109.) clitoris	(113.) perineum	(116.) labia majora
(110.) labia minora	(114.) prepuce	(117.) hymen
(111.) vaginal orifice		

Figure 28.4   Female Reproductive Organs (sagittal section)—p. 534

(118.) uterus

(119.) urinary bladder

(120.) pubic symphysis

(121.) urethra

(122.) clitoris

(123.) labium minora

(124.) labium majora

(125.) cervix

(126.) vagina

(127.) anus

Figure 28.5   Female Reproductive Organs (frontal section)—p. 535

(128.) infundibulum

(129.) ovary

(130.) uterine tube

(131.) myometrium

(132.) cervix

(133.) vagina

[L2]   CONCEPT MAPS—pp. 535–540

I   MALE REPRODUCTIVE TRACT—p. 535

(1.) ductus deferens

(2.) penis

(3.) seminiferous tubules

(4.) produce testosterone

(5.) FSH

(6.) seminal vesicles

(7.) bulbourethral glands

(8.) urethra

II   PENIS—p. 536

(9.) crus

(10.) shaft

(11.) corpus spongiosum

(12.) prepuce

(13.) external urinary meatus

(14.) frenulum

III   REGULATION OF MALE REPRODUCTIVE FUNCTION—p. 537

(15.) anterior pituitary

(16.) FSH

(17.) testes

(18.) interstitial cells

(19.) inhibin

(20.) CNS

(21.) male secondary sex characteristics

IV   FEMALE REPRODUCTIVE TRACT—p. 538

(22.) uterine tubes

(23.) follicles

(24.) granulosa & thecal cells

(25.) endometrium

(26.) supports fetal development

(27.) vagina

(28.) vulva

(29.) labia majora and minora

(30.) clitoris

(31.) nutrients

V   REGULATION OF FEMALE REPRODUCTIVE FUNCTION—p. 539

(32.) GnRH

(33.) LH

(34.) follicles

(35.) progesterone

(36.) bone and muscle growth

(37.) accessory glands and organ

VI   HORMONAL FEEDBACK AND PREGNANCY—p. 540

(38.) anterior pituitary

(39.) ovary

(40.) progesterone

(41.) placenta

(42.) relaxin

(43.) mammary gland development

[L2]   BODY TREK—MALE REPRODUCTIVE SYSTEM—p. 541

(44.) seminiferous tubules and rete testis

(45.) body of epididymis

(46.) ductus deferens

(47.) ejaculatory duct

(48.) urethra

(49.) penile urethra

(50.) external urethral meatus

[L2]   MULTIPLE CHOICE—pp. 542–544

(51.) A

(52.) C

(53.) A

(54.) C

(55.) D

(56.) B

(57.) B

(58.) C

(59.) B

(60.) D

(61.) D

(62.) C

(63.) B

(64.) D

(65.) A

(66.) D

(67.) A

(68.) B

(69.) D

[L2]   COMPLETION—pp. 544–545

(70.) androgens	(78.) smegma	(86.) tunica albuginea
(71.) inguinal canals	(79.) detumescence	(87.) fimbriae
(72.) raphe	(80.) menopause	(88.) cervical os
(73.) cremaster	(81.) zona pellucida	(89.) fornix
(74.) ejaculation	(82.) corona radiata	(90.) hymen
(75.) rete testis	(83.) corpus luteum	(91.) clitoris
(76.) acrosomal sac	(84.) corpus albicans	(92.) menses
(77.) prepuce	(85.) mesovarium	(93.) ampulla

[L2]   SHORT ESSAY—pp. 545–547

94.   (a)   maintenance of the blood-testis barrier

(b)   support of spermiogenesis

(c)   secretion of inhibin

(d)   secretion of androgen-binding protein

95.   (a)   It monitors and adjusts the composition of tubular fluid.

(b)   It acts as a recycling center for damaged spermatozoa.

(c)   It is the site of physical maturation of spermatozoa.

96.   (a)   seminal vesicles, prostate gland, bulbourethral glands

(b)   activates the sperm, provides nutrients for sperm motility, provides sperm motility, produces buffers to counteract acid conditions

97.   Seminal fluid is the fluid component of semen. Semen consists of seminal fluid, sperm, and enzymes.

98.   *Emission* involves peristaltic contractions of the ampulla, pushing fluid and spermatozoa into the prostatic urethra. Contractions of the seminal vesicles and prostate gland move the seminal mixture into the membranous and penile walls of the prostate gland.

Ejaculation occurs as powerful, rhythmic contractions of the ischiocavernosus and bulbocavernosus muscles push semen toward the external urethral orifice.

99.   (a)   Promotes the functional maturation of spermatozoa.

(b)   Maintains accessory organs of male reproductive tract.

(c)   Responsible for male secondary sexual characteristics.

(d)   Stimulates bone and muscle growth.

(e)   Stimulates sexual behaviors and sexual drive.

100.   (a)   Serves as a passageway for the elimination of menstrual fluids.

(b)   Receives penis during coitus; holds sperm prior to passage into uterus.

(c)   In childbirth it forms the lower portion of the birth canal.

101.   (a)   *Arousal*—parasympathetic activation leads to an engorgement of the erectile tissues of the clitoris and increased secretion of the greater vestibular glands.

(b)   *Coitus*—rhythmic contact with the clitoris and vaginal walls provides stimulation that eventually leads to orgasm.

(c)   *Orgasm*—accompanied by peristaltic contractions of the uterine and vaginal walls and rhythmic contractions of the bulbocavernosus and ischiocavernosus muscles giving rise to pleasurable sensations.

102.   Step 1: Formation of primary follicles

Step 2: Formation of secondary follicle

Step 3: Formation of a tertiary follicle

Step 4: Ovulation

Step 5: Formation and degeneration of the corpus luteum

103. Estrogens:

    (a) stimulate bone and muscle growth;

    (b) maintain female secondary sex characteristics;

    (c) stimulate sex-related behaviors and drives;

    (d) maintain functional accessory reproductive glands and organs;

    (e) initiate repair and growth of the endometrium.

104. (a) menses

    (b) proliferative phase

    (c) secretory phase

105. (a) human chorionic gonadotrophin (HCG)

    (b) relaxin

    (c) human placental lactogen (HPL)

    (d) estrogens and progestins

106. By the end of the sixth month of pregnancy the mammary glands are fully developed, and the glands begin to produce colostrum. This contains relatively more proteins and far less fat than milk, and it will be provided to the infant during the first two or three days of life. Many of the proteins are immunoglobulins that may help the infant ward off infections until its own immune system becomes fully functional.

[L3] CRITICAL THINKING/APPLICATION—p. 547

1. (a) The normal temperature of the testes in the scrotum is 1°–2° lower than the internal body temperature—the ideal temperature for developing sperm. Mr. Hurt's infertility is caused by the inability of sperm to tolerate the higher temperature in the abdominopelvic cavity.

    (b) Three major factors are necessary for fertility in the male:

- adequate motility of sperm—30%–35% motility necessary
- adequate numbers of sperm—20,000,000/ml minimum
- sperm must be morphologically perfect—sperm cannot be malformed

2. The 19-year-old female has a problem with hormonal imbalance in the body. Females, like males, secrete estrogens and androgens; however, in females, estrogen secretion usually "masks" the amount of androgen secreted in the body. In females an excess of testosterone secretion may cause a number of conditions such as sterility, fat distribution like a male, beard, low-pitched voice, skeletal muscle enlargement, clitoral enlargement, and a diminished breast size.

3. The contraceptive pill decreases the stimulation of FSH and prevents ovulation. It contains large quantities of progesterone and a small quantity of estrogen. It is usually taken for 20 days beginning on day 5 of a 28-day cycle. The increased level of progesterone and decreased levels of estrogen prepare the uterus for egg implantation. On day 26 the progesterone level decreases. If taken as directed, the Pill will allow for a normal menstrual cycle.

4. In males the disease-causing organism can move up the urethra to the bladder or into the ejaculatory duct to the ductus deferens. There is no direct connection into the pelvic cavity in the male. In females the pathogen travels from the vagina to the uterus, to the uterine tubes, and into the pelvic cavity where it can infect the peritoneal lining, resulting in peritonitis.

Chapter 29:

Development and Inheritance

[L1]   MULTIPLE CHOICE—pp. 549–552

(1.) C	(7.) A	(13.) B	(19.) A	(25.) B
(2.) B	(8.) C	(14.) D	(20.) D	(26.) D
(3.) A	(9.) B	(15.) A	(21.) C	(27.) A
(4.) D	(10.) C	(16.) D	(22.) D	(28.) B
(5.) C	(11.) B	(17.) C	(23.) C	(29.) C
(6.) B	(12.) D	(18.) A	(24.) A	(30.) A

[L1]   COMPLETION—p. 553

(31.) fertilization
(32.) capacitation
(33.) polyspermy
(34.) induction
(35.) thalidomide
(36.) second trimester
(37.) first trimester

(38.) chorion
(39.) human chorionic gonadotropin
(40.) placenta
(41.) true labor
(42.) dilation
(43.) childhood

(44.) infancy
(45.) meiosis
(46.) gametogenesis
(47.) autosomal
(48.) homozygous
(49.) heterozygous

[L1]   MATCHING—p. 554

(50.) H	(54.) C	(58.) G	(62.) Q	(66.) S
(51.) E	(55.) I	(59.) B	(63.) O	(67.) M
(52.) F	(56.) D	(60.) P	(64.) L	(68.) N
(53.) A	(57.) J	(61.) R	(65.) K	

[L1]   DRAWING/ILLUSTRATION—LABELING—p. 555

Figure 29.1   Spermatogenesis—p. 555

(69.) spermatogonia
(70.) primary spermatocyte
(71.) secondary spermatocyte
(72.) spermatids
(73.) spermatozoa

Figure 29.2   Oogenesis—p. 555

(74.) oogonium
(75.) primary oocyte
(76.) secondary oocyte
(77.) mature ovum

[L2]   CONCEPT MAPS—pp. 556–557

I   INTERACTING FACTORS—LABOR AND DELIVERY—p. 556

(1.) estrogen
(2.) relaxin
(3.) ↑ prostaglandin production
(4.) positive feedback
(5.) parturition

II   LACTATION REFLEXES—p. 557

(6.) anterior pituitary
(7.) prolactin
(8.) ↑ milk secretion
(9.) posterior pituitary
(10.) oxytocin
(11.) milk ejection

[L2]   BODY TREK—p. 558

(12.) secondary oocyte
(13.) fertilization
(14.) zygote
(15.) 2-cell stage
(16.) 8-cell stage
(17.) morula
(18.) early blastocyst
(19.) implantation

[L2] MULTIPLE CHOICE—pp. 559–561

(20.) B	(24.) A	(28.) A	(32.) B	(36.) D
(21.) C	(25.) B	(29.) D	(33.) D	(37.) B
(22.) D	(26.) D	(30.) C	(34.) C	(38.) D
(23.) C	(27.) B	(31.) C	(35.) A	(39.) B

[L2] COMPLETION—pp. 561–562

(40.) inheritance	(46.) oogenesis	(52.) corona radiata
(41.) genetics	(47.) spermiogenesis	(53.) hyaluronidase
(42.) chromatids	(48.) alleles	(54.) cleavage
(43.) synapsis	(49.) X-linked	(55.) chorion
(44.) tetrad	(50.) simple inheritance	(56.) differentiation
(45.) spermatogenesis	(51.) polygenic inheritance	(57.) activation

[L2] SHORT ESSAY—pp. 562–564

58. In simple inheritance, phenotypic characters are determined by interactions between a single pair of alleles.

   Polygenic inheritance involves interactions between alleles on several genes.

59. Capacitation is the activation process that must occur before a spermatozoon can successfully fertilize an egg. It occurs in the vagina following ejaculation.

60. (a) cleavage (b) implantation (c) placentation (d) embryogenesis

61. (a) ectoderm (b) mesoderm (c) endoderm

62.
(a)	(b)
• yolk sac	• endoderm and mesoderm
• amnion	• ectoderm and mesoderm
• allantois	• endoderm and mesoderm
• chorion	• mesoderm and trophoblast

63. (a) The respiratory rate goes up and the tidal volume increases.
   (b) The maternal blood volume increases.
   (c) The maternal requirements for nutrients increase.
   (d) The glomerular filtration rate increases.
   (e) The uterus increases in size.

64. (a) estrogens (b) oxytocin (c) prostaglandins

65. (a) Secretion of relaxin by the placenta—softens symphysis pubis.
   (b) Weight of the fetus—deforms cervical orifice.
   (c) Rising estrogen levels.
   (d) Both b and c promote release of oxytocin.

66. (a) dilation stage (b) expulsion stage (c) placental stage

67. infancy, childhood, adolescence, maturity, senescence

68. An Apgar rating represents an assessment of the newborn infant. It considers heart rate, respiratory rate, muscle tone, response to articulation, and color at 1 and 5 minutes after birth. In each category the infant receives a score ranging from 0 (poor) to 2 (excellent), and the scores are totaled. An infant's Apgar rating (1–10) has been shown to be an accurate predictor of newborn survival and the presence of neurological damage.

69. (a) Hypothalamus—increasing production of GnRH.
   (b) Increasing circulatory levels of FSH and LH (ICSH) by the anterior pituitary
   (c) FSH and LH initiate gametogenesis and the production of male or female sex hormones that stimulate the appearance of secondary sexual characteristics and behaviors.

70. (a) Some cell populations grow smaller throughout life.
   (b) The ability to replace other cell populations decreases.
   (c) Genetic activity changes over time.
   (d) Mutations occur and accumulate.

[L3] CRITICAL THINKING/APPLICATION—p. 564

1. Color blindness is an X-linked trait. The Punnett square shows that sons produced by a normal father and a heterozygous mother will have a 50 percent chance of being color blind, while the daughters will all have normal color vision.

Maternal alleles

	X^C	X^c
X^C	$X^C X^C$	$X^C X^c$
Y	$X^C Y$	$X^C Y$ (color blind)

Paternal alleles

2. The Punnett square reveals that 50 percent of their offspring have the possibility of inheriting albinism.

Maternal alleles

	a	a
A	Aa	Aa
a	aa (albino)	aa (albino)

Paternal alleles

3. Both the mother and father are heterozygous-dominant.

T—tongue roller   t— non-tongue roller

The Punnett square reveals that there is a 25 percent chance of having children who are not tongue rollers and a 75 percent chance of having children with the ability to roll the tongue.

Maternal alleles

	T	t
T	TT (yes)	Tt (yes)
t	Tt (yes)	tt (no)

Paternal alleles

4. Amniocentesis is a diagnostic tool to determine the possibility of a congenital condition. To obtain the sample of amniotic fluid, a needle is inserted into position using ultrasound. This represents a potential threat to the health of the fetus and mother. Sampling cannot be safely performed until the volume of amniotic fluid is large enough to avoid injury to the fetus. The usual time is at a gestational age of 14–15 weeks. By the time the results are available, the option of abortion may not be available. Chorionic villus sampling may be an alternative since it analyzes cells collected from the villi during the first trimester.